P9-DWA-226

BioMEMS
Science and Engineering Perspectives

BioMEMS
Science and Engineering Perspectives

Simona Badilescu
Muthukumaran Packirisamy

CRC Press is an imprint of the
Taylor & Francis Group, an **informa** business

CRC Press
Taylor & Francis Group
6000 Broken Sound Parkway NW, Suite 300
Boca Raton, FL 33487-2742

Printed in the United States of America on acid-free paper
10 9 8 7 6 5 4 3 2 1

International Standard Book Number: 978-1-4398-1699-8 (Hardback)

Visit the Taylor & Francis Web site at
http://www.taylorandfrancis.com

and the CRC Press Web site at
http://www.crcpress.com

SB: To Sabina and Iris, with love.

MP: To my wife, Indrani, son, Sudarsan, and parents, Kamatchi and Packirisamy, for their support and love.

Contents

Preface

We are proud to present this book as an attempt to bridge different areas that constitute the field of biomicroelectromechanical systems (BioMEMS), often called biomicrosystems. The field of BioMEMS has been growing rapidly since the early 1990s due to the advancements in microtechnologies that could cater to the vast application requirements of bio areas. The potential of BioMEMS suits this technology for many applications, including clinical and environmental diagnostics, drug delivery, agriculture, nutrition, pharmaceuticals, chemical synthesis, etc. It is foreseen that BioMEMS will have a deep impact on many aspects of the life science operations and functionalities in the near future.

Scientists and students that work in the field of BioMEMS will need to have knowledge and skills at the interface between engineering and biosciences. Development of a BioMEMS device usually involves many scientists and students from various disciplines, such as biosciences, medicine, biochemistry, engineering, physics, etc. One could anticipate many communication and understanding issues that would arise among these people with varied expertise and training. The methods, details, and languages of training are quite different for the students and researchers of engineering and biosciences. As a result, researchers and students involved with multidisciplinary projects like BioMEMS undergo an interesting and refreshing learning on multidisciplinary subjects along the project development. This book aims to support and expedite the multidisciplinary learning involved with the development of biomicrosystems, from both bioscience and engineering perspectives. Due to the variety and intensity of the subject matter that exists with an engineering and biological focus, there are many excellent books available dealing only with the engineering or biological aspect of biomicrosystems. But, this books attempts to cover the subjects that are important from both science and engineering perspectives. This effort of combining both perspectives in a single book presents a challenge of covering in detail the wide spectrum of areas and topics that are associated with those disciplines. The authors sincerely feel it is possible that some important works were unintentionally missed in this book due to the vastness of the subjects covered and other limitations. We suggest that readers refer to other advanced books and publications on the topics of interest that arise after reading this book.

The science perspectives of BioMEMS include an introduction to molecules of biological interest that are the building blocks of cells and viruses, and also to organic molecules that are involved in the formation of self-assembled monolayers (SAMs), linkers, hydrogels, etc., used for making different surfaces biocompatible through functionalization. The presented engineering perspectives include methods of manufacturing bioactive surfaces and devices, microfluidics modeling and experimentation, and also device level implementation of BioMEMS concepts for different applications. As the field of BioMEMS is application driven, the concepts of lab-on-a-chip (LOC) and micro total analysis system (μTAS) are also discussed, along with their pertinence to the emerging point-of-care (POC) and point-of-need (PON) applications.

This book tries to present both engineering and bioscience topics in a more balanced way. It has nine chapters, with four chapters assigned to science and four chapters assigned to engineering, while the first chapter introduces the readers to many aspects of biomicrosystems, including biological, engineering, application, and commercialization. Chapter 2 deals with different materials and platforms that are used for developing biomicrosystems. Biological entities, including pathogens, are introduced in Chapter 3, in the order of increasing complexities. The multidisciplinary aspects of engineering bioactive surfaces are provided in Chapter 4. Different types and methods of characterizing bioactive surfaces are outlined in Chapter 5, while the fundamentals of biosensing along with methods are given in Chapter 6. Chapter 7 presents the engineering aspects of fabricating

BioMEMS devices, and Chapter 8 introduces different aspects of microfluidics. Different life science applications along with case studies are presented in Chapter 9. With the present organization, it is expected that readers will enjoy not only the topics related to engineering, but also the topics related to the physicochemical understanding of biological processes that are involved in various BioMEMS devices.

The authors thank the graduate students of our Optical Bio Microsystems Laboratory at Concordia for their contribution toward this field. The authors are pleased to publish this book through Taylor and Francis Group.

The Authors

Dr. Simona Badilescu is a senior scientist with a background in physical chemistry and a rich experience in teaching and research. She received her PhD degree from the University of Bucharest (Romania) and specialized in molecular spectroscopy, surface science, and analytical applications of infrared spectroscopy. Dr. Badilescu has several years of experience in an industrial environment as director of the spectroscopy department of a petrochemical company in Romania. After three years of teaching in Algeria as associate professor at the University of Blida, she came with her family to Canada and was a research associate at the Universite de Montréal in the chemistry department. Her research interest focused on vibrational spectroscopy of molecules of biological interest. Since 1987, she has been part of an interdisciplinary group, at the University of Moncton in New Brunswick, working on ATR spectroscopy of thin-film systems. Afterward, she was a professor of chemistry at the University of Moncton in New Brunswick, and later a senior scientist in the physics department. Dr. Badilescu joined the electrical and computer engineering department in 2002, and since then, she has been a part of the Nanomaterials and Devices Laboratory and contributed to different projects related to biosensing. Presently, she is part of the Nanomaterials and Nanodevices Laboratory in the mechanical engineering department. Dr. Badilescu has published two books and several chapters, mainly on topics related to spectroscopy. She is author of more than two hundred articles and conference papers. She is a member of several professional societies and a reviewer for journals such as the *Journal of Physical Chemistry*, *Applied Physics Letters*, *Advanced Materials*, etc.

Muthukumaran Packirisamy is a professor and Concordia research chair on optical BioMEMS in the Department of Mechanical and Industrial Engineering, Concordia University, Canada. He is the recipient of the Fellow and I. W. Smith Award from the Canadian Society for Mechanical Engineers, the Concordia University Research Fellow, the Petro Canada Young Innovator Award, and the ENCS Young Research Achievement Award. His research interests include optical BioMEMS, integration of microsystems, and micro-nano integration.

He obtained his PhD from Concordia University, master's from the Indian Institute of Technology, Madras, India, and bachelor's from the University of Madras, India. He has also worked for many MEMS industries in Canada. He is presently involved with developing BioMEMS devices in collaboration with industries.

An author of more than 225 articles published in journals and conference proceedings, he has nine patents in the area of microsystems.

1 Introduction

1.1 INTRODUCTION TO BIOMEMS

The term *microelectromechanical systems* (MEMS) was coined in the 1980s to define devices that were fabricated using microfabrication techniques and whose primary function was not electronic but mainly mechanical. Within a short span of time, the field of MEMS exposed its immense potential to impact every aspect of human life.

A variety of areas, like aerospace, communication, medical, sensing, and actuation, started enjoying the advantages of miniaturization with MEMS technology or microsystem technology. As a result, the field of MEMS evolved into an enabling technology in the mid-1990s and gave rise to the creation of many subdisciplines, such as optical MEMS, radio frequency (RF) MEMS, power MEMS, etc., depending upon the focus of application. In this line, the field of BioMEMS was created with a focus on biological and chemical applications.

The concept of integrating biochemical analysis with microelectromechanical systems (MEMS) is involved in the new field of BioMEMS, which is undergoing tremendous growth in a multitude of applications. The applications spectrum covers from tissue engineering to proteomics. Some of the applications include cell culture, cell sorting, cell manipulation, stem cell growth, separation and mixing of biological and chemical fluids, enzymatic reactions and gene isolation and transformation, DNA purification, antigen-antibody interaction, protein level interaction, drug diagnosis and delivery, therapeutics, chemical and biosensing, etc. The important recent applications include point-of-care (POC) *in vitro* diagnostics, and synthesis of nanoparticles using BioMEMS.

The field of BioMEMS inherits all the advantages of miniaturization, such as small sample volume, scalability, integration of multiple functions and fields, low cost, low power consumption, etc. As a result, the microsystems facilitate the implementation of many laboratory works at the microchips that are millimeters to centimeters in size. Some of the standard laboratory tasks that can be implemented under a microenvironment include sample preparation, mixing, separation, diagnosis, sensing, manipulation, control, delivery, data acquisition, and analysis. When some of the laboratory functions out of a process are integrated at the microchip, the devices are called *lab-on-a-chip* (LOC). The concept of LOC was demonstrated by S. C. Terry et al. in 1979[1] with the development of a silicon-chip-based gas chromatographic analyzer. Manz et al.[2] introduced the concept of *micro total analysis system* (μTAS) in 1990, by developing a device that conducts all the steps of a process. μTAS is generally considered to incorporate all the operations needed in a process, involving sample preparation. Even though μTAS is a subset of LOC, it involves many functions and integrates many domains, such as micromechanical, microfluidic, microelectronics, microphotonics, etc. μTAS can also integrate moving elements, such as micropumps, microvalves, etc. LOC and μTAS will be used interchangeably in this book. Biomedical technologies contribute to the use of LOCs in healthcare of various specialties, ophthalmology, cardiology, anesthesiology, and immunology. For example, such LOCs combine a number of biological functions, such as enzymatic reactions, antigen-antibody conjugation, and DNA/gene probing, in addition to microfluidic function, such as sample dilution, pumping, mixing, metering, incubation, separation, and detection in micron-sized channels and reservoirs. Lab-on-a-chip devices aim to address a wide range of life science applications, including drug discovery and delivery as well as clinical and environmental diagnostics. The integration and automation capabilities of LOC can improve the reproducibility of results, reduce test time, and eliminate preparation errors that may occur in the intermediate stages of an analytical procedure.

BioMEMS devices are defined as the devices or microsystems that are fabricated with methods inspired from micro- and nanotechnologies and are used for many processes involving biological and chemical species.[3]

BioMEMS may involve in-parts or a combination of (1) microchannels, microchambers, etc., for handling a small volume of liquids in the range of microliters to nanoliters; (2) micromechanical elements, such as microvalves, micropumps, microcantilevers, microdiaphragms, etc.; (3) microelectrical elements such as electrodes and heaters; and (4) microphotonic elements, such as waveguides, gratings, interferometers, etc., to handle bio or chemical species with an elaborate integration feasibility. BioMEMS offers the following advantages: portability, scalability, reliability, reduced sample/reagent volume, low power consumption, high throughput, integrability, disposability, batch fabrication, low cost, high sensitivity, reduced test time, etc. BioMEMS devices combine a biological recognition system called a *bioreceptor* with a physical or chemical *transducer* to selectively and quantitatively detect the presence of specific compounds in a given external environment. A bioreceptor is a biological molecular species (e.g., antibody, enzyme, protein, or nucleic acid) or a living biological system (e.g., cells, tissue, or whole organisms) that undergoes a biochemical mechanism for recognition. The interaction of an analyte with a bioreceptor produces an effect to be measured by the transducer, which converts the information into a measurable effect, such as an electrical or optical signal.

Lab on a chip is characterized by some level of integration of different functions, and it can be used to perform a combination of analyses on a single miniaturized device, for biological and clinical assays. Most of the key application areas of this technology belong to the areas of life science research (genomics, pharmacogenomics, and proteomics), drug delivery, and point-of-care diagnostics, and they offer the advantages of integrating sample handling and preparation, mixing, separation, lysing of cells, and detection.

Biosensors and *biochips* can be classified by either their bioreceptor or their transducer type. Comprised of microarrays of genetic, protein, or cellular materials on microfluidic devices, these miniature platforms can perform parallel analysis of large data. Lab-on-a-chip can integrate many operations, including sample preparation, detection, and analysis on a single chip. It is also useful in drug discovery, target identification and validation, toxicology, and clinical drug safety.

1.2 APPLICATION AREAS

Research in BioMEMS covers a wide range, from diagnostics, DNA, and protein microarrays, to novel materials for BioMEMS, microfluidics, surface modification, etc. In addition, BioMEMS also finds many applications in the chemical, healthcare, biotechnological, and manufacturing industries.

The emerging applications for BioMEMS include agricultural and food engineering areas. DNA microarrays have become the most successful example of the integration among microelectronics, biology, and chemistry. Similarly, protein and antibody arrays can identify disease-specific proteins with enormous medical, diagnostic, and commercial potential as disease markers or drug targets. With the recent thrust in genomics and proteomics technologies, many new gene products and proteins are being discovered almost daily, and it has become a difficult task to analyze experimental data. In such situations, array-based integrated BioMEMS chips and microfluidics hold great potential to analyze systematically the proteins and to determine protein-protein or protein-DNA interactions. In the case of protein chips, protein is arrayed into many spots using robots, and each spot is addressed by other affinity proteins. The binding between recognized proteins or antigen-antibody has traditionally been detected by fluorescence-based methods. But it can also be detected by changes in surface plasmon resonance (SPR) due to changes in surface refractive index, or in a mechanical way of detecting the changes in structural properties due to interaction. For example, enzyme-linked immunosorbent assay

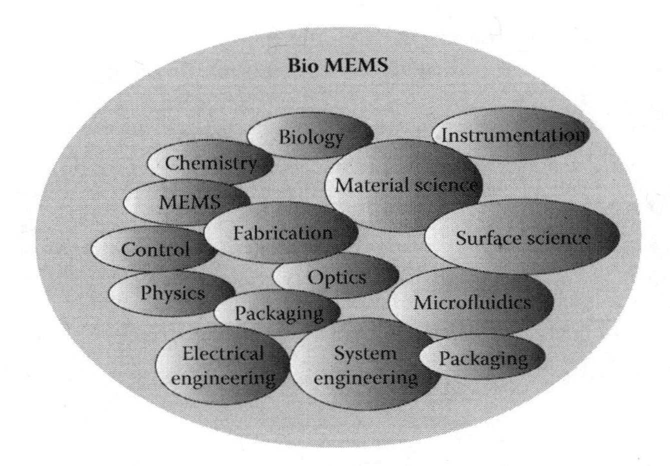

FIGURE 1.1 BioMEMS at the intersection of science and engineering.

(ELISA) type assays use selective bonding to antibodies immobilized on microfabricated surfaces and detect the bioaffinity binding using electrical or optical detectors. Several accomplishments in using BioMEMS with various modes of detection technologies have been reported. A current goal of BioMEMS research is identification and manipulation at the molecular/cellular level. Microfluidics based lab-on-a-chip devices has proved useful for realizing single molecule/cell detection also.

1.3 INTERSECTION OF SCIENCE AND ENGINEERING

The BioMEMS devices are made of silicon, glass, or polymer materials to produce highly functional miniaturized devices. A major driving force behind this is the rapidly emerging biochip market, wherein original MEMS techniques are integrated with advanced techniques from molecular biology, physics, chemistry, and data analysis, as shown in Figure 1.1.

Expansion of the biochip sector is predicted to be one of the key drivers for the growth of the microsystems market, and to have a profound impact on many aspects of life science industries. BioMEMS chips also promise to be among the most important pharmaceutical research and development tools in the postgenomic era.

Diagnostics applications form the largest part of BioMEMS applications. A very large and increasing number of BioMEMS devices for diagnostic applications have been developed and presented in the literature by many groups within the last decade. These devices are used to detect cells, microorganisms, viruses, proteins, DNA and related nucleic acids, and many small molecules of biochemical interest. BioMEMS devices for cell characterization will greatly contribute toward the development of new diagnostics and therapies. Cellular analysis will provide many benefits, including increased sensitivity for early detection of diseased conditions and lab-on-a-chip devices for faster analysis, lower cost, and efficient drug discovery.

1.4 EVOLUTION OF SYSTEMS BASED ON SIZE

One could consider the biological species like protein molecules, cells as tiny machines or mechanisms that produce low signals when they interact with the environment or other bio species. This necessitates the use of structures or devices of similar dimensional scale with the possibility of tuning their properties in order to increase the sensitivity and throughput while reducing the sample volume and cost. Hence, the use of BioMEMS becomes imminent for biological and chemical applications. As a result, BioMEMS devices could serve as a complete laboratory-on-a-chip, combining biochemistry, surface chemistry, microfabrication, microfluidics, microelectronics, and microphotonics. Figure 1.2 shows schematically the trend in the emerging of microsystems with the size of biological systems.[4]

Challenges and opportunities still exist for continuous monitoring and early detection of clinically significant proteins directly from blood and other body fluids. Figure 1.3 shows the research areas that result from the integration of life sciences with micro- and nanoscale systems. The applications of biology to micro- and nanoscale systems are shown on the right side, while the left side of the drawing shows schematically the applications to biomedical problems.

1.5 COMMERCIALIZATION, POTENTIAL, AND MARKET

Figure 1.3 shows the research and application areas of BioMEMS devices for life sciences and chemical sciences.

The emerging field of personalized medicine requires rapid detection technologies. In therapeutic application of BioMEMS, much progress has also been made; for example, the silicon-based implantable devices that can be electrically actuated have an orifice to release preloaded drugs and microcapsules with nanoporous membranes for the release of insulin recognition proteins. Healthcare-related applications include real-time circulatory system monitoring, glucose testing, pacemakers and heart monitoring, nerve and muscle stimulations, and biotechnology analytical systems.

Although a variety of MEMS components and devices, including microreservoirs, micropumps, cantilevers, valves, sensors, and other structures, have been fabricated and tested in the laboratory environment, most devices are designed for *in vitro* diagnostics. However, as MEMS devices can be aseptically fabricated and hermetically sealed, interest in using microfabrication technologies for *in vivo* applications is growing.

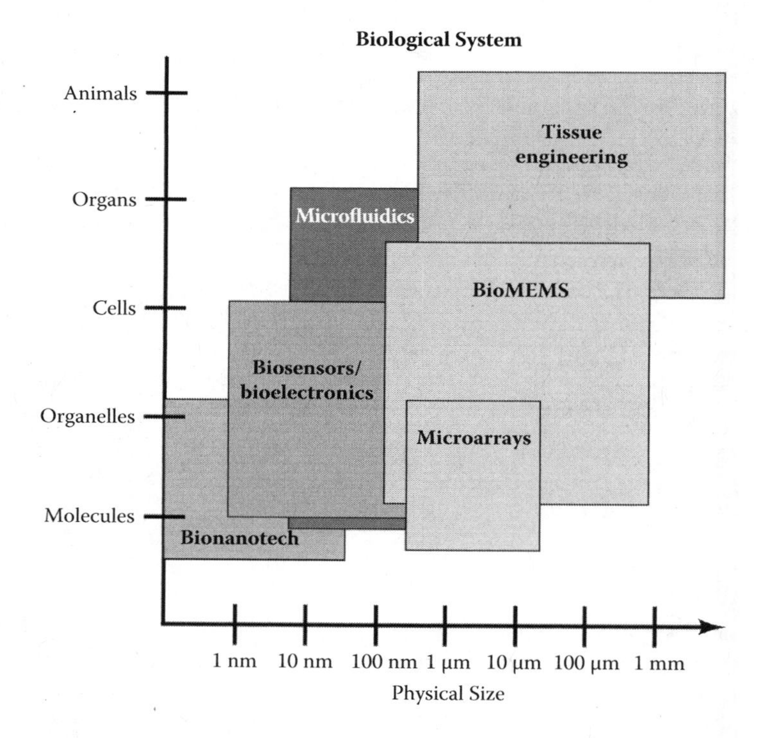

FIGURE 1.2 Evolution of biomicronanotechnologies based on size and complexity. (Adapted from Wikswo et al., *IEE Proc. Nanobiotechnol.*, 153, 81–101, 2006.)

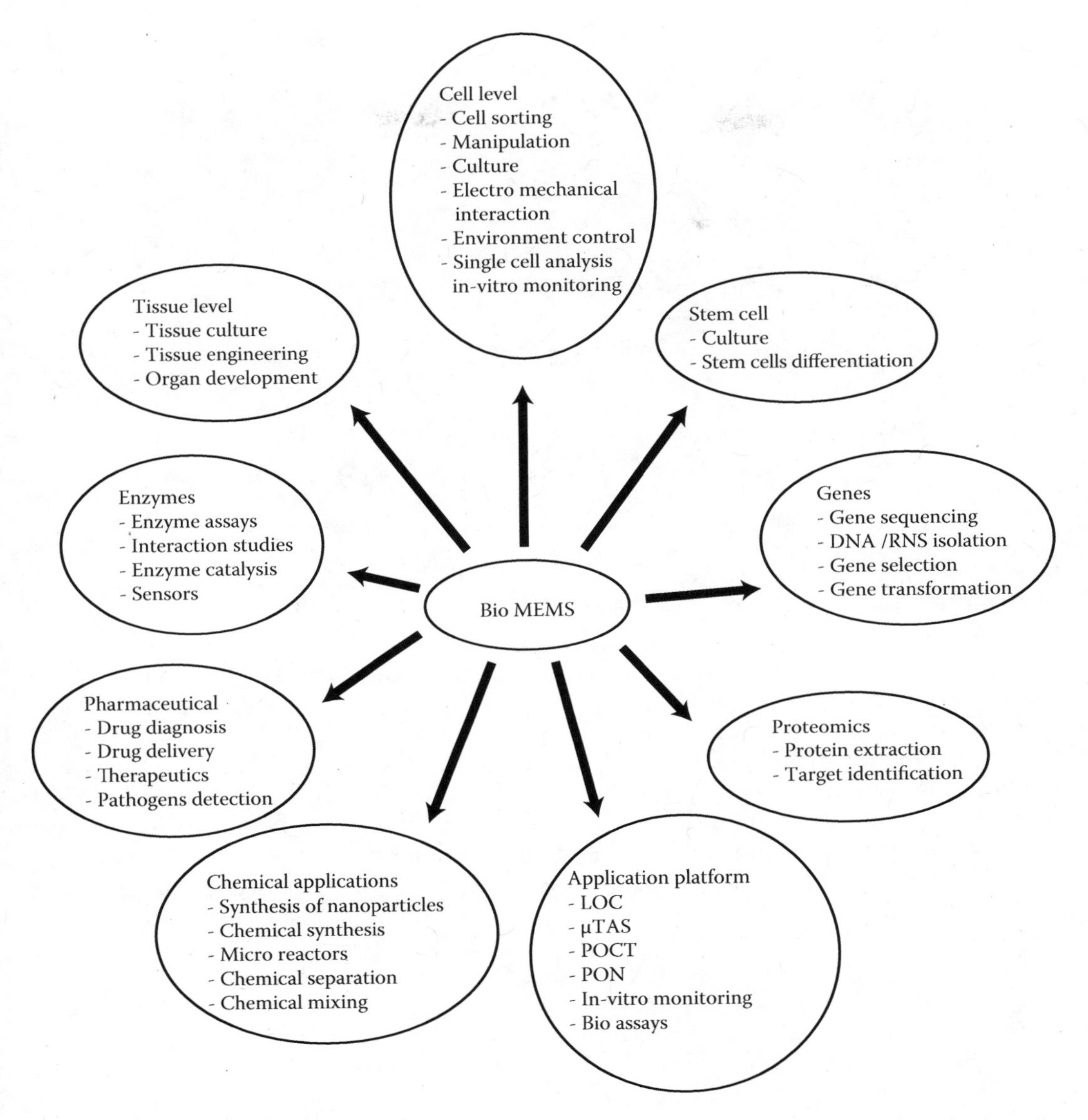

FIGURE 1.3 Research areas resulting from the integration of micro- and nanoscale systems and biomedical sciences.

The most common laboratory techniques, such as immunoassays, nucleic acid amplification tests, blood chemistries (enzymes, gases, electrolytes, etc.) and flow cytometry (counting cells with specific characteristics), can all be carried out in BioMEMS instrumentation. Completely self-contained, battery-operated LOC devices that do not require calibration and are able to be operated at a wide range of ambient temperatures have been developed for the benefit of the developing world.

As we have seen, microfluidic systems could allow miniaturization and integration of complex functions with a capability to process measurements from complex fluids without the need of expert operators to create portable point-of-care (POC) medical diagnostic tools well suited for the developing world. As an example, Figure 1.4 shows a system that uses a "reader" with single use. It was designed for monitoring small-molecule analytes such as hormones and drugs in saliva, and it is based on surface plasmon resonance (SPR) imaging.

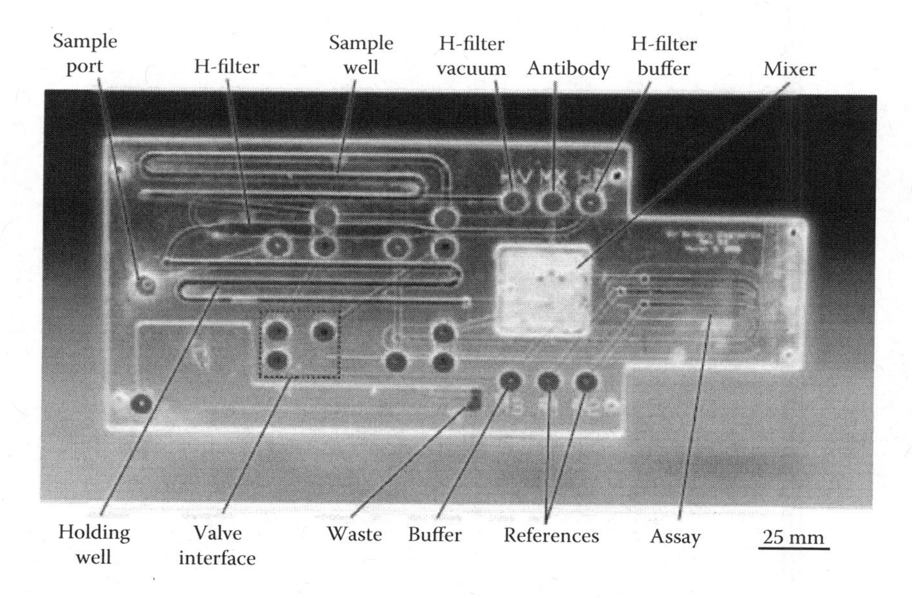

FIGURE 1.4 (See color insert.) POC microchip.

Similarly, disposable single-use lab-on-a-chip platforms were developed for the identification of pathogens as well.

The sensitivity and specificity of the microfluidic diagnostic assay are comparable to those obtained by conventional laboratory microbiological assays. As a result, microchips will have a significant impact on all aspects of diagnostic testing, and miniaturization will facilitate clinical testing to be performed in nonlaboratory settings, closer to the location of the need or patient using point-of-care testing (POCT) and point-of-need (PON) technologies, instead of large specialized central laboratories.[6] Because of the small size and weight, BioMEMS devices are portable, and as a consequence, they can serve as personal laboratories where all the analytical steps can be performed automatically. With recent growth in BioMEMS areas, many of the classical analytical techniques, such as gas chromatography, spectrometry, cytometry, capillary electrophoresis, etc., have been miniaturized and are available commercially.

The applications of BioMEMS have great commercial potential and value. Some of the business areas include microarrays, lab-on-chip, and μTAS for POCT and PON diagnostic applications.

A recent technical market research report from BCC Research[7] estimates that the global market value for biochip products is expected to increase to nearly $6 billion in 2014 from $2.6 billion in 2009. As can be seen in Figure 1.6, the largest segment of the BioMEMS market is DNA microarray, which is expected to grow to $2.7 billion in 2014 from $1.3 billion in 2009. The report stresses the commercialization of biochips from primary research and applications, especially in clinical diagnostics.

BioMEMS is a multidisciplinary field involving expertise from the science disciplines of biology, physics, biochemistry, etc., and the engineering disciplines of fluid mechanics, micromechanics, microelectronics, micro- and nanofabrication, packaging, integration, etc. Hence, this book attempts to introduce the science concepts involved in BioMEMS devices, with engineering perspectives. The initial chapters are devoted to the background knowledge of materials for microfabrication in Chapter 2, classes of biomolecules in Chapter 3, surface manipulation techniques on materials such as silicon, glass and ceramics, and polymers in Chapter 4, and methods of surface characterization in Chapter 5. Although the content of the initial chapters was formed with the aim of revisiting the basic science, sections devoted to biosensing strategies and various applications are included as well.

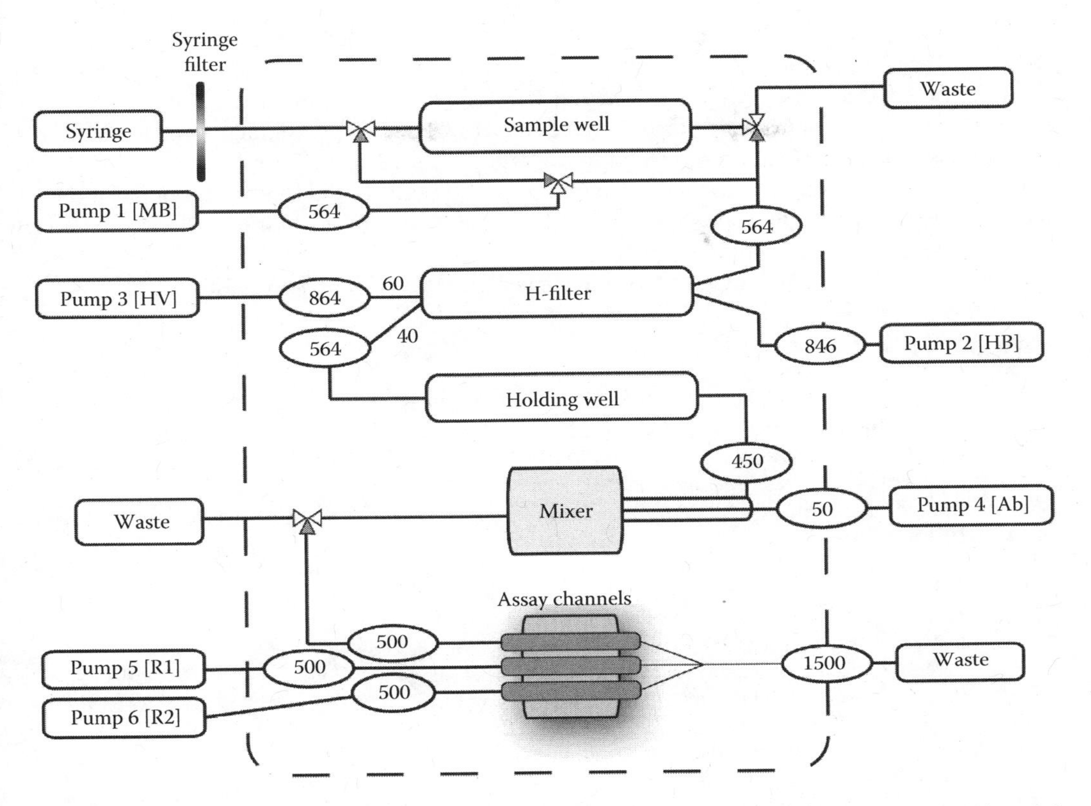

FIGURE 1.5 An integrated disposable diagnostic LOC. Schematic layout of the chip functionalities of the card. Syringe-pump-filtered saliva through an H-filter for further sample conditioning, a herringbone mixer for mixing antibodies with the sample, and channels with gold-coated surfaces for detection of analyte in the sample using an SPR imaging-based immunoassay. Numbers in ovals indicate flow rates in nl/s. Percentages of flow are indicated by the numbers at the H-filter four ports. MB, mixing buffer; HB, H-filter buffer; HV, H-filter vacuum; R1 and R2, reference solutions, typically a positive and negative control. (Reproduced from Yager et al., *Nature*, 442, 412–18, 2006. With permission.)

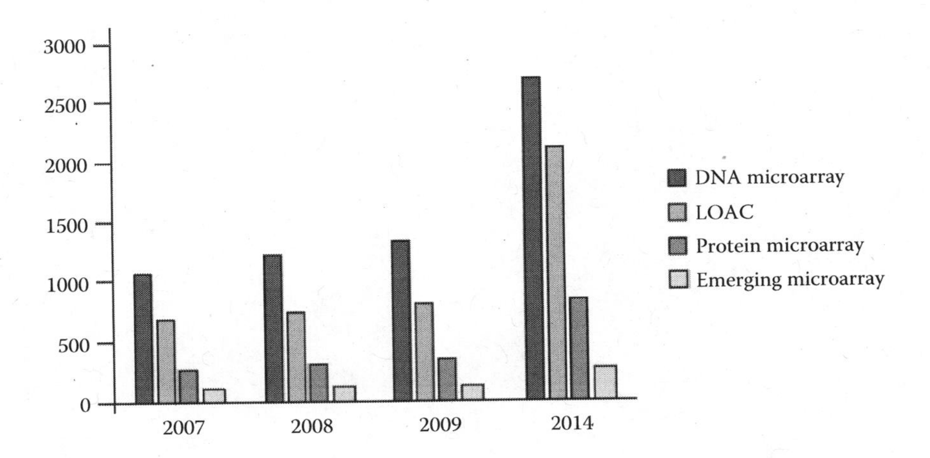

FIGURE 1.6 Global value of biochip products by type. (Adapted from BCC Research, www.bccresearch.com.)

The fundamentals of biosensing are discussed in Chapter 6 and are illustrated with many case studies describing enzyme-, DNA-, and antibody-based sensors. This chapter makes the transition from the science perspective to the engineering aspects of the BioMEMS devices. The later chapters focus on the engineering perspective of BioMEMS devices, with micro- and nanofabrication methods in Chapter 7, an introduction to microfluidics in Chapter 8, and life science applications in Chapter 9. Chapter 9 presents a detailed survey of some of the life science applications of BioMEMS, for example, DNA, protein, and cell-based microarrays and their importance in clinical chemistry along with case studies.

References are provided at the end of each chapter, together with review questions.

REFERENCES

1. Terry, S. C., Jerman, G. H., Angell, J. B. 1979. A gas chromatographic air analyzer fabricated on a silicon wafer. *IEEE Trans. Electron Devices ED26* 12:1880–86.
2. Manz, Z., Graber, N., Widmer, H. M. 1990. Miniaturized total chemical analyses systems. A novel concept for chemical sensing. *Sens. Actuators B Chem.* B1:244.
3. Bashir, R. 2004. BioMEMS: State-of- the-art in detection, opportunities and prospects. *Adv. Drug Deliv. Rev.* 56:1565–86.
4. Wikswo, J. P., Prokop, A., Baudenbacher, F., Cliffel, D., Csukas, B., Velkowski, M. 2006. Engineering challenges of BioNEMS: The integration of microfluidics, micro- and nanodevices, models and external control for systems biology. *IEE Proc. Nanobiotechnol.* 153:81–101.
5. Yager, P., Edwards, T., Fu, E., Helton, K., Nelson, K., Tam, M.R., Weigl, B.H. 2006. Microfluidic diagnostic technologies for global public health. *Nature* 442:412–18.
6. Kricka, L. J. 2001. Microchips, microarrays, biochips and nanochips: Personal laboratories for the 21st century. *Clin. Chim. Acta* 307:219–23.
7. BCC Research. www.bccresearch.com.

2 Substrate Materials Used in BioMEMS Devices

2.1 INTRODUCTION

Since microfluidic systems first came into existence, silicon and glass have been used extensively as substrate materials. Microfabrication methods such as photolithography, wet chemical etching, and fusion bonding adopted from the manufacturing of microelectronic devices were used to structure these materials.[1,2] Deep reactive ion etching (DRIE) is used as an alternative to wet chemical etching when deeper channels are fabricated. These processes have proved to be methods of choice for electronic and mechanical devices at the micrometer and millimeter scales, but are expensive, requiring clean room conditions and intensive labor. Devices were initially fabricated by using only one material. For example, a microfluidic channel was etched into a glass plate and sealed with another glass plate, to produce a monolithic glass chip. In the last decade, both materials and methods have changed, other materials and fabrication technologies have come into play, and polymeric materials are being explored as more versatile alternatives for the fabrication of microfluidic devices. They have become the base material for microfluidic devices due to a wide range of available polymers with a variety of material properties. The polymers that were used in the beginning were mostly polymethylmethacrylate (PMMA) and polycarbonate (PC). The development of polymer microfabrication methods opened the door for the production of disposable microfluidic devices suitable for commercialization. In the case of polymers, a large number of devices can be made from a mold by using techniques such as casting, injection molding, or hot embossing with no need for clean rooms. However, hybrid microfluidic devices are more often used today, especially polymer-metal, polymer-glass, silicon-glass, and polymer-silicon combinations. The general trend is the integration of more functional elements made of other materials into polymer microfluidic devices. In the following sections, a short discussion of the materials used for microfluidic devices and their most important properties is briefly given. The materials and their corresponding microfabrication methods are discussed in detail in the relevant chapters.

2.2 METALS

Metals are used in some microfluidic devices that require high-volume processing of chemicals, high temperature, pressure, chemical resistance, and high thermal conductivity. Metal microfluidic devices are likely to be more expensive than polymer devices, and they will be used in applications where a permanent mold device is used for manufacturing of high volumes of copies. Metals are used in microfluidics, principally for the fabrication of microelectrodes in sensors and actuators that involve electrical potentials. Fabrication of stable microelectrode arrays on polymer wafers introduces serious problems due to low adhesion between metal layers and polymer substrates. This problem can be partially solved by applying a surface treatment to polymers before metal deposition. But, the sealing process that follows electrode patterning becomes difficult with the integration of thick metallic electrodes, due to the creation of a gap between the structured layer and the cover plate. To avoid leakage, the electrodes must be inserted into the microchannel without a step between

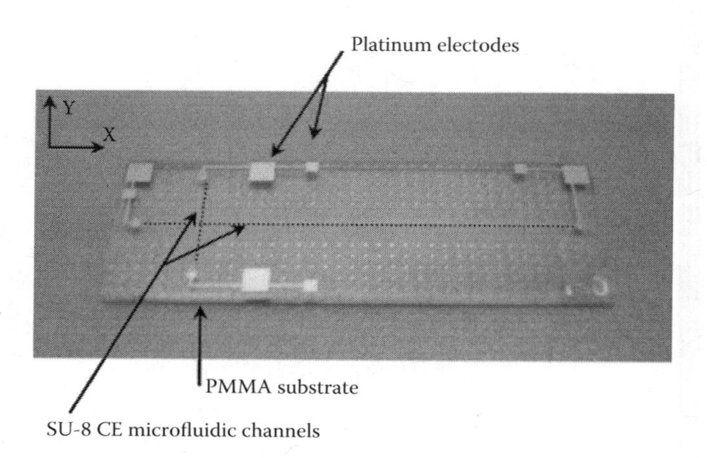

FIGURE 2.1 Polymer microfluidics chip integrated with metal electrodes. (Reproduced from Arroyo et al., *J. Micromech. Microeng.*, 17, 1289–98, 2007. With permission.)

the channel and the substrate. Figure 2.1 shows a polymer microfluidic device with integrated platinum microelectrodes. Figure 2.1 shows a metal-coated microchannel in a silica platform.

Titanium is a potential material for BioMEMS applications as it is a mechanically stable implant material known for its biocompatibility, and it is used in orthopedics due to its osteointegrative properties. However, surface coatings and functionalization of titanium are necessary for cell culture and tissue engineering.

2.3 GLASSES AND CERAMICS

Glass is a robust material, with good chemical and temperature resistance. However, glass microfluidic devices have a higher cost per function when manufactured in volume compared to polymer chips. Glass is often favored when sensitive chemical reactions are involved. Besides silicon, glass is a widely used substrate material in microsystem technology, in particular in the manufacture of microfluidic devices for biological analyses, as it provides beneficial structural and functional material properties. In comparison to silicon, the use of glass in micro total analysis systems (μTAS) applications is advantageous with regard to its optical transparency, which allows for visual inspection and online optical detection. Glasses and ceramics have good fluorescence and dielectric properties that allow them to withstand high voltages used in electrokinetically driven flows. Other beneficial properties of glass are its good chemical resistance, high thermal stability, chemical inertness, and established protocols for surface modification and functionalization (silane modification), which make glass the most used substrate for the fabrication of DNA arrays. The use of glass substrates may also improve the long-term chemical stability of the devices in comparison with silicon-based systems. Many applications would also exploit high mechanical strength and the good mechanical stability of glass.

Glass is a brittle material and it is not a good conductor. But it is readily amenable for metallization, while it is resistant to many harsh chemicals; it has good optical properties as well. It has proved to be a better substrate than silicon for applications involving high electric fields, namely, electroosmosis. As it is transparent in visible and near-infrared (NIR) wavelength ranges, it is a perfect material for optical-based detection and flow visualization. High endurance to high-temperature sterilization and excellent biocompatibility make this material suitable for developing biochips. Both silicon and glass possess excellent chemical stability and solvent compatibility. However, microfabrication of glass is difficult because of its amorphous nature. Moreover, the process used for bonding glass chips requires a clean room environment. Devices were first manufactured in common glass, Pyrex, and quartz, into which channels were etched by using similar methodologies as those used for silicon.

In summary, glass material is less preferred than silicon and polymer due to their limitations in microfabrication, in spite of being an excellent optical and compatible material.

There exist a variety of glass materials based on silica-activated oxide, such as soda-lime glass, borosilicate glass, and pure silica glass (quartz glass). A special variety of glass is amenable to anisotropic photostructuring, so it does not require an intermediate photoresist layer for patterning. It is commercially available through various suppliers and is patterned by photolithography using a mask. Schott Glass Corporation manufactures a photosensitive glass (trade name Foturan) that is UV sensitive in the wavelength range of 290 to 330 nm, with mechanical, thermal, and electrical properties similar to those of conventional glass. It is composed of alkali (lithium)-aluminosilicate glass doped with ions of cerium as a photosensitizer and silver, which makes the glass photosensitive. The process consists of a UV exposure, followed by a heat treatment and wet etching. Exposure to UV results in the formation of a three-dimensional latent image within the glass confined to the zone where the UV is absorbed. The image is formed by the silver atoms. The subsequent heat treatment promotes good crystallization within the exposed areas. The microcrystallites (lithium silicate) are grown around the silver nuclei during an annealing process and are subsequently dissolved in an HF solution. Microchambers and channels fabricated in Forturan glass are shown in Figure 2.2.

As the chemical etching process is anisotropic and the etch rate for the crystallized parts is up to twenty times higher than the amorphous regions, almost vertical sidewalls can be obtained. It is possible for a fine structure down to 25 μm with a high aspect ratio to be fabricated in glass.

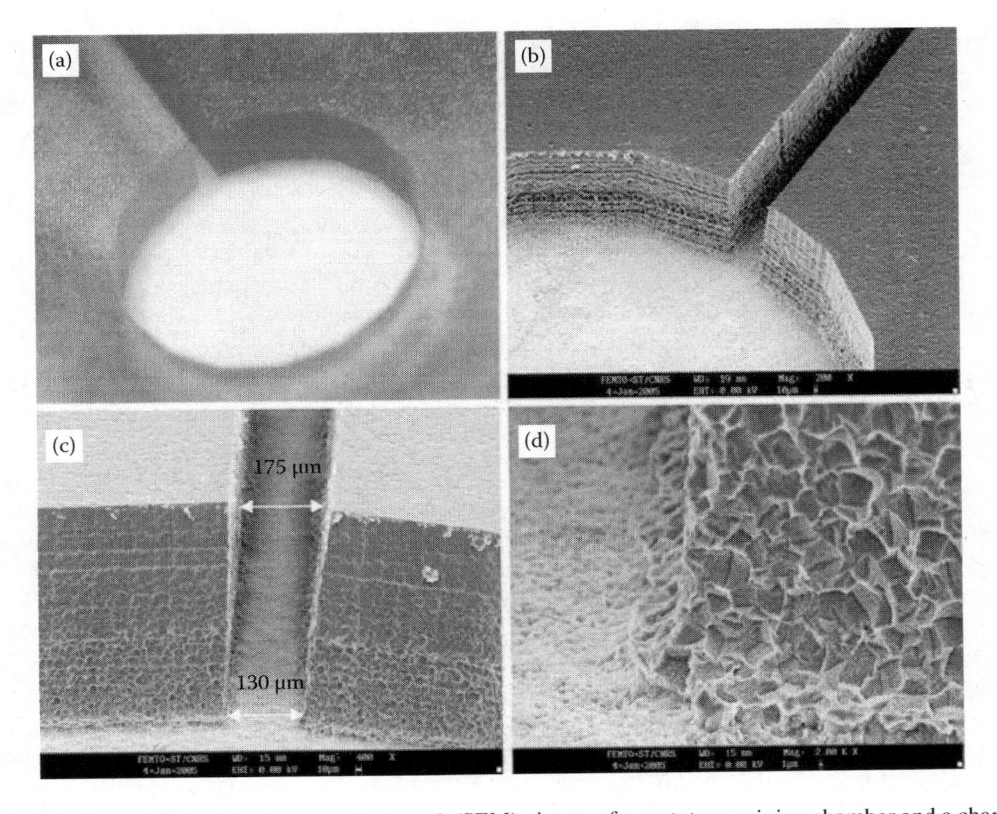

FIGURE 2.2 (a) Scanning electron micrograph (SEM) picture of a prototype mixing chamber and a channel etched in Foturan glass. (b) SEM picture of the close-up channel and reservoir. (c) SEM picture of the channel entrance showing the difference in width of the microchannels as a function of depth (130 to 175 μm wide, 415 μm deep). (d) Close-up of the bottom wall showing glass grains. (Reproduced from Khan Malek et al., *Microsyst. Technol.*, 13, 447–53, 2007. With permission.)

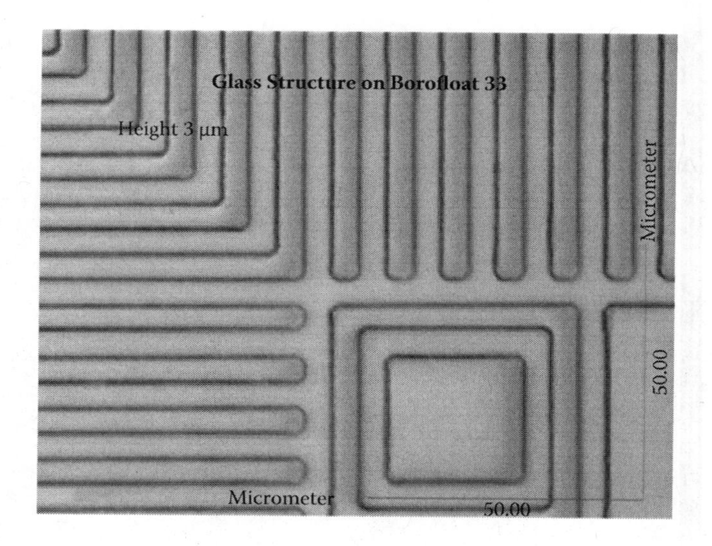

FIGURE 2.3 MSG microfluidic structure. (Reproduced from Munds and Leib, "Novel Microstructuring Technology for Glass on Silicon and Glass Substrates," in *Electronic Components and Technology Conference*, Las Vegas, NV, 2004, pp. 939–42. With permission.)

Microstructuring glass (MSG) is an exciting new technology that has the potential for solving demands in microelectronic and microfluidic applications, including hermetic wafer level packaging. MSG offers many advantages, such as chemical-resistant borosilicate glass layers, layered structuring with photoresist, and high glass deposition rates at low temperatures. A microfluidic chip manufactured using MSG technology is shown in Figure 2.3. The chip has many microchannels of 3 μm in height.

The enormous progress made in the field of medicine over the past few decades has been due to the introduction of new instruments and the use of new materials. It is impossible to imagine modern medicine without metals, alloys, sintered corundum, organic high polymers (also as composite materials), glassy carbon, etc. The biocompatibility of these glasses has led them to be investigated extensively for use as implant materials in the human body to repair and replace diseased or damaged bone. *Bioceramics* and bioglasses are ceramic materials that are biocompatible. Bioceramics are an important subset of biomaterials. Bioceramics range in biocompatibility from the ceramic oxides, which are inert in the body, to the other extreme of resorbable materials, which are eventually replaced by the original materials that are used for repair. Bioglass ceramics opens up new possibilities for medical treatment and constitute a new area of research in the natural sciences and medicine. Bioglass ceramics can be more easily adapted to suit medical requirements than customary implants due to their tunable properties. Two properties of bioglass ceramics are of primary importance: *biocompatibility*, i.e., acceptance of the material by the tissues of the human body without irritation, rejection reactions, or toxic effects; and *bioactivity*, i.e., the ability to establish firm intergrowths with tissues of the human body. This property is not shared by any of the classical biomaterials. In addition, *bioadaptability*, the capacity of the material to collaborate with the surrounding tissue in replacing the removed body part in the best way, is an essential requirement.

Composites containing polymer or metal with various bioceramics that can give enhanced mechanical properties and bioactivities have been developed. Polymers such as HDPE, PMMA, polysulfone, polylactides, and their copolymers with bioglasses have been found to be highly suitable for biomedical applications. Most of the bioactive glasses are based on the SiO_2-P_2O_5–CaO-Na_2O system. Other glass and glass ceramics include ZnO, Ag_2O, and Al_2O_3. The production of the materials using the sol-gel process allows the tailoring of textural characteristics, such as size and structure of pores, in order to enhance the bioactive behavior.

2.4 SILICON AND SILICON-BASED SURFACES

Silicon is an excellent material with good mechanical, electrical, optical, and biocompatibility properties. Silicon is favored for the availability of a large variety of high-precision microfabrication and micromachining methods dedicated to it. The manufacturing methods that exist for silicon substrate are not only precise but also capable of producing chips in batch production with less cost. Silicon opens up an immense possibility of integration with microelectronics and microphotonics elements. The transparency of silicon in the near-infrared wavelength range makes the biochips integrate monolithically with silicon photonics elements such as waveguides, couplers, gratings, interferometers, etc. The well-established silicon-based microelectronics also makes silicon an excellent material for biochips with electrical controls for electroosmosis, electrowetting, etc.

Silicon is used to provide low-volume prototypes where direct integration of electronics and photonics adds value. The availability of lithography, etching, and other fabrication methods from the microelectronics industry, which have been extensively developed over the last decades, makes silicon interesting. As an excellent mechanical material, silicon has a high Young's modulus and a thermal conductivity that is suitable for applications involving deflection and heat transfer. Even though it is basically a brittle material, it has a high yield strength and an elastic regime compared to metals. In essence, it can have more elastic deformation than metals. The disadvantages of using silicon are its high cost, elaborate steps of fabrication (cleaning, resist coating, photolithography, etc.), and a long fabrication time. The requirement of a high-quality clean room facility necessary for the fabrication of silicon-based microfluidic devices makes it unsuitable for many research groups to produce chips in few numbers. In addition, harmful chemicals such as SiH_4, NH_3, solvents, HF, and KOH are usually involved in the process. The microfluidic chip fabricated on silicon is shown in Figure 2.4.

Geometries of the cross section of microchannels and their aspect ratio (height vs. width) are limited due to the isotropic nature of silicon. Silicon is optically opaque in the visible/ultraviolet region of the spectrum (200 to 800 nm). As a result, it is suitable for optical-based detection only in the NIR range and not in the visible range. In spite of microelectronic compatibility, the semiconductivity of silicon may cause problems in the commonly used electroosmosis pumping. Its intrinsic stiffness is not beneficial for making devices with moving parts. As silicon is hydrophobic, adsorption of biomolecules is possible only after surface functionalization, and the surface chemistry is difficult to achieve with high temperatures required to bond silicon and glass. It is not biodegradable and has limited biocompatibility, and therefore is not a suitable biomaterial for a tissue engineering scaffold. In addition, as mentioned, the semiconductive properties can be limiting in applications that involve electroosmotic and electrophoresis flows. Glass is more suitable for these kinds of applications because of the surface charge. The cost of silicon is likely to be uncompetitive

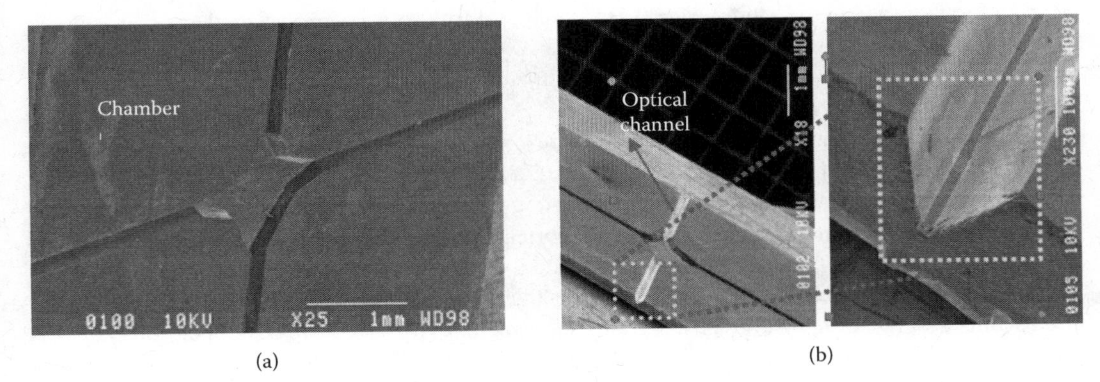

FIGURE 2.4 SEM images of (a) fabricated microfluidic chip on silicon. (b) Close-up of the channel. (Reproduced from Chandrasekaran et al., *Sensors*, 7, 1901–15, 2007. With permission.)

in the long run. Finally, silicon circuitry is sensitive to temperature, moisture, and magnetic fields. The more likely approach is that silicon sensors will be more and more hybridized with glass and polymer microfluidic devices, to create advanced sensors. However, such systems can be technically demanding and expensive. Silicon is a well-suited material for integrating electrical connections, but integrating fluidic interconnections is more difficult and needs hybrid integration with other materials for interconnection.

2.5 POLYMERS

Today, polymers are likely to be the most widely used substrate materials because of their low cost. This makes them suitable for applications where either a microfluidic device is used once and disposed of or cost is critical. Polymeric lab-on-a-chip devices are easy to fabricate without the use of advanced micromachining technology. However, polymers are not robust, under high temperatures or harsh chemical environments. Polymers have good optical and mechanical properties. However, it is difficult to deposit metal onto them, and so it is problematic to have electrical lines into a polymer system, especially when high voltage needs to be applied to the device. Polymers, generally, have low thermal conductivity. In a high-temperature operating environment, cooling mechanisms have to be used to dissipate heat. Some polymers are transparent in both visible and NIR ranges and suitable for optical-based detection and flow visualization. They provide a variety of material choices with various physicochemical properties to meet different applications. Polymers are suitable for electrophoresis and other electrokinetic applications.

In spite of the above advantages, polymer chips suffer from some drawbacks. The fabrication methods are not fully developed. In addition to being thermally unstable, they are not suitable for applications involving heat transfer due to their poor thermal conductivity. At high temperatures, they tend to deform due to the residual molding stress. One has to be careful in optical applications, as most polymers have strong absorption at certain wavelengths, depending upon their chemistry. Commercial manufacturers of microfluidic devices see many benefits in using plastics that include reduced cost and simplified manufacturing procedures when compared to glass and silicon. There is a wide range of available plastic materials, which allows the manufacturer to choose materials' properties based on their suitability for specific applications. For many applications, plastics instead of pure polymers are used to fabricate microfluidic devices. Plastics are polymers with a number of additives, such as fillers, plasticizers, heat, and UV stabilizers.

When hot embossing and injection molding fabrication methods are used for polymers, there are many critical parameters, such as glass transition temperature, melting temperature, and coefficient of thermal expansion, to be considered. Glass transition temperature is the temperature range at which the polymer changes from a rigid glassy material to a soft material. The degree to which a thermoplastic material softens is dependent on its crystallinity. The melt temperature is the temperature at which the polymer melts. The melting temperature is much higher than the glass transition temperature. The coefficient of thermal expansion refers to the change in length per unit length due to the change in temperature. It is a critical parameter to be considered in fabrication and bonding where different materials are thermally bonded. In addition, their good isolating properties make polymers suitable for several microfluidic applications. Different polymeric materials and fabrication processes have been developed in order to fabricate microfluidic devices. Thermoplastic, thermosetting, and elastomer materials have been tested for different microfluidic applications. In order to create microchannels with different polymers as the structural materials, technologies such as hot embossing, laser ablation, soft lithography, and injection molding have been developed. The key feature to obtain robust polymeric microfluidic devices is to obtain an effective seal between the structured layer and a cover plate. Table 2.1 provides a list of some common polymer-based materials and their properties that are important for fabrication. As more micro- and nanofluidic methodologies are developed for a growing number of diverse applications, it becomes increasingly apparent that

TABLE 2.1
Material Properties of Polymers

	Polymethyl methacrylate (PMMA)	Polycarbonate (PC)	Polyimide	Polystyrene (PS)	Cycloolefine Copolymer	Polydimethyl siloxane (PDMS)
Density (× 10^3 kg/m^3)	1.16	1.2	1.39	1.05	1.02	1.23
Glass transition temp., T_g (°C)	106	150	285	100	90–136	−120
Useful temp. range (°C)	−70–100	−150–130	−73–240	−40–70	−73–80	−40–150
Thermal conductivity (W/mK)	0.186	0.21	0.2	0.18	0.16	0.17–0.3
Transmission (%) in visible range	92	89	87	90	92–94	91
Solvent resistance	Poor	Poor	Fair	Poor	Fair to poor	Poor

Source: Adapted from Becker and Gartner, *Electrophoresis*, 21, 12–26, 2000.

the choice of substrate material can have a profound effect on the eventual performance of a device. This is mostly due to the high surface-to-volume ratio that exists within such small structures.

Among the polymers that are used as materials for the fabrication of microfluidic devices, polydimethylsilaxane (PDMS) has received the most attention due to its ease of preparation, low cost, good transparency, and nontoxicity. In addition, it becomes an elastomeric material after cross-linking, with a low Young's modulus of ~750 kPa, which enables it to form reversible seals. The UV-visible spectrum of PDMS, together with that of glass and other polymers, is shown in Figure 2.5.

Optical and mechanical properties of PDMS strongly depend on the amount of cross-linking agent used for the fabrication of polymer. Figure 2.6 shows the dependency of UV-visible transparency on the amount of cross-linking agent.

PDMS has a distinctive mechanical property: the Young's modulus can be tuned over two orders of magnitude by controlling the amount of cross-linking between polymer chains. The mechanical properties and gas permeability make it highly suitable for manufacturing devices supporting living cells. The fabrication of PDMS chips is simple and can be achieved in any laboratory by soft lithography, hot embossing, casting, etc. PDMS is easily cast against the master to yield a polymeric replica containing the network of microchannels. The master in metal can be made through standard machining, while masters in photoresist or silicon are made through lithography and micromachining.

Among the limitations with polymers, hydrophobicity is important, as it leads to nonspecific adsorption of some hydrophobic compounds on its surface when the surface is not oxidized. Oxidized PDMS has a short lifetime, as short as 1 day. As a result, PDMS needs surface treatment for fabricating devices. A lot of research focus is presently directed on the methods used for the modification of the PDMS surface in order to make it more hydrophilic. Another limitation of PDMS is its low compatibility for electroosmotic flow and for many organic solvents. Solvents such as dichloromethane, acetone, toluene, etc., swell the PDMS-based microfluidic devices, and they cannot be flown inside the channels. Swelling causes distortion of the fluidic structure, and even channel closure if the channel width is not large enough. It also causes pattern distortion in

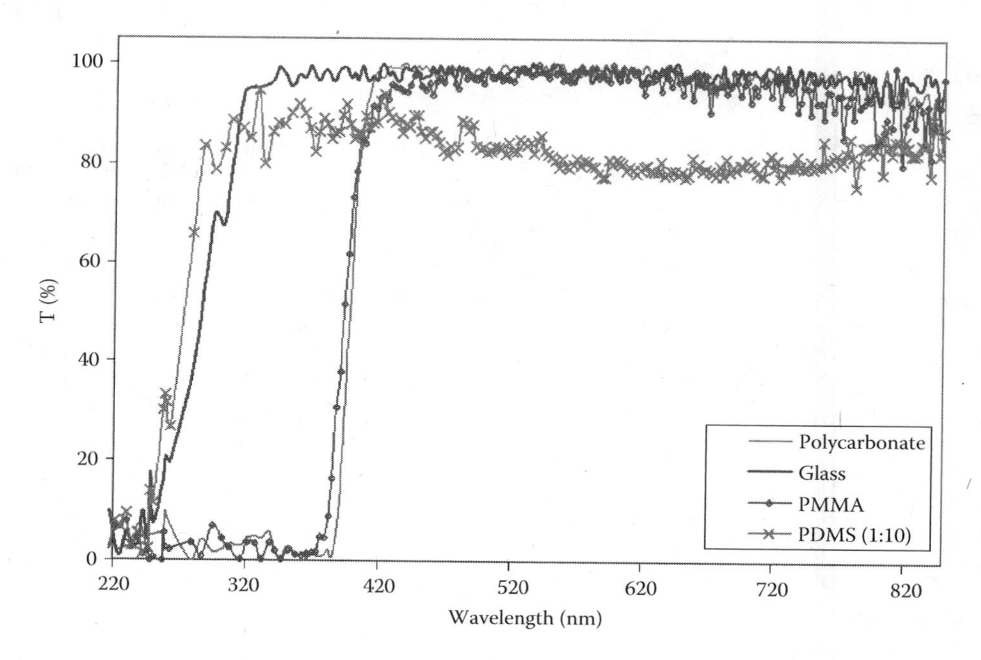

FIGURE 2.5 (See color insert.) UV-visible spectrum of PDMS, glass, polycarbonate and polymethylmethacrylate. (Unpublished data from the authors' laboratory, Acsharia, "Modeling, Fabrication and Testing of Oled and Led Fluorescence Integrated Polymer Microfluidic Chips for Biosensing Applications," MASc thesis, 2007.)

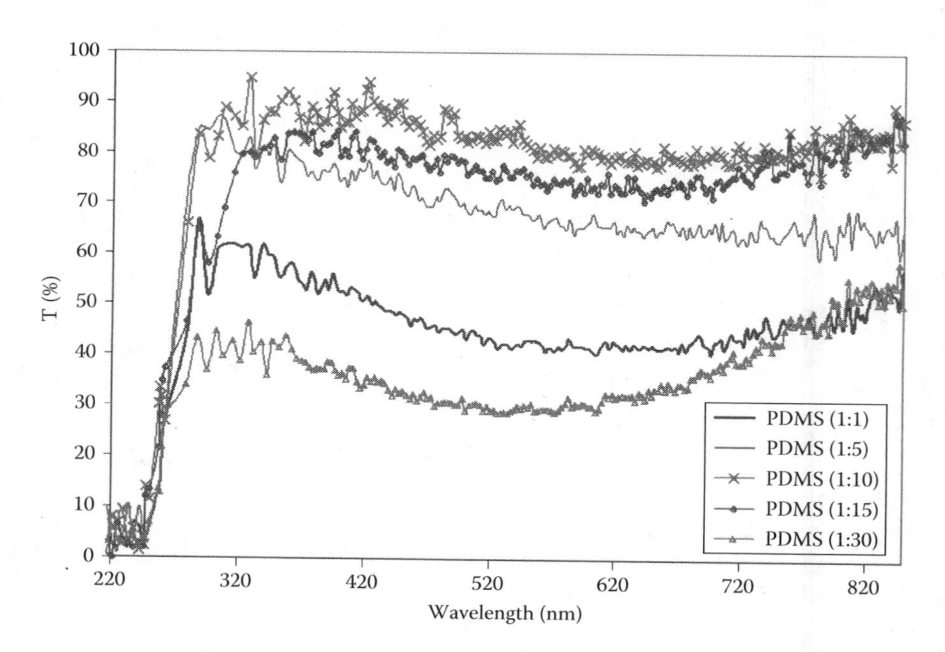

FIGURE 2.6 (See color insert.) UV-visible spectra of PDMS fabricated with different amounts of cross-linking agents. (Unpublished data from the authors' laboratory, Acsharia, "Modeling, Fabrication and Testing of Oled and Led Fluorescence Integrated Polymer Microfluidic Chips for Biosensing Applications," MASc thesis, 2007.)

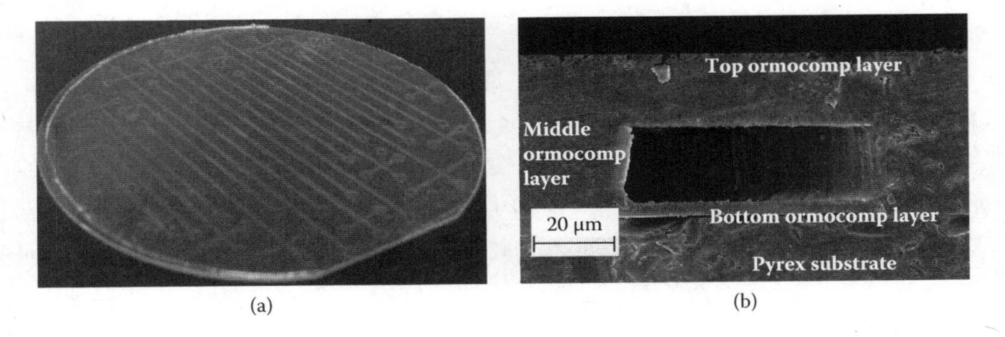

(a) (b)

FIGURE 2.7 (a) A photograph of a full-wafer (diameter 100 mm) array of enclosed Ormocomp separation microchips. (b) A scanning electron micrograph of the cross section of an Ormocomp separation channel (on Pyrex substrate), demonstrating the good bonding quality between the different Ormocomp layers. (Reproduced from Sikanen et al., *Anal. Chem.*, 1004053, 2010. With permission.)

lithography, and even mold pattern collapse. To overcome the swelling problem, a passivation layer of 0.2 to 5 μm thick polytetrafluoroethylene (PTFE) is formed on the channel surface. PDMS also lacks the ability to be easily functionalized with biologically active components due to the processing requirements (solvents, thermal processing). Recently, polymethylhydrosiloxane (PMHS) has been investigated as a functional material for microfluidic chips.[9] PDMS is also used for adhesive bonding of various materials. Cycloolefin polymers and copolymers, a group of relatively hard plastics, have also received attention because of their high chemical stability and optical transparency, which make them promising materials for applications in analytical chemistry and molecular biotechnology. Zeonor resins, for example, are cyclic olefin polymers and copolymers that exhibit low water absorption, high transparency, and good chemical resistance to polar solvents, making them well suited for microchip designs involving organic solvents. The wide UV-visible transmittance is beneficial for on-chip optical detection.

Polyimides are nontoxic materials, stable in physiological environment, and used for neural implants. They are flexible and have an insulation resistance similar to that of silicon. Hybrid microsystems for flexible neural interfaces in case of neural disorders have been developed.[10]

A new, commercial hybrid material, Ormocomp, was developed by Microresist Technology GmbH for microfluidic bioanalytical applications. Ormocomp belongs to the family of organically modified ceramics (ORMOCERs), the properties of which can be tailored to different applications, such as optical devices or antistatic and antiadhesive coatings. ORMOCERs were first developed by the Fraunhofer Institute in the 1980s. The manufacturing process includes a sol-gel preparation of the inorganic backbone (typically Si-O-Si), followed by cross-linking of the organic side chains. By varying the amount of inorganic and organic elements, one can adjust the properties of ORMOCERs in a wide range. The original idea was to combine the properties of organic polymers (functionalization, ease of processing at low temperatures, toughness) with those of glass-like materials (hardness, chemical and thermal stability, transparency).

Ormocomp was introduced to the fabrication of microfluidic separation chips (Figure 2.7) using two independent techniques, UV lithography and UV embossing. Both fabrication methods provided Ormocomp chips with stable cathodic electroosmotic flow, which enabled examination of its biocompatibility by means of microchip capillary electrophoresis (MCE) and (intrinsic) fluorescence detection. The hydrophobic and hydrophilic properties of Ormocomp were examined by screening its interactions with bovine serum albumin and selected amino acids of varying hydrophobicity. The results show that the ceramic, organic-inorganic polymer structure natively resists biofouling on microchannel walls, so that the Ormocomp microchips can be used in protein analysis without prior surface modification. Ormocomp was shown to be suitable for optical fluorescence detection even in the near-UV range (for example, in the 355 nm range), with detection limits at a nanomolar level (approximately 200 nM) for selected inherently fluorescent pharmaceuticals.

Many applications of polymer-drug systems were considered. By using polymers as carriers, drugs can be continuously released for very long periods of time (over a year in some cases). By using different polymeric systems or altering polymer-drug incorporation procedures, one can obtain widely varying release rates.

The most used system is a reservoir, a core of drug surrounded by a swollen or nonswollen polymer film, and the drug diffuses through the polymer. These systems include membranes, capsules, microcapsules, liposomes, and hollow fibers, as shown in Figures 2.8 and 2.9. The polymers used clinically are mostly hydrogels, such as hydron, and ethylene-vinyl acetate copolymer. These polymers are relatively inert, do not readily biodegrade, have good tissue biocompatibility, and are generally permeable only to low molecular weight solutes (MW < 600).

Stimuli-responsive polymers, also called "smart" polymers or environmentally sensitive polymers, are being used more and more in the field of controlled and self-regulated drug delivery. Small changes in the environment may trigger rapid changes in the microstructure of these polymers, from a hydrophilic to a hydrophobic state. Potential stimuli and responses of stimuli-responsive polymers are shown in Figure 2.10.

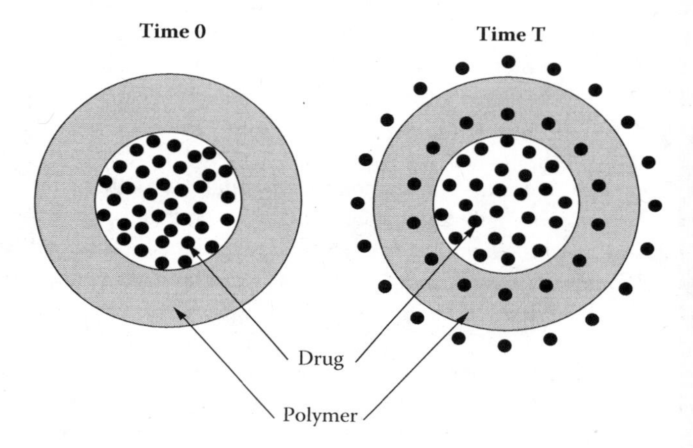

FIGURE 2.8 Idealized diffusion-controlled reservoir release system. (Adapted from Langer and Peppas, *Biomaterials*, 2, 201–14, 1981.)

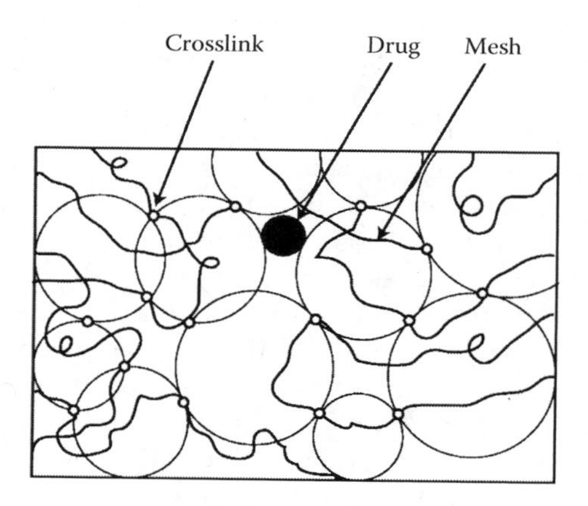

FIGURE 2.9 Screening effect of macromolecular cross-linked structure on drug diffusion through membranes. (Adapted from Langer and Peppas, *Biomaterials*, 2, 201–14, 1981.)

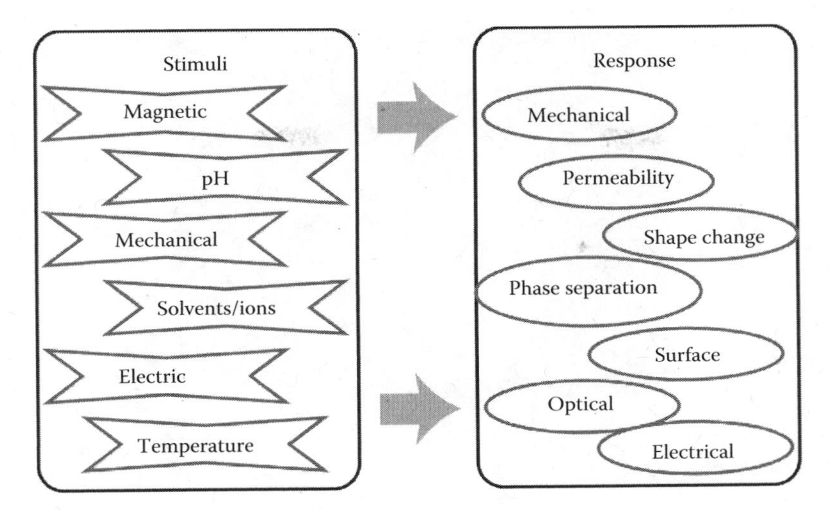

FIGURE 2.10 Common stimuli and the changes triggered in the structure of polymers. (Adapted from Bawa et al., *Biomed. Mater.*, 4, 1–15, 2009.)

Hydrogels that swell in aqueous media, for example, are able to respond rapidly to small changes in pH and temperature, allowing for pulsatile drug delivery to specific sites. Block copolymers with hydrophobic polymers such as polystyrene, polybutylmethacrylate, or poly-lactic acid form polymeric micelles and show a change in the hydration and dehydration properties upon a change in the temperature. Thermoresponsive polymers can be utilized for drug delivery to tumor sites.

2.6 BIOPOLYMERS

Applications such as tissue engineering, drug delivery, and other biomedical applications need precise control of the interface between the synthetic and living materials. New versatile materials have been developed, and they can be easily processed and functionalized for bioapplications.

An ideal biomaterial for BioMEMS fabrication would have the following material properties: (1) it can be processed using mild conditions to facilitate protein or growth factor incorporation, (2) it naturally promotes adhesion and normal function of seeded cells, (3) it contains moieties for potential chemical modification of the surface, (4) it exhibits slow and predictable degradation rates to maximize the duration of functional implanted devices, (5) it has robust, yet flexible mechanical properties, and (6) it is relatively inexpensive. One class of *natural biomaterials* that could potentially meet these material requirements is silk fibroin from the *Bombyx mori* silkworm, which has been used in medicine for a wide variety of applications, including surgical, drug delivery, and tissue engineering. Silk fibroin exhibits *in vitro* and *in vivo* biocompatibility, robust mechanical properties, including high mechanical modulus and toughness, and relatively slow proteolytic biodegradation.

Other biopolymers, such as collagen, chitosan, gelatine, etc., are also naturally occurring, and inherently biocompatible for *in vitro* and *in vivo* situations. They can be processed using microfabrication techniques (micromolding and sealing) and can be used for biomolecules and cell cultures.[14] Biopolymers are used as bulk and film materials in microfluidics. Some materials, for example, gelatine (melting point at 35°C), have to be cross-linked in order to be stabilized. The enzymatically cross-linked material is thermally stable up to 95°C, and the material shows low autofluorescence (Figure 2.11).

Another biomaterial used for cell culture in microfluidic channels is silk, which has proved to be suitable for biologically interfaced microdevices.[15]

When selecting the most suitable material for a particular application, the physical, chemical, and biological properties of the material have to be carefully considered. In addition, especially in the

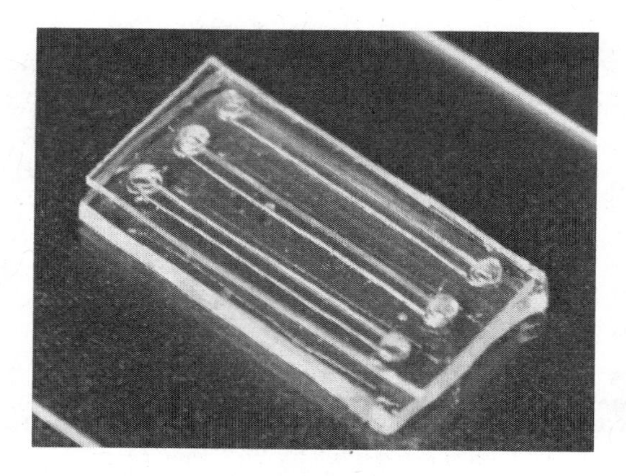

FIGURE 2.11 Microfluidic device produced from transglutaminase cross-linked gelatine with a PDMS top layer for sealed connection ports. (Reproduced from Paguirigan and Beebe, *Lab Chip*, 6, 407–13, 2006. With permission.)

case of biological applications, the wetting properties of the material have to be considered. The compatibility with the detection methods has to be taken into account as well in selecting the materials.

In addition to natural biopolymers, there is a wide range of polymeric materials that are highly biocompatible, such as medical grade silicone (Silastic), polyvinylalcohol (PVA), polyethyleneoxide (PEO), and polyimide (PI). In order to pattern these materials at a small scale and tailor their surface features and porosity, new methods of microfabrication are being developed.[17]

2.7 ORGANIC MOLECULES (FUNCTIONAL GROUPS) INVOLVED IN THE FORMATION OF SELF-ASSEMBLED MONOLAYERS

Self-assembled monolayers (SAMs) are formed on the surface of gold or other substrates, such as glass, metal oxides, etc. They are discussed in detail in Chapter 4, dedicated to the functionalization of surfaces, in order to make them compatible with biomolecules. The most important classes of organic molecules adsorbed as SAMs on various substrates are as follows:

- *Silanes* ($RSiX_3$, R_2SiX_2, or R_3SiX, where R is alkyl chain and X is alkoxy or Cl) on hydroxylated surfaces, such as silicon dioxide on silicon (SiO_2 on Si), aluminum oxide on aluminum (Al_2O_3 on Al), and glass
- *Alkanethiols* (shortly thiols) on Au, Ag, and Cu
- *Dialkyl sulfides* on Au
- *Alcohols* and *amines* on Pt
- *Carboxylic acids* (C_nH_{2n+1} COOH) on Ag and Al_2O_3

In most of the systems, the self-assembly of thiols on gold surfaces leads to the formation of Au-S covalent bonds. Because alkanethiols and silanes are the most important classes of organic compounds that form SAMs, their structural features and general physical and chemical properties will be briefly discussed. Their characteristic functional groups are thiol (SH) and Si-OR (silane) groups. The *functional group* is a group of atoms attached to a carbon backbone formed of C-C and C-H bonds. It represents the reactive part of the molecule and distinguishes one organic molecule from another. A functional group determines not only the physical properties and chemical reactivity of a class of compounds, but also the type and strength of intermolecular forces. *Thiols* (general formula: R-SH, where R is an organic radical formed of C and H) are the sulfur analogues of alcohols (general formula: R-OH). However, the S-H bonds are less polar than the O-H bonds since sulfur is less electronegative than oxygen. Thiols have lower boiling points than the corresponding alcohols and phenols because they show less association by hydrogen bonding. Thiols are also stronger acids than similar alcohols, while their pKa is around 10 to 11, compared to 16 to 19 for R-OH. The *thiol* group is also referred as a *sulfhydryl group*, or *mercaptan*. Examples of thiols are *methanethiol* (IUPAC nomenclature) or *methyl mercaptan* (CH_3SH) and *benzenethiol* or *thiophenol* (C_6H_5SH). The deprotonated form, RS^-, is called a *thiolate*. Oxidation of thiols leads to disulfides (R-S-S-R), as follows:

$$2R\text{-}SH + Br_2 \rightarrow R\text{-}S\text{-}S\text{-}R + 2HBr$$

This reaction, which occurs easily with oxygen or bromine, is biologically important (cysteine → cystine in proteins).

The *silane* molecule (SiH_4) is the silicon analogue of *methane* (CH_4). In general, silane is a silicon analogue of an alkane hydrocarbon and can be represented by the general formula Si_nH_{2n+2}. Silicon is more electropositive than carbon, and so silicon-based chemicals exhibit important physical and chemical differences compared to analogous carbon-based chemicals. An *organosilane* is a molecule that contains at least one carbon-silicon bond (CH_3-Si-). Chlorsilanes, silyl amines, and alkoxy silanes are molecules that have the chlorine, nitrogen, or methoxy groups, respectively, attached directly to silicon. These molecules are very reactive, and they react readily with water to form silanols and siloxane bonds (-Si-O-Si-), which are very stable structures. Generally, silanes are used to bond glass, an inorganic material, to organic materials or compounds.

In the case of glass, the hydrolyzed silane that has OH groups is adsorbed on the surface OH groups of glass and forms an O-Si-O- network, as shown in Figure 2.12.

For example, a silane molecule that has an amino (NH_2) group at one end will bind to gold nanoparticles through the lone pair of the nitrogen atom, and hence adhere to the glass substrate, as shown in Figure 2.13. The silicon atom forms a covalent bond with the OH (hydroxyl) groups adsorbed on glass. The nanoparticle is shown in gray; the black part shows the citrate groups adsorbed on the gold (citrate comes from the synthesis of gold and has carboxyl groups). The carboxyl groups have a negative charge, and they stabilize the gold by repulsions.

The organosilane compounds that have at least two reactive groups of different types, bonded to the silicon atom, are called *silane coupling agents*. One of the groups, for example, methoxy, ethoxy, or silanolic hydroxy group, reacts with an inorganic material, such as glass, glass fibers, silica, etc., while the other reactive groups (amino, mercapto, etc.) may form a bond with a metal such as gold or with an organic compound or polymer, as shown in Figure 2.12. In applications where silanes are used to promote adhesion or biocompatibility, the main goal is to form a uniform layer. In addition to this, one would like to use as little of the silane as needed. Aminoalkane-substituted silanes are extensively used for the chemical modification of glass for subsequent colloidal gold adsorption. In this process, aminosilanes react with silanol groups on the glass surface to form the -Si-O-Si-R-NH_2 structure, leaving the amine functionality on the surface. The formation of SAMs based on silanes is discussed in detail in Chapter 4.

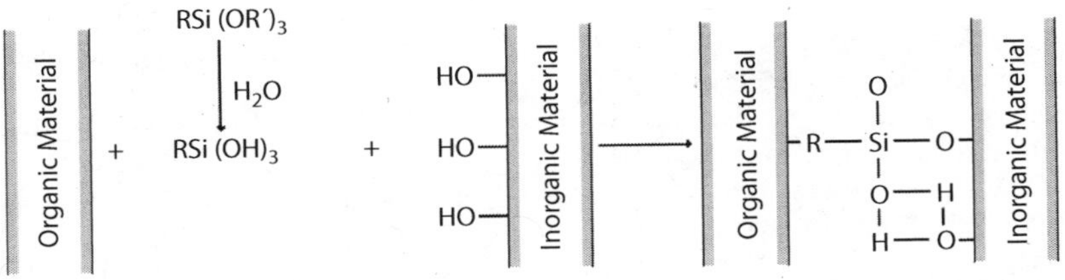

FIGURE 2.12 Schematic of adsorption of silane on glass and coupling to an organic compound (R and R' are organic radicals).

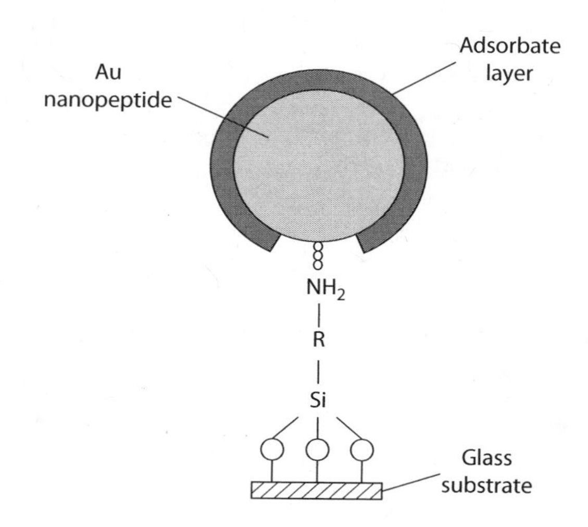

FIGURE 2.13 Adhesion of gold to a silanized glass substrate.

REFERENCES

1. Manz, A., Graber, N., Widmer, H. M. 1990. Miniaturized chemical analysis systems: A novel concept for chemical sensing. *Sensors Actuators B* 1:244–48.
2. Harrison, D. J., Fluri, K., Seiler, K., Fan, Z. H., Effenhauser, C. S., Manz, A. 1993. Micromachining a miniaturized capillary electrophoresis-based chemical analysis system on a chip. *Science* 261:895–96.
3. Arroyo, M. T., Fernández, L. J., Agirregabiria, M., Ibañez, N., Aurrekoetxea, J., Blanco, F. J. 2007. Novel all-polymer microfluidic devices monolithically integrated within metallic electrodes for SDS-CGE of proteins. *J. Micromech. Microeng.* 17:1289–98.
4. Khan Malek, C., Robert, L., Boy, J.-J., Blind, P. 2007. Deep microstructuring in glass for microfluidic applications. *Microsyst. Technol.* 13:447–53.
5. Munds, D., Leib, J. 2004. Novel microstructuring technology for glass on silicon and glass substrates. In *Electronic Components and Technology Conference*, Las Vegas, 2004, pp. 939–42.
6. Chandrasekaran, A., Acharya, A., You, J. L., Soo, K. Y., Packirisamy, M., Stiharu, I., Darveau, A. 2007. Hybrid integrated silicon microfluidic platform for fluorescence based biodetection. *Sensors* 7:1901–15.
7. Becker, H., Gartner, C. 2000. Polymer microfabrication methods for microfluidic analytical applications. *Electrophoresis* 21:12–26.
8. Acsharia, A. L. 2007. Modeling, fabrication and testing of Oled and led fluorescence integrated polymer microfluidic chips for biosensing applications. MASc thesis, Concordia University, 2007.
9. Lee, S. J., Goedert, M., Matyska, M. T., Ghandehari, E. M., Vijay, M., Pesek, J. J. 2008, Polymethylhydrosiloxane (PMHS) as a functional material for microfluidics chips. *J. Micromech. Microeng.* 18:025026.
10. Stieglitz, T., Beutel, H., Schuettler, M., Meyer, J.-U. 2000. Micromachined, polyimide-based devices for flexible neural interfaces. *Biomed. Microdevices* 2:283–94.
11. Sikanen, T., Aura, S., Heikkilä, L., Kotiaho, T., Franssila, S., Kostiainen, R. 2010. Hybrid ceramic polymers: New, nonbiofouling, and optically transparent materials for microfluidics. *Anal. Chem.* 1004053.
12. Langer, R. S., Peppas, N. A. 1981. Present and future applications of biomaterials in controlled drug delivery systems. *Biomaterials* 2:201–14.
13. Bawa, P., Pillay, V., Choonara, Y. E., du Toit, L. C. 2009. Stimuli-responsive polymers and their applications in drug delivery. *Biomed. Mater.* 4:1–15.
14. Domachuk, P., Tsioris, K., Omenetto, F. G., Kaplan, D. L. 2010. Bio-microfluidics: Biomaterials and biomimetic designs. *Adv. Mater.* 22:249–60.
15. Bettinger, C., Cyr, K., Matsamoto, A., Langer, R., Borenstein, J., Kaplan, D. 2007. Silk fibroin microfluidic devices. *Adv. Mater.* 19:2847–50.
16. Paguirigan, A., Beebe, D. J. 2006. Gelatin based microfluidic devices for cell culture. *Lab Chip* 6:407–13.
17. Korivi, N. S., Ajmera, P. K. 2009. Micropatterning of biocompatible materials. Paper presented at ICMEMS Conference, Madras, India.

REVIEW QUESTIONS

1. What are the most important properties of glass that make it suitable for microfluidic applications?
2. Give an example of a silica-based oxide glass material.
3. What is the typical composition of a bioactive glass?
4. What are the advantages and disadvantages of the silicon-based materials for microfluidic applications?
5. What is the glass transition temperature of a polymer?
6. What properties make PDMS a very suitable material for microfabrication?
7. What kind of material is ORMOCOMP, and what properties make it suitable for microfluidic applications?
8. What are the principal criteria that a biopolymer should fulfill in order to be used for BioMEMS fabrication?
9. Give two examples of biopolymers.

3 Biomolecules and Complex Biological Entities

Structure and Properties

BioMEMS devices deal with biomolecular species from the molecular level to the cellular level, as seen in Chapter 1. The interaction between the devices and biospecies differs so much, depending upon their size and nature. Hence, it is important to understand the structure and nature of different biospecies that would interact with microsystems.

3.1 AMINO ACIDS

Amino acids play central roles both as building blocks of *proteins* and as intermediates in metabolism. Proteins are created by polymerization of amino acids, a process in which amino acids are joined to form chains called *peptides* or *polypeptides*. The primary structure of a protein refers to the nature of the sequence of amino acids that makes up the protein. The biological activity of proteins is determined by the properties of amino acids that constitute protein composition. The formation of proteins from amino acid building blocks is shown in Figure 3.1.

Proteins are important, not only because they catalyze all of the reactions in living cells, but also due to their role in controlling virtually all the cellular processes. Proteins act as catalysts, regulate transport across cell membranes, and control expression of genes and replication of genetic material. In addition, proteins contain the information regarding the folding of protein into a three-dimensional structure, as well as the stability of this structure within their amino acid sequences. The protein's specific function in the body is determined by its three-dimensional shape, which results from the sequence of amino acids. In addition to their function in the formation of proteins, amino acids have many other biologically important roles. For example, amino acids such as *glycine* and *glutamate*, which are found in most proteins, also function as neurotransmitters that inhibit and amplify, respectively, the transmission of nerve impulses in cells. A nonstandard amino acid called *carnitine* transports fatty acids into muscle cells for energy production. Many amino acids are used to synthesize molecules other than proteins. For example, *tryptophan*, an essential amino acid found in proteins such as *casein* and *fibrin*, is a precursor of *serotonin*, which is a neurotransmitter important for the regulation of mood. On the other hand, glycine takes part in the biosynthesis of porphyrins such as *heme*, a component of the hemoglobin molecules found in red blood cells.

Substances derived from amino acids also have important uses in medicine and the food industry. For example, aspartame (aspartyl-phenylalanine-1-methyl ester) is an artificial sweetener, and monosodium glutamate is a food additive that enhances flavor. Other amino acids, such as 5-hydroxytryptophan and L-hydroxy phenylalanine, are used to treat neurological problems.

As shown in Figures 3.2 and 3.3, all amino acids have an amino (-NH_2) and a carboxylic acid (-COOH) functional group attached to the same tetrahedral carbon atom, the α-carbon atom. The amino acids having this type of structure are referred to as α-amino acids. Twenty types of α-amino acids that are used to form proteins in the human body are called standard or *proteinogenic* amino acids. The side chains (R) are specific to each amino acid, and they distinguish one amino acid from the other. The R group, which is an organic radical, can vary from a hydrogen atom in glycine to large aliphatic or aromatic groups. Depending on the nature of the R groups, the twenty amino acids

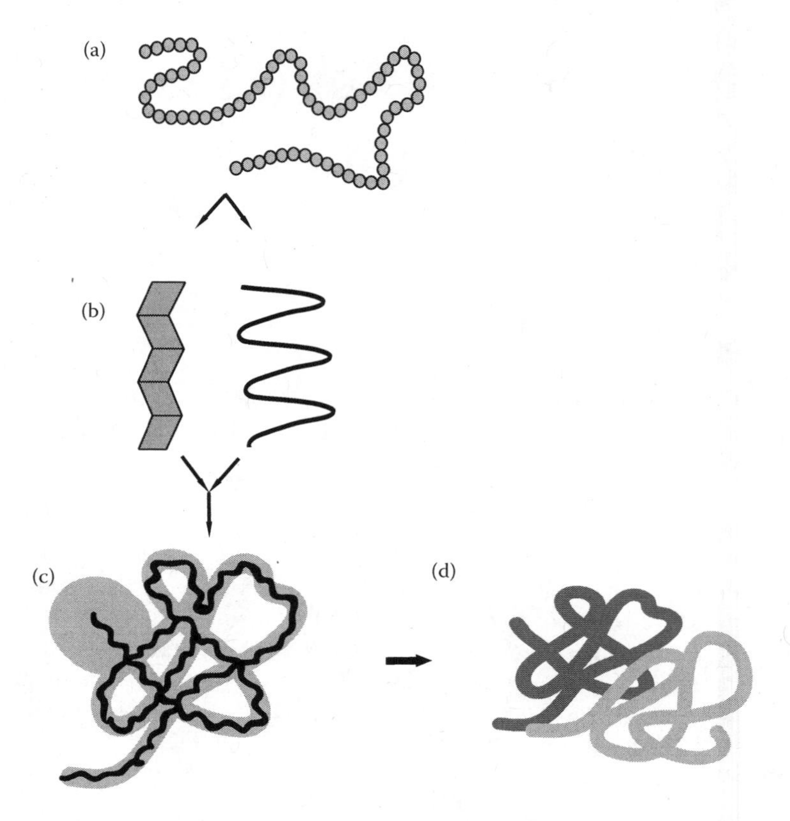

FIGURE 3.1 Formation of proteins from amino acid building blocks. (a) Primary protein structure is the sequence of a chain of amino acids. (b) Secondary protein structure is the structure formed when the sequence of amino acids is linked by hydrogen bonds. (c) Tertiary protein structure occurs when α-helices and pleated sheets are attracted through intermolecular forces. (d) Quaternary protein structure occurs when more than one amino acid chain is interacting. (Courtesy of National Human Genome Research Institute.)

H H O
H—N—C—C—OH
Amino group
R
Side chain
Carboxylic acid group

FIGURE 3.2 Structure of an amino acid.

can be classified into two broad groups: hydrophobic or hydrophilic. The hydrophobic amino acids, which repel water, will be found in the interior of proteins, shielded from being in direct contact with water, while the hydrophilic ones are predominantly found on the exterior surfaces of proteins and interact with water through hydrogen bonds.

From an engineering point of view, the hydrophilic or hydrophobic characteristics of different amino acids are of paramount importance for understanding the adsorption of proteins onto solid surfaces. Indeed, the response of a particular protein to a solid substrate is determined not only by the surface chemistry, but also by the structure and polarity of the side chains in the amino acids. For example, glycine, alanine, and leucine are considered nonpolar, while serine is polar neutral and aspartic and glutamic acid are polar acid.

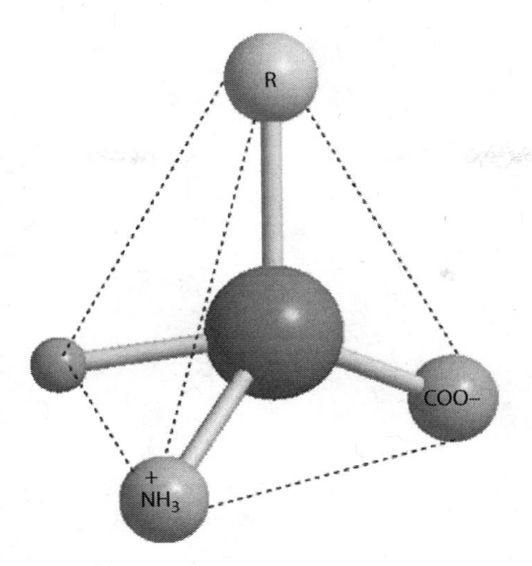

FIGURE 3.3 Tetrahedral structure of amino acids.

There is another way to classify the properties of amino acid side chains (R groups): using the Venn diagram shown in Figure 3.4, which has the advantage of providing a comparative view of their polarity.

Nonpolar groups are mostly found to be buried in the hydrophobic central core of proteins, while charged groups, which are the most polar, are mostly on the surface of globular protein, where they can strongly interact with water through hydrogen bonds or with other charged groups through electrostatic interactions. It is important to mention one of the very important sulfur-containing amino acids, cysteine, which has the unique property to cross-link covalently with an adjacent cysteine group in the protein and form disulfide bridges (-S-S-), as shown in Figure 3.5. The structure of cysteine is shown in Figure 3.6.

In the folded state of the protein, the disulfide bonds are formed between cysteine groups at a distant part of the protein, and they have a stabilizing effect, as shown in Figure 3.5.

When the basic amino group of one amino acid reacts with the acidic carboxyl group of a second amino acid, a *peptide bond* forms and the two amino acids are linked together, as shown in Figure 3.7.

A molecule of water formed from a H^+ and an OH^- group is lost during this condensation reaction. The peptide bond is rigid, planar, and polar due to a partial double-bond character. The double-bond character imposes an energy barrier to free rotation, and hence a planarity of the peptide bond, as shown in Figure 3.8.

Due to the resonance between the two structures, the peptide bond has a considerable double-bond character (approximately 40%), which prevents rotation around this bond. These resonance structures lead to a planar arrangement of the peptide carbonyl C and amide N, as shown in Figure 3.9.

Figures 3.10 and 3.11 show the two rotation angles, φ and ϕ, that determine the conformation of the protein.

The adsorption of amino acids on solid surfaces is a good model system for protein adsorption. For example, the adsorption of glycine, the simplest of amino acids, can be investigated by photoelectron diffraction measurements.

Figure 3.12 shows the adsorption of glycine on a Cu(110) surface. X-ray emission spectroscopy has been used to study the bonding and chemistry of glycine chemisorbed on Cu (110). The results show that glycine deprotonates upon adsorption. The adsorbate exhibits a rich surface chemistry leading to several intermediate adsorption structures. The most stable geometry has been found to involve both carboxylic and amino functional end groups in the bond.

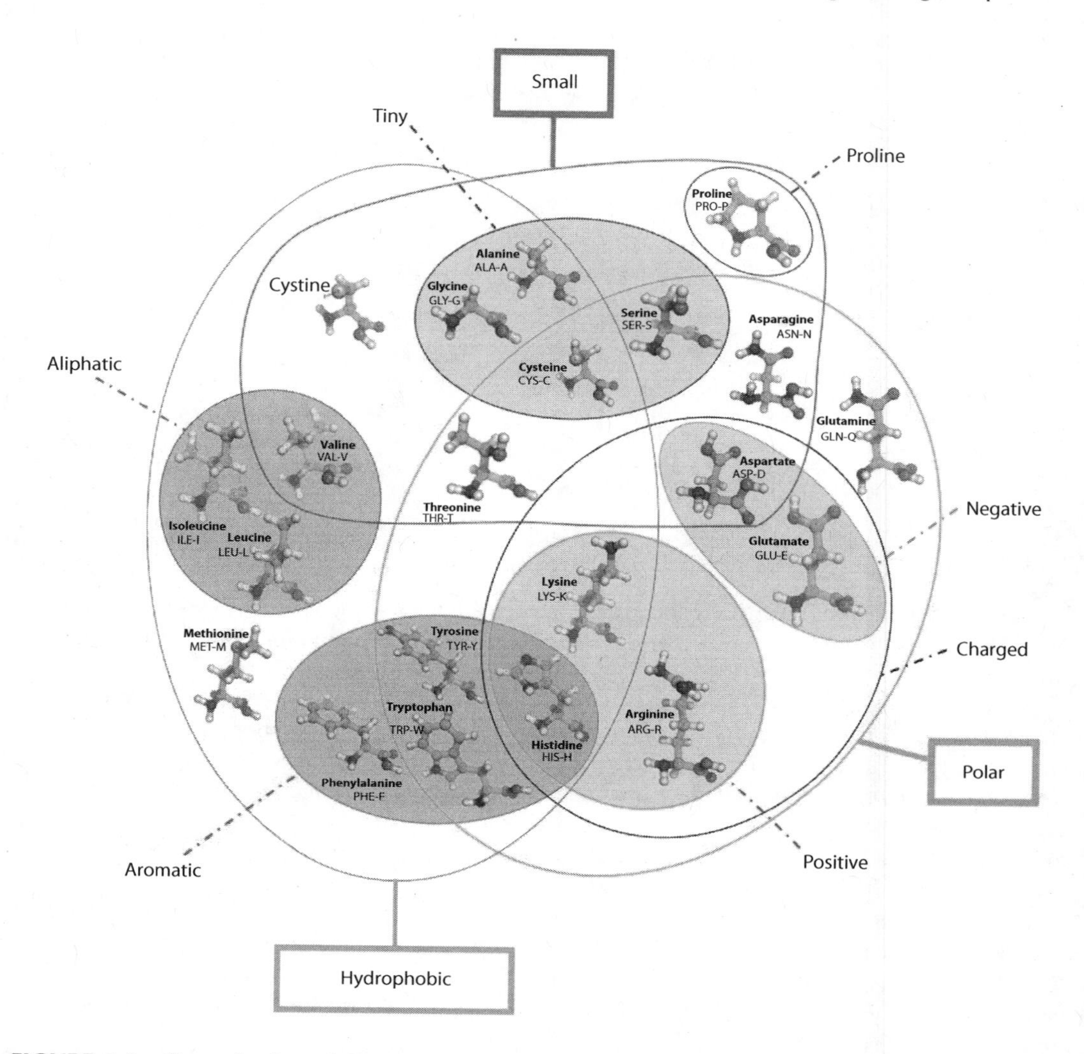

FIGURE 3.4 (See color insert.) Venn diagram of amino acids. (Reproduced from sites.google.com/site/apodtele/nutshell.)

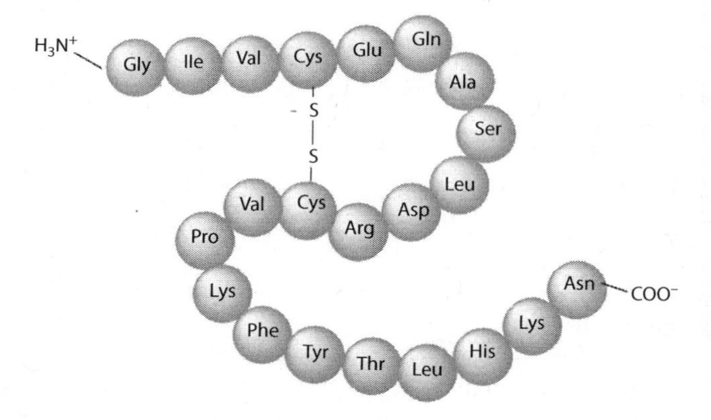

FIGURE 3.5 Formation of a disulfide bond between distant amino acids that is necessary to fold the chain structure.

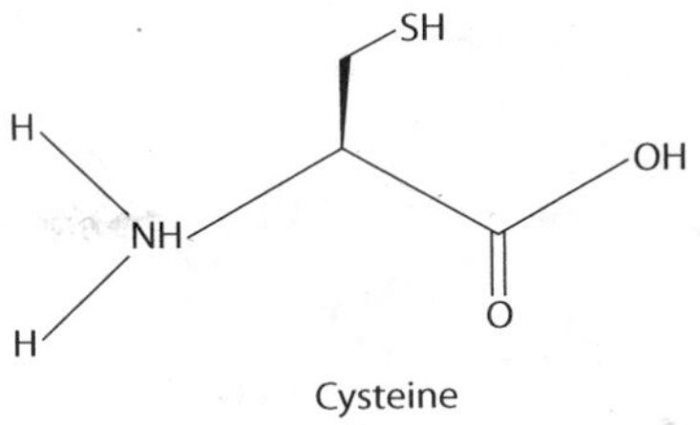

Cysteine

FIGURE 3.6 Structure of cystein.

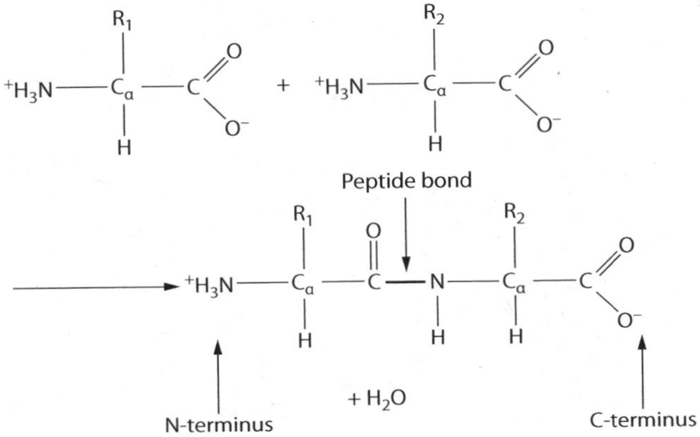

FIGURE 3.7 Formation of a peptide bond between two amino acids. The forward reaction is called condensation, while the reverse reaction is called hydrolysis.

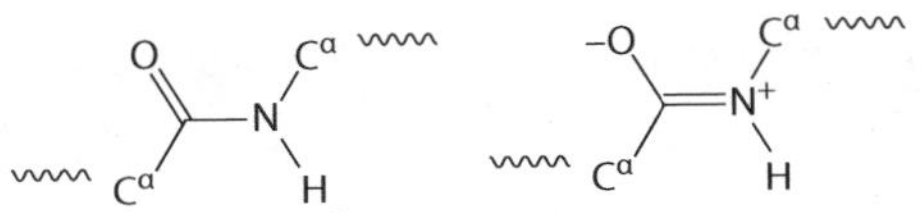

FIGURE 3.8 Resonance structures corresponding to the peptide bond.

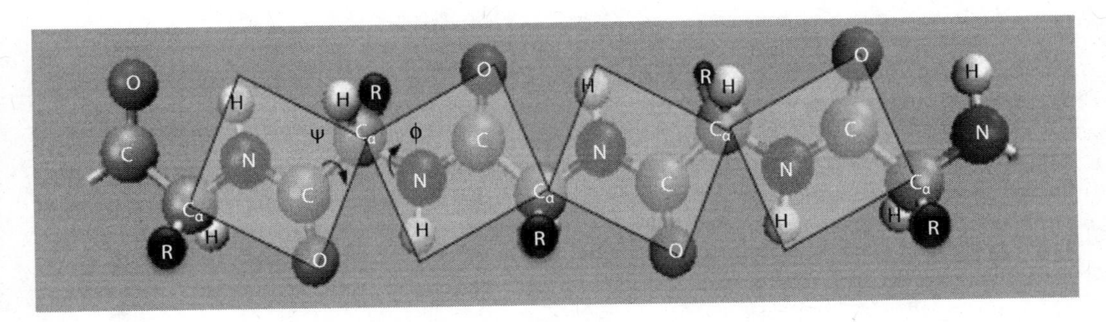

FIGURE 3.9 (See color insert.) Extended polypeptide chain with planar amide bonds. (Reproduced from http://employees.csbsju.edu/Hjakubowski/classes/ch331/protstructure/olunderstandconfo.html. With permission.)

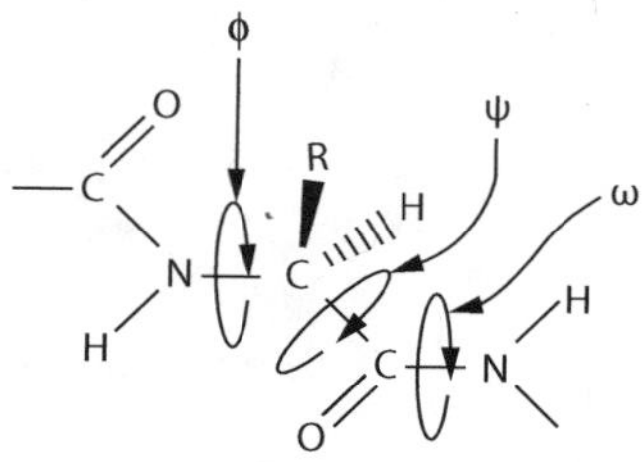

FIGURE 3.10 Two rotation angles that determine the conformation of a polypeptide or protein (angle ϕ-rotation around nitrogen-carbon bond and ψ-rotation around carbon-carbon bond), shown here for a simple peptide.

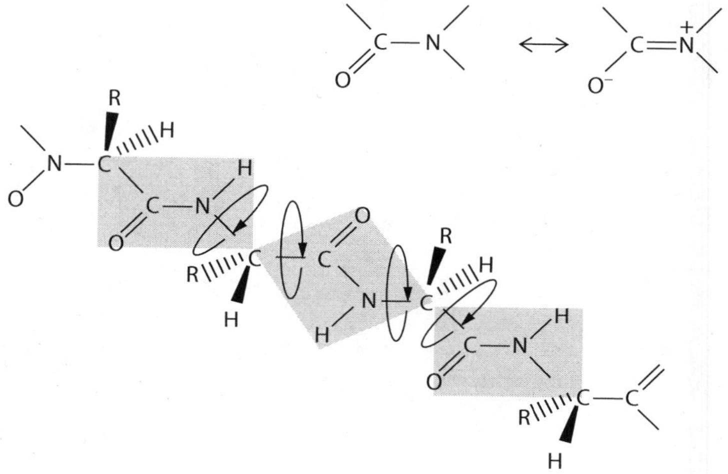

FIGURE 3.11 (See color insert.) Two rotation angles shown for a peptide formed of three amino acids.

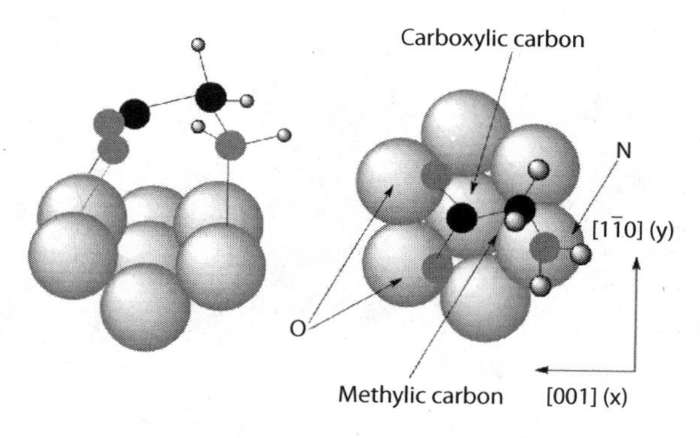

FIGURE 3.12 Local bonding geometry of glycine adsorbed on Cu (110). (Reproduced from Nyberg et al., *J. Chem. Phys.*, 112, 5420, 2000. With permission.)

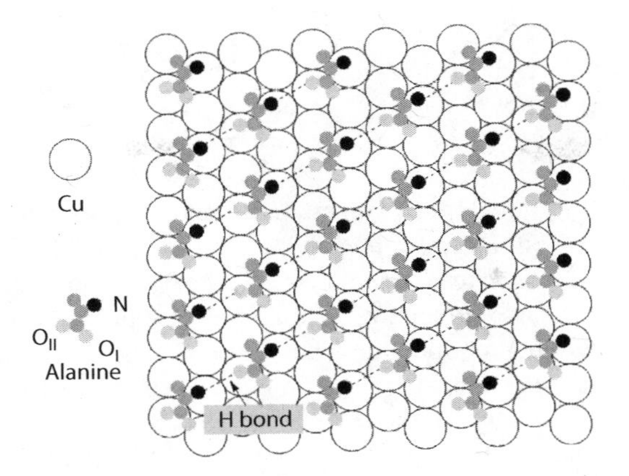

FIGURE 3.13 Cu(111)-alanine superstructure. (Reproduced from Ge et al., *Chin. Phys. Lett.*, 23, 1558, 2006. With permission.)

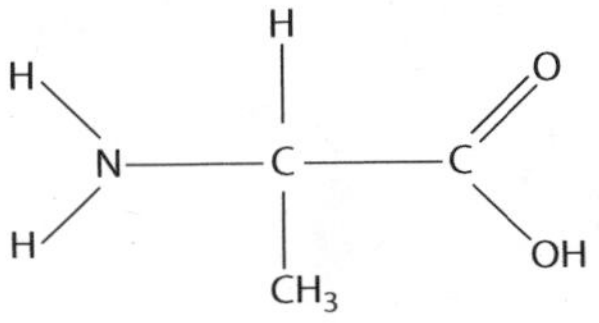

FIGURE 3.14 The structure of alanine.

The superstructure formed through adsorption of alanine on Cu(111) is shown in Figure 3.13.

The adsorption of alanine in the neutral form has been studied by reflection absorption infrared spectroscopy (RAIRS). The results show that both carboxylate oxygen groups are equidistant from the surface when the exposure time is short. For reference, the molecular structure of alanine is shown in Figure 3.14.

3.2 POLYPEPTIDES AND PROTEINS

Different biomedical applications, such as biosensors, drug delivery schemes, immunological tests, etc., are based on protein adsorption and cell adhesion at solid-liquid interfaces. The adsorption process depends mostly on various intermolecular forces that exist between protein molecules and surfaces, and also on intramolecular forces that may produce conformational changes upon the adsorption. In order to understand and be able to control the adsorption of proteins for various biosensor applications, it is important to have a good basic knowledge of the structure and properties of proteins.

Proteins are large biological molecules made of amino acids linked into linear chains by peptide bonds. The structures built from two to one hundred amino acids are usually called *polypeptides*, while longer polypeptide structures are called *proteins*. The origin of the word *protein* in Greek is *prota*, meaning "of primary importance," and was introduced by the well-known chemist Berzelius in 1838 to design large organic molecules, important for animal nutrition.

Proteins carry many functions in the living cells. The most important group of proteins is called *enzymes*, and they are catalysts of metabolic reactions in cells. Other forms of proteins are the hormones, transport protein, structural protein, storage protein, motor protein, antibodies, etc. According to their location in the living cell, proteins are classified as membrane proteins, internal, external, and virus proteins.

The *primary structure* of peptides and proteins is defined by the amino acid sequence, independent of their spatial arrangement in the polypeptide chain. The convention for the designation of the order of amino acids is that the N-terminal end is to the left and the C-terminal end containing a free α-carboxyl group is to the right, as shown, for example, in Figure 3.11. The *secondary structure* is the local spatial arrangement of the main chain atoms. The organization of the polypeptide into *regular repetitive patterns* over short segments is without considering the conformation, i.e., the spatial arrangement of its side chains.

In general, depending on the way the proteins fold, there are two broad classes of structures, termed *globular proteins* and *fibrous proteins*. Globular proteins are compactly folded and coiled compared to fibrous proteins, which are more elongated. The polypeptide chains may assume different conformations within the same protein, depending on the sequence of amino acids. Other proteins may be membrane spanning, multidomain, etc., or can be a member of more than one class.

The *α-helix* shown in Figures 3.15 and 3.16 is a common secondary structure encountered in proteins of globular class. The formation of the α-helix is spontaneous and is stabilized by H-bonding between amide nitrogens and carbonyl carbons of peptide bonds spaced four residues apart. This orientation of H-bonding produces a helical coiling of the peptide backbone such that the R groups lie on the exterior of the helix and perpendicular to its axis (Figure 3.17). The average number of amino acids in a helix is 11, which corresponds to three turns. The α-helix is a more compact structure than a fully extended polypeptide chain, with ϕ and φ angles of 180°.

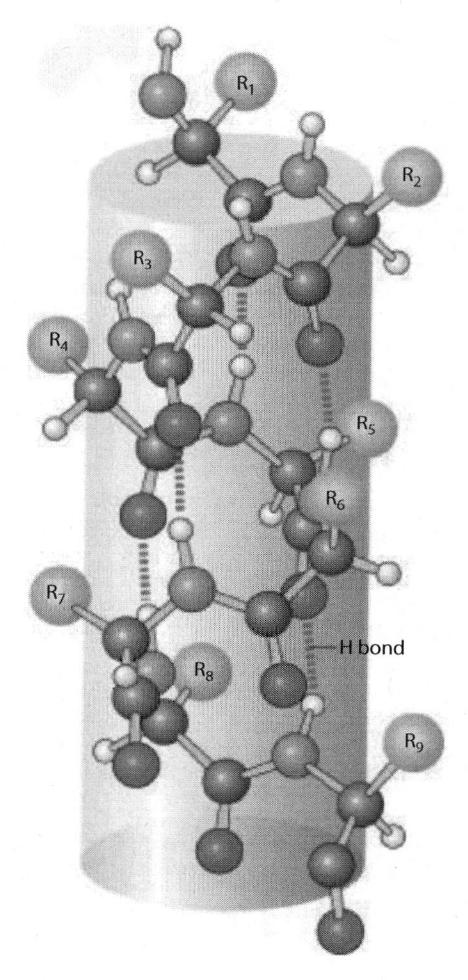

FIGURE 3.15 (See color insert.) Typical α-helix. (Reproduced from themedicalbiochemistrypage.org. With permission.)

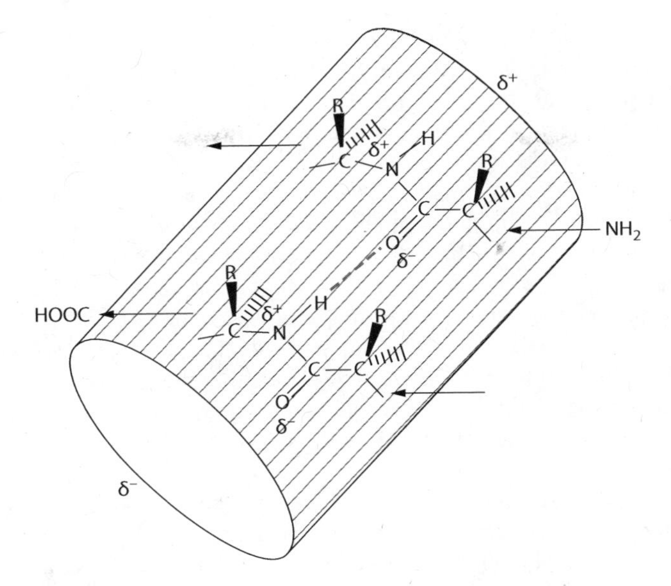

FIGURE 3.16 (See color insert.) α-Helix structure showing the charge distribution along the peptide bond (δ is the charge on the atoms).

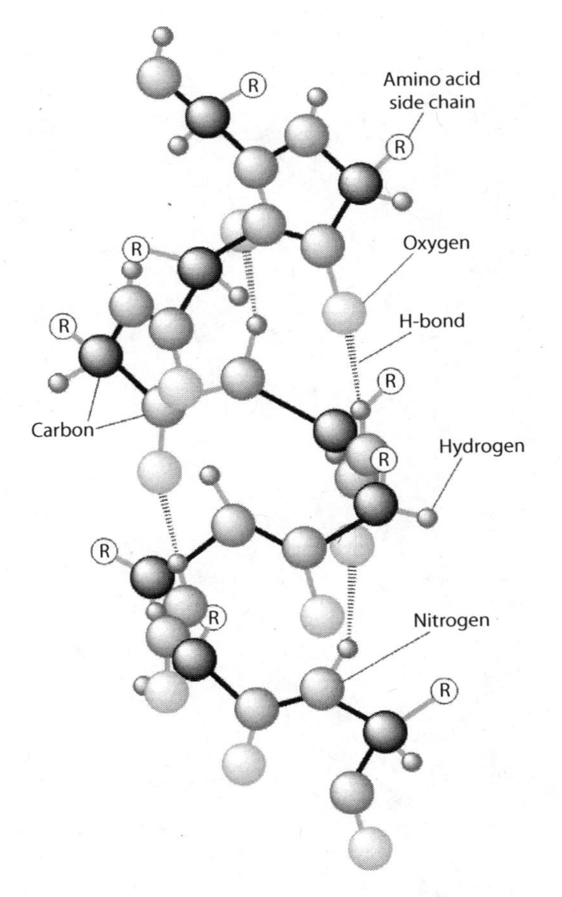

FIGURE 3.17 Regular conformation of the polypeptide backbone observed in the α-helix.

β-Strands are an extended form in which the side chains alternate on either side of the extended chain. The backbones of β-strands hydrogen bond with the backbone of an adjacent β-strand to form a β-sheet structure. The strands in a β-sheet can be either parallel or antiparallel, and the hydrogen bonding pattern is different between the two forms.

The α-helix is composed of a single linear array of helically disposed amino acids; β-sheets are composed of two or more different regions of stretches of at least five to ten amino acids. As shown in Figure 3.18, the folding and alignment of stretches of polypeptide backbone side by side to one another to form β-sheets is stabilized by H-bonding between amide nitrogens and carbonyl carbons. The H-bonding residues are present in adjacently opposed stretches of the polypetide backbone, as opposed to a linearly contiguous region of the backbone in an α-helix. β-Sheets are said to be pleated due to the positioning of α-carbons of a peptide bond, which alternates above and below the plane of the sheet, as shown in Figure 3.19. β-Sheets are either *parallel* or *antiparallel*, as shown in Figures 3.18 and 3.19. In parallel sheets, adjacent peptide chains proceed in the same direction (i.e., the direction of N-terminal to C-terminal ends is the same), whereas in antiparallel sheets adjacent chains are aligned in opposite directions.

Tertiary structure is the overall folding of the whole polypeptide. Tertiary structure refers to the complete three-dimensional structure of the polypeptide units of a given protein. Therefore, tertiary structure also describes the relationship between different domains within a protein. The tertiary structure is governed by hydrogen bonding and hydrophobic interactions.

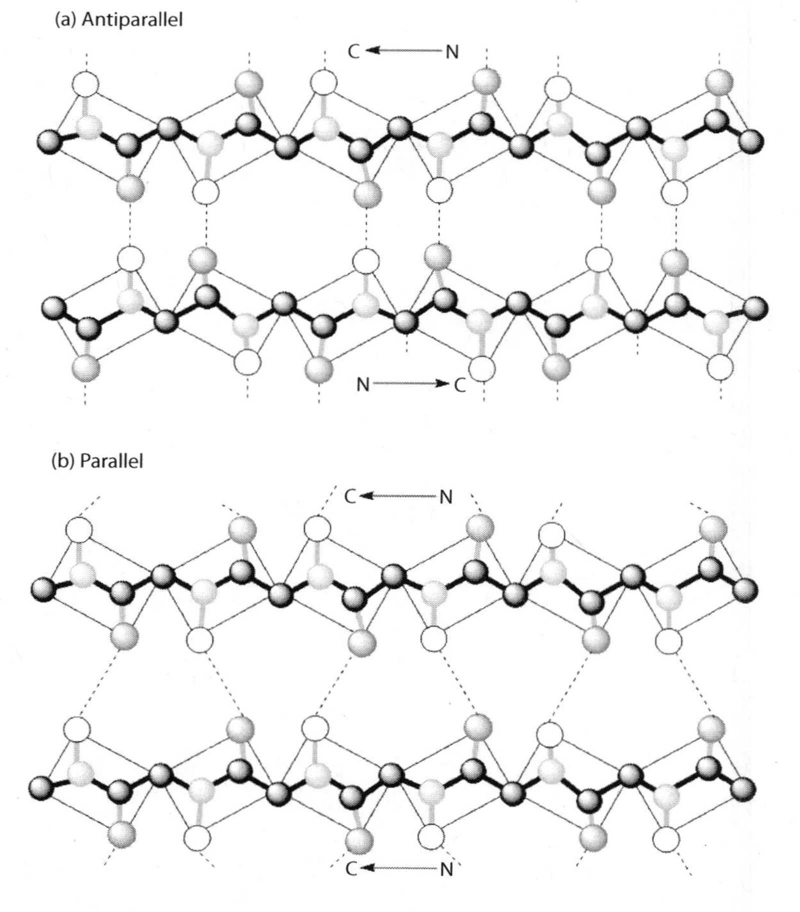

FIGURE 3.18 Parallel and antiparallel β-sheets showing the pleated β-structure.

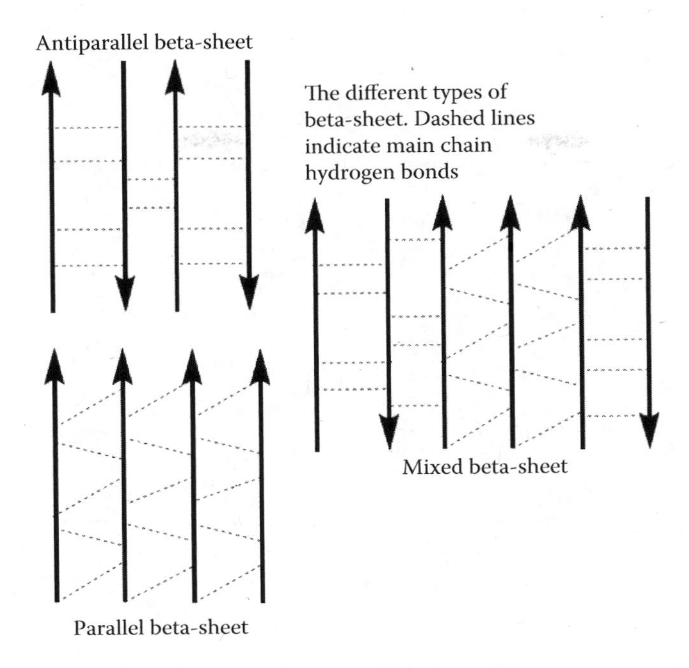

FIGURE 3.19 Conformation of the polypeptide backbone observed in the β-sheet structure.

Quaternary structure is the grouping of several protein molecules into a single larger entity that acts cooperatively with others to yield special properties to grouping, not possessed by the single subunit. Not all proteins have a quarternary structure; e.g., a single myoglobin molecule functions by itself, but its close relative, hemoglobin, is a tetramer of four globin subunits. Each globin is structurally very similarly to myoglobin.

3.3 LIPIDS

Lipids are naturally occurring organic molecules that are all soluble in organic solvents. The group of lipids encompasses compounds of large structural diversity. The only property they have in common is their hydrophobicity. In spite of their different general properties, compounds such as fatty acids, waxes, phospholipids, terpenes, and steroids are all considered lipids. Lipids are an important component of blood and of other biological fluids that are adsorbed on the biomaterials surfaces for various biomedical applications. Understanding the interfacial interactions between biologically significant lipid molecules and synthetic biomaterials requires a good knowledge of the structure and properties of various lipids and their interactions with proteins. In addition, physiological lipids such as phospholipids, cholesterol, and triglycerides are important as major components of lipid-based control drug delivery vehicles. Different types of lipid-based carriers, such as liposomes, lipid microspheres, microtubules, and microcylinders, are currently under development, as they can serve as carriers for fragile and highly vulnerable molecules such as pharmaceutical proteins and peptides, as well as for vaccines and adjuvants. These systems have been proposed as a new type of lipid-based encapsulation system, and their structure and physicochemical properties will be discussed in more detail in the subsection devoted to phospholipids.

There are four main groups of lipids: fatty acids, glycerides, nonglyceride lipids, and complex lipids (lipoproteins), as shown in Figure 3.20. Other systems are based on their chemical properties. In addition, lipids are classified as saponifiable and nonsaponifiable. A more general scheme of classification of lipids is shown in Figure 3.20.

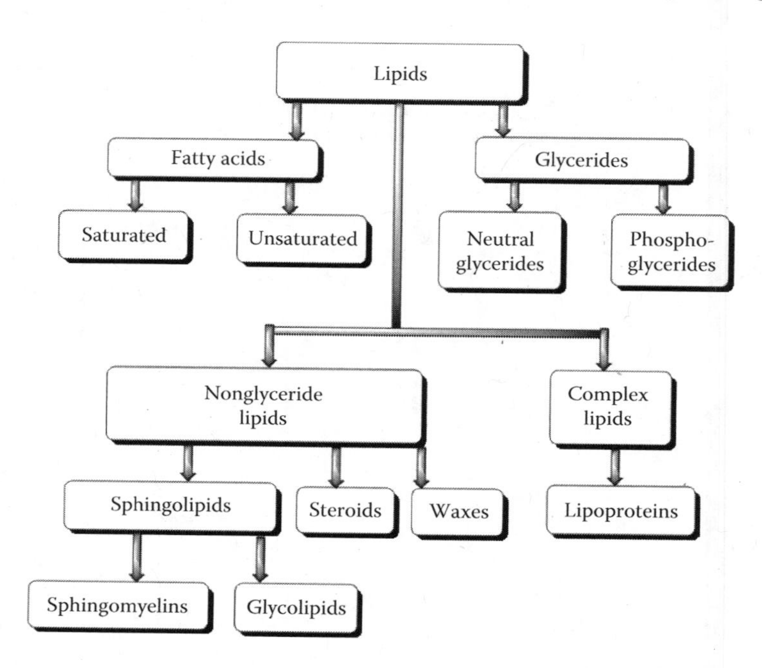

FIGURE 3.20 Classification of lipids. (Adapted from Denniston et al., *General Organic Chemistry and Biochemistry*, 6th ed., McGraw-Hill Higher Education, 2008, p. 581.)

3.3.1 Fatty Acids and Their Esters

The naturally occurring fats, oils, and waxes are all esters formed between long-chain fatty acids and glycerol. *Fatty acids* are long-chain carboxylic acids (saturated or unsaturated) and are generally referred by their common names. Examples of *saturated fatty acids* are lauric acid, $CH_3(CH_2)_{10}COOH$; myristic acid, $CH_3(CH_2)_{12}COOH$; and stearic acid, $CH_3(CH_2)_{16}COOH$, with melting points increasing with increasing number of carbon atoms (45, 55, and 69°C, respectively). The presence of one or more double bonds in unsaturated fatty acids lowers the melting point relative to saturated acids because their chains are bent and cannot be packed closely together as in the saturated acids. A typical *unsaturated fatty* acid with eighteen carbons in the chain and one double bond is oleic acid, $CH_3(CH_2)_7CH{=}\,CH(CH_2)_7COOH$, with a melting temperature of 13°C and a *cis* configurated double bond. It is the long hydrocarbon chains that make the fatty acids hydrophobic.

Figure 3.21 shows the *cis* and *trans* configurational isomers of 9-octadecenoic acid. The Latin prefixes *cis* and *trans* describe the orientation of the hydrogen atoms with respect to the double bond: "on the same side" (*cis*) and "on the other side" (*trans*). Naturally occurring fatty acids generally have the *cis* configuration. The natural form of 9-octadecenoic acid (oleic acid) found in olive oil has a V shape due to the *cis* configuration.

All fats and oils are composed of esters of fatty acids with glycerol called glycerides or acylglycerols. Glycerol (1,2,3-trihydroxy propane) is an alcohol that contains three hydroxyl groups (OH). Glycerol can combine with up to three fatty acids to form monoglycerides, diglycerides, and triglycerides. The reaction is called esterification. Animal fats are solid because of the large percentage of saturated fatty acids, while oils originate mostly in plants, consist of unsaturated fatty acids, and are liquid at room temperature. The formula of a mixed triglyceride formed from two molecules of unsaturated acids (oleic) and one saturated one (palmitic) is shown in Figure 3.22.

By hydrolysis in the presence of NaOH (saponification), triglycerides produce fatty acid salts (soaps) and glycerol. The C=C double bond in oils can be hydrogenated to give saturated fats similar to those found in animal fat.

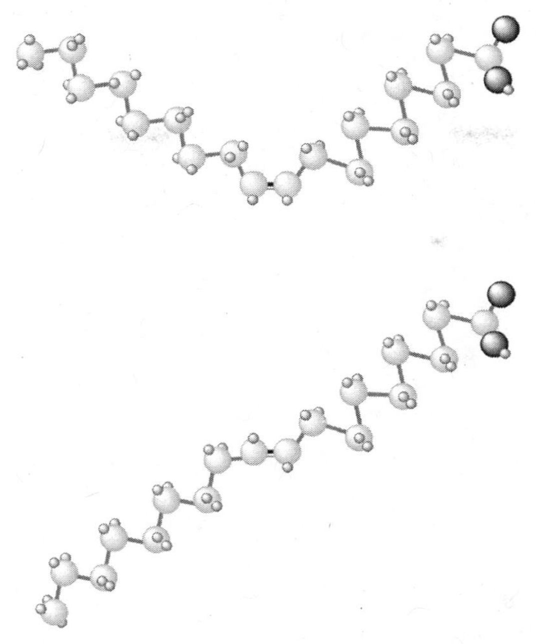

FIGURE 3.21 Three-dimensional molecular projections of two configurational isomers.

$$\begin{array}{l} CH_2{-}CH(CH_2)_7C(O)O = CH_3(CH_2)_7CH \\ | \\ CH{-}CH(CH_2)_7C(O)O = CH_3(CH_2)_7CH \\ | \\ CH_2{-}CH_3(CH_2)_4C(O)O \end{array}$$

FIGURE 3.22 Triglyceride (triacylglycerol).

3.3.2 Phospholipids

Phospholipids are the main constituents of cell membranes. They are fatty acyl esters of glycerol that consist of two long-chain fatty acid molecules (C_{14-18}) joined by ester bonds to the hydroxyl groups of glycerol phosphate, as shown in Figure 3.23. Different small groups, such as choline, serine, and ethanolamine, can be bound to glycerol through phosphoester bonds.

Phospholipids are ionic *amphiphiles* (or *amphipathic*); i.e., they have dual properties: *hydrophilic* (polar) due to the phosphate and substituent groups and *hydrophobic* (nonpolar) due to the long chains of the fatty acid residues. When phospholipids are dispersed in water, they aggregate spontaneously and form phospholipid bilayers or vesicles called *liposomes*.

Bilayers form spontaneously in aqueous environments due to the amphipathic nature of phospholipids. When surrounded by water on all sides, they aggregate to protect their hydrophobic tails in the interior of the bilayer or micelle. The bilayer is actually the more favorable structure since it protects the tails on all sides from interaction with water. These bilayer structures are conveniently employed in nature to partition individual cells from one another as well as organelles within the cells themselves. The fluidity of the structure is also advantageous to its functions within and among the cells. As mentioned, the fluidity of a specific bilayer can vary. The type of phopholipid bilayer surrounding a specific cell or cell compartment (i.e., lysosome, Golgi, or ER) is governed by its function and surrounding temperature. The lipid bilayer is usually not just phopholipids, but contains cholesterol and glycolipids, in addition to the proteins already mentioned. The structures of liposome, micelle, and bilayer are shown in Figure 3.24.

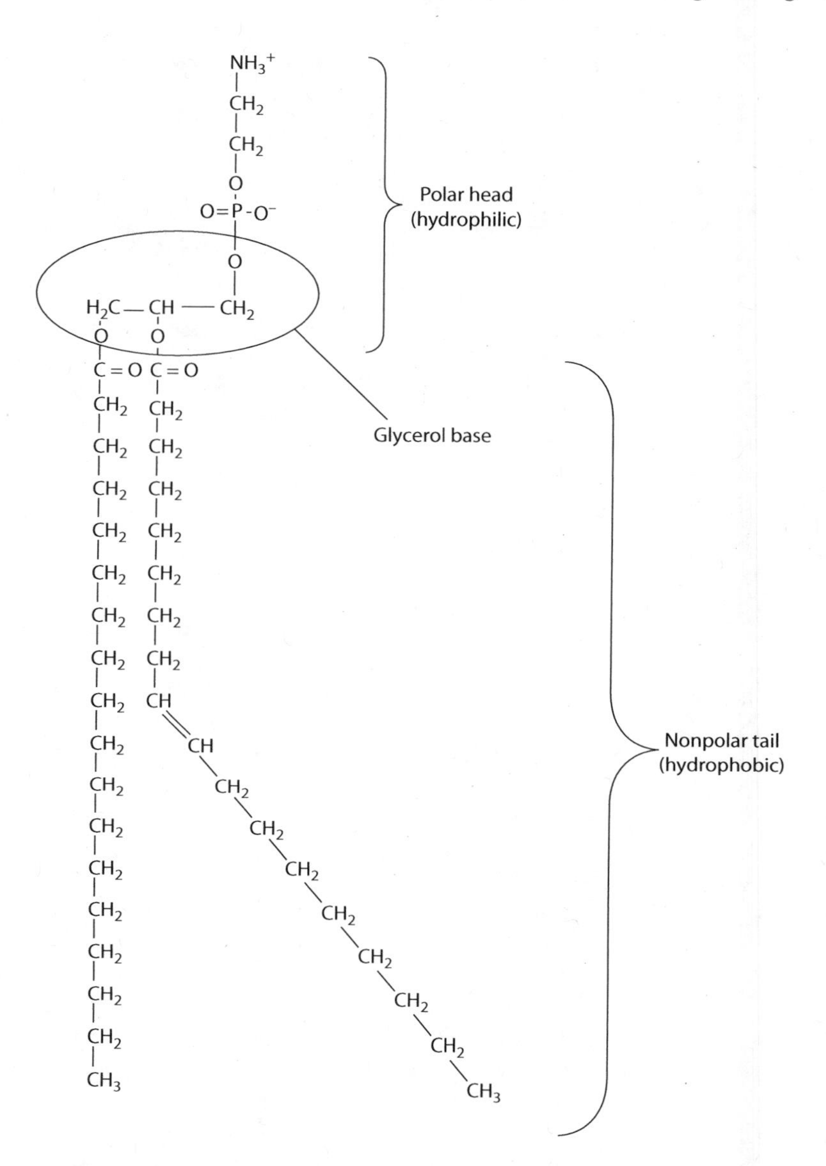

FIGURE 3.23 Simplest structure of a phospholipid.

Liposomes are used for drug delivery by introducing the drug into the free space inside the liposome. The *phospholipid bilayer* is composed of two parallel sheets of lipid molecules arranged tail to tail; i.e., the hydrophobic tails in each layer face each other, and the hydrophilic heads are exposed to water, as shown in Figure 3.24.

Liposomes are well-recognized drug delivery vehicles. They have been shown to enhance the therapeutic activity of several anticancer drugs. Liposomes have long been used as carrier systems for the delivery of vaccines, therapeutic drugs, and hormones because of easy preparation, good biocompatibility and biodegradability, low toxicity, and commercial availability. The delivery system of liposome is given in Figure 3.26. The efficient functioning of enzymes inside liposomes opens up new possibilities of applications in biocatalysis and bioanalytical tools. For example, enzyme-containing vesicles can serve as nanoreactors for biospecific reactions. In such reaction systems, specific substrates, which permeate across the vesicle membrane lipid bilayer(s), are converted to

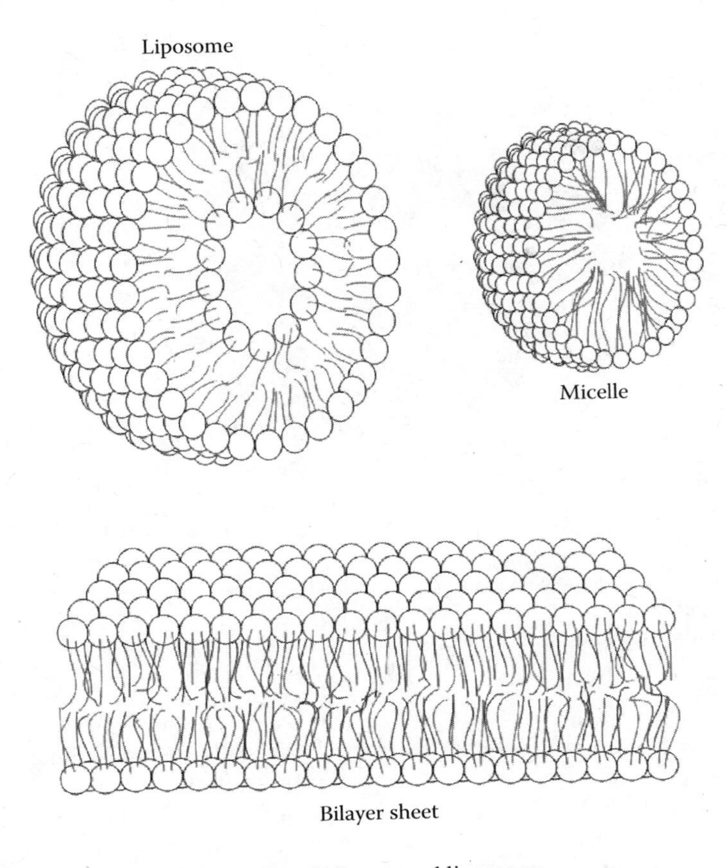

FIGURE 3.24 (See color insert.) Formation of bilayers and liposomes.

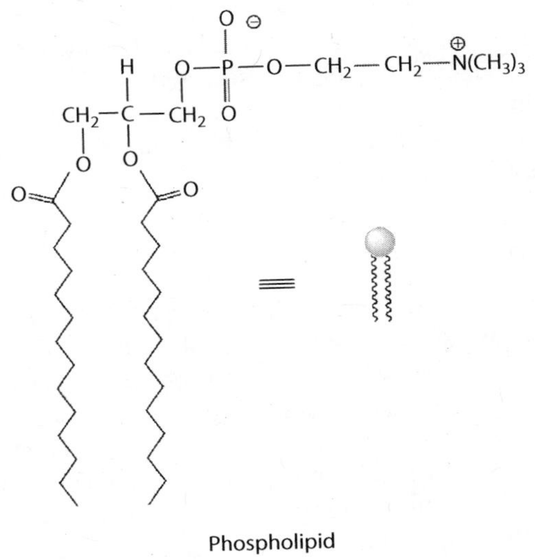

Phospholipid

FIGURE 3.25 Structure of a phospholipid molecule.

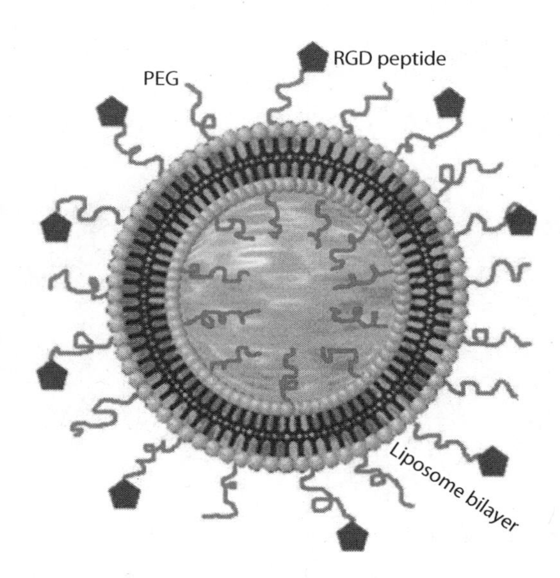

FIGURE 3.26 (See color insert.) Schematic representation of the targeted liposome delivery system. Cyclic RGD peptides coupled to the distal end of maleimide-PEG-DSPE in the liposome bilayer of PEG-grafted LCL. (Reproduced from Nallamothu et al., *AAPS PharmSciTech*, 7(article 32), 2008. With permission.)

products by the entrapped enzymatic catalyst. *Liposomal enzyme electrodes* have been developed with glucose oxidase encapsulated within liposomes. Figure 3.27 shows the lipid bilayer in a cell membrane. Surface-active liposomes are used for targeted cancer therapy as well.

The polar head groups are oriented toward the aqueous environment making up the external surfaces of the bilayer, while the nonpolar tails are directed to the center of the aggregate and are shielded from the surrounding water. As can be seen from the figure, the membrane contains proteins. Their hydrophobic amino acids associate with the lipids in the membrane, and the hydrophilic amino acids are oriented toward the water content outside of the membrane.

Supported lipid bilayers present a simple model of cell membranes in configurations that can be easily investigated with common quantitative surface-sensitive tools. They are of particular interest as components of future generations of biosensors based on transmembrane proteins.

Recent years have seen an increased understanding of the interaction of lipid bilayers with solid surfaces. However, many fundamental problems still await answers and further investigations. Among these is the question of how the liposome-surface interaction is affected by nano- and microscale chemical and topographical patterns, and how the biologically relevant lipid compositions influence these interactions.

3.3.3 Lipoproteins

Lipoproteins, shown in Figure 3.28, are complex aggregates of lipids and proteins that make the lipids compatible with the aqueous environment of body fluids and enable their transport throughout the body of all vertebrates and insects. Lipoproteins are synthesized mainly in the liver and intestines. Within the circulation, these aggregates are in a state of constant flux, changing in composition and physical structure as the peripheral tissues take up the various components before the remnants return to the liver. The most abundant lipid constituents are triacylglycerols, free cholesterol, cholesterol esters, and phospholipids (phosphatidylcholine and sphingomyelin), though fat-soluble vitamins and antioxidants are also transported in this way. Free (unesterified) fatty acids and lysophosphatidylcholine are bound to the protein albumin by hydrophobic forces in plasma.

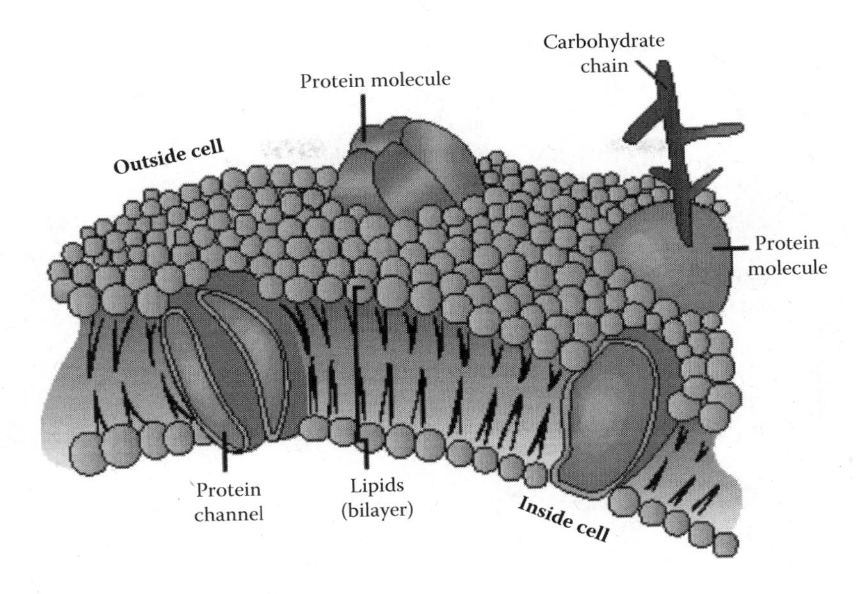

FIGURE 3.27 (See color insert.) Architecture of cell membrane.

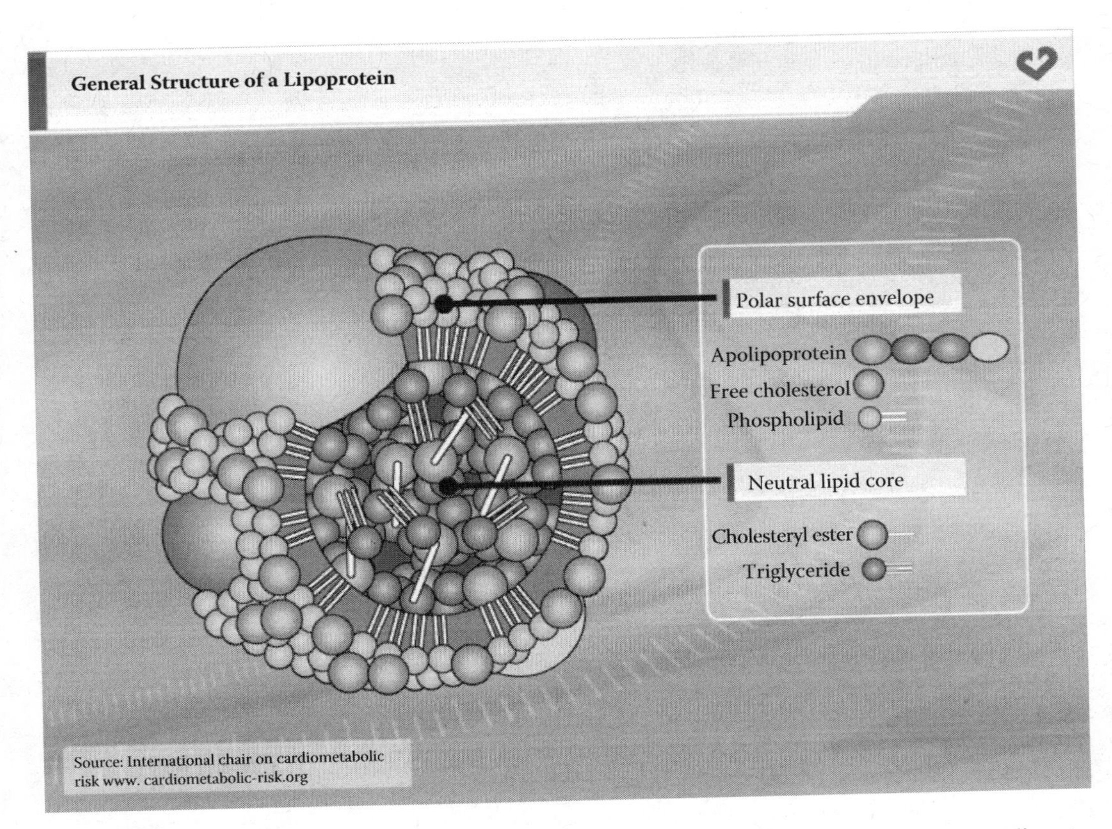

FIGURE 3.28 (See color insert.) General structure of a lipoproteine. (Reproduced from www.cardiometabolic-risk.org. With permission.)

3.4 NUCLEOTIDES AND NUCLEIC ACIDS

Due to its unique structure and composition, *deoxyribonucleic acid* (DNA) serves as a genetic (hereditary) information-carrying molecule in living organisms. DNA is the molecule responsible for both the storage and transmission of information to the next generation. The discovery of the structure and function of DNA (Watson and Crick, 1953) is considered the most important of the twentieth century and had a tremendous impact on science and medicine. For replication, DNA acts as a template for the production of RNA and proteins. The discovery of DNA has contributed to a better understanding of genetic and infectious diseases, to the creation of new drugs, and to the development of gene therapy. Because of the importance of DNA as a biological template, enormous efforts have been focused on the complete understanding of its functions and contribution in biological systems, as well as on the isolation, detection, and analysis of DNA.

Most DNAs exist in the famous form of a *double helix* called double-stranded DNA (dsDNA), in which two linear strands of DNA are wound around one another. As will be shown below, the major force promoting the formation of this helix is *complementary base pairing* through hydrogen bonds. The base pairs, i.e., adenine (A), thymine (T), cytosine (C), and guanine (G), are nucleotides that follow the Watson-Crick rule of base pairing. These pairs form the basis for the high specificity of the biorecognition process. Most of the DNA detection techniques are based on the DNA hybridization mechanism. Hybridization works on the principle of base pairing interaction, which will be discussed in detail in Chapter 9. The affinity pairing occurs either between A and T or between G and C. In DNA hybridization, unknown single-stranded DNA (ssDNA), called the target, is identified by a probe that is nothing but a complementary ssDNA. As a result, a double-stranded (dsDNA) helix structure is formed. In order to understand how DNA biosensors, called *genosensors*, work, understanding the structure and properties of nucleic acids, as well as of the nucleotides, will be useful and will be discussed in this section.

3.4.1 Nucleotides

Nucleotides are the building blocks of nucleic acids. They have three components: a five-carbon sugar called ribose or 2'-deoxyribose, a heterocyclic base (derivatives of pyrimidine and purine), and a phosphate group. The general structure of nucleotides is shown in Figure 3.29, and its components are shown in Figure 3.30.

The combination of only a base and pentose without the phosphate is called a *nucleoside.*

The carbon and nitrogen atoms in the heterocyclic bases are conventionally numbered, and in the pentose ring the carbon atoms are designed by prime numbers and the phosphate group is esterified to the 5' carbon of the pentose ring. There are five different bases in a nucleotide: adenine, cytosine, guanine, thymine, and uracil (Figure 3.31).

3.4.2 Nucleic Acids

DNA (deoxyribonucleic acid) and RNA (ribonucleic acid) are synthesized in cells by *DNA polymerases* and *RNA polymerases*, which are enzymes, while short fragments of nucleic acids can be produced without enzymes. They are polymers of nucleotides linked through *phosphodiester bonds*. As shown in the introduction, DNA serves as an information-carrying molecule.

Because of the presence of hydrogen-bond-forming groups in their molecules, there are two types of base pairs. Usually, the bases are identified by the first letter in their names, like A, T, G, C. The possible base pairs are A-T and G-C, as shown in Figure 3.32. Most DNAs exist in the form of a double helix, in which two linear strands are wound around each other, as seen in Figure 3.32. In double-stranded nucleic acids, base pairs are always formed between a purine and a pyrimidine. Thymine (T) is found only in DNA, while uracil (U) is found only in RNA, as seen in Figure 3.33. There is a phosphodiester linkage between the 5'-phosphate group of one nucleotide unit and the

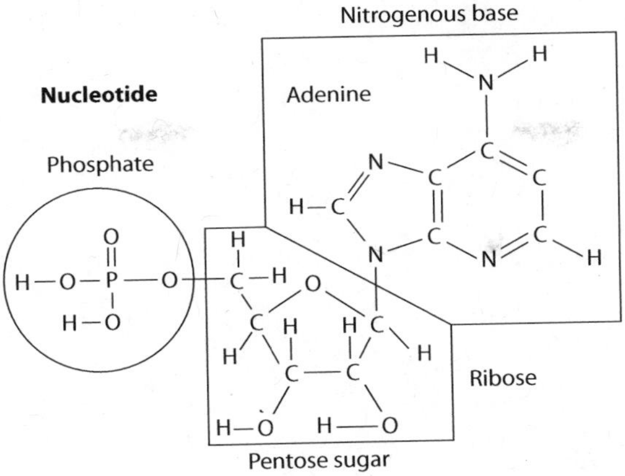

Nitrogenous base

Nucleotide

Adenine

Phosphate

H—O—P—O (=O, H—O)

H—C, N, C, C, C, N, C, N, C, H (with NH_2: H—N—H)

Ribose

Pentose sugar

H—O H—O

FIGURE 3.29 Structure of nucleotides showing the phosphate, pentose, and heterocyclic base moieties.

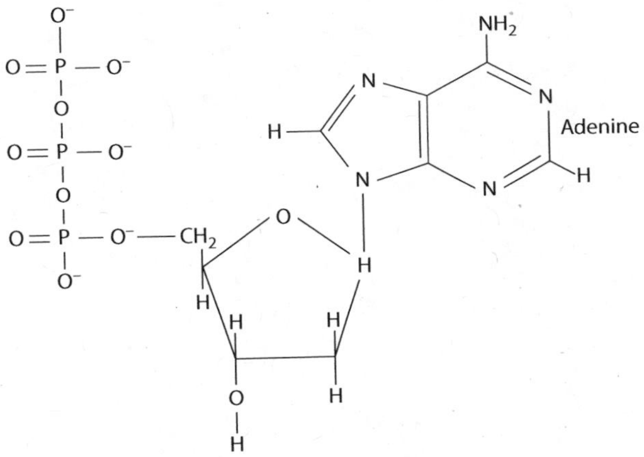

O^-

$O=P—O^-$

O

$O=P—O^-$

O

$O=P—O^-—CH_2$

O^-

NH_2

N N

H— N N

Adenine

H

O

H H H H

O H

H

Deoxyadenosine triphosphate = dATP

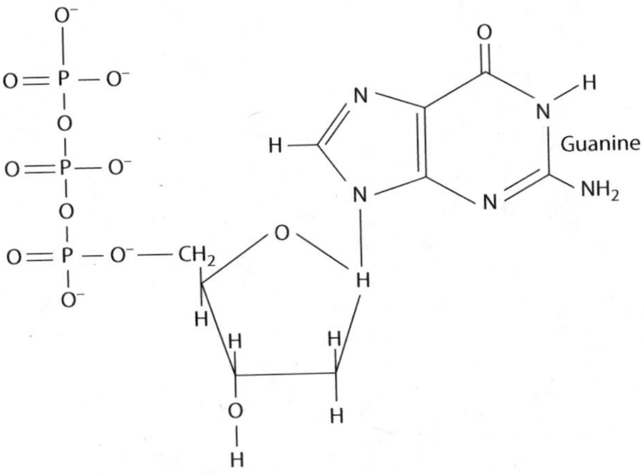

O^-

$O=P—O^-$

O

$O=P—O^-$

O

$O=P—O^-—CH_2$

O^-

O

N N—H

H— Guanine

N N NH_2

O

H H H H

O H

H

Deoxyguanosine triphosphate = dGTP

FIGURE 3.30 Components of nucleotides (four examples of deoxynucleoside triphosphates). *(Continued)*

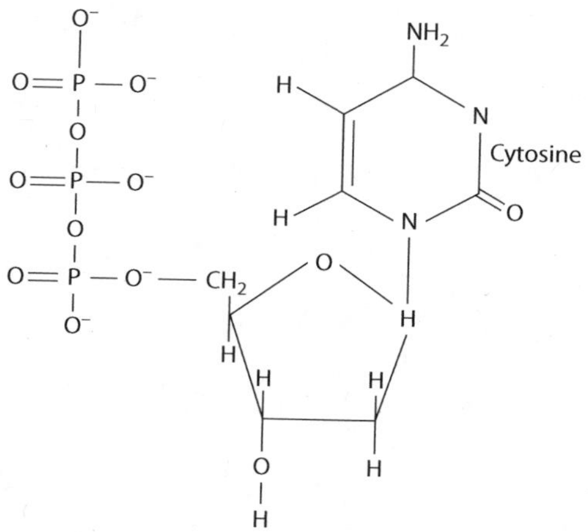

Deoxycytidine triphosphate = dCTP

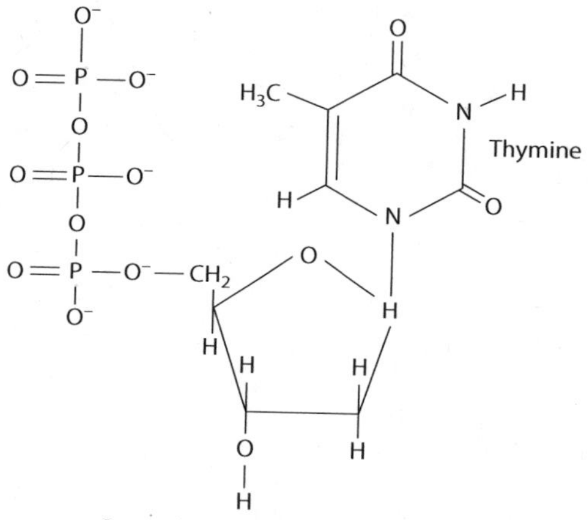

Deoxythymidine triphosphate = dTTP

FIGURE 3.30 **(Continued)** Components of nucleotides (four examples of deoxynucleoside triphosphates).

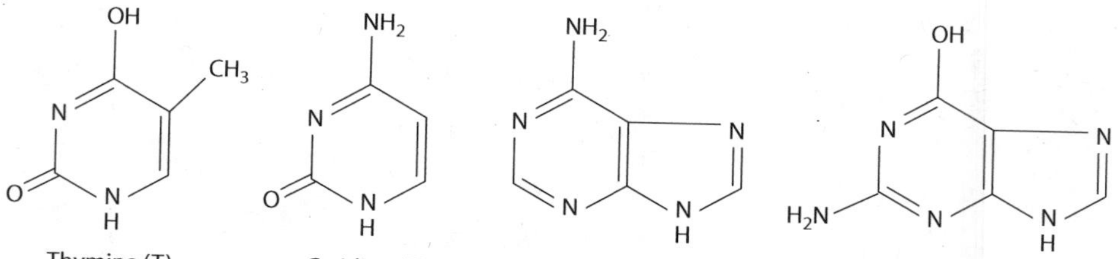

Thymine (T) Cytidine (C) Adenine (A) Guanine (G)

FIGURE 3.31 Purine and pyrimidine bases of nucleic acids.

3'-hydroxyl group of the next one. The backbone of nucleic acids is formed of alternating pentose and phosphate residues, which are both hydrophilic, while at near-neutral pH, the bases are hydrophobic. The bases may be regarded as side groups of the backbone, as can be seen in Figure 3.32. Nucleic acids are synthesized in a 5' to 3' direction. During the synthesis in cells, the enzymes (DNA and RNA polymerases) add nucleotides to the 3' end of the previously incorporated base. By convention, a short nucleic acid, with fifty or less nucleotides, is called an *oligonucleotide*.

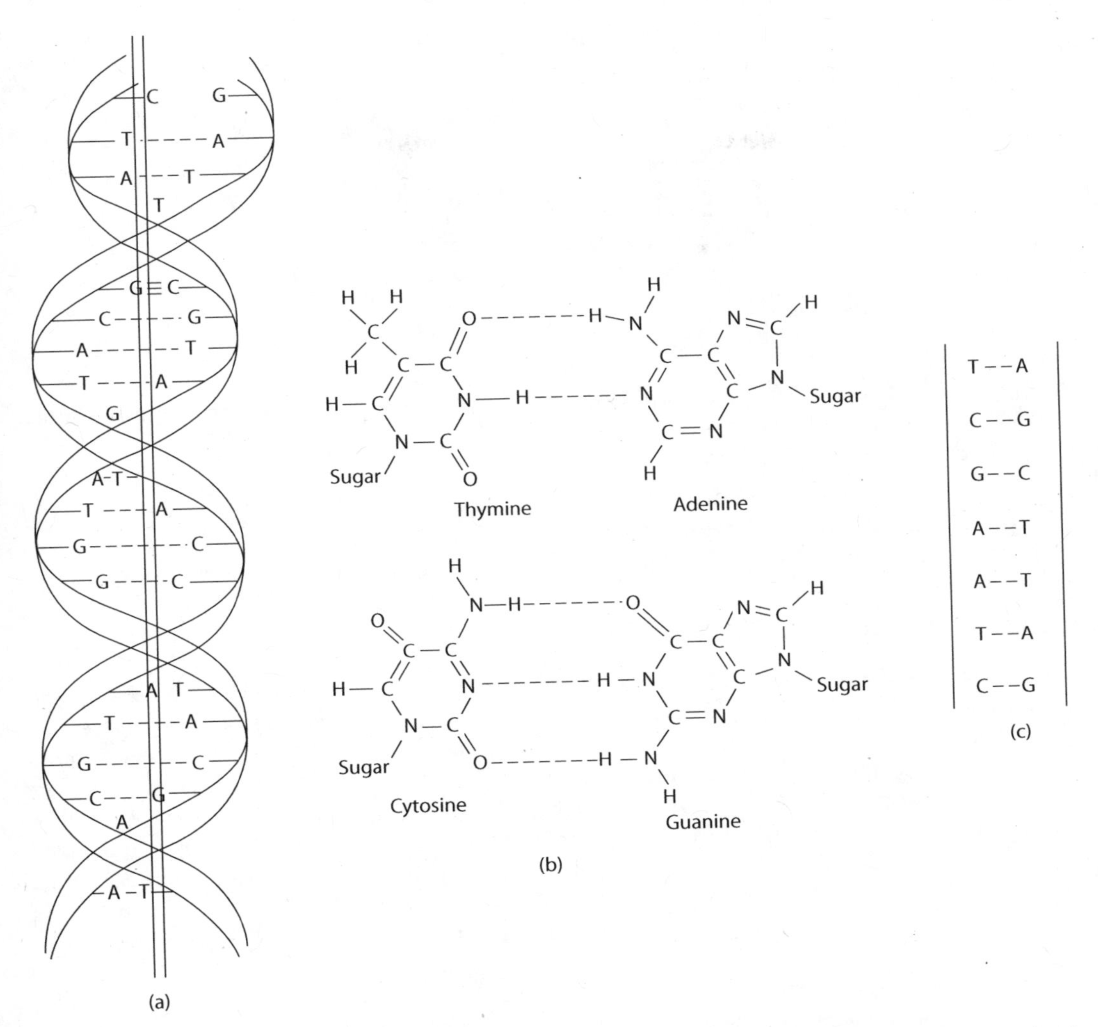

FIGURE 3.32 The double helix. (a) Deoxyribonucleic acid (DNA) is a double-stranded, helical molecule composed of nucleotide units. Replication of DNA occurs through the pairing of nucleotide bases to form new strands. (b) Diagram of the molecular structure of the base pairs of DNA. "Sugar" indicates the bond to deoxyribose. Phosphates connect the sugars to make the "backbone" of DNA. A nitrogen base plus deoxyribose and a phosphate group constitute a nucleotide. Adenine (A) always pairs with thymine (T), and cytosine (C) always pairs with guanine (G). (c) Diagram showing the pairing of the nucleotide bases in a short DNA segment. (Reproduced from the Education Center of the American Phytopathological Society, http://www.apsnet.org/education/K12PlantPathways/TeachersGuide/Activities/DNA_Easy/top.htm. With permission.)

Figure 3.32 shows that G-C base pairs have three hydrogen bonds, whereas A-T base pairs have only two hydrogen bonds; i.e., the energy necessary to disrupt GC-rich DNA is larger than that for AT-rich DNA. *Denaturation* of double-stranded DNA is generally achieved by heating a solution of DNA to a temperature that disrupts the hydrogen bonds. The energy required to separate two complementary DNA strands depends on the base composition, but also on the strand length and the chemical environment (presence of Na^+ ions).

Base pairing in RNA follows the same principles as with DNA, and the base pairs that form are A-U and G-C. Although DNA stores the information for protein synthesis, RNA carries out the instructions encoded in DNA. There are three kinds of RNA molecules that perform different functions in protein synthesis: *messenger RNA* (mRNA), which carries the genetic information from

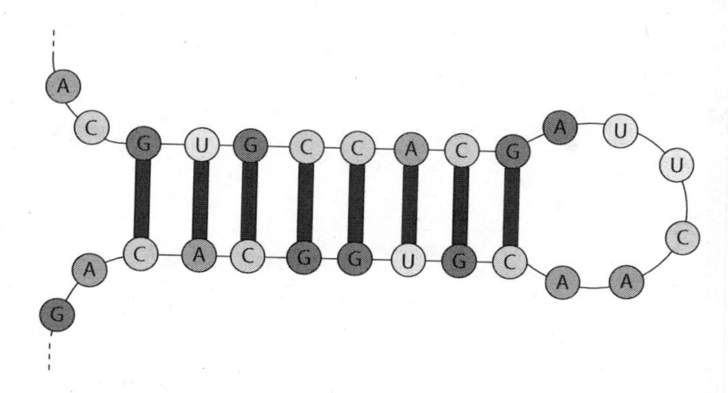

FIGURE 3.33 Single strand of RNA with intramolecular loops formed by complementary base pairing. (Adapted from http://en.wikipedia.org/wiki/Stem-loop.)

DNA; *transfer RNA* (tRNA), which is the key to deciphering the code in mRNA; and *ribosomal RNA* (rRNA), which associates with a set of proteins to form ribosomes. *Ribosomes* are made from proteins and ribosomal RNAs and are the sites of protein synthesis/translation in the cell.

3.4.3 DNA Sensing Strategies

DNA biosensors and gene chips are based on nucleic acid recognition processes called *hybridization*, and they are used in DNA diagnostics, involving the testing of genetic and infectious diseases.

DNA hybridization is based on the ability of single-stranded DNA (ssDNA) to recognize a strand with a complementary nucleotide sequence (ss probe DNA) and form the DNA duplex (double-stranded DNA (dsDNA)). Hybridization in biosensing is based on the immobilization of the ssDNA probe onto the transducer surface. In the "sandwich" strategy the target nucleic acid is immobilized onto the sensor surface by the capture probe and detected by hybridization with a detector probe coupled to the label, such as ferrocene derivatives. The redox signals of the sensor electrodes are detected amperometrically. The sandwich sensor shown in Figure 3.34 is based on an indirect detection of DNA.

Optical, electrochemical, and mass-sensitive DNA hybridization biosensors have also been developed for obtaining sequence-specific information. Many aspects of genetic analysis have been significantly advanced by the implementation of DNA microarrays. These devices, integrated on

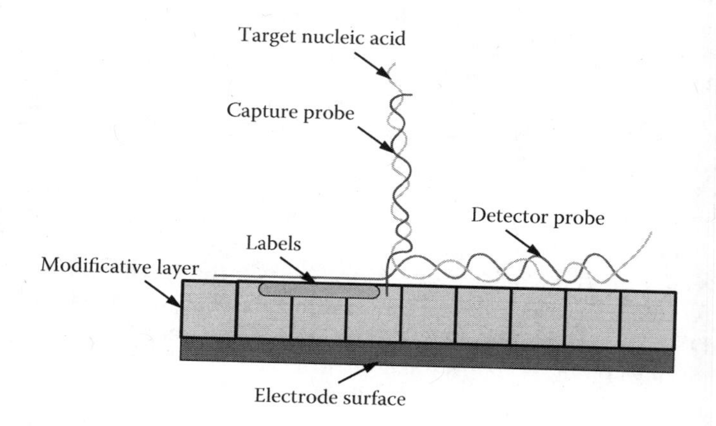

FIGURE 3.34 Schematic of a "sandwich" DNA sensor. (Adapted from Wang et al., *Sensors* 8, 2043–81, 2008.)

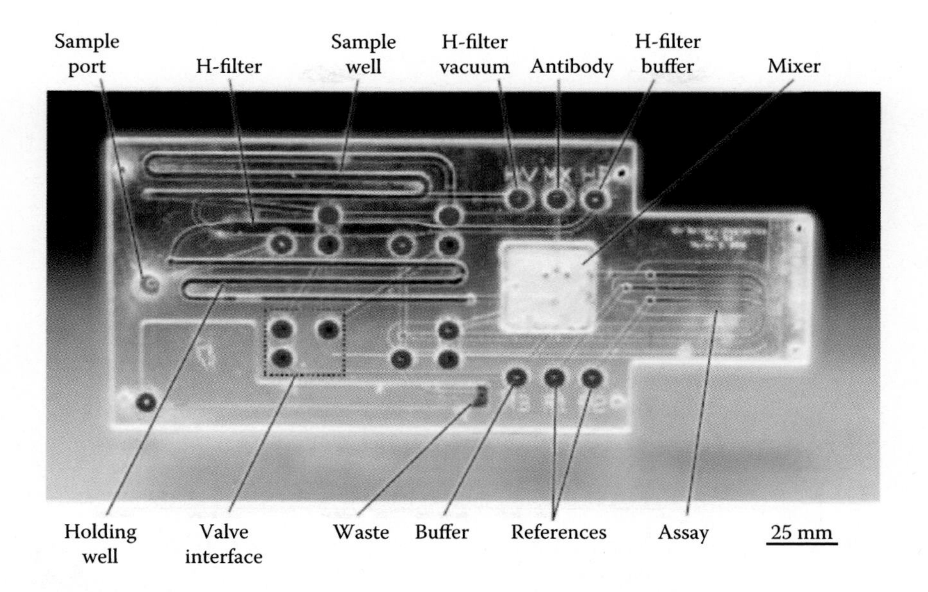

FIGURE 1.4 POC microchip.

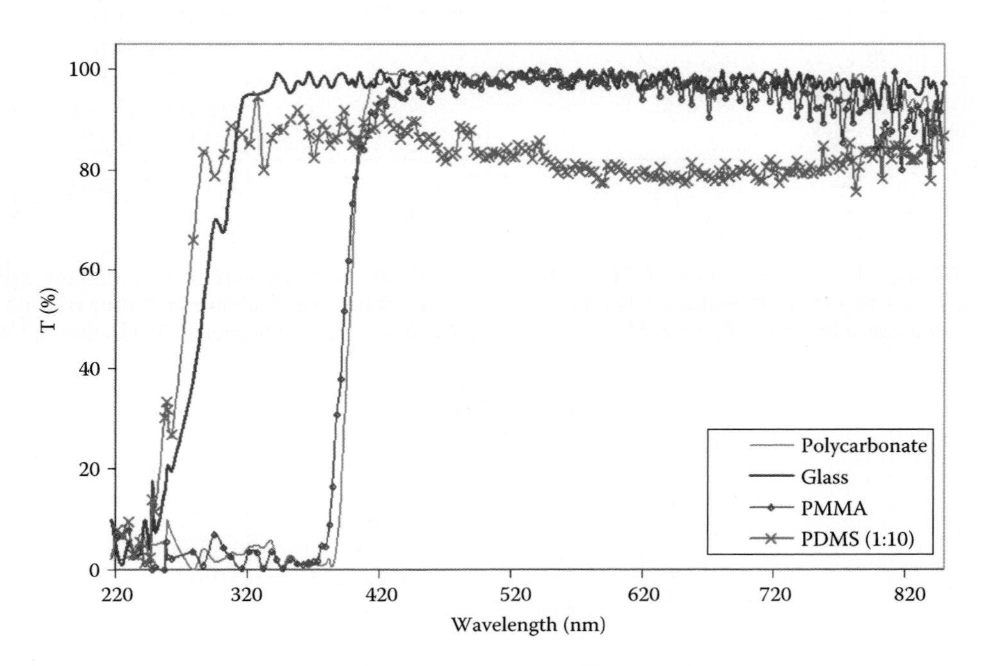

FIGURE 2.5 UV-visible spectrum of PDMS, glass, polycarbonate and polymethylmethacrylate. (Unpublished data from the authors' laboratory, Acsharia, "Modeling, Fabrication and Testing of Oled and Led Fluorescence Integrated Polymer Microfluidic Chips for Biosensing Applications," MASc thesis, 2007.)

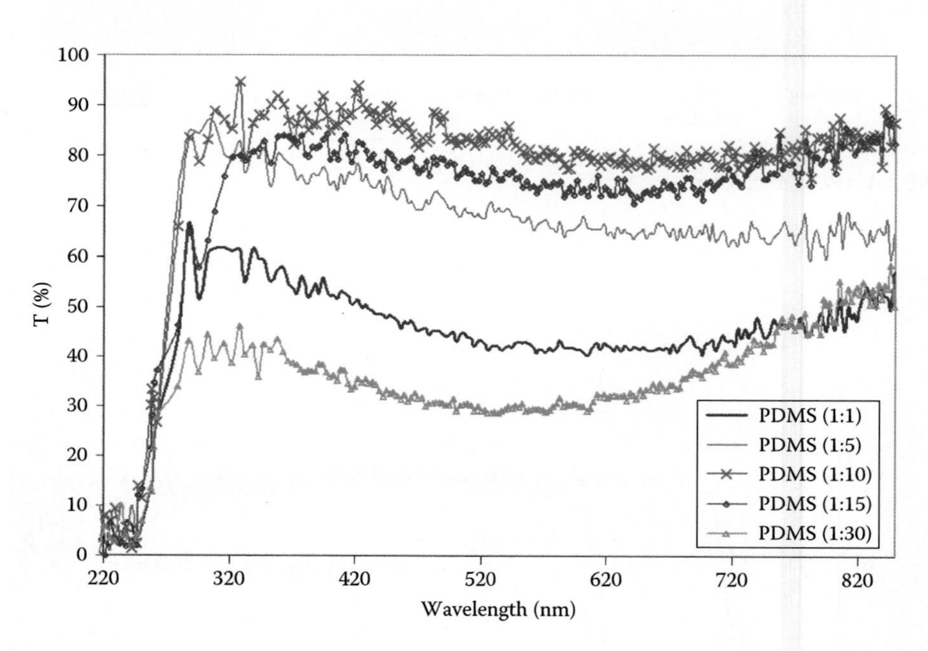

FIGURE 2.6 UV-visible spectra of PDMS fabricated with different amounts of cross-linking agents. (Unpublished data from the authors' laboratory, Acsharia, "Modeling, Fabrication and Testing of Oled and Led Fluorescence Integrated Polymer Microfluidic Chips for Biosensing Applications," MASc thesis, 2007.)

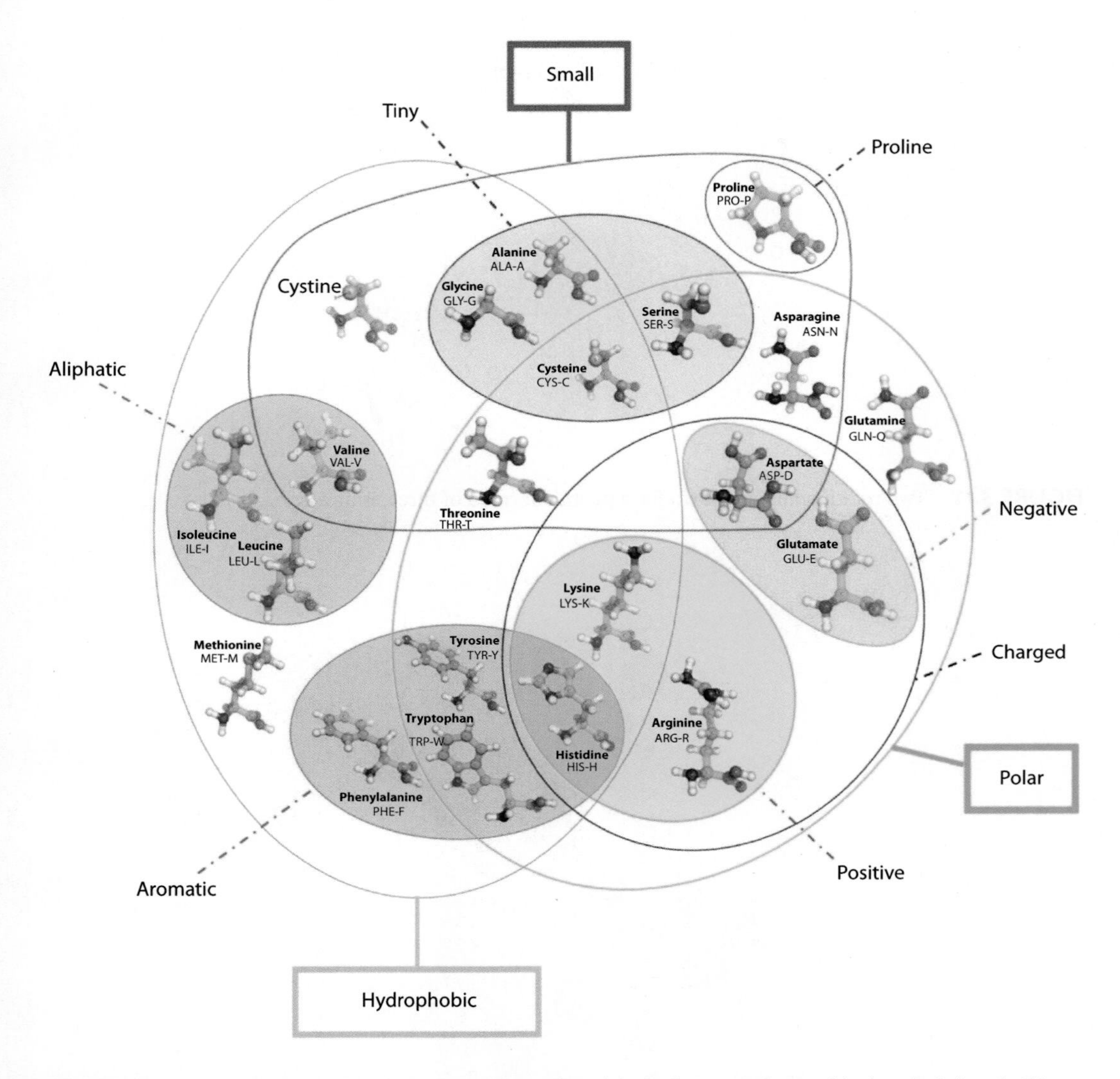

FIGURE 3.4 Venn diagram of amino acids. (Reproduced from sites.google.com/site/apodtele/nutshell.)

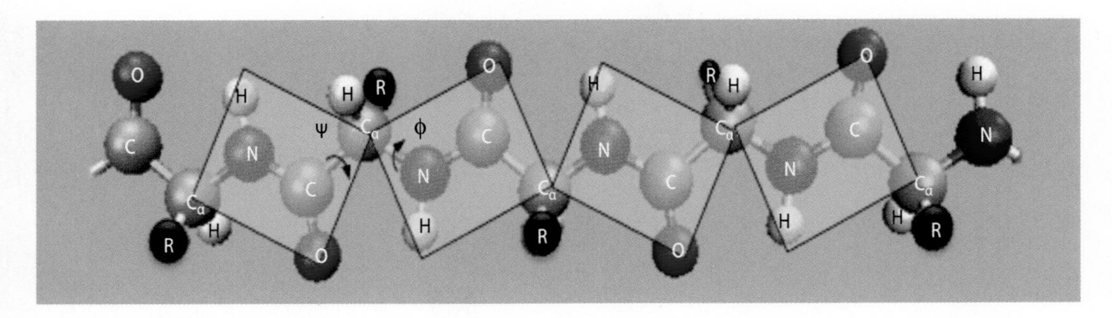

FIGURE 3.9 Extended polypeptide chain with planar amide bonds. (Reproduced from http://employees.csbsju.edu/Hjakubowski/classes/ch331/protstructure/olunderstandconfo.html. With permission.)

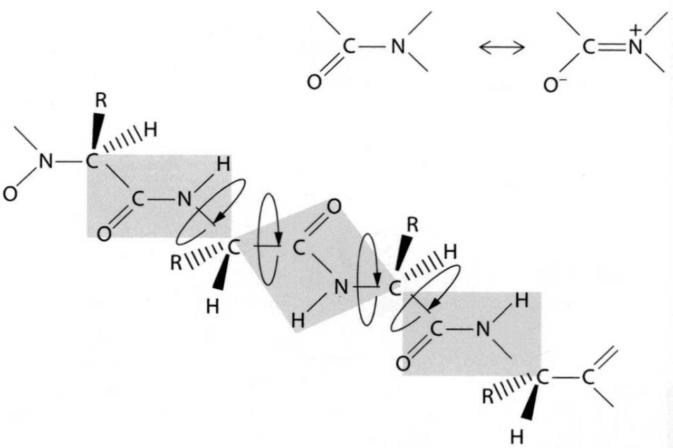

FIGURE 3.11 Two rotation angles shown for a peptide formed of three amino acids.

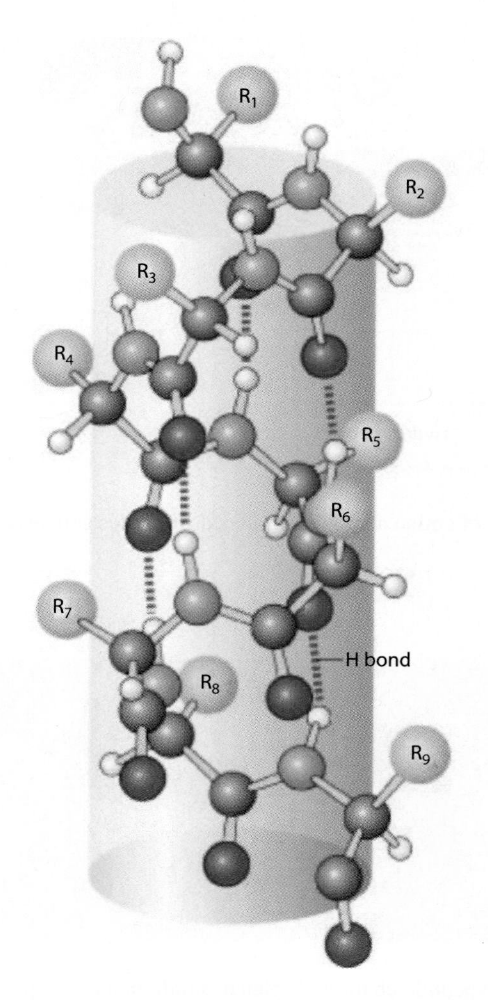

FIGURE 3.15 Typical α-helix. (Reproduced from themedicalbiochemistrypage.org. With permission.)

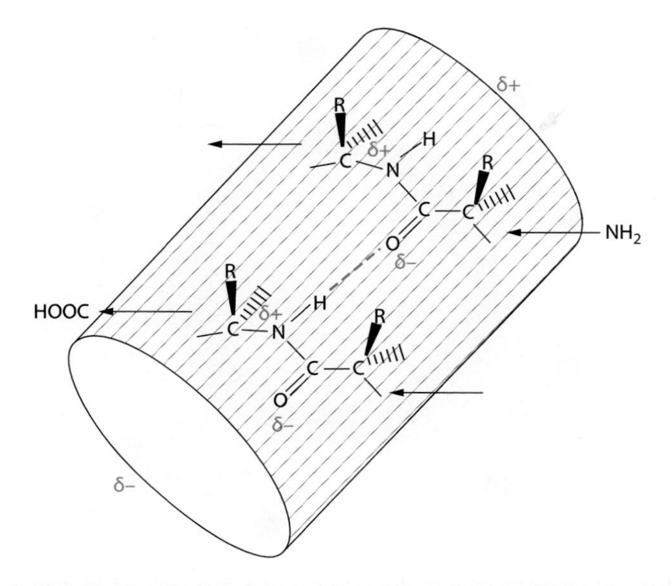

FIGURE 3.16 α-Helix structure showing the charge distribution along the peptide bond (δ is the charge on the atoms).

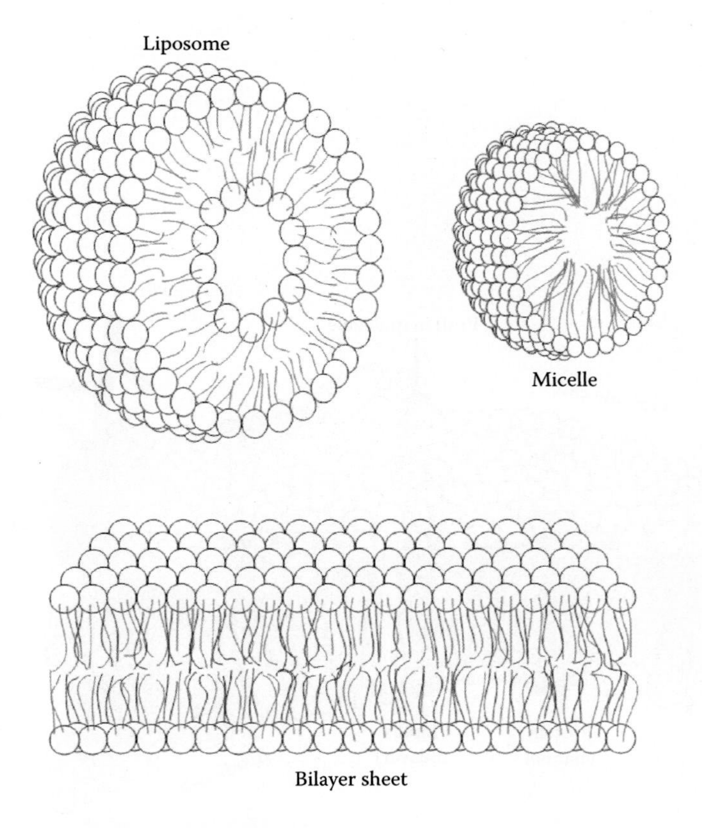

FIGURE 3.24 Formation of bilayers and liposomes.

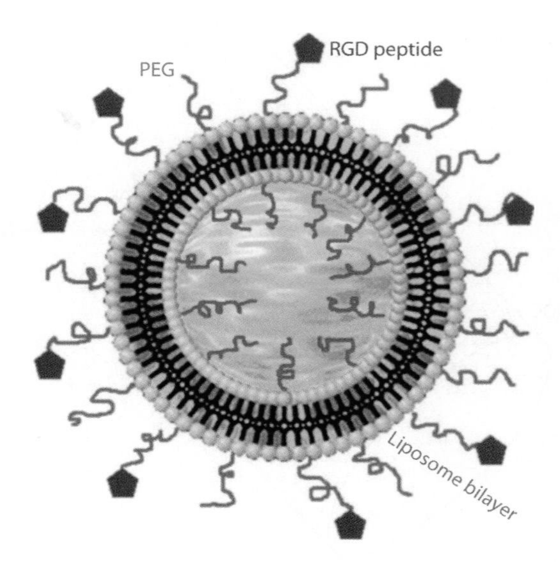

FIGURE 3.26 Schematic representation of the targeted liposome delivery system. Cyclic RGD peptides coupled to the distal end of maleimide-PEG-DSPE in the liposome bilayer of PEG-grafted LCL. (Reproduced from Nallamothu et al., *AAPS PharmSciTech*, 7(article 32), 2008. With permission.)

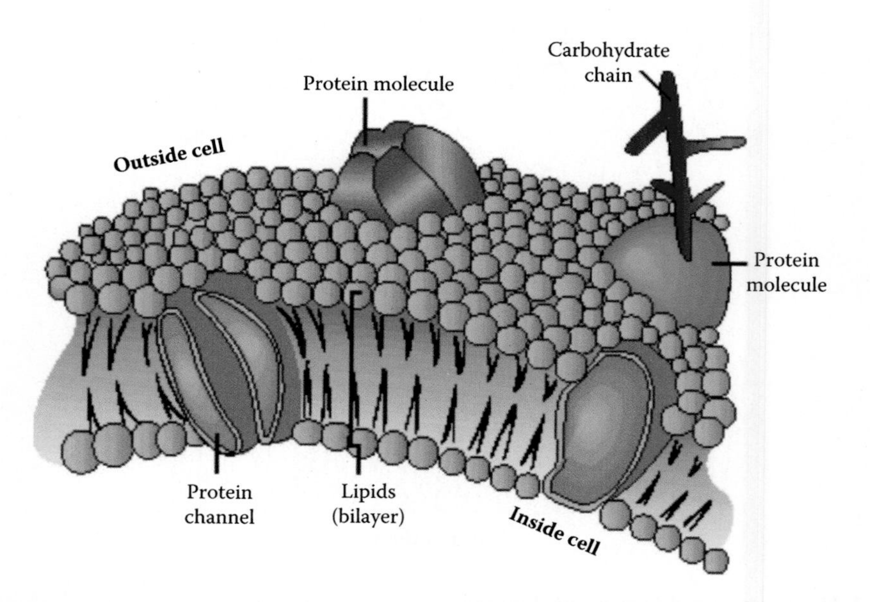

FIGURE 3.27 Architecture of cell membrane.

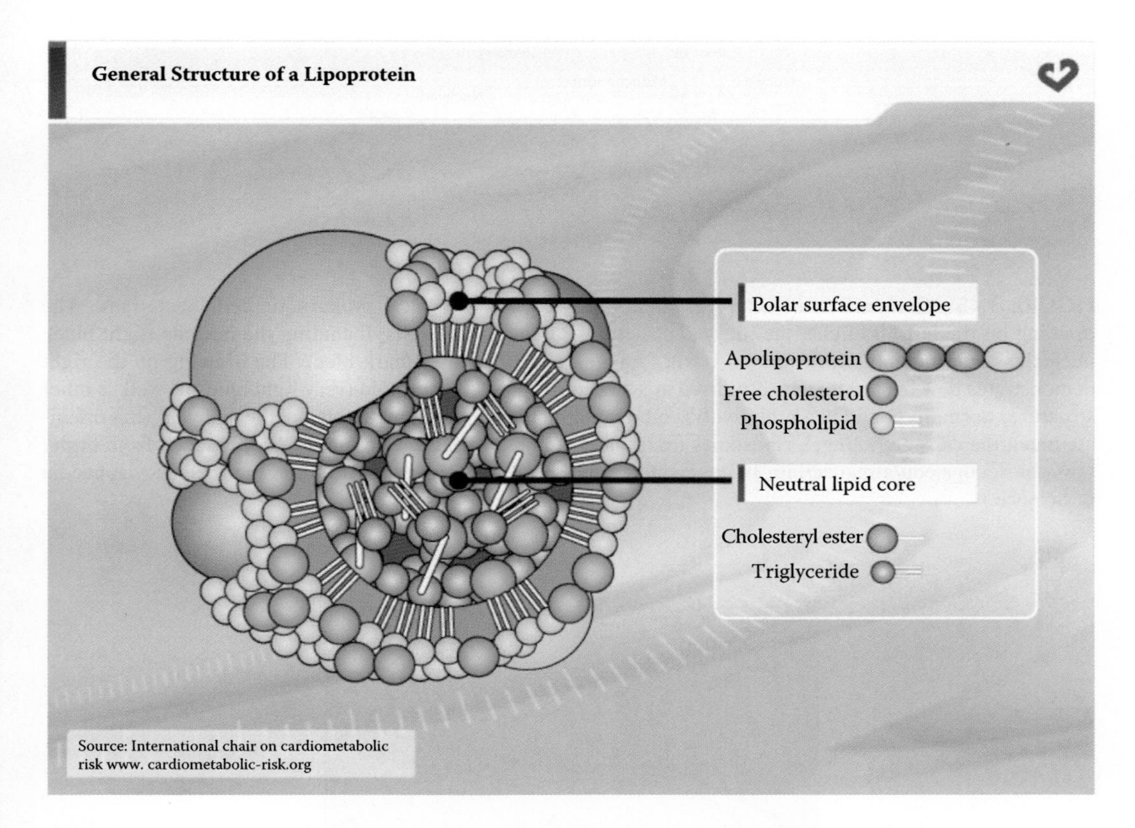

FIGURE 3.28 General structure of a lipoproteine. (Reproduced from www.cardiometabolic-risk.org. With permission.)

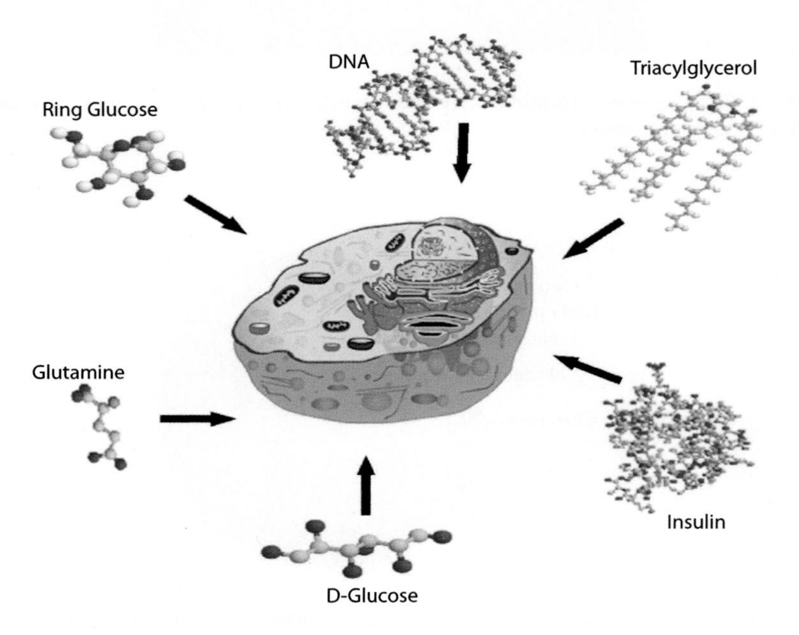

FIGURE 3.44 Molecular components of cells.

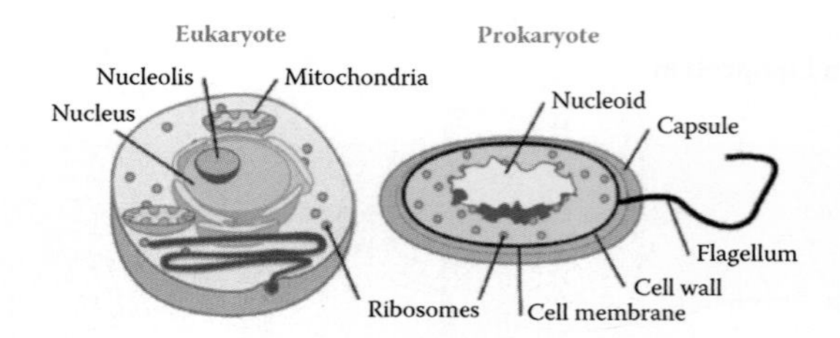

FIGURE 3.45 Comparison of a typical eukaryotic cell with a typical prokaryotic cell (bacterium). The drawing on the left highlights the internal structures of eukaryotic cells, including the nucleus (light blue), nucleolus (intermediate blue), mitochondria (orange), and ribosomes (dark blue). The drawing on the right demonstrates how bacterial DNA is housed in a structure called the nucleoid (very light blue), as well as other structures normally found in a prokaryotic cell, including the cell membrane (black), the cell wall (intermediate blue), the capsule (orange), ribosomes (dark blue), and a flagellum (also black). (Reproduced from http://www.biosci.uga.edu/almanac/bio_103/notes/may_15.html. With permission from the Internet Encyclopaedia of Science.)

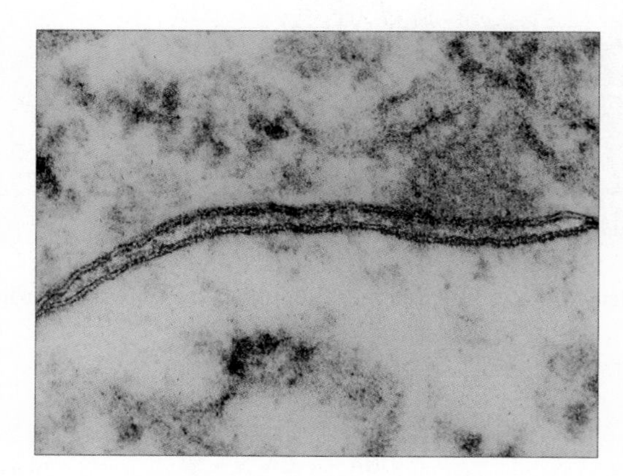

FIGURE 3.46 Cell cytoplasmic membrane. (From www.DennisKunkel.com, copyright Dennis Kunkel Microscopy, Inc. With permission.)

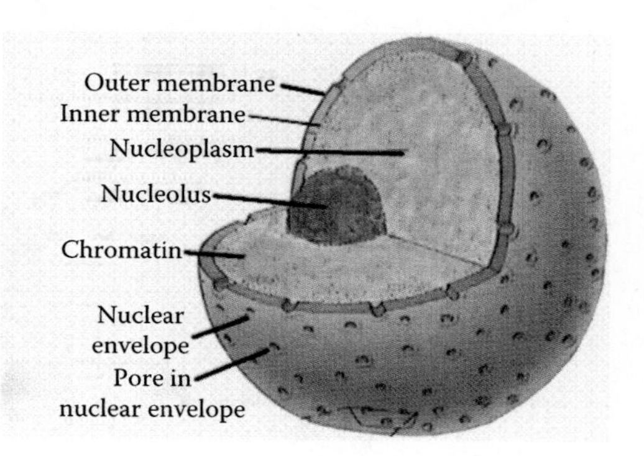

FIGURE 3.47 Structure of the nucleus. (Reproduced from Purves et al., *Life: The Science of Biology*, 4th ed., Sinauer Associates and WH Freeman, 1995. With permission.)

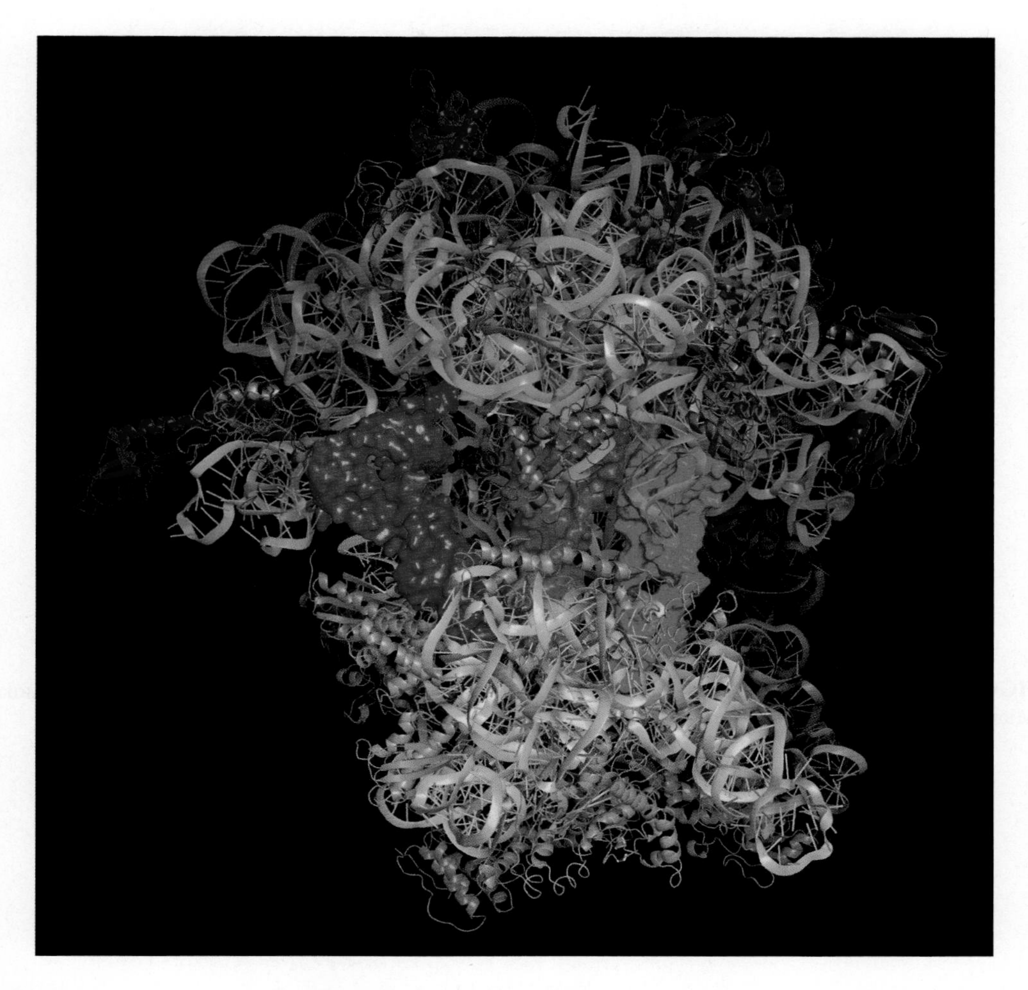

FIGURE 3.49 Ribosome. (Reproduced from http://scienceblogs.com/transcript/ribosome.06-09-18.jpg. With permission.)

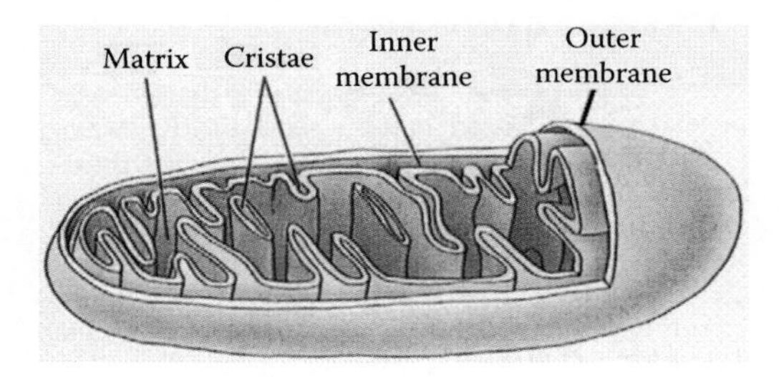

FIGURE 3.50 Structure of a mitochondrion.

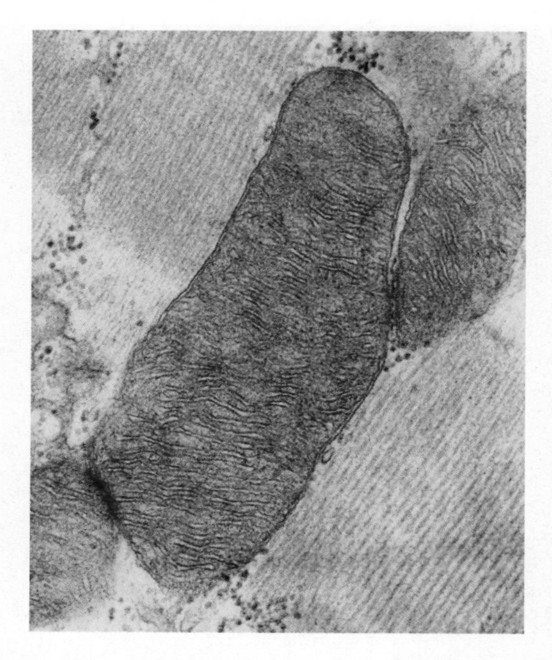

FIGURE 3.51 Muscle cell mitochondrion. (From www.DennisKunkel.com. Copyright Dennis Kunkel Microscopy, Inc. With permission.)

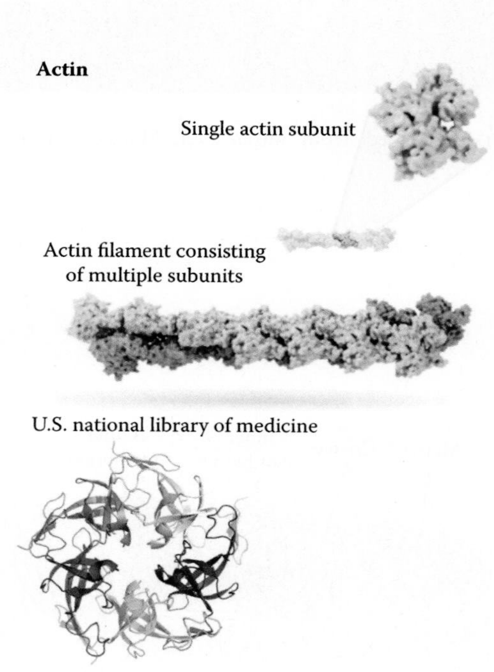

FIGURE 3.52 Structure of actin filaments.

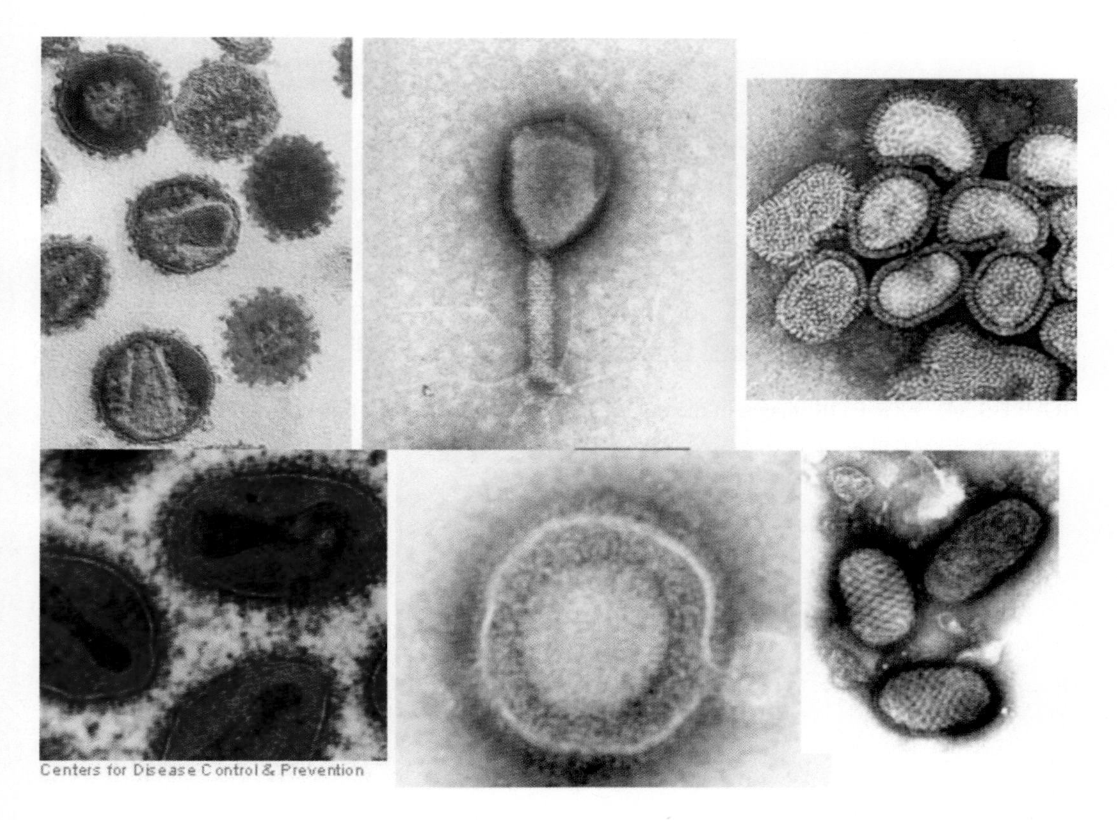

FIGURE 3.56 Gallery of electron micrographs of viruses illustrating diversity in form and structure. Clockwise: Human immunodeficiency virus (HIV), Aeromonas virus 31, influenza virus, Orf virus, herpes simplex virus (HSV), smallpox virus. (Courtesy of Centers for Disease Control and Prevention, Gallery of electron micrographs, textbookofbacteriology.net/.../Phage.html.)

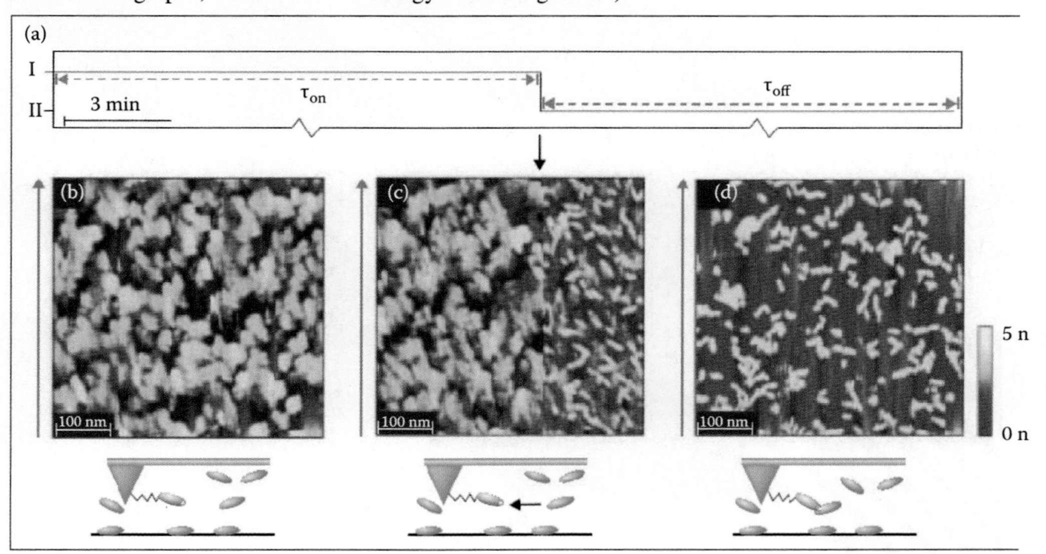

FIGURE 5.23 Force recognition imaging of antibody-antigen interaction. (Reprinted from Allison et al., *Curr. Opin. Biotechnol.*, 13, 47–51, 2002; original figure in Raab et al., *Nat. Biotechnol.*, 17, 901–5, 1999. Copyright 2005, with permission from Elsevier.)

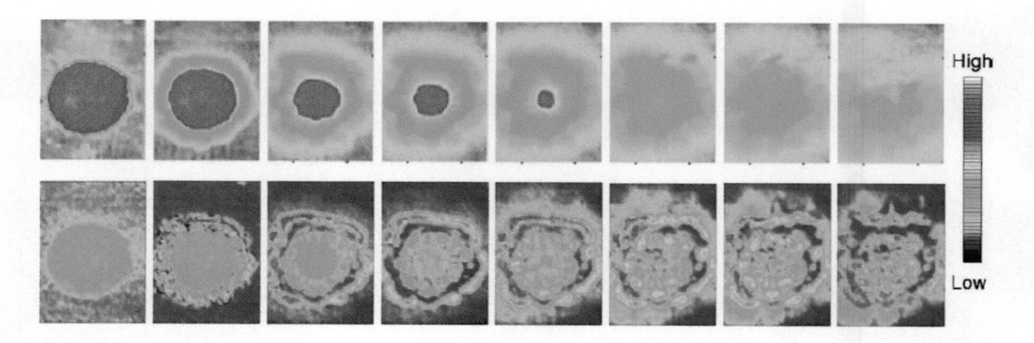

FIGURE 5.34 ATR-FTIR images of the PEG/ibuprofen formulation as a function of time.

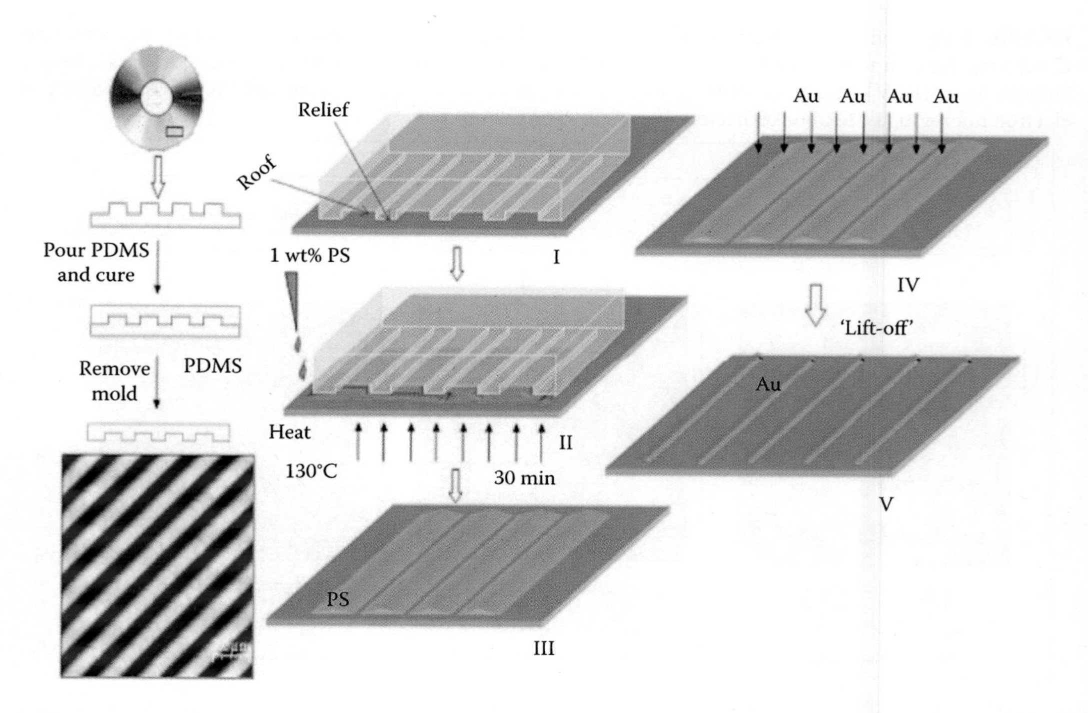

FIGURE 7.30 Preparation of the PDMS stamp from a CD and the AFM image of the stamp; molding of polystyrene using the PDMS stamp (steps I to III) and the metal deposition (steps IV and V). (Reprinted from Radha et al., *Appl. Materials & Interfaces* 1, 257-260, 2009.)

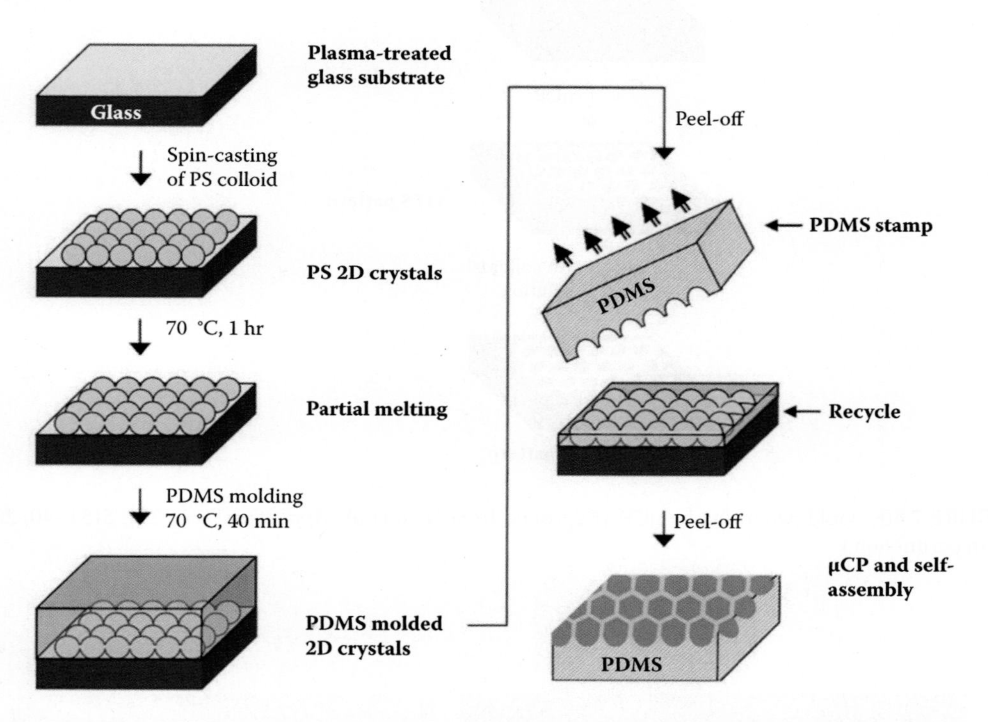

FIGURE 7.59 Schematics of two-dimensional nanopatterning by PDMS relief structures of polymeric colloidal crystals. (Adapted from Nam et al., *Appl. Surf. Sci.*, 254, 5134–40, 2008.)

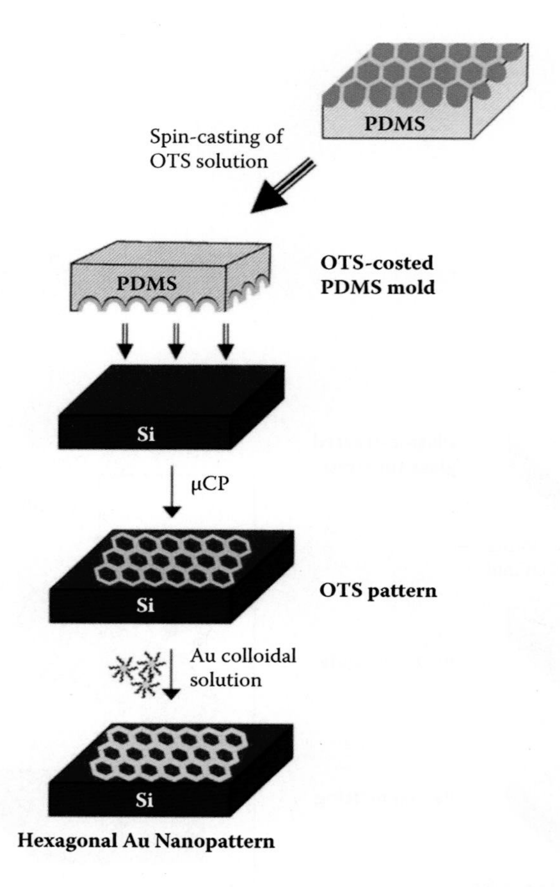

FIGURE 7.60 Gold patterning by μCP. (Reprinted from Nam et al., *Appl. Surf. Sci.*, 254, 5134–40, 2008. With permission.)

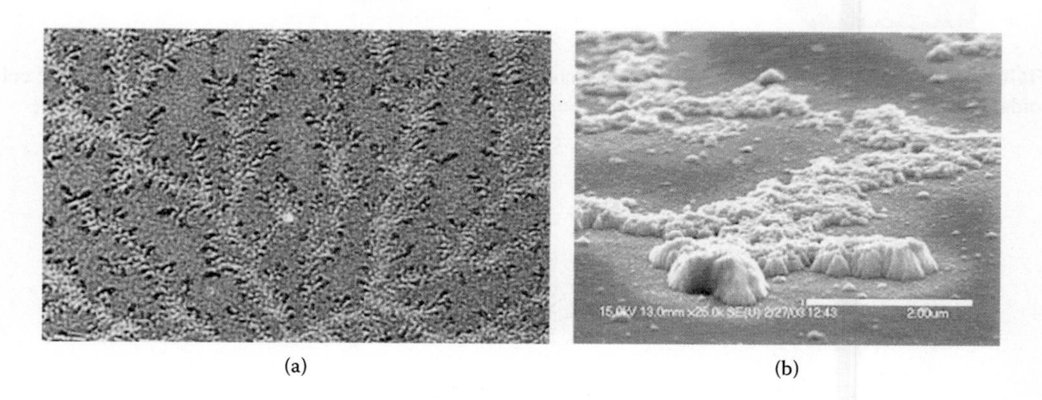

(a) (b)

FIGURE 8.32 (a) An optical micrograph of a silver electrodeposit on Ag-Ge-Se solid electrolyte. The field of view is approximately 200 μm width. (b) A scanning electron micrograph of one branch of the electrodeposit. The size bar is 2 μm. (From Przekwas and Makhijani, "Mixed-Dimensionality, Multi-Physics Simulation Tools for Design Analysis of Microfluidic Devices and Integration Systems," in *Technical Proceedings of the 2001 International Conference on Modeling and Simulation of Microsystems*, 2001, www.cr.org.)

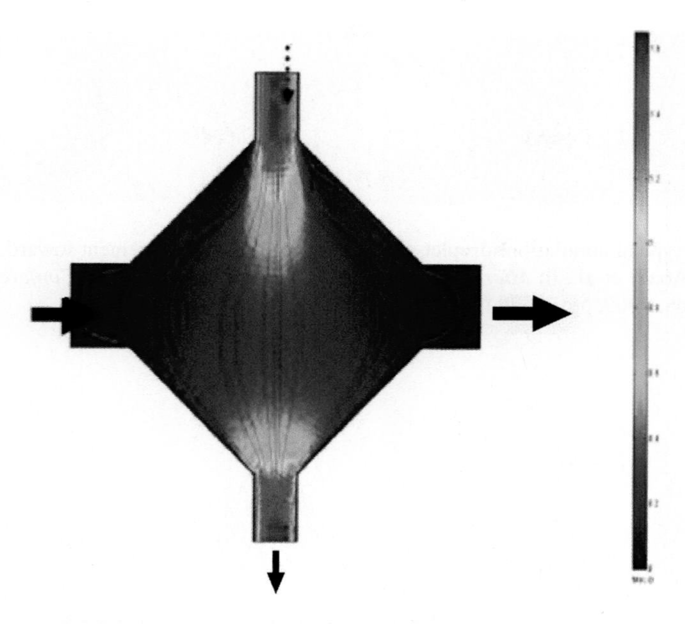

FIGURE 8.35 Predicted flow behavior in the chamber using FEM simulation.

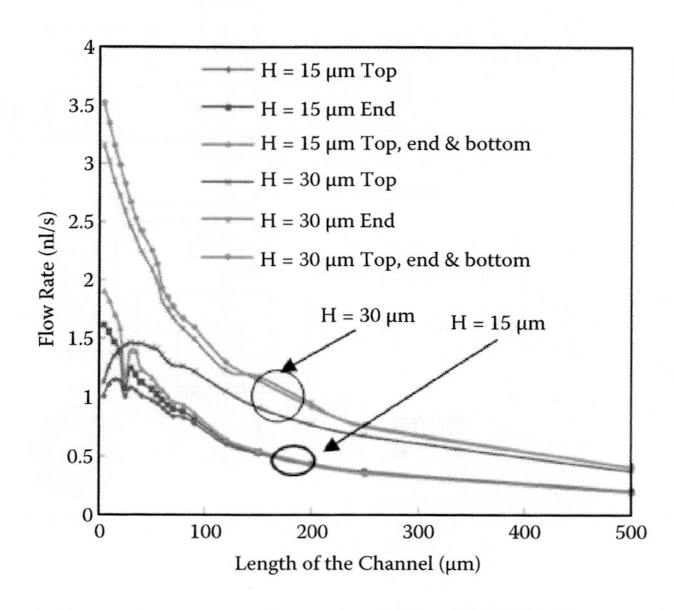

FIGURE 8.37 Influence of electrode position on flow rate with varying the length of the pump. (From Ozhikandathil et al., "Modeling and Analysis of Low Voltage Electro-Osmotic Micropump." paper presented at Proceedings of ASME 2010 3rd Joint US-European Fluids Engineering Summer Meeting and 8th International Conference on Nanochannels, Microchannel and Minichannels, FEDSM ICNMM2010-31213, Montreal, Canada, 2010.)

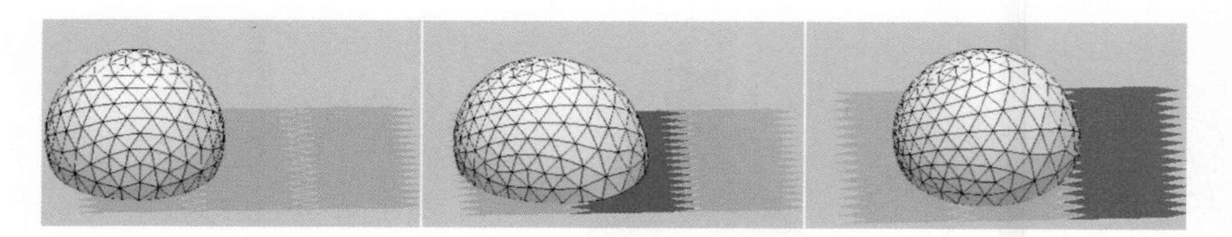

FIGURE 8.42 A typical simulation: droplet at rest (left) and during movement toward activated electrode (right). (From SadAbadi et al., in *Microsystems and Nanoelectronics Research Conference, 2009*, MNRC 2nd, Ottawa, Canada, 2009, pp. 76–79.)

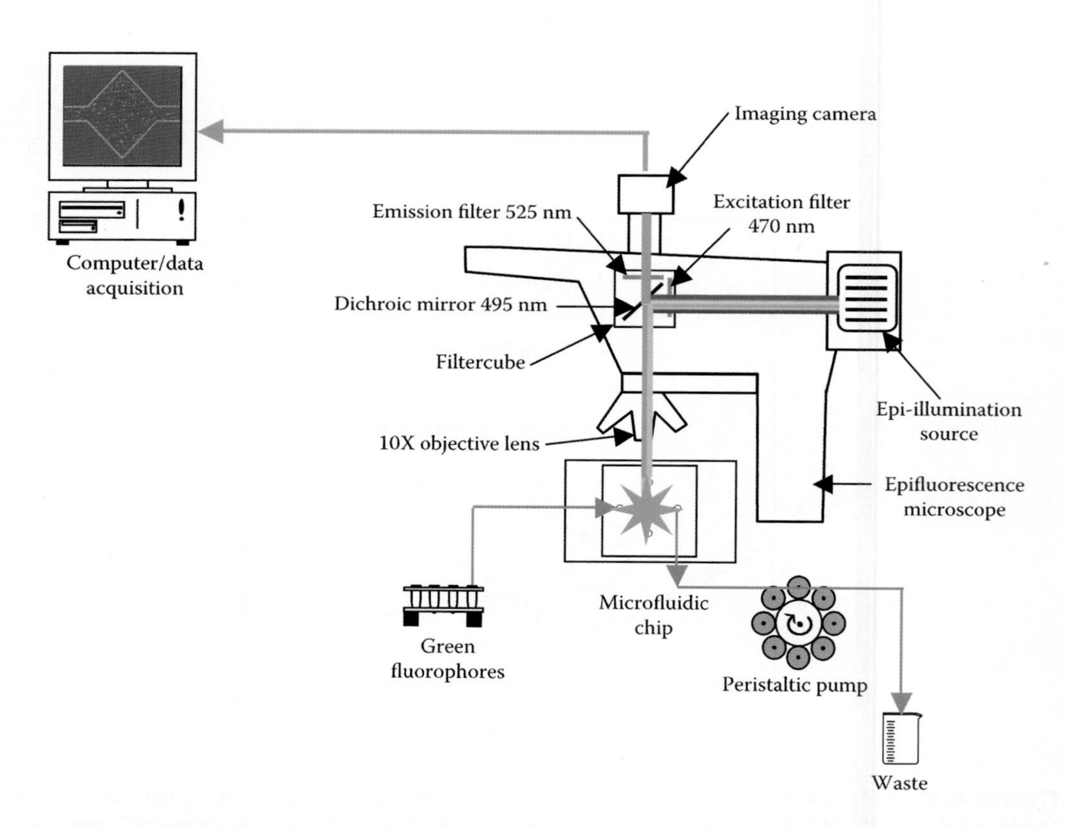

FIGURE 8.44 Schematic of epifluorescence microflow visualization setup. (From Acharya, "Modeling, Fabrication and Testing of OLED and LED Fluorescence Integrated Polymer Microfluidic Chips for Biosensing Applications," MASc thesis, Concordia University, 2007.)

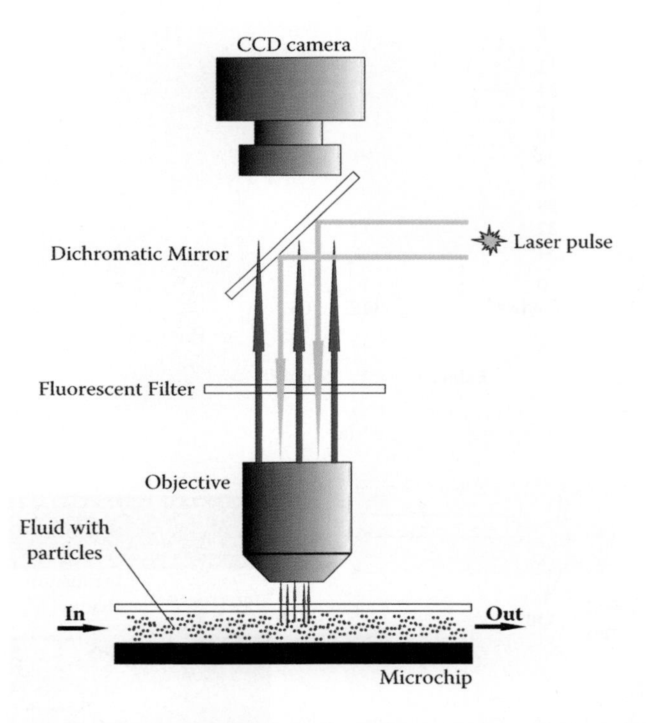

FIGURE 8.48 Optical detection setup used in other studies by μPIV. (From Fang, Q., "Optical Fiber Coupled Low Power Micro-PIV Measurement of Flow in Microchambers: Modeling, Fabrication, Testing and Validation," MASc thesis, Concordia University, 2009.)

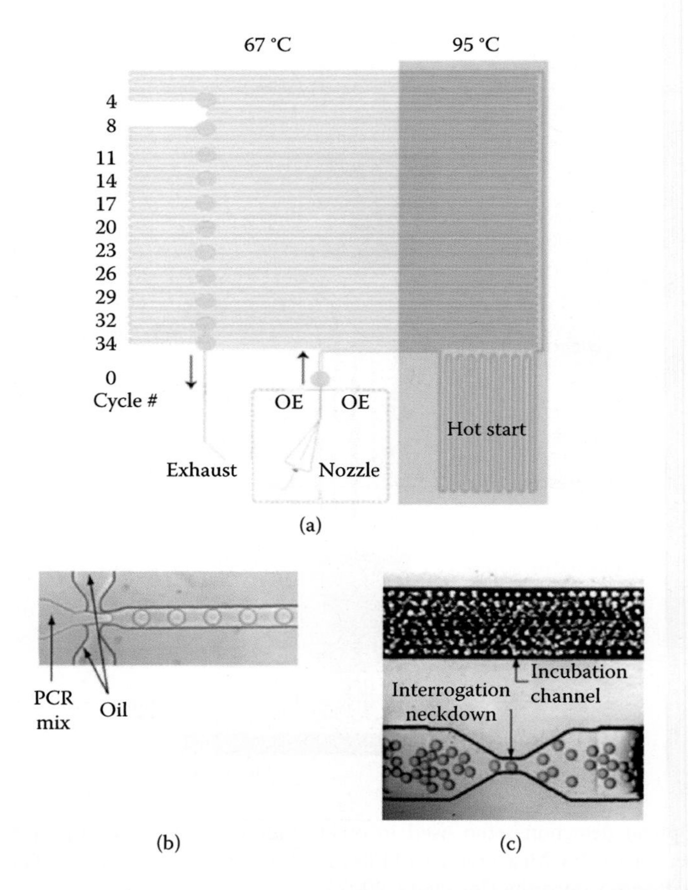

FIGURE 9.18 The layout of PCR chip. (a) With different temperature zones for thermal cycling. (b) The PCR mixture is injected into the main nozzle, while oil is injected through the side nozzles for droplet generation at the nozzle. (c) Generated uniform picoliter droplets are seen in the downstream channel and neck areas. (Reprinted from Kiss et al., *Anal. Chem.*, 80, 8975–81, 2008. With permission.)

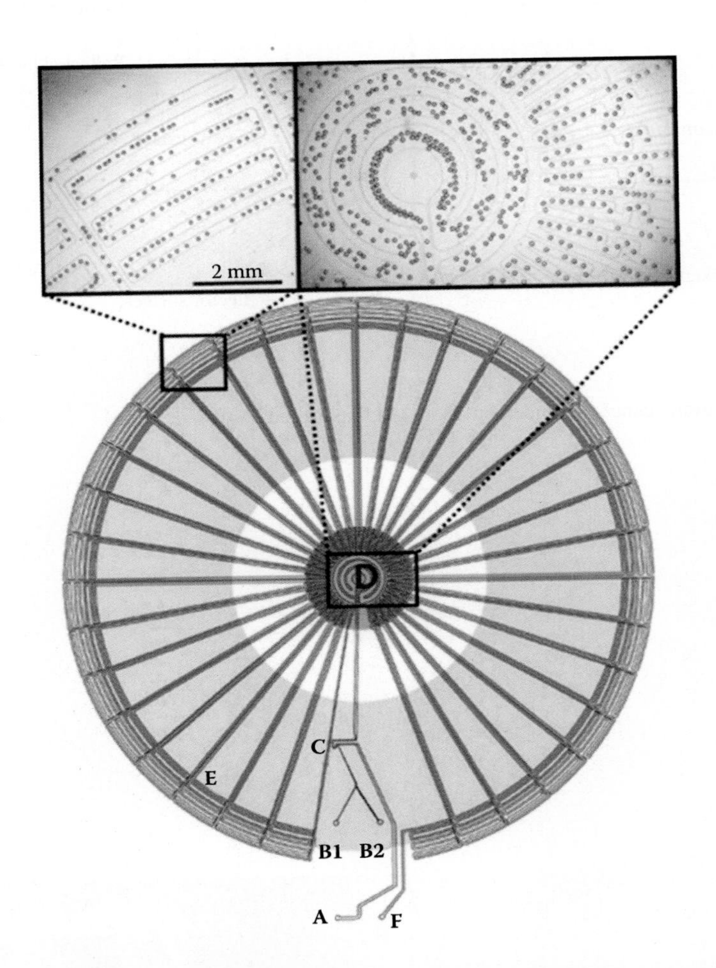

FIGURE 9.20 Schematic diagram of a radial PCR device. Droplets are generated by sending carrier fluid oil at inlet A, and the aqueous phase at two inlet channels, B1 and B2. Droplets are generated at the T-junction denoted by C. Initial denaturation is implemented at location D, while annealing and extension are implemented in the peripheral zone, E, where primer annealing and template extension occur. The droplets are collected at exit F after thirty-four PCR cycles. The heating is implemented using the Peltier heat module. (Reprinted from Zhang and Ozdemir, *Anal. Chim. Acta*, 638, 115–25, 2009. With permission.)

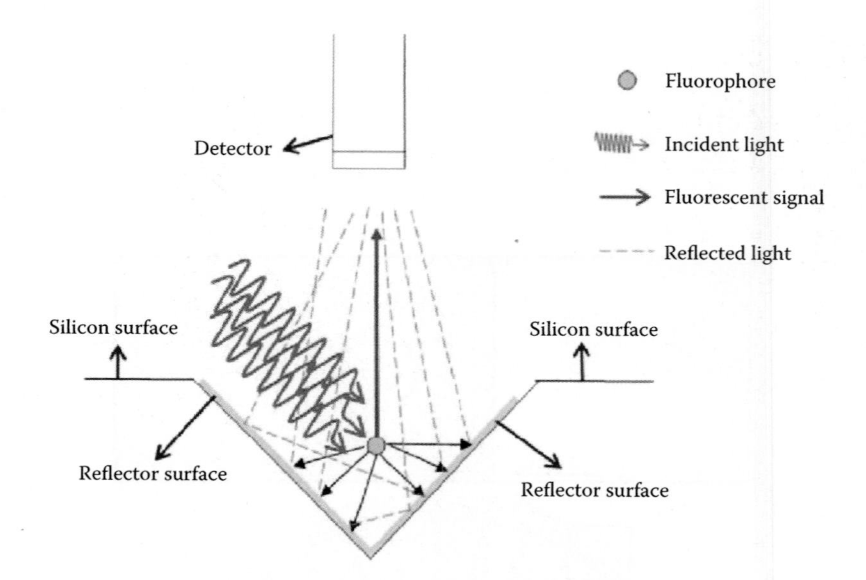

FIGURE 9.67 Fluorescence enhancement from gold-sputtered V-groove.

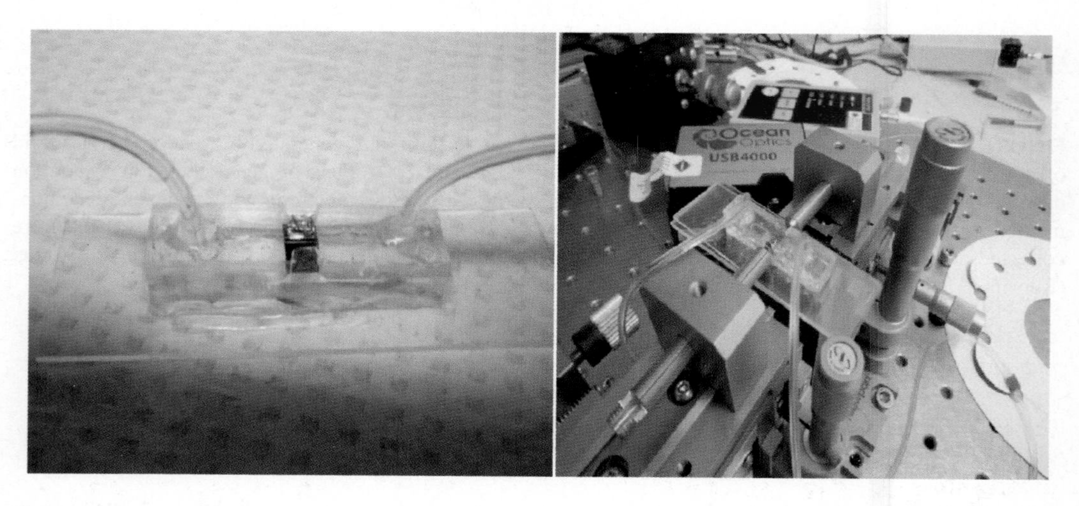

FIGURE 9.68 SOS-PDMS platform for the fluorescence detection of QD[48] and optical characterization setup.

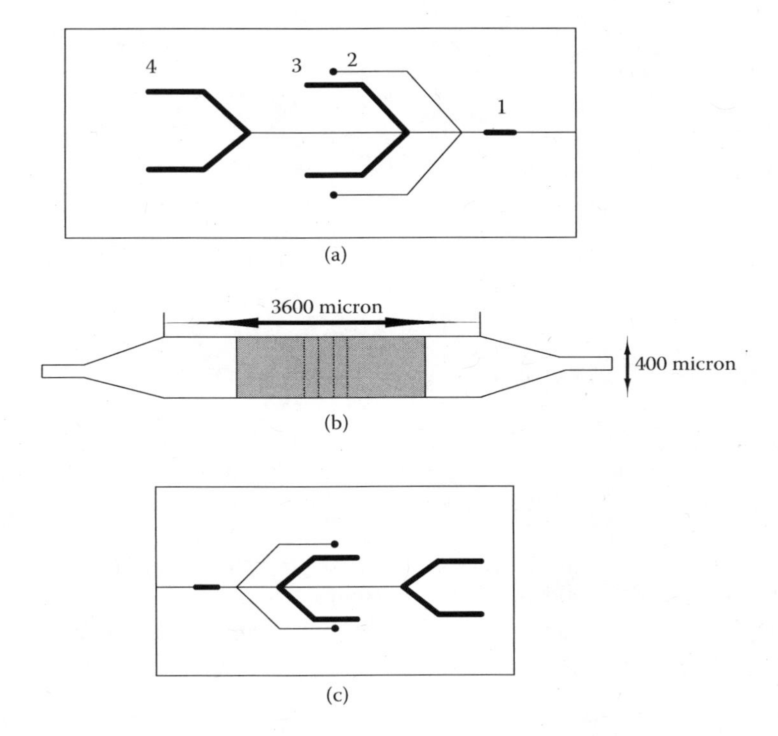

FIGURE 3.35 Microfluidic system based on microarray hybridization. (a) Molded PDMS microfluidic unit on a glass slide consisting of chambers, microchannels for reagent flow, and hybridization chamber 1. (b) View of the hybridization chamber that can accommodate up to 150 nucleic acid capture probes spotted onto a glass slide. (c) PDMS chip applied onto a glass slide on which the capture probes are arrayed. The glass slide is placed on a CD support that can hold up to five slides overlaid with PDMS. The hybridization reagents are positioned to be pumped sequentially through the hybridization chamber by centrifugal force, starting with chamber 2. (Reproduced from Peytavi et al., *Clin. Chem.*, 51, 1836–44, 2005. With permission.)

microfluidic platforms, conduct many processes, such as sample collection with DNA extraction, amplification, hybridization, and detection. The schematic of such a device is shown in Figure 3.35.

DNA microarrays offer great promise for monitoring sequence-specific information in a faster and cheaper manner than traditional hybridization assays.

3.5 CARBOHYDRATES

3.5.1 Introduction

Carbohydrates or *saccharides* are the most abundant biomolecule. They serve as a major source of metabolic energy (storage and transport), as structural material (cellulose), and are one of three essential components of nucleic acids (deoxyribose). Carbohydrates linked to lipid molecules (*glycolipids*) and proteins covalently linked to carbohydrates (*glycoproteins*) are the common components of biological membranes. These two classes of biomolecules, called *glycoconjugates*, are important components of cell walls and extracellular structures in plants, animals, and bacteria. For example, *muramic acid* and *neuraminic acid*, derivatives of polysaccharides, are components of cell membranes of higher organisms and also of bacterial cell walls. In addition to the structural roles such molecules play, they also serve in a variety of processes involving recognition events. Recognition events are important in normal cell growth, transformation of cells, and other processes. Members of this large and highly varied family act as ligands for complementary binding

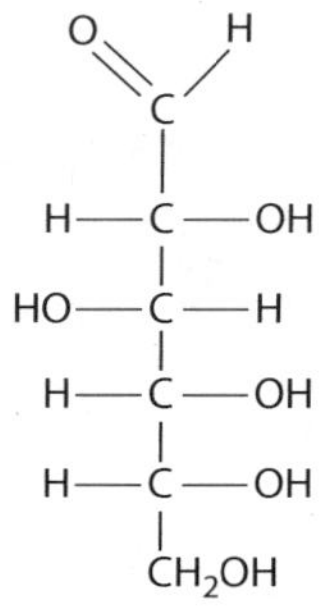

FIGURE 3.36 Glucose molecular structure.

proteins called lectins, present on nearby cells. Lectin-carbohydrate interactions mediate intercellular communication in cell-cell recognition systems in the brain and elsewhere. The study of cell surface complex carbohydrates, lectins, and their roles in cell physiology is part of the rapidly emerging discipline called *glycobiology.*

The term *carbohydrate* is a generic one that refers primarily to carbon-containing compounds that contain hydroxyl, keto, or aldehydic functionalities. The name *carbohydrate* arises from the basic molecular formula $(CH_2O)_n$, which can be rewritten $(C{\cdot}H_2O)_n$, to show that these substances are hydrates of carbon, where $n = 3$ or more.

3.5.2 Monosaccharides

Carbohydrates can range in size from a simple *monosaccharide* (simple sugar) to an *oligosaccharide* or *polysaccharide.* A simple monosaccharide, called *glucose*, which is a building block for larger structures, called oligosaccharides or polysaccharides, is shown in Figure 3.36. A ketose has a ketone functionality, while an aldose has an aldehyde functionality associated with it.

The names of carbohydrates often end in the suffix *-ose*. The monosaccharides are classified according to the number of carbons they contain in their backbone structures. The major monosaccharides contain four to six carbon atoms. *Hexoses* (aldoses with six carbon atoms) are the most abundant sugars in nature. For any given monosaccharide, there are a number of stereoisomers because of the presence of asymmetric (chiral) carbons. For example, glucose has four asymmetric carbon atoms, resulting in $2^4 = 16$ possible stereoisomers. The predominant form in nature is the D- form, as shown in Figure 3.37.

The aldehyde and ketone moieties of the carbohydrates with five and six carbons that have the alcohol and carbonyl functionalities in the same molecule will spontaneously react to produce intramolecular *hemiacetals* or *hemiketals*, respectively. This results in the formation of five- or six-membered rings, as shown in Figure 3.37 for the case of glucose.

Because the five-membered ring structure resembles the organic molecule *furan*, derivatives with this structure are termed *furanoses*. Those with six-membered rings resemble the organic molecule *pyran* and are termed *pyranoses***.** In the case of pyranose rings, the two favored structures are the *chair conformation* and the *boat conformation***.** Straight and ring structures of glucose are given in Figure 3.38 for reference.

3.5.3 Oligosaccharides and Polysaccharides

Most of the carbohydrates found in nature occur in the form of high molecular weight polymers called *polysaccharides.* The monomeric building blocks used to generate polysaccharides can be varied. In all cases, however, the predominant monosaccharide found in polysaccharides is D-glucose. When polysaccharides are composed of a single monosaccharide building block, they

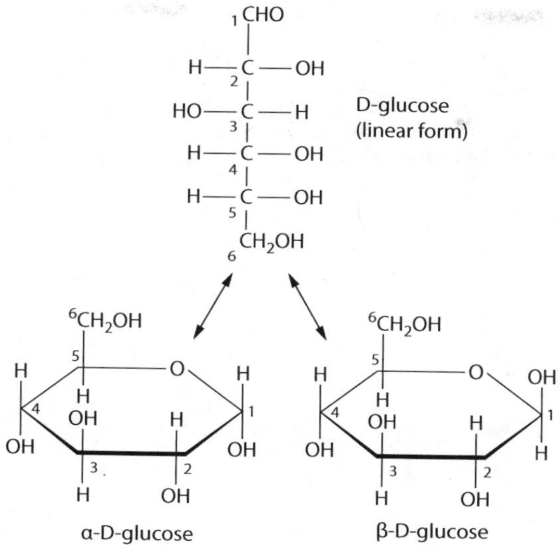

D-glucose (linear form)

α-D-glucose

β-D-glucose

FIGURE 3.37 Glucose: Linear and ring structures.

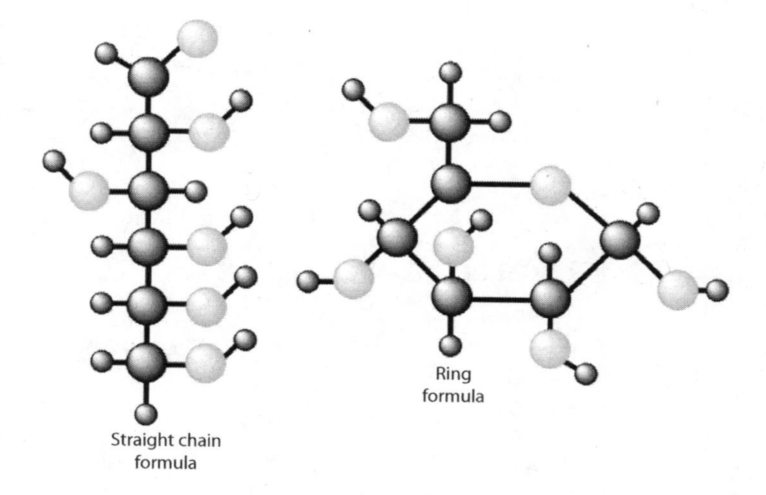

FIGURE 3.38 Different formulas for representing glucose.

are termed *homopolysaccharides*. Polysaccharides composed of more than one type of monosaccharide are termed *heteropolysaccharides*.

Polysaccharides or *glycans* are polymers of monosaccharides linked by glycosidic linkages. The simplest polysaccharide formed from two joined monosaccharides is called a *disaccharide*. *Saccharose* is composed of D-glucose and D-fructose ($C_{12}H_{22}O_{11}$) bound together by a *glycosidic linkage*. Another important disaccharide is *lactose*, composed of one D-galactose molecule and one D-glucose molecule. When the number of monosaccharide units is over ten, the corresponding molecule is called a *polysaccharide*.

Examples are *starch* (amylose or amylopectin) used as a storage polysaccharide in plants, while cellulose is a structural polysaccharid used in the cell walls. Oligosaccharides are often attached to proteins or lipids to form glycoproteins or glycolipids, respectively.

The structures of three common disaccharides (sucrose, lactose, and maltose) are shown in Figure 3.39.

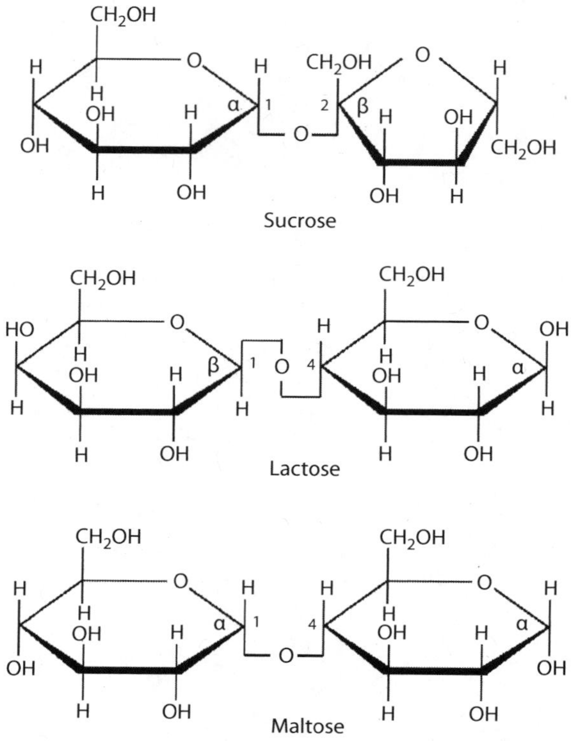

FIGURE 3.39 The structures of sucrose, lactose, and maltose.

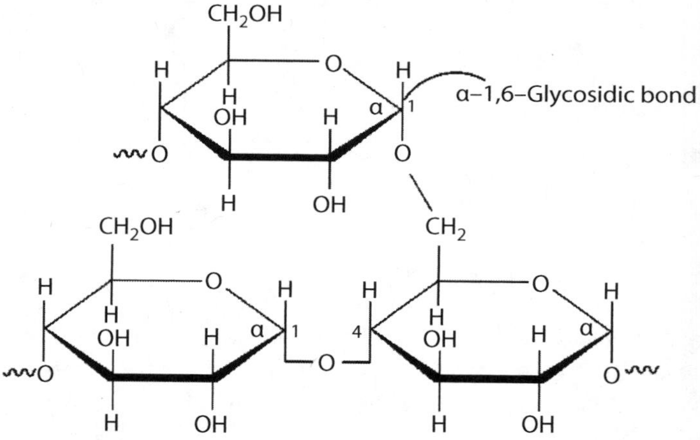

FIGURE 3.40 The structure of glycogen.

Glycogen is a polysaccharide composed of several branches, each of them having around ten molecules of glucose. It is stored in liver and muscles.

Oligosaccharides are important components of cell walls of bacteria. Starch is the major form of stored carbohydrate in plant cells. Its structure is identical to that of glycogen, shown in Figure 3.40, except for a much lower degree of branching (about every twenty to thirty residues). Unbranched starch is called *amylose*; branched starch is called *amylopectin*.

3.5.4 Biosensing Applications

Because diabetes is a world health problem, the reliable measurement of glucose in blood is an important biomedical analysis issue. Because of the high sensitivity and selectivity to glucose, enzymes (glucose oxidase) have been widely used for the amperometric detection of glucose. However, due to the sensitivity of enzymes to temperature and humidity, their lifetime is short and the integration of the sensor and microfluidic devices is difficult.

An integrated microfluidic system for automatic glucose sensing and insulin injection has been reported recently,[13] but the detection is still enzyme based. Nonenzymatic electrochemical glucose sensing based on a nanoporous platinum thin film has also been developed. A microfluidic system contains a Pt electrode and a Ag/AgCl reference electrode integrated on a chip. The transfer of sample and buffer solution to the cell is realized by programmed electroosmotic flow (EOF). The sensitivity of the system was found to be 1.65 $\mu A\ cm^{-2}\ mM^{-1}$ in the concentration range 1 to 10 mM in PBS. A detailed discussion of glucose sensing devices is given in Chapter 6.

3.6 ENZYMES

3.6.1 Definition and Nomenclature

Enzymes are complex protein structures with a large and globular three-dimensional structure that catalyze all biochemical reactions needed by the living cell. They are very efficient catalysts that mediate all synthetic and degradative reactions carried out by living organisms. A living cell needs thousands of biochemical reactions, and each of them is mediated by a unique enzyme. The activity of the cell is altered by removing specific enzymes or synthesizing new enzymes.

Like all catalysts in chemistry, enzymes reduce the activation energy of a reaction, but enzyme catalysts are highly selective for a particular substrate, and the speed of catalysis is generally much higher. In addition, enzymes work under mild conditions, and as a result, several steps are involved in an enzyme-catalyzed reaction. Enzymes are named by adding the suffix -ase, for example, *deoxyribonuclease*, or DNAse, for the enzyme that breaks down DNA, and *lactase* for the enzyme that attacks the disaccharide sugar lactose. A few familiar enzymes retain their older names, such as *trypsin*, *chymotrypsin*, etc.

The International Union of Biochemistry and Molecular Biology has introduced a nomenclature for enzymes, based on the EC numbers; each enzyme is described by a sequence of four numbers preceded by EC, the first number being related to its mechanism. For example, EC 1 *oxidoreductases* catalyze oxidation/reduction reactions, EC 3 *hydrolases* catalyze hydrolysis reactions of various bonds, etc.

Because the molar amount of the enzyme is often not known, its amount can be expressed in terms of the activity observed. The International Commission on Enzymes defined *one international unit* of enzyme as the amount that catalyzes the formation of 1 μM product in 1 min. Another definition for units of enzyme activity is the *katal*, the amount of enzyme catalyzing the conversion of 1 M substrate to product in 1 sec.

The *turnover number* of an enzyme is a measure of its maximal catalytic activity and is defined as the number of substrate molecules converted into product per enzyme molecule per unit time when the enzyme is saturated with substrate.

Another example of enzymes is the *catalase*, which catalyzes the decomposition of hydrogen peroxide into water and oxygen. Each second, one molecule of catalase can break 40 million molecules of hydrogen peroxide, and *carbonic anhydrase*, found in red blood cells, catalyzes the reaction

$$CO_2 + H_2O \leftrightarrow H_2CO_3$$

acetylcholinesterase, which acts as a catalyst for breakdown of acetylcholine (neurotransmitter).

3.6.2 Mechanism of the Enzymatic Catalysis

In order for a reaction to occur, reactant molecules must contain sufficient energy above a threshold potential energy barrier known as the *activation energy*, which is the energy input needed to bring about the reaction.

Catalysts are able to accelerate the reaction by reducing the size of the activation energy barrier. The reduction in free energy is more important for an enzyme-catalyzed reaction than an acid-catalyzed reaction. The activation energy is reduced by the enzymes in the following way. In an aqueous solution, there is a site within the three-dimensional structure of the enzyme known as the *active site* that contains the hydrophobic regions of protein. At this site, the substrate is bound to the enzyme through hydrophobic interactions. According to lock-and-key analogy first postulated in 1894 by Emil Fischer, the active site has a unique geometric shape that is complementary to the geometric shape of a substrate molecule, as shown in Figure 3.41. This means that enzymes specifically react with only one or a very few similar compounds. Different enzymes have differently shaped active sites.

It is thought that the amino acid side chains at the enzyme active site are of crucial importance for reducing the activation energy barrier. as shown in Figure 3.42. ΔG_c, standard free energy of a catalyzed reaction, is less than ΔG_u, the standard free energy ofan uncatalyzed reaction. The figure also indicates the possible transition states of reaction. They can be involved with hydrogen bonds with stabilized transition states.

$$E + S \rightleftharpoons ES \rightleftharpoons EP \rightleftharpoons E + P \tag{3.1}$$

Many enzymes require coenzymes or cofactors that can be small organic molecules or single metal ions, such as Mg, Zn, Co, Mn, etc. Examples of such enzymes include alcohol dehydrogenase, peroxidase, catalase, xanthine oxidase, etc., which contain sites for binding metal ions. Many of the important dietary vitamins are coenzymes used by certain types of enzymes. Many enzymes can be inhibited by molecules with a similar shape to the substrate, called a competitive inhibitor, that may block the substrate if its concentration is high enough. *Inhibitors* are compounds that combine with enzymes and prevent an enzyme and substrate from forming the ES complex. They may cause a reduction in the rate of the enzyme-catalyzed reaction or bring about a loss of activity. Two broad classes of enzyme inhibitors generally recognized are reversible and irreversible inhibitors, depending on whether the inhibition can be reversed or not. Heavy metal ions (e.g., mercury and lead) should generally be prevented from coming into contact with enzymes, as they usually cause such irreversible inhibition by binding strongly to the amino acid backbone. It has been shown that enzymes function best within a narrow range of temperature and pH. For human intracellular enzymes the maximum enzyme action was found at 37°C and pH 7.

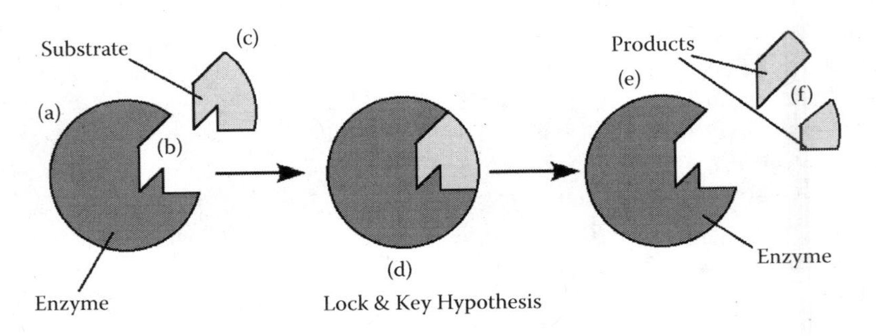

FIGURE 3.41 Schematic representation of an enzyme active site and of the formation of an enzyme-substrate complex. After the reaction is complete, the enzyme dissociates from the product and the active enzyme is regenerated. (Adapted from http://click4biology.info/c4b/3/Chem3.6.htm#one.)

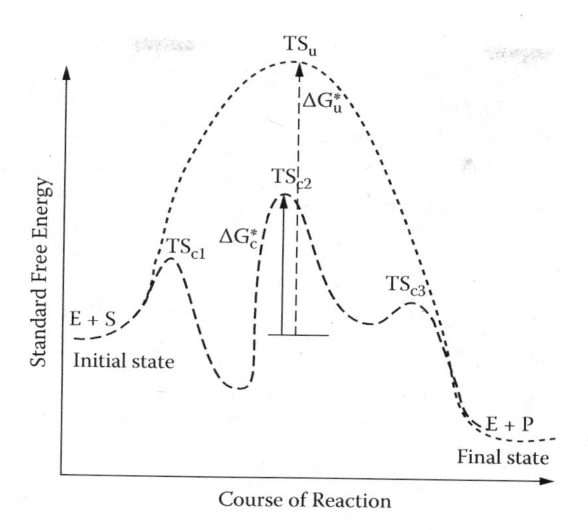

FIGURE 3.42 Free energy profile for an enzyme-catalyzed reaction involving the formation of enzyme-substrate (ES) and enzyme-product (EP) complexes. S = substrate; P = product; TS_{c1}, TS_{c2}, TS_{c3} = transition states of catalyzed reaction; TS_u = transition state of uncatalyzed reaction; ΔG_c = standard free activation energy of the uncatalyzed reaction; ΔG_c = standard free activation energy of catalyzed reaction.

3.6.3 Catalysis by RNA

It was long assumed that all enzymes are proteins. However, in recent years, more and more instances of biological catalysis by RNA molecules have been discovered. It was shown that a small piece of RNA, in the absence of any protein, was able to splice itself out of a long piece of RNA. These catalytic RNAs, or *ribozymes*, are substrate specific. They enhance the reaction rate, and they emerge from the reaction unchanged, like any enzyme. It is thought that because RNA is capable of both information storage and biological catalysis, it may have been a "forerunner" to DNA in the evolution of life. The catalytic activity of RNA is still under study.

3.6.4 Applications of Enzymes in Biotechnology and Biosensing

The biotechnology industry uses enzyme catalysis for commercial applications, for example, the production of semisynthetic penicillins using an enzyme called *penicillin acylase*. *Glucose isomerase* is used for the production of fructose from glucose, while an enzyme called *sucrase* brings about the hydrolysis of sucrose with the production of glucose and fructose. For bioprocessing, enzymes are immobilized in gel beds and held in suspension in the nutrient medium in a sterile bioreactor. Enzymes immobilized in a solid support are also used as biosensors. Selective enzyme inhibition is successfully used for drug discovery as well.

The most widespread biosensor today is the *glucose biosensor*, used to measure the glucose concentration in blood. There are quite a few enzymes that could be used as glucose detectors. The following example shows the glucose enzyme reaction with *glucose oxidase*, one of the most stable enzymes for glucose detection. By using this enzymatic reaction, the glucose concentration is determined by measuring either the oxygen consumption or the hydrogen peroxide production. The presence of the second enzyme (*catalase*) enhances the first reaction by the decomposition of the hydrogen peroxide. Enzyme-based biosensors use semipermeable membranes. Target analytes diffuse toward the immobilized enzyme through this membrane. For example, *L-glutamate oxidase* is adsorbed onto a phosphatidylethanolamine-coated platinum electrode for the measurement of the neurotransmitter glutamate.[15] The amperometric current due to oxidation of H_2O_2 is measured by a potentiostat. Various enzymatic biosensors for organophosphorous, carbamate pesticides, and other potentially harmful pollutants in the environment have been developed.[16] These biosensors are based on the activity of the *choline oxidase* and the inhibition of cholinesterase enzymes by several toxic chemicals. *Cholinesterases* are important enzymes that hydrolyze acetylcoline in the nervous system. Enzyme-based biosensors will be described and discussed in Chapter 6. An example of using microfluidics to monitor cellular secretions by online fluorescence-based enzyme assay, as shown in Figure 3.43, is described in Clark et al.[17]

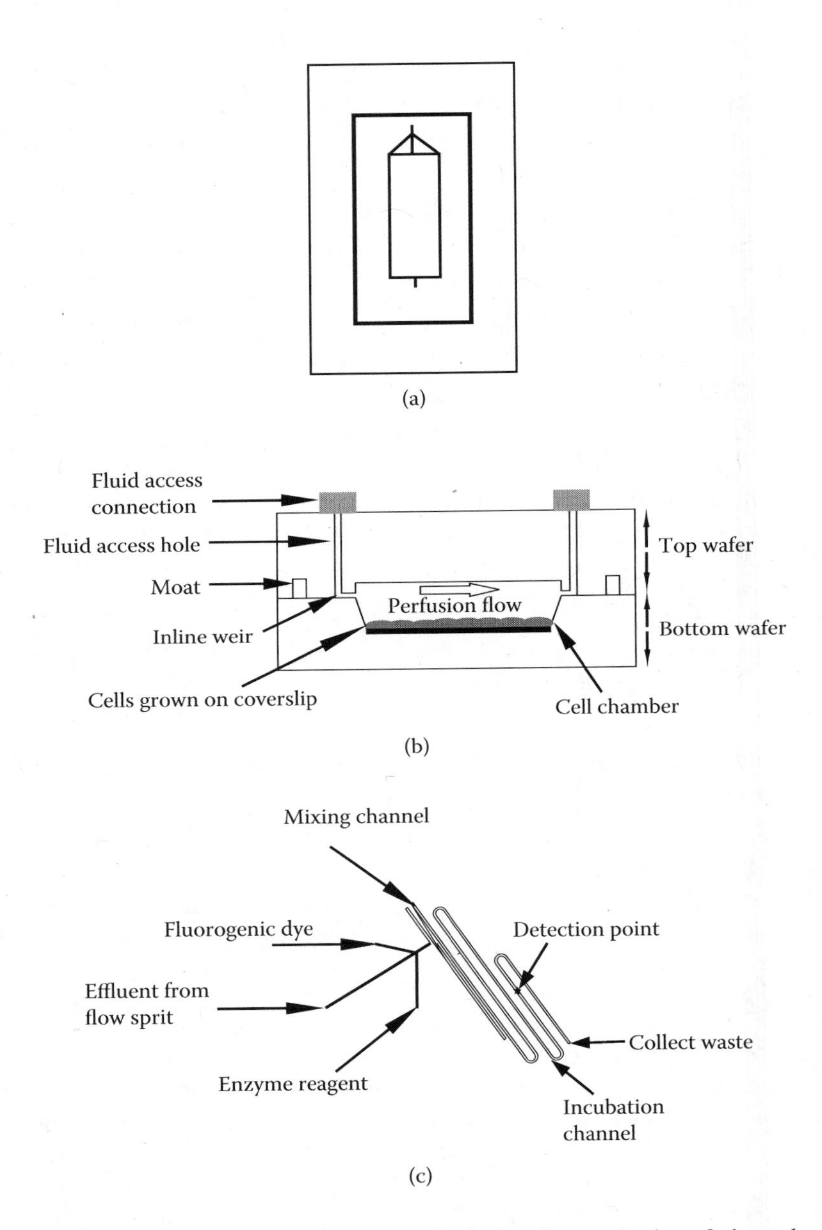

FIGURE 3.43 Schematics of the dual-chip microfluidic devices for monitoring of glycerol secretion from an adipocyte cell line using a continuous flow enzyme assay. (a) Top view of the two wafers. (b) Side view of the perfusion chip. (c) On-line mixing of solutions. (Reproduced from Clark et al., *Anal. Chem.*, 81, 2350–56, 2009. With permission.)

Metabolic secretions from the cell, for example, glycerol, can be monitored in real time by the enzyme assay chip. The enzymatic assay scheme is as follows:

1. Glycerol + ATP → glycerol-1-phosphate + ADP, in the presence of glycerol kinase (GK)
2. Glycerol-1-phosphate + O_2 → dihydroxyacetone + H_2O_2, in the presence of glycerol phosphate oxidase (GPO)
3. H_2O_2 + Amplex UltraRed Reagent → Resonufin

The use of a microfluidic device for mixing and detection reduced the consumption of costly reagents and labor.

3.7 CELLS

3.7.1 Cellular Organization

A cell is a complex system formed of many different building blocks enclosed in a membrane bag, as shown in Figure 3.44. In a human body there are about 6×10^{13} cells of hundreds of different types. The size of a cell may vary, depending on the type of cell. The diameter of animal and plant cells is estimated to be between 10 and 100 microns. There are two types of cells corresponding to the two types of organisms: *eukaryotes* and *prokaryotes*, which have different building blocks. Prokaryotes are single cellular organisms, smaller, and have a simpler structure than eukaryotes. *Bacteria* and *protozoa* belong to the prokaryotes.

A model of a *eukaryotic cell* is shown in Figure 3.45. In the center of the cell is a nucleus, which contains the *chromosomes* – the carrier of the genetic material. The nucleus controls all cellular activities. The contents of the cell are called protoplasm, and are further subdivided into *cytoplasm* (a fluid-filled space inside cells) and *nucleoplasm* (all of the material within the nucleus). The centrioles, lysosomes, mitochondria, etc., are called organelles (major subcellular structures), which are specialized for different processes. The mitochondria, for example, are specialized for energy production, while chloroplasts (organelles in plant cells) are specialized in photosynthesis. The membranes containing lipid bilayers act as a barrier to the environment and also regulate the flow of information in and out of the cell. Proteins are synthesized in different parts in the cell, for example, in cytoplasm cells, mitochondria, and chloroplasts in plants.

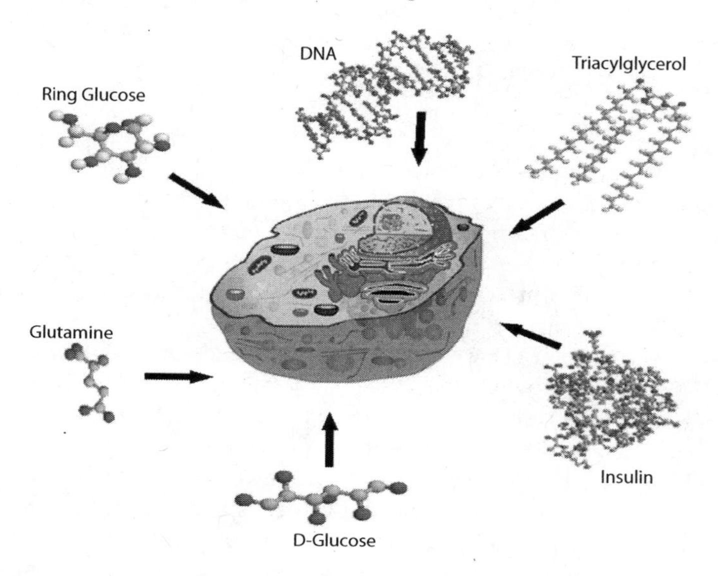

FIGURE 3.44 (See color insert.) Molecular components of cells.

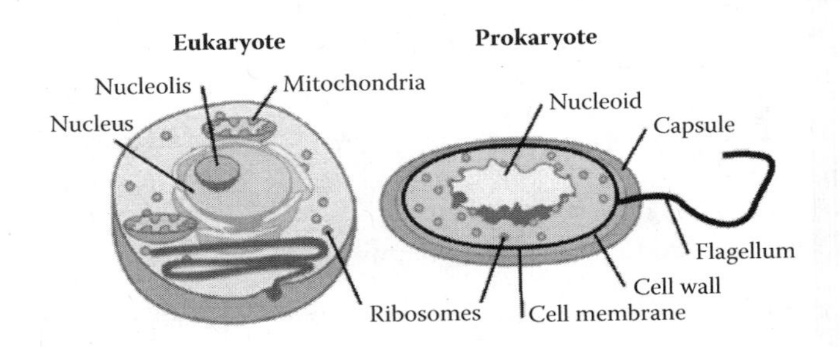

FIGURE 3.45 (See color insert.) Comparison of a typical eukaryotic cell with a typical prokaryotic cell (bacterium). The drawing on the left highlights the internal structures of eukaryotic cells, including the nucleus (light blue), nucleolus (intermediate blue), mitochondria (orange), and ribosomes (dark blue). The drawing on the right demonstrates how bacterial DNA is housed in a structure called the nucleoid (very light blue), as well as other structures normally found in a prokaryotic cell, including the cell membrane (black), the cell wall (intermediate blue), the capsule (orange), ribosomes (dark blue), and a flagellum (also black). (Reproduced from http://www.biosci.uga.edu/almanac/bio_103/notes/may_15.html. With permission from the Internet Encyclopaedia of Science.)

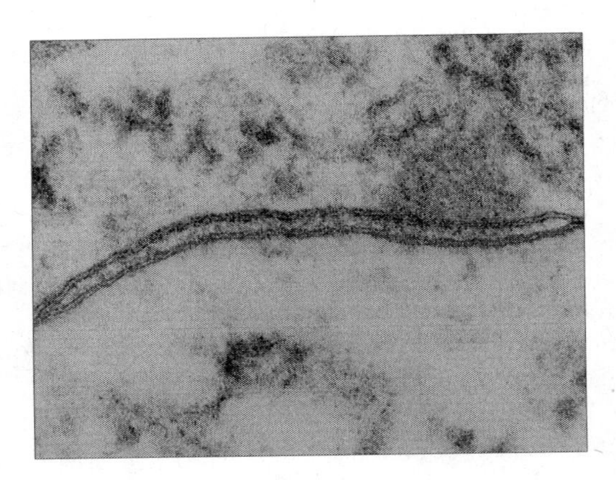

FIGURE 3.46 (See color insert.) Cell cytoplasmic membrane. (From www.DennisKunkel.com, copyright Dennis Kunkel Microscopy, Inc. With permission.)

The cell membrane shown in Figure 3.46 allows very few molecules across it and keeps the majority of organically produced molecules inside the cell. The scanning electron microscope (SEM) study of cell membranes has led to the lipid bilayer model. The phospholipids that form the bilayer have a polar (hydrophilic) head and two nonpolar (hydrophobic) tails. They are aligned tail to tail so the nonpolar areas form a hydrophobic region between the hydrophilic heads on the inner and outer surfaces of the membrane. The bilayer in the cell membrane is shown in Figure 3.46.

Cell membranes also contain cholesterol in the hydrophobic areas. *Cholesterol* is thought to enhance the flexibility of a cell membrane. Proteins, in the inner surface of the membrane called *integral* or *gateway proteins*, allow certain molecules to move through the protein channel. Carbohydrates are also attached to integral proteins. The outer surface of the membrane is rich in glycolipids. Their hydrophobic tails are embedded in the hydrophobic region of the membrane, and their hydrophilic heads exposed outside the cell. Plant cells have walls located outside the membrane that contain cellulose (a polysaccharide), and some plants have lignin and other chemicals in their wall as well. Because animal cells lack a cell wall, their membrane maintains the integrity of the cell. The nucleus, shown in Figure 3.47, exists only in eukaryotic cells.

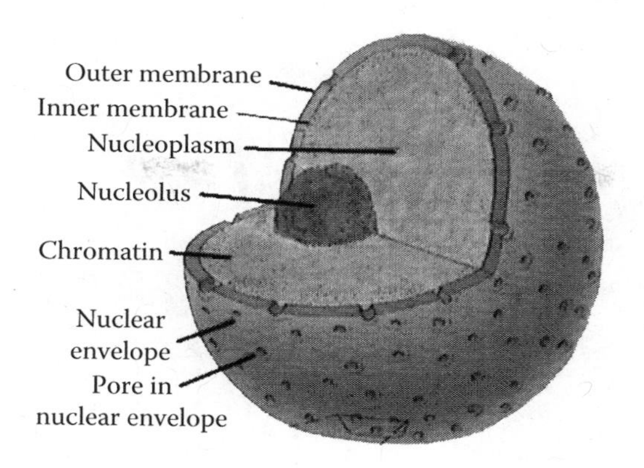

FIGURE 3.47 (See color insert.) Structure of the nucleus. (Reproduced from Purves et al., *Life: The Science of Biology*, 4th ed., Sinauer Associates and WH Freeman, 1995. With permission.)

Deoxyribonucleic acid (DNA) is the physical carrier of inheritance. All DNA, with the exception of DNA found in the chloroplast and mitochondrion, is found in the nucleus. Ribonucleic acid (RNA) is formed in the nucleus using the DNA base sequence as a template, and then it moves out into the cytoplasm. The nucleolus is the area of the nucleus where ribosomes are constructed.

As has been defined earlier, the material between the cell membrane and the nuclear envelope is the cytoplasm. In it, there are fibrous proteins called cytoskeleton that maintain the shape of the cell. The cytoskeleton is a network of connected microtubules and actin filaments that both function in cell division.

Ribosomes are the sites of protein synthesis in both prokaryotes and eukaryotes. Eukaryotic ribosomes are slightly larger than prokaryotic ones. The structure of a ribosome is shown in Figures 3.48 and 3.49.

It can be seen that the ribosome consists of a small and a larger subunit. Biochemically, the ribosome consists of ribosomal RNA (rRNA) and structural proteins.

Mitochondria contain their own DNA (termed mDNA) and function as the sites of energy release that occurs due to the glycolysis in the cytoplasm and ATP formation that occurs by chemiosmosis. The mitochondrion has been termed the powerhouse of the cell. Mitochondria are bounded by two membranes, as seen in Figure 3.50. The inner membrane folds into a series of cristae, which are the surfaces on which adenosine triphosphate (ATP) is generated. The matrix is the area of the mitochondrion surrounded by the inner mitochondrial membrane. Ribosomes and mitochondrial DNA are found in the matrix. Muscle cell mitochondrion is shown in Figure 3.51 for reference.

3.7.2 Cell Movement

Cell movement or *motility* is a highly dynamic phenomenon that is essential to a variety of biological processes.[22] Although cell movement was observed by van Leeuwenhoek as early as 1675, the molecular mechanisms behind cell movement have been studied only in the past few decades. Experimental techniques such as fluorescence microscopy and advances in molecular biology have enabled the discovery of the processes underlying motility. Biophysical studies helped identify regions where different force-generating proteins are located and measured the forces associated with movement.

It has been found that cell movement is the internal movement of organelles and chloroplasts. They are governed by actin filaments and other cytoskeleton components that allow the cell and its organelles to move. Internal movement is known as cytoplasmic streaming. An actin filament consists of two chains of globular actin monomers twisted to form a helix, as seen in Figures 3.52 and 3.53.

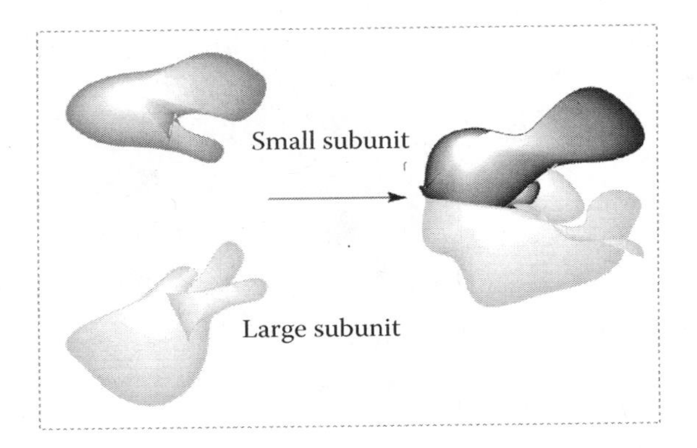

FIGURE 3.48 Structure of ribosome.

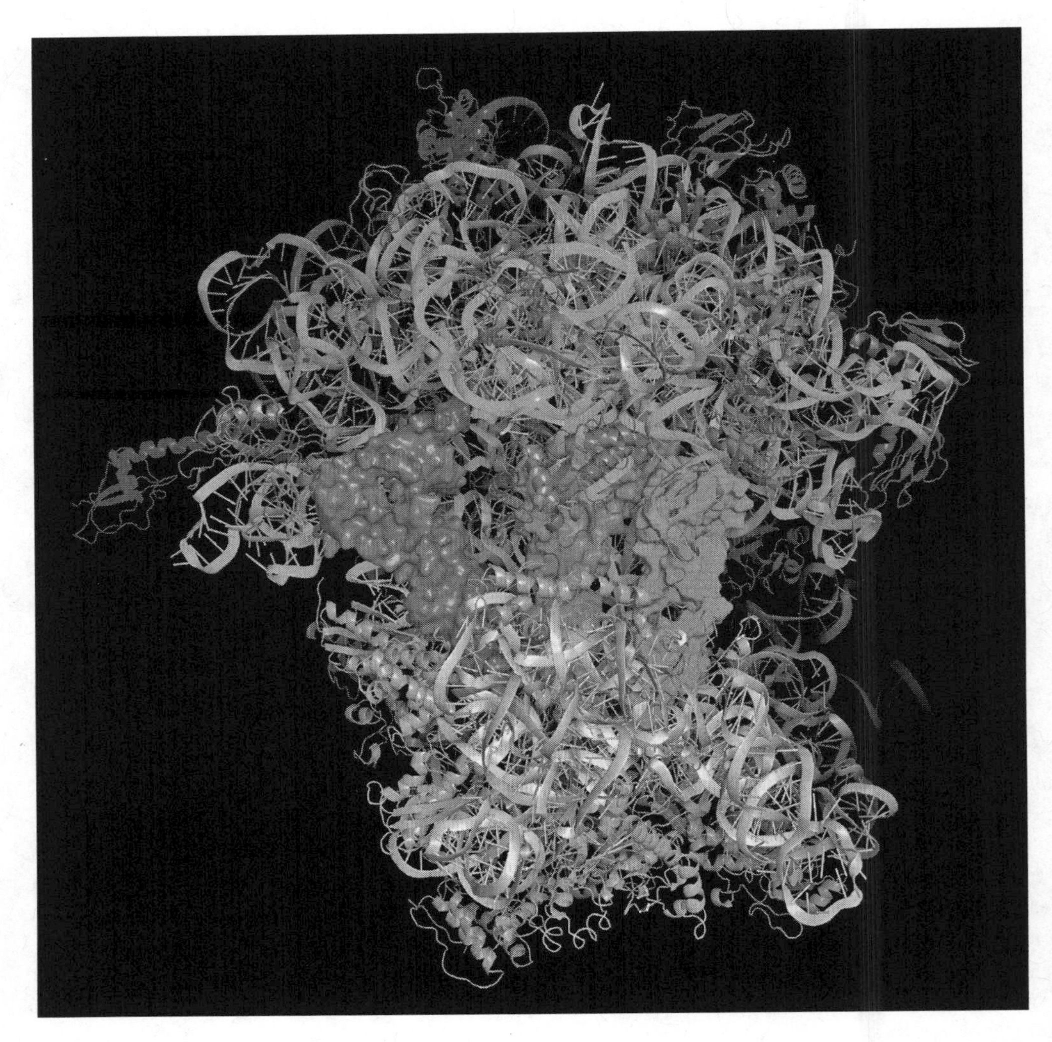

FIGURE 3.49 (See color insert.) Ribosome. (Reproduced from http://scienceblogs.com/transcript/ribosome.06-09-18.jpg. With permission.)

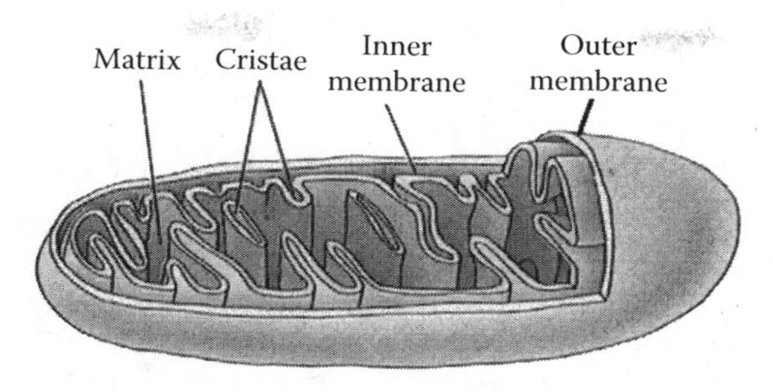

FIGURE 3.50 (See color insert.) Structure of a mitochondrion.

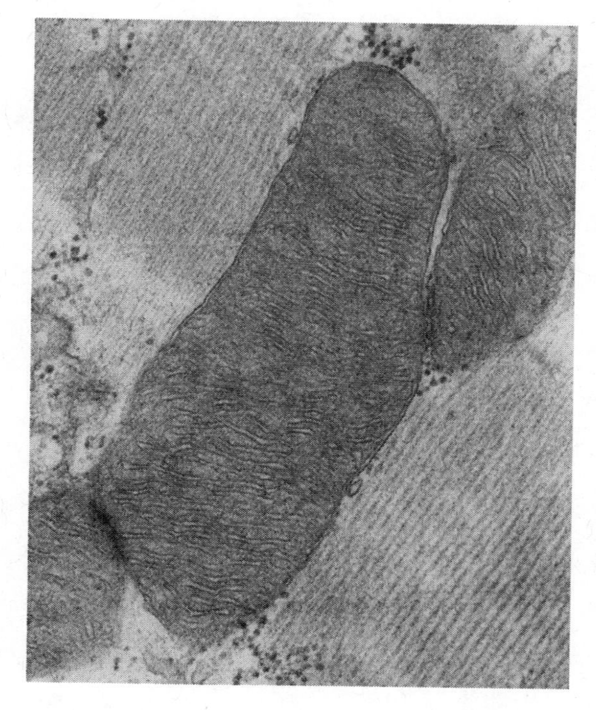

FIGURE 3.51 (See color insert.) Muscle cell mitochondrion. (From www.DennisKunkel.com. Copyright Dennis Kunkel Microscopy, Inc. With permission.)

Actin filaments are dynamic polymers whose ATP-driven assembly in the cell cytoplasm drives shape changes, cell locomotion, and chemotactic migration. Actin filaments also participate in muscle contraction.

The external movement of cells is determined by special organelles for locomotion. The crawling movements of cells on a surface represent a basic form of cell locomotion, employed by a wide variety of different kinds of cells. All these movements are complex phenomena based on the dynamic properties of the actin cytoskeleton, but the origins of the forces that drive locomotion remain to be fully understood. Examples include the movements of amoebas, the migration of cells involved in wound healing, the elongation of pollen tubes, and the spread of cancer cells during the metastasis of malignant tumors. The movement can be divided into several components: protrusion of the leading edge of the cell, adhesion of the leading edge and adhesion at the cell body and rear, and cytoskeletal contraction to pull the cell forward. Each of these steps is driven by physical forces generated by unique segments of the cytoskeleton.

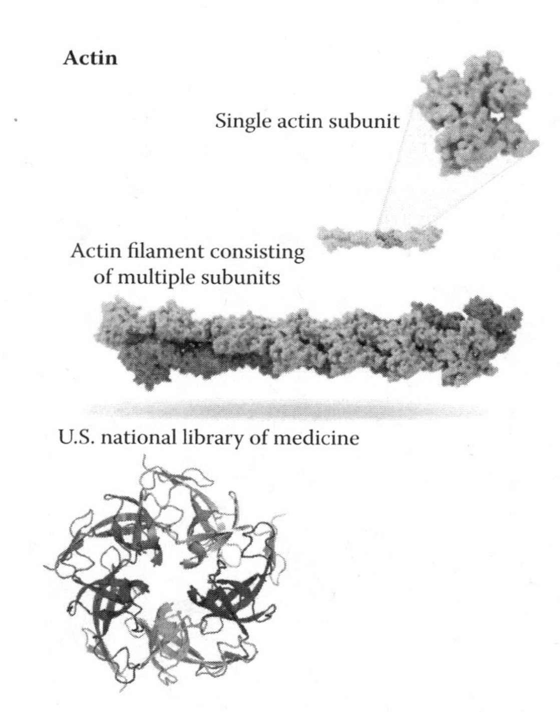

FIGURE 3.52 (See color insert.) Structure of actin filaments.

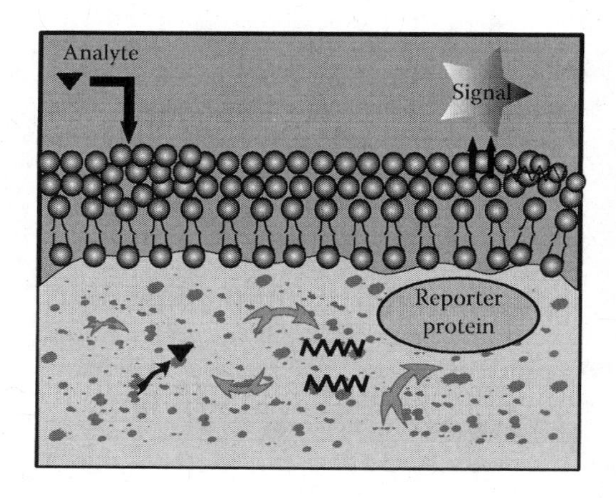

FIGURE 3.53 Schematic depicting a whole cell sensing system based on reporter genes. (Adapted from Daunert et al., *Chem. Rev.*, 100, 2705–38, 2000.)

3.7.3 Whole Cell Biosensors: Applications

Due to their enhanced sensitivity to the environment, living cells are becoming increasingly popular as biosensors, as they host a large number of enzymes and bioreceptors. However, in recent years, the development of sensors for environmental monitoring has received impetus from the increasing public awareness of "green" issues. Living cells have been used for a variety of applications in environmental analysis, such as water quality testing, metals monitoring, clinical diagnostic testing, etc. The greatest advantage of cell-based systems is their ability to provide physiologically relevant data in response to the analyte. The whole cell approach is extremely useful, as there is no need for extraction of enzymes.

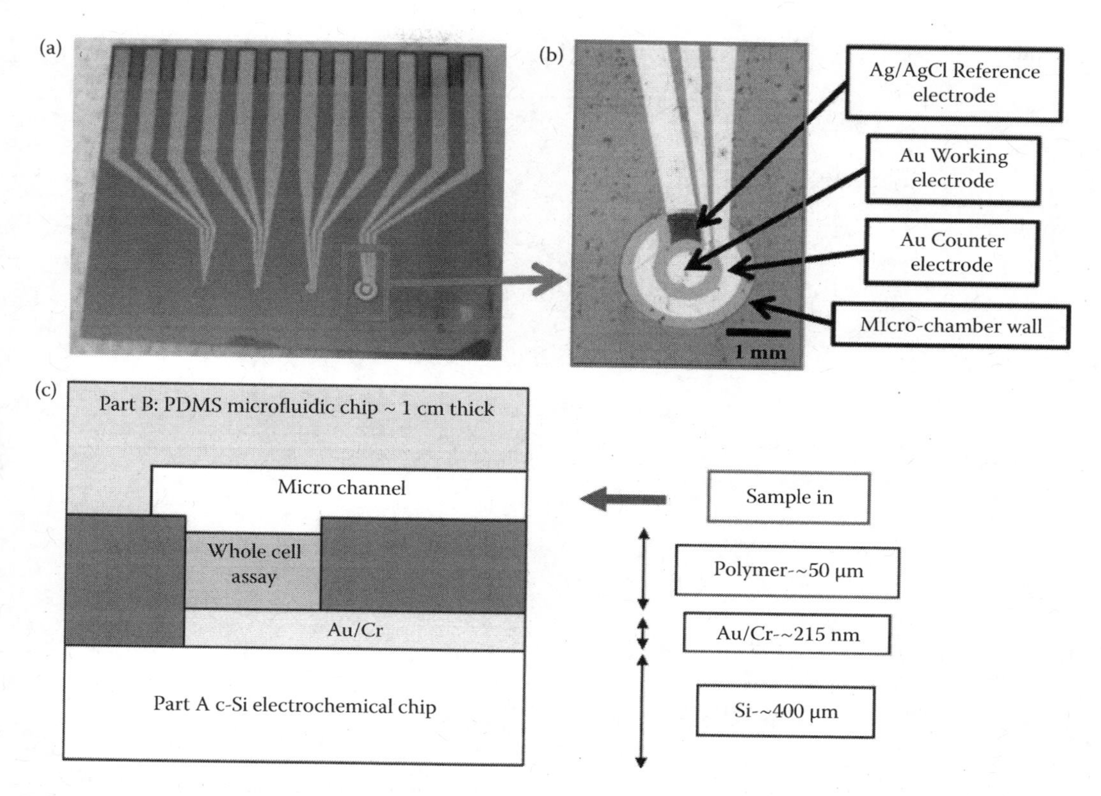

FIGURE 3.54 Microfluidic whole cell biosensor based on electrochemical detection. (a) Electrochemical chambers in silicon. (b) Inside view of the three-electrode cell. (c) Layout of the microfluidic mold. (Reproduced from Ben-Yoav et al., *Electrochim. Acta*, 54, 6113–18, 2009. With permission.)

In a sensing system with genetically engineered whole cells based on reporter genes, they are coupled to a sensing element, which recognizes an analyte and thus confers selectivity to the system. The reporter protein produces a detectable signal, which determines the system sensitivity. Reporter proteins or genes help to explain the role of a particular gene by identifying when and where the proteins are produced. Reporter genes allow the study of genes at various stages of development. Commonly used reporter genes that induce visually identifiable characteristics usually involve fluorescent and luminescent proteins. It has been demonstrated that a whole cell biosensing system can be adapted to a miniaturized microfluidic platform with the capability of performing quantitative analysis.[25] A microfluidic whole cell biosensor based on electrochemical detection of water toxicity is shown in Figure 3.54.

The bacteria were manipulated genetically to produce the enzyme alkaline phosphatase. A bacterial suspension was deposited on the ITO electrodes, and the current generated by the bacteria in the presence of a toxic material (nalidixic acid) was measured.

In addition to biosensing, microfluidics can provide small structures, actually mimicking the *in vivo* environment of cells.

3.8 BACTERIA AND VIRUSES

The emerging use of BioMEMS is the detection of pathogens such as bacteria, viruses, and parasites that create many diseases, such as HIV, diabetes, yellow fever, etc. The understanding of these pathogens will be useful to develop lab-on-a-chip (LOC) for their detection.

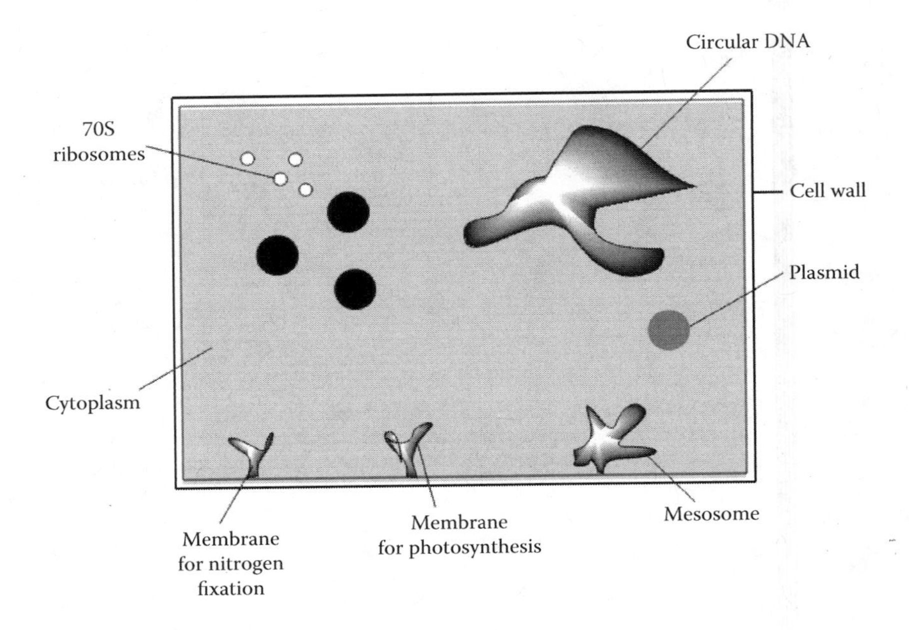

FIGURE 3.55 Prokaryotic cells, i.e., bacteria and cyanobacteria.

3.8.1 Bacterial Cell Structure

Bacteria is one of the oldest living organisms on earth. Evidence has shown that bacteria were in existence 3.5 billion years ago. They are microscopic (usually 0.3 to 2.0 μm in diameter) and mostly unicellular, with a relatively simple cell structure lacking a cell nucleus, cytoskeleton, and organelles (Figure 3.55). For example, *Escherichia coli*, which is an average sized bacterium, has a cell length of approximately 1 μm. Cell shape depends on the bacterial species. Some bacteria have elongated shapes, such as rods, and others are crescent shaped. The structural components of a prokaryotic cell, as they were evidenced by SEM in the 1950s, are *ribosomes*, *cell membrane*, and *cell wall* (Figure 3.56). Structurally, there are attachments to the cell surface called *flagella* and *pili*, composed predominantly of proteins. *Prokaryotic* structural components consist of nucleic acids (DNA, RNA), proteins, polysaccharides, phospholipids, etc. For example, the wall contains peptidoglycan, a molecule composed of a protein and a polysaccharide. The presence of peptidoglycan makes the cell walls rigid and determines eventually the shape of the cell. The term *gram negative* or *gram positive* refers to the staining procedure used to determine the cell wall composition of unknown bacteria, which helps to determine the appropriate antimicrobial treatment. It does not refer to the electrical charge of the bacteria. Both gram-negative and gram-positive bacteria are negatively charged. The staining and washing technique devised by the Danish physician Hans Christian Gram (1884) allows differentiation between two types of bacteria, those with a thick cell wall high in peptidoglycan and those with a thin cell wall with a negligible amount of peptidoglycan. Gram-positive bacteria appear dark blue or violet due to the crystal violet stain. Gram-negative bacteria appear red or pink because they do not retain the crystal violet stain. This difference between bacteria is due to the structure of their cell wall. Gram-positive bacteria have a thick cell wall high in peptidoglycan and techoic acid. Lipoteichoic acids are linked to lipids within the *cytoplasmic membrane*, and they give the wall an overall negative charge. Gram-negative bacteria have a thinner cell wall composed mainly of lipids. In addition, Gram-negative bacteria have an outer membrane and a periplasmic space, which contains peptidoglycan, different protein constituents, and metabolites. Due to the presence of *lipopolysaccharides*, the cell wall has an overall negative charge. The bacterial cytoplasmic membrane having a phospholipid bilayer is acting as a permeability barrier and provides sites for the transport of solutes. The functions for cell growth, metabolism,

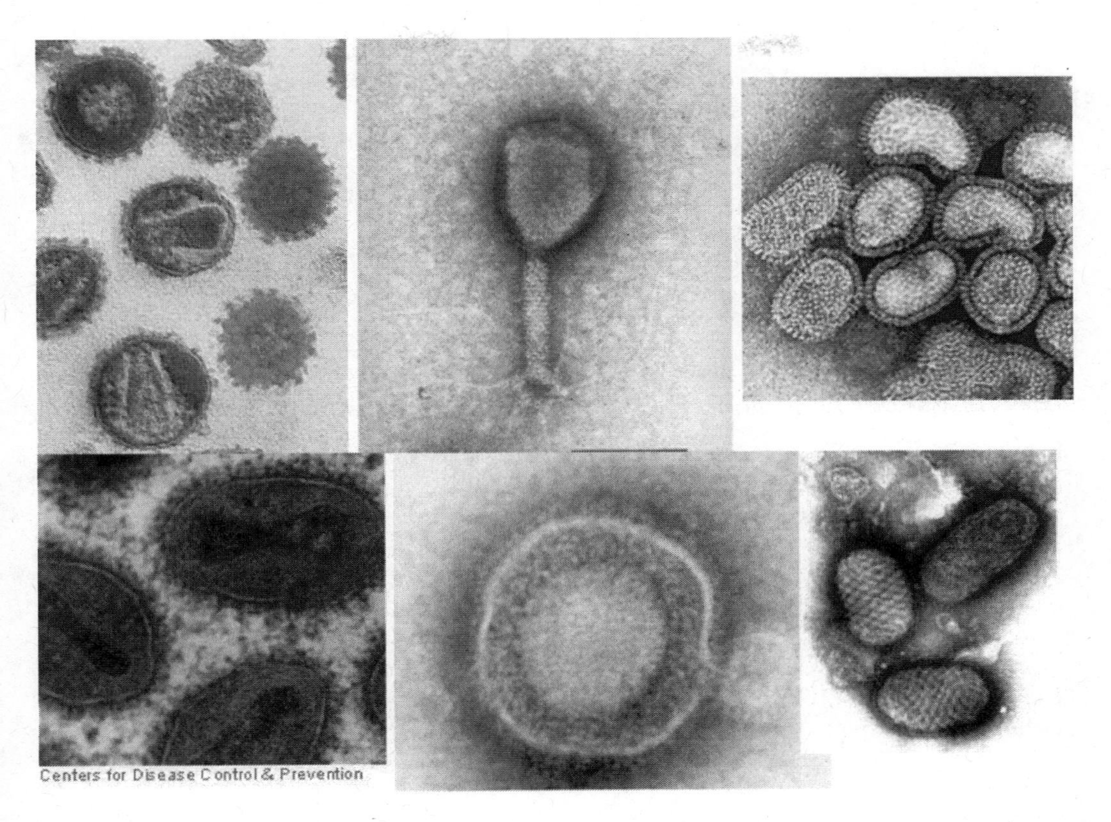

FIGURE 3.56 (See color insert.) Gallery of electron micrographs of viruses illustrating diversity in form and structure. Clockwise: Human immunodeficiency virus (HIV), Aeromonas virus 31, influenza virus, Orf virus, herpes simplex virus (HSV), smallpox virus. (Courtesy of Centers for Disease Control and Prevention, Gallery of electron micrographs, textbookofbacteriology.net/.../Phage.html.)

and replication are all carried out in the cytoplasm. Bacteria contain within their membrane many kinds of saturated and unsaturated fatty acids that maintain an optimum fluidity of the membrane. *Plasmids* are extrachromosomal genetic material made of a small circular piece of DNA, but they are not involved in reproduction. The genetic material of the cell that contains DNA is found in the *chromosome*. Plasmids contain only a few genes and can move freely around. Plasmids are important in bacteria for resistance to changes in the environment.

Flagella are usually found at the extremities of bacteria, and they help bacteria to move toward nutrients, or toward the light in the case of photosynthetic cyanobacteria. *Ribosomes* are the sites of protein synthesis, and they translate the genetic code from the molecular language of nucleic acids to that of amino acids, which are the building blocks of proteins. Bacterial ribosomes are smaller than those in eukaryotes and are distributed throughout the cytoplasm.

The cell chromosome (DNA), ribosomes, and various sorts of inclusions are found in the cytoplasmic region.

3.8.2 Virus Structure

Viruses, shown in Figure 3.56, are noncellular entities composed of a genetic molecule surrounded by a protein coating called a *capsid*, which contains the genetic molecule of the virus and sometimes a membraneous "envelope." The capsid is composed of protein molecules called *capsomeres*, and they may have different shapes, such as helical, polyhedral, etc. They may be as small as 20 nm, which is smaller than a ribosome. The largest virus may reach up to 1,000 nm. Viruses are considered *obligate intracellular parasites* because they require living organisms as hosts because they are not self-replicating; i.e., they cannot reproduce by themselves. Viruses need other cells. The host cell may be any form of eukaryote or prokaryote.

An isolated virus cannot synthesize proteins, as it has no metabolic enzymes. They would enter a host cell and use this cell's DNA polymerases to duplicate the viral DNA and make viral copies. After the new viruses are assembled, the host cell is destroyed. Viruses that recognize specific receptors on the bacterium surface and bind to them are called *bacteriophages* (or phage$). The phages inject their genetic material in the host cell. As will be shown later in this section, the specificity of this recognition opens important possibilities for biosensor development.

3.8.3 Biosensors and BioMEMS Sensor Systems for the Detection of Pathogenic Microorganisms and Bacterial Toxins

Biosensors have several potential advantages over other methods of analysis, in terms of sensitivity in the range of ng/ml for microbial toxins and <100 colony-forming units/ml for bacteria. Two examples of sensing systems will be briefly described below; the first is an impedance biosensing system for the detection of a chemically attached bacteriophage T4 structure onto the sensor surface.[28] The second is an array-based technique where bacterial toxins are attached to carbohydrates.

Figure 3.57a shows the structure of the bacteriophage when it approaches the host bacterium surface. The functional receptors that recognize the target bacteria are located on the extremity of the tail. The stability of the sensor platform is improved by chemical biotinylation of the phage and attachment to the streptavidin immobilized on the sensor surface, a gold electrode (Figure 3.57b

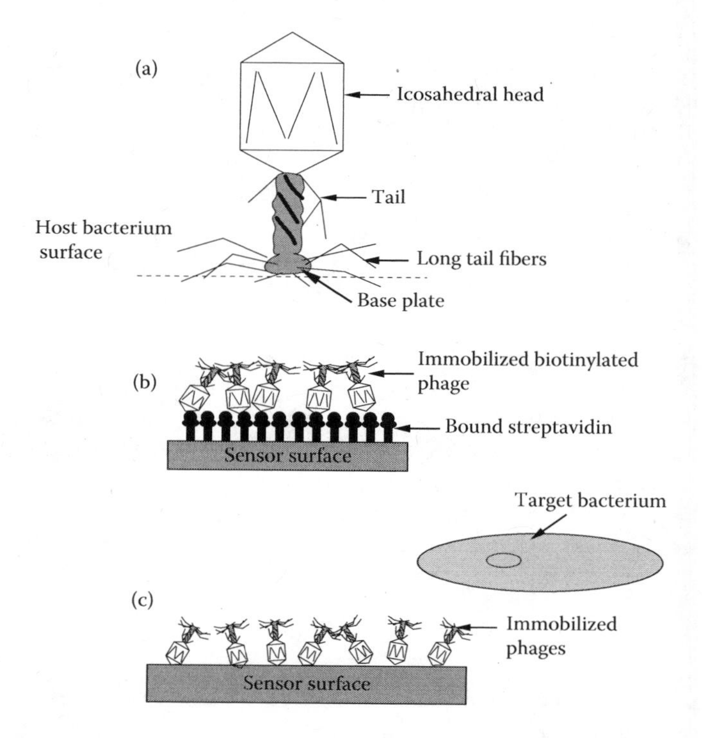

FIGURE 3.57 Attachment of phages in biosensing platforms for bacterial detection. (a) Structure of the bactereophage. (b) Immobilization of biotinylated phage to bound streptovidin. (c) Immobilization of phages on the sensor surface. (Adapted from Gervais et al., *Sensors Actuators B*, 125, 615–21, 2007.)

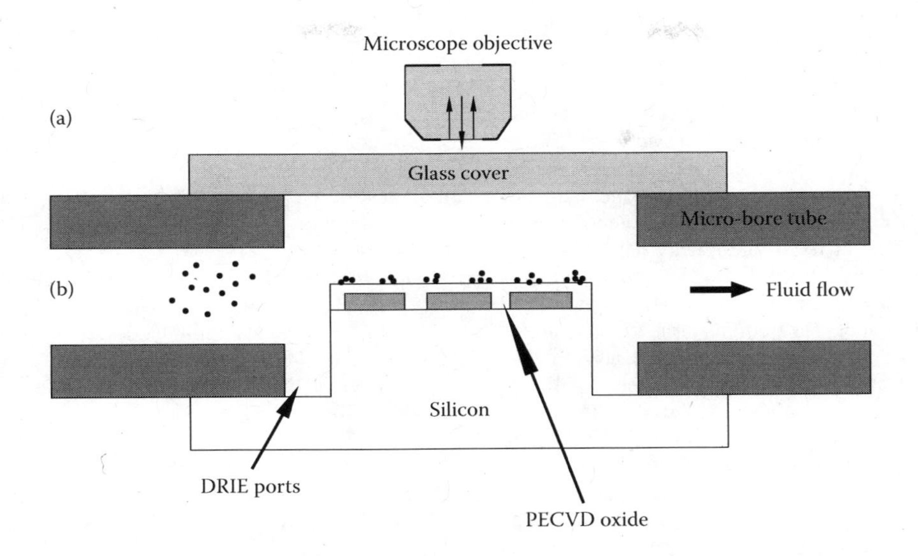

FIGURE 3.58 Dielectrophoretic capture and epifluorescence microscopic visualization of Vaccinia virus particles. (a) Cross-sectional drawing of the microfluidic device. (b) Dielectrophoretic capture. (Reproduced from Akin et al., *Nano Lett.*, 4, 257–59, 2004. With permission.)

and c). Biotin is bound on the capsid protein of the virus, leaving free the tail fibers to capture the bacteria (an *E. coli* solution).

The increase of impedance caused by the growth of the bacteria host (incubated into a culture medium) in the presence of the phages immobilized on gold was monitored. It has been concluded that the use of biotin-streptavidin interaction chemistry improved significantly the attachment of viruses into the gold surface. To detect bacterial toxins, monosaccharide derivatives were immobilized in the channels of the polydimethylsilaxane (PDMS) template, and interactions between the monosaccharides and fluorescently labeled protein toxins and bacterial cells were studied. Toxins such as cholera toxin and tetanus toxin were detectable at 100 nmol/ml. Monosaccharides were chosen because bacterial toxins target carbohydrate derivatives on the cell surface to attach and enter into the cell.

Although biosensors are not commonly used for food microbial analysis, it has been demonstrated that they have great potential for the detection of microbial pathogens and their toxins in food. Several applications have been developed for microbial analysis of food pathogens, including *E. coli* O157:H7, *Staphylococcus aureus*, *Salmonella*, and *Listeria monocytogenes*, as well as various microbial toxins, such as staphylococcal enterotoxins and mycotoxins. Carbohydrate microarrays combine the benefits of immobilized format assays with the capability of detecting thousands of analytes simultaneously. At the same time, the sensor platform can mimic the interactions at cell-cell interfaces. Figure 3.58 shows the dielectrophoretic capture of Vaccinia virus in a silicon microfluidic platform.

REFERENCES

1. sites.google.com/site/apodtele/nutshell.
2. http://employees.csbsju.edu/Hjakubowski/classes/ch331/protstructure/olunderstandconfo.html.
3. Nyberg, M., Hasselström, J., Karis, O., Wassdahl, N., Weinelt, M., Nilsson, A. 2000. The electronic structure and surface chemistry of glycine adsorbed on Cu(110). *J. Chem. Phys.* 112:5420.
4. Ge, S.-P., Lu, C., Zhao, R.-G. 2006. Adsorption of L-alanine on Cu(111) studied by scanning tunneling microscopy. *Chin. Phys. Lett.* 23:1558.
5. themedicalbiochemistrypage.org.

6. Denniston, K., Topping, J., Caret, R. 2008. *General organic chemistry and biochemistry*, 581. 6th ed. McGraw-Hill Higher Education.
7. Nallamothu, R., Wood, G. C., Pattillo, C. B., Scott, R. C., Kiani, M. F., Moore, B. M., Thomas, L. A. 2008. A tumor vasculature targeted liposome delivery system for combretastatin A4: Design, characterization, and *in vitro* evaluation. *AAPS PharmSciTech* 7(article 32).
8. www.cardiometabolic-risk.org.
9. The Education Center of American Phytopathological Society. http://www.apsnet.org/education/K12PlantPathways/TeachersGuide/Activities/DNA_Easy/top.htm.
10. http://en.wikipedia.org/wiki/Stem-loop.
11. Wang, Y., Zhang, J., Li, G. 2008. Electrochemical sensors for clinical analysis. *Sensors* 8:2043–81.
12. Peytavi, R., Raymond, F. R., Gagne, D., Picard, F. J., Jia, G., Zoval, J., Madou, M., Boissinot, K., Boissinot, M., Bissonnette, L., Ouellette, M., Bergeron, M. G. 2005. Microfluidic device for rapid (<15 min) automated microarray hybridization. *Clin. Chem.* 51:1836–44.
13. Huang, C.-J., Chen, Y. H., Wang, C.-H., Chou, T.-C., Lee, G.-B. 2007. Integrated microfluidic systems for automatic glucose sensing and insulin injection. *Sensors Actuators B* 122:461–68.
14. http://click4biology.info/c4b/3/Chem3.6.htm#one.
15. Pancrazio, J. J., Whelan, J. P., Borkholder, D. A., Ma, W., Stenger, D. A. 1999. Development and application of cell-based biosensors. *Ann. Biomed. Eng.* 27:697–711.
16. Rodriguez-Mozaz, S., López de Alda, M. J., Marco, M.-P., Barceló, D. 2005. Biosensors for environmental monitoring. *Talanta* 65:291–97.
17. Clark, A., Sousa, K. M., Jennings, C., MacDougald, O. A., Kennedi, R. T. 2009. Continuous flow enzyme assay on a microfluidic chip for monitoring glycerol secretion from cultured adipocytes. *Anal. Chem.* 81:2350–56.
18. http://www.biosci.uga.edu/almanac/bio_103/notes/may_15.html.
19. www.DennisKunkel.com.
20. Purves et al. *Life: The science of biology*. 4th ed. Sinauer Associates and WH Freeman.
21. http://scienceblogs.com/transcript/ribosome.06-09-18.jpg.
22. Ananthakrishnan, R., Ehrlicher, A. 2007. The forces behind cell movement. *Int. J. Biol. Sci.* 3:303–17.
23. US National Library of Medicine.
24. Daunert, S., Barrett, G., Feliciano, J. S., Shetty, R. S., Shrestha, S., Smith-Spencer, W. 2000. Genetically engineered whole-cell sensing systems: Coupling biological recognition with reporter genes. *Chem. Rev.* 100:2705–38.
25. Rothert, A., Deo, S. K., Millner, L.,Puckett, L. G., Madou, M. J. 2005. Whole-cell-reporter-gene-based biosensing systems on a compact disk microfluidics platform. *Anal. Biochem.* 342:11–19.
26. Ben-Yoav, H., Biran, A., Pedahzur, R., Belkin, S., Buchinger, S., Reifferscheid, G., Shacham-Diamanda, Y. 2009. A whole cell electrochemical biosensor for water genotoxicity bio-detection. *Electrochim. Acta* 54:6113–18.
27. Centers for Disease Control and Prevention. Gallery of electron micrographs. textbookofbacteriology.net/.../Phage.html.
28. Gervais, L., Gel, M., Allain, B., Tolba, M., Brovko, L., Zourob, M., Mandeville, R., Griffiths, M., Evoy, S. 2007. Immobilization of biotinylated bacteriophages on biosensor surfaces. *Sensors Actuators B* 125:615–21.
29. Akin, D., Li, H., Bashir, R. 2004. Real-time virus trapping and fluorescent imaging in microfluidic devices. *Nano Lett.* 4:257–59.

REVIEW QUESTIONS

1. What are the functional groups that characterize the amino acids?
2. What information about the amino acids can be found in Venn's diagram?
3. How can the primary and secondary structures of a protein be defined?
4. Give two examples of lipids.
5. Describe briefly the structure of a phospholipid bilayer.
6. What are the three important moieties in a nucleotide?
7. Give two examples of nucleic acids.
8. Explain how DNA hybridization is used in biosensing.

9. What is a carbohydrate and how can carbohydrates be classified?
10. Why is it important to analyze the concentration of glucose in blood?
11. What is an enzyme and what is its role in the living organism?
12. How are enzymes used in biosensing of glucose?
13. What are the most important parts of a eukaryotic cell?
14. Where does the synthesis of proteins take place?
15. What are viruses?
16. How can bacteria be detected?

4 Engineering of Bioactive Surfaces

4.1 INTRODUCTION

Surfaces and interfaces play an important role in biology and medicine, as most of the biological reactions are surface phenomena. Because the molecules at the surface of a material have different free energies, reactivities, and structures from those in the bulk, surfaces can even be considered a fourth state of matter.[1] The design of high-performance biosensors and medical devices is only possible if the interactions between the biomolecules and surfaces are fully controlled. The characteristics of sensors such as the *sensitivity* and *selectivity* of detection are determined by the quality of the biointerface. Modifying the surface with adequate chemical groups allows improvement of the biological response. Surfaces may also be functionalized to produce antifouling coatings that will reduce the adsorption of proteins and cell adhesion from physiological fluids on implants, medical devices, etc. The ultimate goal of surface modification is the maximization of the performance of sensing devices (biochips, BioMEMS), that is, the ability to achieve targeted sensitivity and maximum specificity. The specificity implies minimal nonspecific adsorption and an adequate reaction of the ligand with its target molecule. Biochips and biosensors use biological molecules that are immobilized onto solid surfaces in the proximity of a transducer to detect the interactions with an analyte. Up to date, a variety of surface modification methods have been used, such as self-assembled monolayers (SAMs), poly-L-lysine coatings, nitrocellulose adlayers, etc.

Organic functionalization of surfaces is emerging as an important area in the development of semiconductor-based materials and devices. The direct covalent attachment of organic layers to a semiconductor interface provides many new properties, including biocompatibility. Surfaces are called *biocompatible* if they do not cause allergic reactions or rejections when the device is introduced in the organism. Choosing the correct surface is a critical step in bioassay development as well. The general objective is to produce interfaces using cost-effective, robust techniques that allow the elimination of nonspecific protein adsorption (nonfouling surfaces). Fouling of a surface implies adsorbing or interacting with nonspecific biospecies. The bioligands of controlled surface density and molecular conformation have to be added in order to maximize the biological response. The ability to design, fabricate, and optimize surfaces tailored for a particular application is of crucial importance for the development of new biosensors.

Molecules may be immobilized into the surface, either passively through hydrophobic or ionic interactions, or by covalent attachment with activated surface groups. Immobilization is often necessary for binding molecules (especially proteins) that do not adsorb on a particular surface, or adsorb with an improper orientation and conformation.

Covalent immobilization may result in better bioactivity, reduced nonspecific adsorption, and better stability. In order to ensure the successful immobilization of a biomolecule, the reaction should occur rapidly, and the process should be selectively and easily monitored. Molecules should be immobilized in such a way that they preserve their original recognition specificity. For example, the loss of specificity, called *denaturation*, may occur when biomolecules are bonded too strongly on metal surfaces.

There are a number of ways to modify solid supports for the covalent immobilization of biomolecules. Among them, spontaneous assembly of multifunctional molecules at surfaces has become a useful technique to design hybrid interfaces for biosensors, model surfaces for cell biological studies, and drug carrier surfaces for medical applications. Today, BioMEMS offer opportunities for astounding medical advances, such as transplants of therapeutic cells, "magic bullets" that deliver controlled cancer-fighting drugs on target, and implantable biochips for online diagnostics.

These microdevices are constructed from many materials, such as silicon, glass, polymers, and hybrid materials. A patterned surface is the basis of many biosensing devices, where the patterned areas serve as sensing elements that interact directly with analytes, such as proteins, viruses, bacteria, or cells, while the background of the substrate remains passive to the analyte.

An important consideration in biochip fabrication is how to minimize nonspecific adsorption and adhesion of biomolecules.

When a microdevice has to be in contact with a biological environment, the issue of biocompatibility becomes a major concern. For blood-contacting devices, the device performance is mediated by surface-induced thrombosis initiated by the adsorption of blood plasma proteins. For tissue-contacting devices, the adsorption of plasma proteins occurs first, followed by adhesion of macrophages that secrete mediators of inflammation and tissue destruction. It is essential to improve the device biocompatibility by creating surfaces with controlled functionalities.

Silicon is a commonly used substrate in microimplants, biosensors, and therapeutics. As a substrate in microimplants, it may have undesired interactions with the immune system. Sometimes, protein coatings are used on silicon-based surfaces as a passivation layer in order to minimize an inflammatory response and to mimic biological surface. Promising methodologies for the construction of novel, well-defined biointerfaces are based on the deposition of uniform self-assembled monolayers (SAMs) of alkane thiols on gold surfaces. The surface chemistry of many materials, such as gold, alumina, mica, and oxidized silicon, can be modified by SAMs. A hydrophilic surface such as glass can be made hydrophobic through the self-assembly of molecules with hydrophobic tail groups such as alkylsilanes. Similarly, SAMs with chemically reactive tail groups can be formed and further chemically modified, thus allowing for the formation of multilayers. Mixed SAMs have been applied for the immobilization of bioreceptor molecules, for example, antibodies on gold. One of the most successful approaches for producing surfaces that are able to resist protein adhesion was proved to be the use of polyethyleneglycol (PEG) as a surface protector. Because of its unique properties, such as hydrophilicity and high exclusion volume in water, PEG increases the attractive forces between solid surfaces and proteins, and it also makes the surfaces biocompatible. Hemo-compatible and tissue-compatible sensor platforms have been fabricated by coupling PEG and silane on silicon and gold-patterned silicon. Sometimes, PEG is grafted on silane coupling agents adsorbed on oxide layers such as glass or aluminum oxide. PEG is also used to functionalize gold nanoparticles when they are used as cellular probes and delivery agents.

Stable and well-controlled surfaces with tailored properties such as hydrophobic and hydrophilic surfaces are interesting from a microfluidic system's point of view. For instance, in capillarity-driven flow systems, wettability-dependent surface tensions of the channel surfaces are of high importance. SAMs in combination with multistream laminar flow or photolithography have been used to pattern the surfaces inside the channels, to direct the liquid flow. It has been shown that aqueous solutions flow only along the hydrophilic pathway. Microfluidic devices are becoming powerful tools for performing chemical or biological assays due to their increased speed and reliability at reduced sample consumption. Most of the commercially available polymers for microfluidic applications (for example, polydimethylsiloxane, polycarbonate, polymethylmethacrilate, etc.) are inherently hydrophobic, which could be become an issue for liquid handling. Indeed, the surfaces of microfluidic devices should be hydrophilic, so that the analyte can flow smoothly through the microchannels. The continuing progress in microfluidics will partly rely on the development of surface modification technologies to control the adsorption and ensure the optimized biocompatibility.

Different strategies have been used to functionalize the surface inside the microchannels. One of the most efficient methods was proved to be the covalent attachment of PEG to the surface of PDMS by UV graft polymerization. The presence of PEG generates a hydrophilic and antifouling surface with stable electroosmotic flow compared to the native substrate. Biomimetic surfaces have also been fabricated by introducing phosphorylcholine functional groups into the polymethylmethacrylate (PMMA) microfluidic channel.

Noncovalent and covalent approaches have been used to functionalize the surface of nanomaterials, as illustrated schematically in Figure 4.1. In the noncovalent approach, the ligand is adsorbed on the surface by physisorption, and the interaction forces are electrostatic, hydrogen bonding, and hydrophobic. Polymers can be adsorbed on the surface of nanoparticles and subsequently derivatized with different ligands for bioconjugation.

Adsorption involving exclusively intermolecular interactions between a ligand and the surface is called *physisorption*. When a covalent bond is formed between the surface and ligand, the phenomenon is called *chemisorption*. Covalent bond formation offers the advantage of stability and robust linkage. Typical examples of chemisorptions include thiol on metal and semiconductor surfaces, silanes on oxides, and phosphates on metal oxides.

The thiol-Au system is the most utilized, and it provides stable SAMs. Usually, the capping layer of Au nanoparticles is displaced by the thiol ligand.

The *Langmuir-Blodgett* (LB) *technique* is frequently used to make monolayers at air-water or air-oil interfaces. Solid films of particles are transferred from the water surface onto solid supports. Hydrophobic surfaces can also be prepared by LB deposition of silica particles on glass and the subsequent formation of a SAM of alkylsilane on the outer surface of the particulate film. The silica and the alkylsilane are supposed to provide the substrate with surface roughness and low surface energy, respectively.

One of the major issues is to achieve prolonged *biofunctionality* of the device. For chemically modified surfaces, the patterns would degrade over time as a result of the biofouling or decomposition of the surface coating. This time span is strongly influenced by the surface chemistry of the substrate and the nature of the chemical bonding between the substrate and the surface coating. The long-term stability and nonbiofouling property of patterned surfaces remain great challenges that require extensive further research.

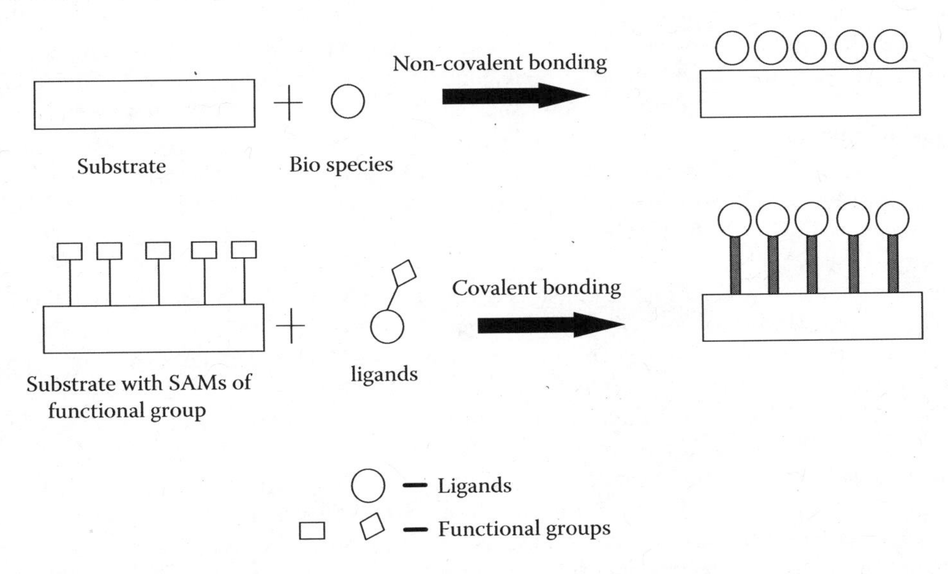

FIGURE 4.1 Surface modification by noncovalent and covalent methods. (Adapted from Wang et al., *Exp. Biol. Med.*, 234, 1128–39, 2009.)

In this chapter, the most important techniques of surface engineering, especially for BioMEMS applications, will be described in more detail. These techniques are as follows:

- *Plasma-mediated surface modification*, where reactive gases are employed for assemblage of predesigned functionalities on glass or polymer surfaces
- *Self-assembled monolayer-mediated surface modification* using highly organized monolayers of alkylsilanes or gold alkanethiol to tailor the interfacial properties of materials
- *Langmuir-Blodgett monolayers*, which are self-assembled films prepared by a controlled transfer from the surface of water onto a solid substrate
- *Layer-by-layer assembly*, a method that takes advantage of electrostatic forces between charged surfaces and charged polyelectrolytes
- *Biosmart hydrogels*, which are hydrophilic polymer networks that facilitate the immobilization of proteins
- *Chemisorption* on functionalized gold nanoparticles by using linkers for immobilization of biomolecules
- *Biomimetic surface engineering* using peptides in order to promote cell adsorption

Sections are also devoted to surface engineering for protein adsorption and tissue engineering applications.

4.2 PLASMA TREATMENT AND PLASMA-MEDIATED SURFACE MODIFICATION

Plasma treatment employing reactive gases is considered one of the most precise methods of fabrication of surfaces with desired functionalities. The use of cold-plasma techniques is becoming an increasingly attractive approach for surface modification of both organic and inorganic substrates. The plasmas contain both neutral and charged active species, such as ions, electrons, radicals, excited species, and photons. Sources of plasmas include glow discharges, radio frequencies, and gas arcs. Molecules are ionized by the electromagnetic fields formed in a corona discharge as well. By using this method, the excited species (ions, radicals, electrons, etc.) are produced at atmospheric pressure. The advantage of plasma chemistry is that the energies of these species are comparable with the bond energies of organic compounds. A good understanding of plasma-induced reaction mechanisms and control of internal plasma parameters allows the assemblage of predesigned functionalities from the molecular fragments created by plasma. Depending on the intended application, oxidizing or reducing plasma atmospheres are created by choosing the feed gas. In order to render a surface hydrophilic, oxidizing atmospheres created by oxygen, H_2O, O_2/O_3, and O_2/H_2O_2 can be used. Under the influence of plasma, oxygen is covalently bonded to the surface, allowing the formation of polar functional groups such as hydroxyl (OH), carbonyl (C=O), ether (C-O-C), aldehyde (CHO), carboxyl (COOH), C-O-OH, and carbon-carbon double bond (C=C).

It is shown that oxygen plasma treatment is a method of precise surface modification for polymethylmethacrylate (PMMA) and SU8 waveguide materials.[3] Oxygen groups are analyzed by contact angle measurements, x-ray photoelectron spectroscopy (XPS), chemical derivatization XPS, and fluorescence labeling.

As shown in Figure 4.2, more than 50% of all oxygen functional groups can be converted in OH groups, resulting in a density of up to 11 OH per 100 carbon atoms at the surface of polyolefins.

When the substrate is a polymer, carbon radicals that originate from the polymer chains at the surface are generated by the plasma treatment. The carbon radicals may combine with the different species present in the plasma to form oxidized species at the surface of polymers.

The formation of functional groups by plasma treatment does not exceed a depth of a few hundred nanometers, and it is only temporary. Although these functional groups improve strongly the adhesion properties of polymers to metals, it is more suitable for more advanced applications to

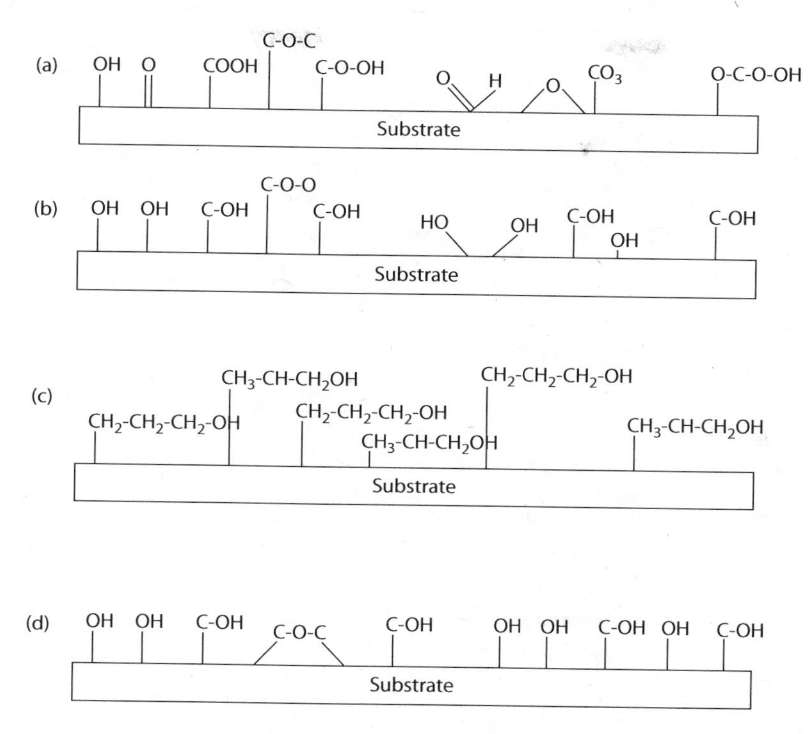

FIGURE 4.2 Polymer surface modification by exposure to an oxygen water vapor plasma followed by reduction with $LiAlH_4$ or B_2H_6, or grafting of a monolayer of allyl alcohol onto the plasma-activated polymer surface. (a) Oxygen plasma polymerization. (b) Plasma functionalization and reduction of oxygen functional groups by diborane/H_2O_2. (c) Grafting of allyl alcohol. (d) Plasma polymerization of thicker layers of allyl alcohol. (Adapted from Kühn et al., *Surface Coating Technol.*, 116–19, 796–801, 1999.)

produce only one type of group at the surface of the polymers. For example, in order to have only OH groups, different functionalities have further to be reduced by using diborane as a reducing agent after exposing the polymer surface to an oxygen plasma.

Reducing plasmas created with a mixture of hydrogen, argon, ammonia, etc., have to be used in order to activate fluorocarbon polymers such as polytetrafluoroethylene (PTFE), a material well suited for implantable medical devices. Other polymeric materials, such as polydimethylsiloxane (PDMS), used in microfluidic devices, are oxidized by gas plasma treatment to make them hydrophilic. It has also been found that plasma treatment may charge the surface of microchannels, therefore supporting electrophoretic or electroosmotic flows. Besides creating hydrophilic channels that can support electroosmotic flow, plasma treatment helps in the binding of molded PDMS to the flat PDMS pieces when making the chip.[4]

Plasma polymer modification is widely used in biomedical applications to improve the biocompatibility of surfaces. It has been demonstrated that polar groups, formed by plasma-based oxidation on the inner surfaces of microfluidic devices, tend to be covered by hydrophobic material. For this reason, a coating with SiO_2 as a barrier material can be used to achieve permanent hydrophilic properties.[5] By using a microplasma-based treatment, as shown in Figure 4.3, the inner surfaces of microfluidic components already sealed were successfully functionalized. For the deposition of silicon dioxide coatings, the precursor molecule, i.e., tetramethyl orthosilicate (TMOS), can be used with the feed gas. By using this technique, surfaces with permanent wettability can be easily obtained.

PDMS is widely used in the fabrication of microfluidic devices, and its surface properties have been thoroughly studied. In addition to oxygen plasma treatment, 2-hydroxyethylmethacrylate (HEMA) was subsequently grafted on its surface in a reactive ion etching system. The mechanism

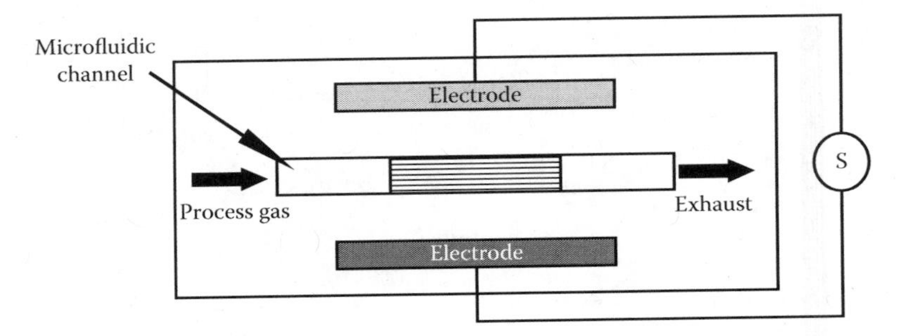

FIGURE 4.3 Microplasma-based treatment of inner surfaces of microfluidic devices. (Adapted from Klages et al., *Contrib. Plasma Phys.*, 47, 49–56, 2007.)

of surface functionalization of PDMS is given in Figure 4.4. Plasma polymerization of acrylic acid not only alters the surface properties, but also allows the deposition of the coatings at low temperatures. During the coating process, a thin layer of deposited polymer becomes strongly attached to the surface of the substrate.

Plasma-polymerized coating may increase considerably the fluid velocity, up to 450 μm/sec in a 150 μm channel, in this particular application.[7] Excellent adhesion to silicon, oxides, metals, and polymers was obtained by coatings deposited by three-phase low-energy plasma polymerization of p-xylene, 1-vinyl-2-pyrrolidinone, and polyvinyldifluoride.[8] The surface functionalization of PMMA using acid-catalyzed hydrolysis and air plasma corona treatment allowed the fabrication of PDMS-embedded valves within PMMA substrates.[9]

One of the limitations of plasma technologies is the multitude of chemical reactions that occur in the plasma, reactions that lead to a diversity of functional groups. This happens because of the hemolytic bond fissions, ionization processes, and fragmentations. Reactive surfaces with carboxyl, hydroxyl, aldehyde, and amine groups can also be used for grafting enzymes, proteins, and antibodies because of their compatibility. The criteria that should be satisfied for interfacial immobilization are the following:

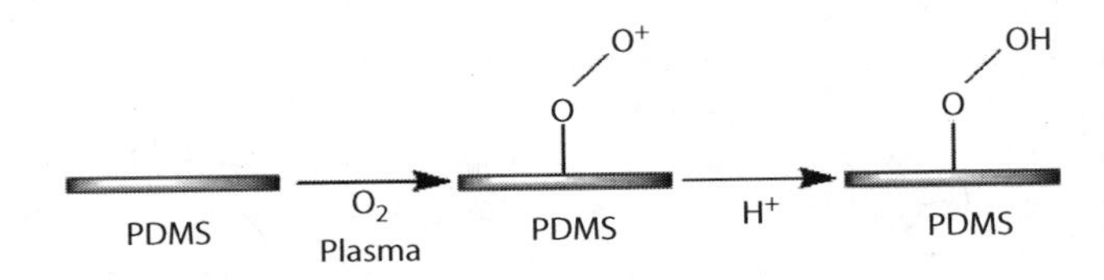

FIGURE 4.4 Surface functionalization of PDMS with oxygen plasma and HEMA treatments. (Adapted from Bodas and Khan-Malek, *Sensors Actuators B*, 123, 368–73, 2007.)

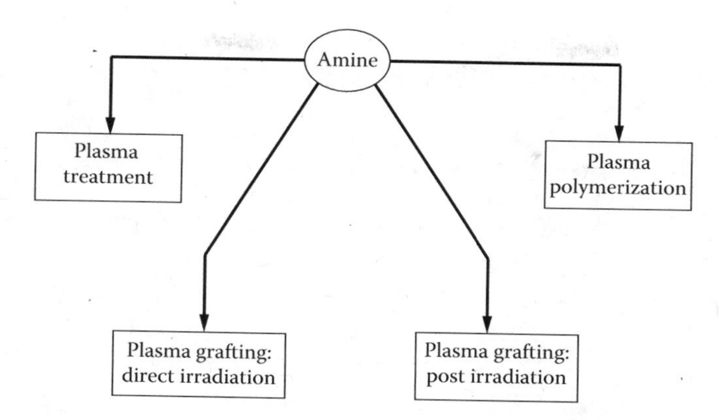

FIGURE 4.5 Approaches used to create amine groups on biomaterial surfaces. (Adapted from Siow et al., *Plasma Process. Polym.*, 3, 392–418, 2006.)

1. Active sites of bioactive molecules should face the biological medium, and the binding site of the bioactive molecules should be situated away from the biologically active area.
2. The linkage should be a covalent bond in order to be stable for the duration of the application.
3. The configuration of the bioactive molecule should not change upon binding.

Active sites are sites that are more accessible, and therefore more reactive, than others. The specific site(s) for binding can be located far away from the active sites, and they are called *attachment sites*.

The methods used to form amino groups on biomaterial surfaces are shown in Figure 4.5. One of the most used methods to introduce NH_2 groups is ammonia plasma treatment because of the ease of fabrication and compatibility of these substrates with cells. However, the surface may revert partly to the original state. For this reason, other techniques are used, such as carbodiimide chemistry, used for the formation of interfacial amide bonds. The reverse reaction has also been used; that is, carboxylated surfaces were used to react with amino groups on the protein.

To create carboxylated surfaces, plasma treatment with CO_2 and CO has been used as shown in Figure 4.6.

The functionalization of a surface by using a plasma process can be achieved with either nitrogen or oxygen, that is, a nonpolymerizable gas, or with unsaturated monomers containing labile groups. The monomers undergo polymerization called *plasma polymerization* through a free radical initiation process. Plasma-induced grafting, called *plasma grafting*, uses inert gas plasma to activate the polymer surface. The formation of free radicals is followed by the introduction of an unsaturated

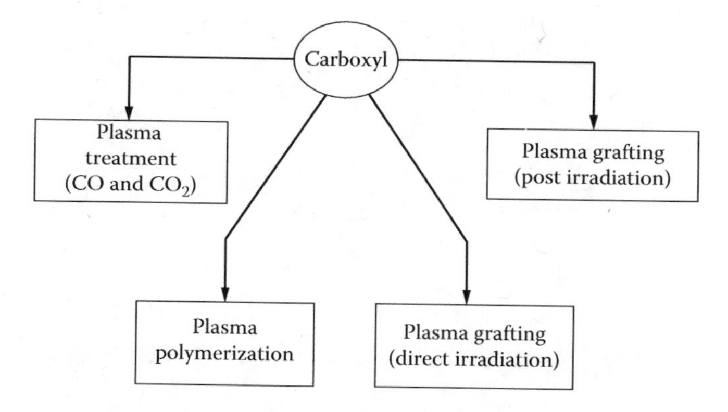

FIGURE 4.6 Plasma-based methods for producing carboxyl groups on substrates. (Adapted from Siow et al., *Plasma Process. Polym.*, 3, 392–418, 2006.)

monomer that will react with the free radicals to yield a grafted polymer. By *direct irradiation* radicals are created on the substrate in the presence of grafting molecules.

CO_2 plasma treatment produces groups such as ketones and esters in addition to COOH groups. Carboxylated surfaces can be produced by plasma polymerization with different carboxy monomers in the presence of CO_2. The monomers used are acrylic acid, propanoic acid, etc., and grafting techniques are used to increase the density of carboxyl groups on the surface.

Plasma treatment in oxygen plasma and plasma polymerization with alcohols have also been used to create OH groups.

It has been demonstrated that the surface chemistry of different plasma- treated surfaces varies with aging, which may affect very much the biological responses.

4.3 SURFACE MODIFICATIONS MEDIATED BY SELF-ASSEMBLED MONOLAYERS (SAMS)

Self-assembled monolayers are highly ordered molecular assemblies formed by the adsorption of a surfactant onto a solid surface.[11] The term *self-assembly* implies the spontaneous adsorption of molecules in a monolayer, onto a solid substrate. Two of the most widely studied systems of SAMs are gold alkanethiol monolayers and alkylsilane monolayers.

The self-assembled monolayer has two terminal groups: interface groups that point away from the substrate, toward the biomolecules that have to be immobilized, and head groups oriented toward the substrate, as shown in Figure 4.7.

The head group, that is, the functional group of the molecule that forms SAMs, has a high affinity for the surface of gold. The interfacial properties of SAMs are, however, determined by the chemical properties of the terminal group (interface group).

There has been tremendous growth in SAM research since the first gold alkylthiolate monolayer was produced by Nuzzo and Allara at Bell laboratories in 1983.[12] SAMs are a powerful research tool that is relatively easy to prepare and manipulate. They offer a unique combination of physical properties that allow fundamental studies of interfacial chemistry and also serve as model surfaces to study the interaction of synthetic materials with biologically relevant systems, especially proteins and cells. SAMs of alkanethiolates on gold provide a flexible system to tailor the interfacial properties of metals, metal oxides, and semiconductor materials.

To prepare a SAM, a dilute solution of the alkanethiol is prepared in ethanol and the gold-coated substrate is immersed in the solution for at least 1 h. The self-assembly of the monolayer takes place almost immediately, but ordering and packing of chains occurs over several hours. A tilt angle of approximately 30° from the normal surface would maximize the van der Waals interactions between the carbon chains.

Generally, for biological applications, two or more alkanethiols with different terminal groups are "mixed," while one of them will further be functionalized. As has already been mentioned, mixed SAMs with different functionalities based on oligo- or polyethyleneglycol are able to prevent nonspecific binding of proteins.[13] To the contrary, in protein adsorption studies, mixed SAMs with hydrophobic groups

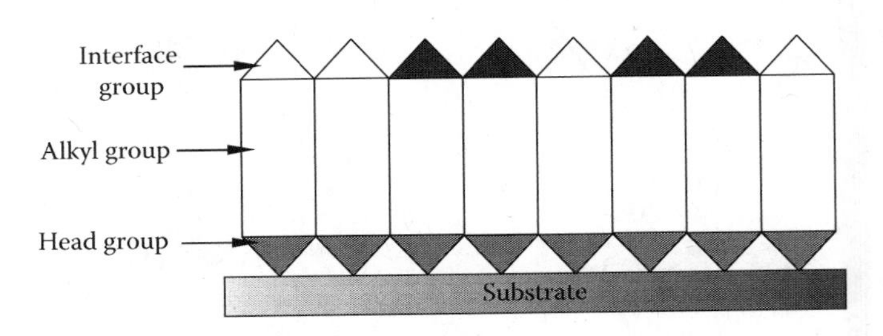

FIGURE 4.7 Monolayer assembled on a substrate.

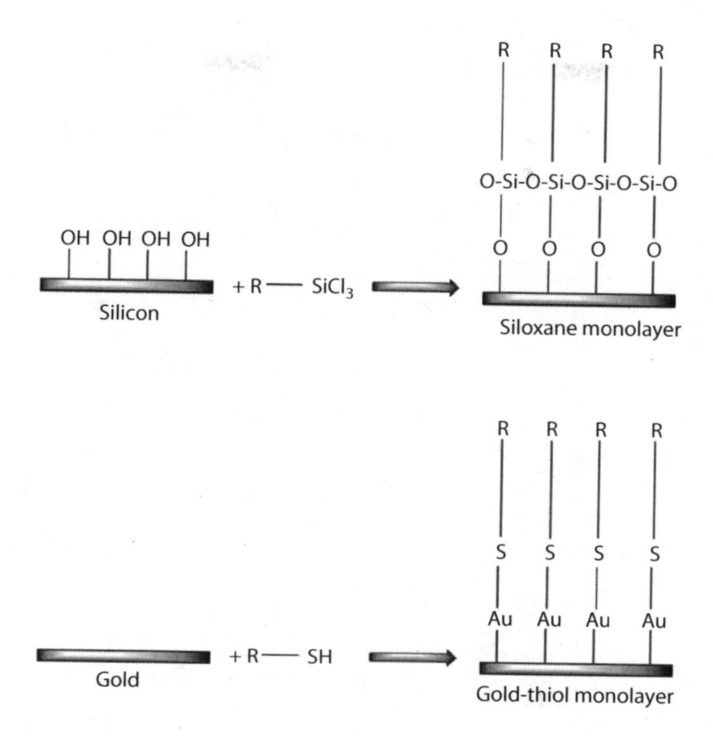

FIGURE 4.8 The structure of a siloxane monolayer and of alkanethiol on gold.

have to be used. Monolayers of organosilicon derivatives can be successfully prepared on hydroxylated surfaces, such as silicon oxide, aluminum oxide, quartz, glass, mica, and gold, as shown in Figure 4.8.

The figure shows the formation of a siloxane monolayer through coupling of a chlorosilane to the hydroxyl (OH) groups of a hydrated silicon substrate. In the case of a gold substrate, silane molecules react covalently with gold with the formation of a Au-S covalent bonding.

The mechanism of formation of the monolayers includes the binding of surface silanol groups (-SiOH) to polysiloxane chains with Si-O-Si bonds. It has been stressed that the quality of the polysiloxane monolayers depends strongly on the conditions of preparation, especially the temperature and the amount of water. Silanes on hydroxylated surfaces can be successfully used as active layers for attachment of biomolecules. While chemical modification techniques of glass are well established, silanization protocols for polymers are not yet fully developed. These techniques are important for the development of BioMEMS technology. Recently, however, some approaches were used to change the surface properties of PDMS and impart biospecificity to the microchannels. Silanization was used to immobilize polyethyleneglycol on a network of plasma-oxidized microfluidic channels used for patterning proteins on a PDMS substrate. The microchannels functionalized with PEG showed better ability to resist protein adsorption than the microchannels treated by oxygen plasma. A series of molecular and polymeric approaches for generating adherent thin films that impart antifouling properties to glass surfaces have been developed. Antifouling polymers, for example, polyacrylates, polymer mimics of phospholipids, etc., can be immobilized by adsorption to surfaces of presynthesized polymer chain end functionalized with chemical anchoring groups. This approach is called a *graft to* method. In *graft from* approaches, a polymer is grown *in situ* from the surface of the substrate through a surface-adsorbed initiation group.

Gold nanoparticle-sol-gel composite structures are also effective for retaining the bioactivity of immobilized biomolecules. Figure 4.9 shows the synthesis of such a composite material by hydrolyzing (3-mercaptopropyl)-trimethoxysilane (MPS) and attaching the hydrolized network on the surface of an electrode. Gold nanoparticles are attached through the thiol group.

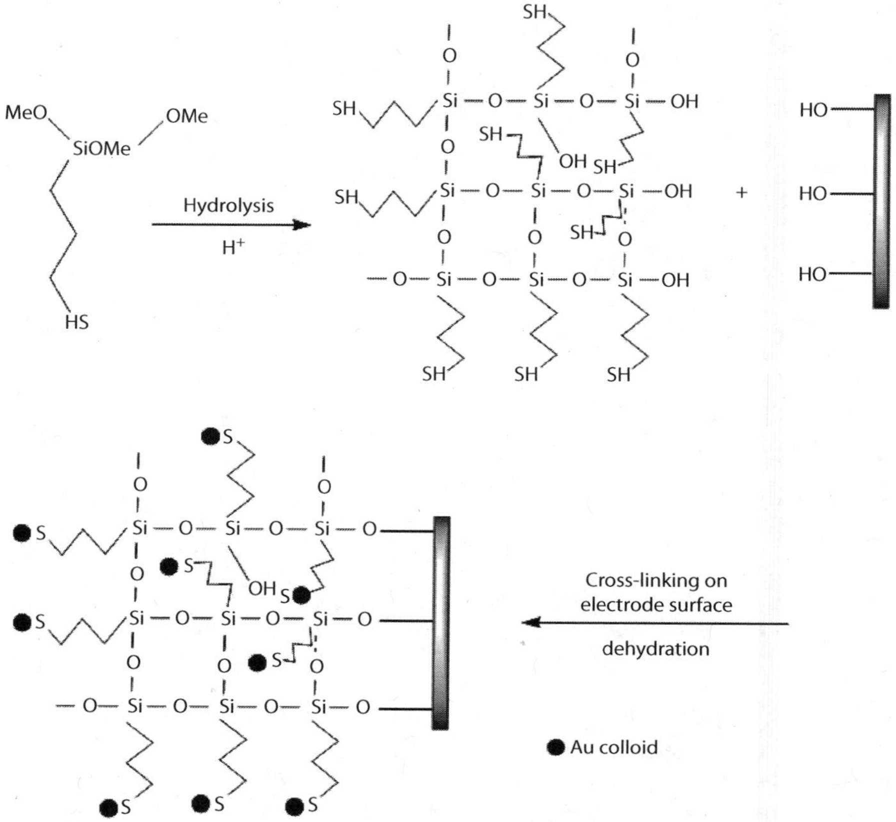

FIGURE 4.9 Preparation of gold nanoparticles doped ((3-mercaptopropyl)-trimethoxysilane membrane (MPS)) on glassy carbon electrode. (Adapted from Chen et al., *Electroanalysis*, 18, 1696–702, 2006.)

Among the methods used to passivate the glass against the nonspecific adsorption of cells, proteins, and other biological species is the use of alkylchlorosilanes (typically n-$C_{18}H_{37}SiCl_3$) with the production of a low-energy hydrophobic surface. The introduction of ethyleneglycol units into the SAMs of the silane proved to retard considerably the nonspecific adsorption of proteins such as insulin, lyzozyme, albumin, etc. The glass substrates modified by the methyl-capped oligo (ethyleneglycol) can be used with very good results in applications that require sterilized glassware, as it maintains its properties and structure when heated to 140°C.

4.4 LANGMUIR-BLODGETT AND LAYER-BY-LAYER ASSEMBLY

Langmuir-Blodgett (LB) films are monolayers, formed by the two-dimensional compression of surface-active molecules (surfactants) and subsequent transfer from the surface of water onto a solid substrate. As shown in Figure 4.10, by moving the substrate slowly, the film can be transferred onto the solid substrate from the water surface. LB films are less stable than SAMs because they are maintained only by weak van der Waals interactions. However, SAMs are more versatile and their preparation does not need special equipment.

The LB film is a quasi-two-dimensional solid with the molecules closely packed and aligned with the hydrophilic head groups on the surface of water. The monolayer attaches to the substrate when a substrate such as silicon or glass is passed through it. The method requires the assembly components to be *amphiphilic*, also called *amphipatic*, that is, with both hydrophilic and hydrophobic groups

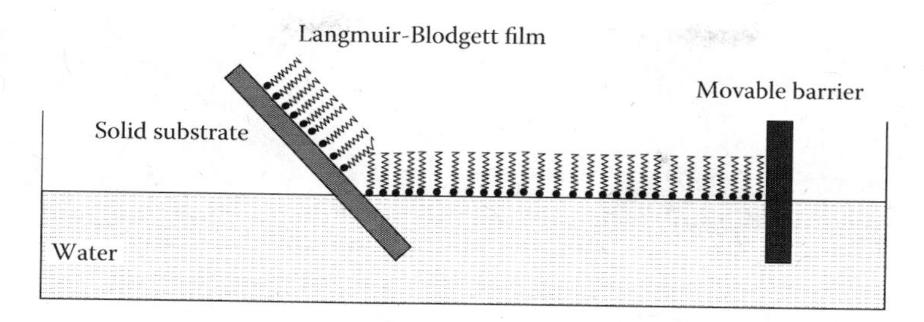

FIGURE 4.10 Deposition of a LB film onto water and transfer onto substrate by dipping.

on the same molecule. Hydrophilic groups are polar groups, while hydrophobic groups are typically hydrocarbon structures. For this reason, there are only a few types of biomolecules that can be embedded in the film. The *layer-by-layer* (LBL) *assembly* was introduced by Decher and coworkers in 1992, and this method proved to be especially adequate for biological applications.[15] It takes advantage of the strong attraction between oppositely charged polyelectrolytes, as shown in Figure 4.11.

The deposition is made on metals, glass, and other substrates that have a net negative charge due to surface oxidation and hydrolysis. Positively charged polyelectrolytes such as polyethyleneimine (PEI) or polyallylamine hydrochloride (PAH) will readily adsorb on the surface, and the surface charge will reverse at each subsequent step because of the excess of positively charged polyelectrolytes. Films can be prepared on substrates of any shape, and the method results in very stable films that allow higher loading of biomolecules than LB or SAM films.

In addition, thin films with biological activity can be easily assembled, as many biomolecules are usually charged. For example, different types of proteins can be incorporated into multilayer LBL films and used as biosensors. An enzyme biosensor using thin films of oxyreductase enzymes prepared on electrode surfaces through LBL assembly is shown in Figure 4.12.

The sensor is immersed in a solution that contains specific substrates and mediators, while the enzyme will catalyze the conversion of the substrates from one redox state to the other. The redox changes of the mediators are detected electrochemically, and then the concentration of the reactants in the solution can be calculated. Immunosensors, protein sensors, as well as nucleic acid and DNA sensors can also be prepared by the LBL method. The LBL assembly technique has been used for biomimetics and tissue engineering as well. Amphiphilic polyelectrolytes can also be used to stabilize LB films.

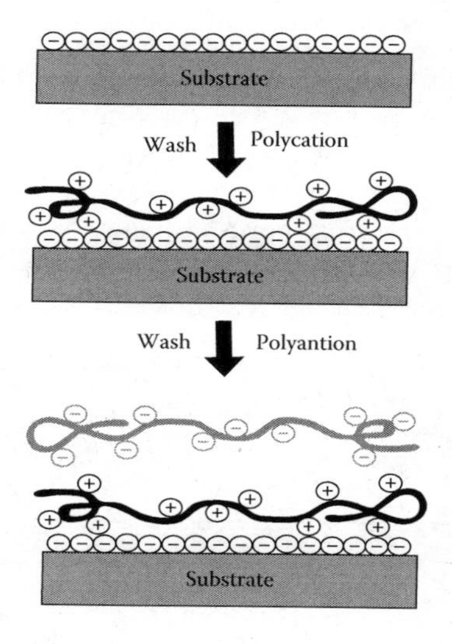

FIGURE 4.11 Scheme of the LBL film deposition process.

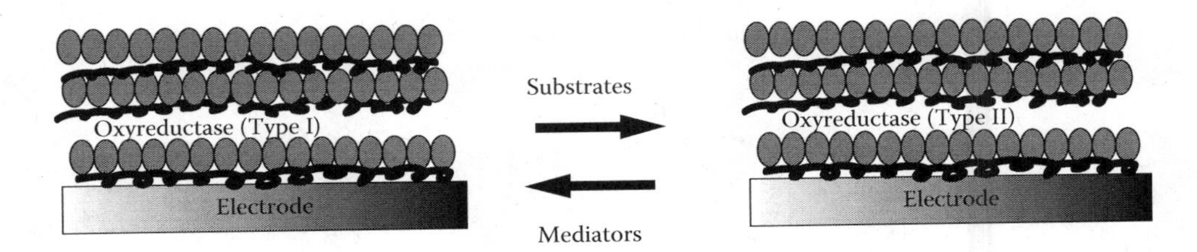

FIGURE 4.12 Mechanism of detection for oxyreductase biosensors using LBL. (Adapted from Tang et al., *Adv. Mater.*, 18, 3203–24, 2006.)

4.5 BIOSMART HYDROGELS

Hydrogels are cross-linked hydrophilic polymer networks that swell due to their high affinity for water. Because of their biocompatibility and high water content, hydrogels are used in biology and medicine for applications such as biosensors, cell encapsulation, and drug delivery. Enzyme-containing PEG hydrogels have been fabricated inside of microchannels for potential use in micro total analysis systems (μTAS).[17]

Carboxymethylated dextran hydrogels on gold surfaces have also been used to develop a platform for protein microarrays.[18] Homogeneously distributed protein ligands have been realized by piezo-dispensing the proteins at high humidity. The increased immobilization of proteins, as compared to monolayer-based coatings, is facilitated by the open structure of dextran, as shown in Figure 4.13. It has been shown that protein loading up to 50 ng.mm^{-2} can been obtained by using a hydrogel with a thickness of 100 nm.[17] Hydrogels that sense and respond to changes in environmental pH were utilized to functionalize the surface of microcantilevers. As a result, there is a change in surface stress that deflects the cantilever, as shown in Figure 4.14.

PEG-based hydrogels are widely used for biomedical applications due to their nontoxic and non-immunogenic nature. The applications encompass tissue engineering, controlled drug delivery, as well as diagnostic devices and medical and biological sensors.

4.6 IMMOBILIZATION AND DETECTION OF BIOMOLECULES BY USING GOLD NANOPARTICLES: CASE STUDIES

4.6.1 Gold Nanoparticles Functionalized by Dextran

Optical sensing based on localized surface plasmon resonance (LSPR) absorption of gold nanoparticles is widely used. Gold nanoparticles have been used in combination with dextran for various sensing applications, for example, in the detection of the sugar binding protein called lectin. Dextran is a linear polysaccharide that can be attached to gold nanoparticles and subsequently activated with peptide coupling reagents (EDC/NHS), as shown in Figure 4.15. Usually, carboxylate groups are introduced in the dextran chains. The synthesis of this material consists of functionalizing the

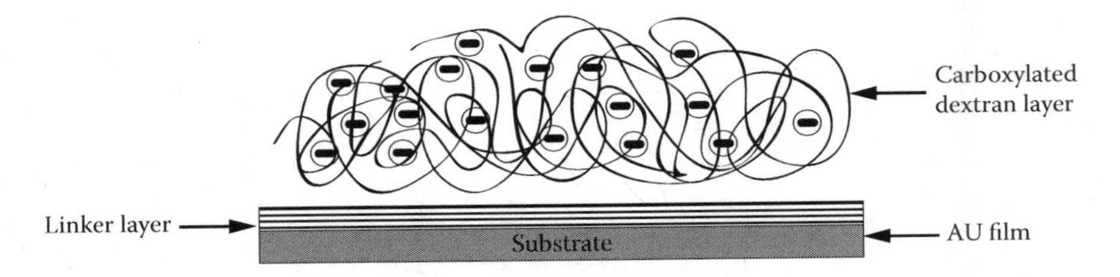

FIGURE 4.13 Schematical illustration of the gold surface modified with carboxymethyldextran matrix. (Adapted from Davies and Higson, *Biosensors Bioelect.*, 21, 1–20, 2005.)

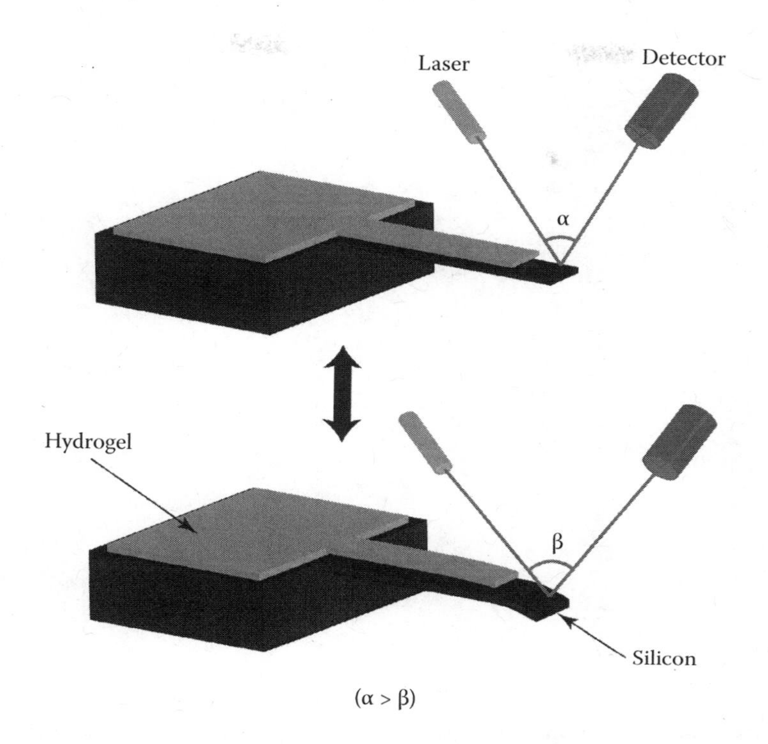

FIGURE 4.14 Schematic of the MEMS sensor platform based on a microcantilever coated with a hydrogel.

surface of gold nanoparticles through a series of steps that lead to epoxy-functionalized nanoparticles. These are subsequently reacted with hydroxyl moieties of the α-d-glucopyranosyl groups of dextran to produce the hybrid material.

4.6.2 Gold Nanoparticles in Hybridization Experiments

Because gold nanoparticles have a good biocompatibility, they can be functionalized by DNA molecules, which later can self-assemble onto functionalized solid surfaces, as shown in Figure 4.16.

Oligonucleotides are used for improved DNA detection on silicon surfaces for diagnostic applications. Experiments onto amino-modified surfaces showed better efficiency than the thiol-modified surfaces. It has been demonstrated that for systems using DNA-modified nanoparticles, the hybridization efficiency and the sensitivity limit increased by a factor ten to one hundred, compared to DNA immobilization on a planar surface.

4.6.3 Enhanced Biomolecular Binding Sensitivity by Using Gold Nanoislands and Nanoparticles

Gold nanoparticles are used in conjunction with gold nanoislands (NIs). Nanoislands are made on glass substrates by evaporation and heat treatment.[22] Streptavidin (STA) and biotinylated bovine serum albumin (Bio-BSA) are used as a model receptor and model analyte, respectively. The addition of gold nanoparticles enhanced the sensitivity of monitoring the biomolecular binding. It has been demonstrated that the combination of gold NI and nanoparticles enhanced the sensitivity more than ten times.

The results show that sensing through a combination of gold nanoislands and nanoparticle conjugates allows the detection of 7 nM Bio-BSA (Figure 4.17).

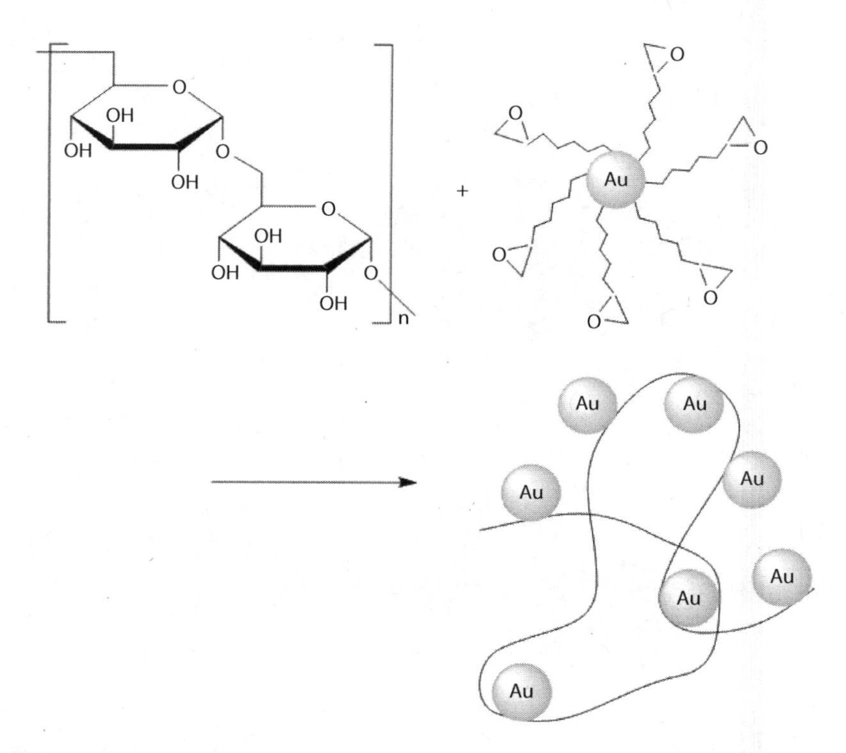

FIGURE 4.15 Preparation of dextran-attached gold nanoparticles for localized surface plasmon resonance (LSPR) sensing. Dextran reacts with epoxy-functionalized gold nanoparticles and yields a hybrid material. The ligands on the backbone of dextran can interact by means of specific biomolecular recognition with proteins that can induce the aggregation of the gold nanoparticles attached to the dextran chains. (Adapted from Lee and Pérez-Luna, *Anal. Chem.*, 77, 7204–11, 2005.)

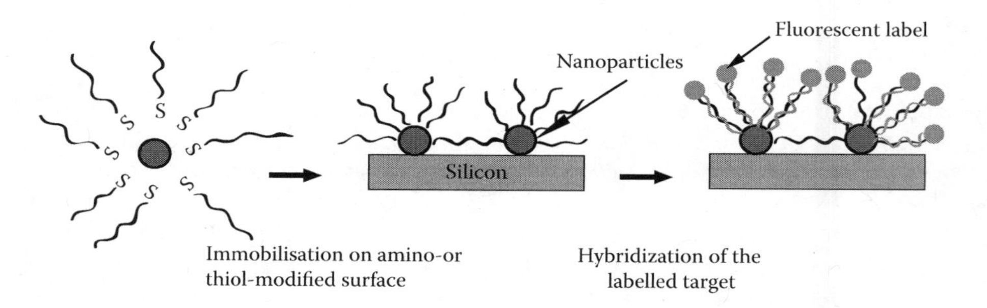

FIGURE 4.16 Immobilization scheme of the modified nanoparticles. (Adapted from Menard-Basquin et al., *IEEE Proc. Nanobiotechnol.*, 152, 97–103, 2005.)

4.6.4 Study of Antigen-Antibody Interactions by Gold Nanoparticle Localized Surface Plasmon Resonance Spectroscopy

Often, glass substrates are functionalized with a silane molecule before the immobilization of gold nanoparticles. As shown in Figure 4.18, gold nanoparticles adsorbed on the silanized glass are further modified by the formation of a self-assembled monolayer of 11-mercaptoundecanoic acid. After activation of a COOH terminal of the SAM through the usual EDAC/NHS system, bovine serum albumin (BSA) is covalently immobilized to SAM, as shown in Figure 4.18.

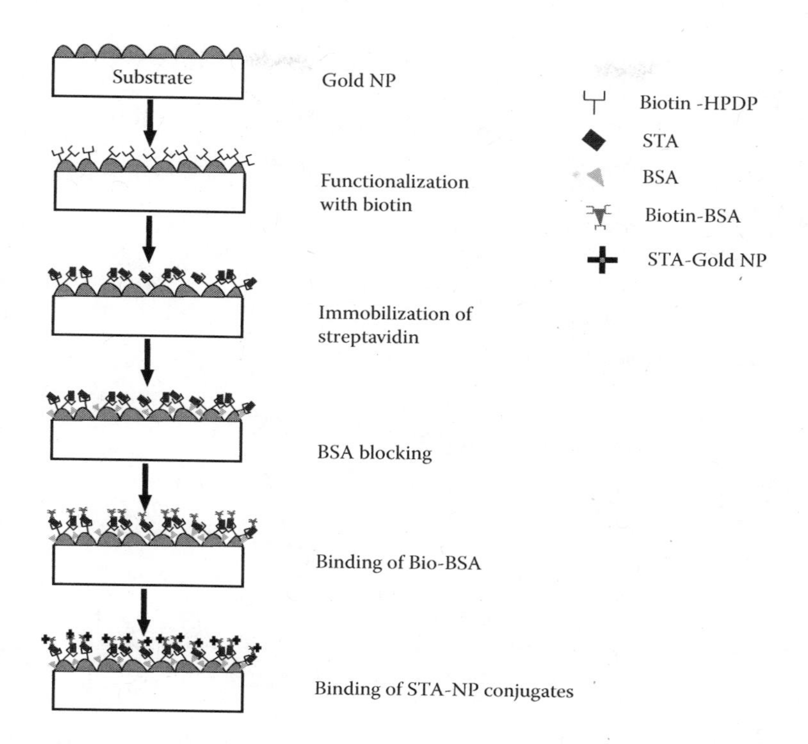

FIGURE 4.17 Schematic illustration representing the modification procedure of gold nanoparticle surfaces and binding of biomolecules. (Adapted from Kim et al., H. M., *Sensors*, 9, 2334–44, 2009.)

4.6.5 Array of Gold Nanoparticles for Binding of Single Biomolecules

The authors present a technology for the positioning of single biomolecules in nanopatterns of gold nanoparticles. This method enables the application of a large number of different biomolecules oriented in a nanoscaled array. The principle of the experimental concept is presented in Figure 4.19. The method permits the exact determination of the number, spatial positioning, and orientation of molecules at the interface.

4.7 BIOMIMETIC SURFACE ENGINEERING

The tripeptide RGD (R, arginine; G, glycine; D, aspartic acid) was found in 1984 to promote cell adhesion and to control effectively the interaction between synthetic polymers and cells. It has been shown that in order to promote strong cell adhesion, RGD peptides should be covalently attached to the polymer through functional groups like hydroxyl, amino, or carboxyl. For polymers that do not have these groups, they can be introduced by different methods, such as blending, chemical, and physical treatments (alkaline hydrolysis, oxidation, plasma deposition, etc.).

An activated surface carboxylic acid group is reacted with the N-(amino) group of the peptide (Figure 4.20) to link RGD peptides to the polymer. In order to reach the binding sites, the RGD peptide must extend out from the polymer surface. Oligoglycine spacers on Polyacrylonitrile beads have been used to enhance cell adhesion. The number of attached cells increases with the RGD surface density and surface distribution. Not only polymer substrates, but also glass, silicon, and titanium, coated with star-shaped polyethyleneglycol prepolymers, were modified with linear RGD peptides to allow specific cell adhesion. Even though star PEG coating prevents protein (peptide) adsorption, cell adhesion, spreading, and survival were observed when modified with linear RGD peptides. The star PEG/RGD-coated substrates may have interesting applications for regenerative medicine.

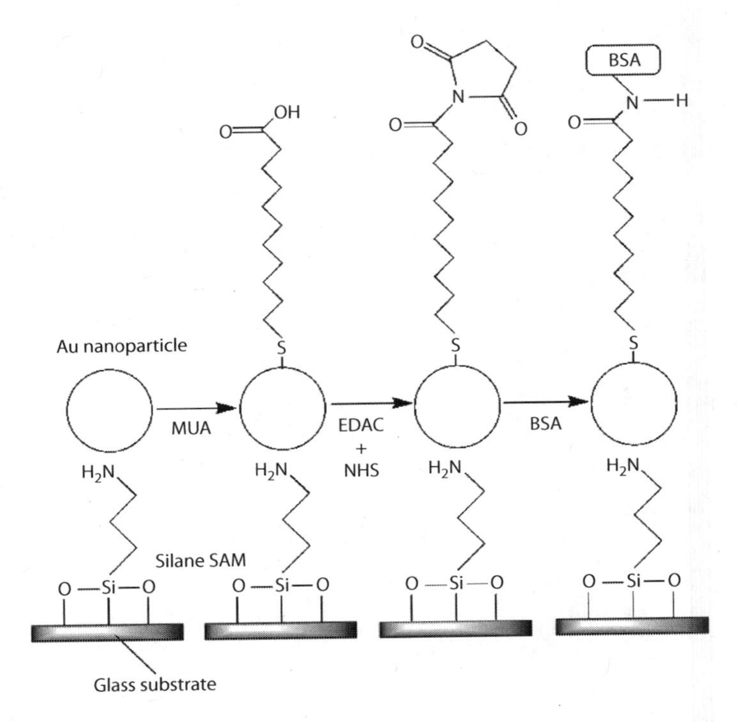

FIGURE 4.18 Schematic drawing showing BSA immobilization. MUA, 11-mercaptodecanoic acid; NHS, N-hydroxisuccinimide; EDAC, carbodiimide derivative; BSA, bovine serum albumin. (Adapted from Fujiwara et al., *Anal. Bioanal. Chem.*, 386, 639–44, 2006.)

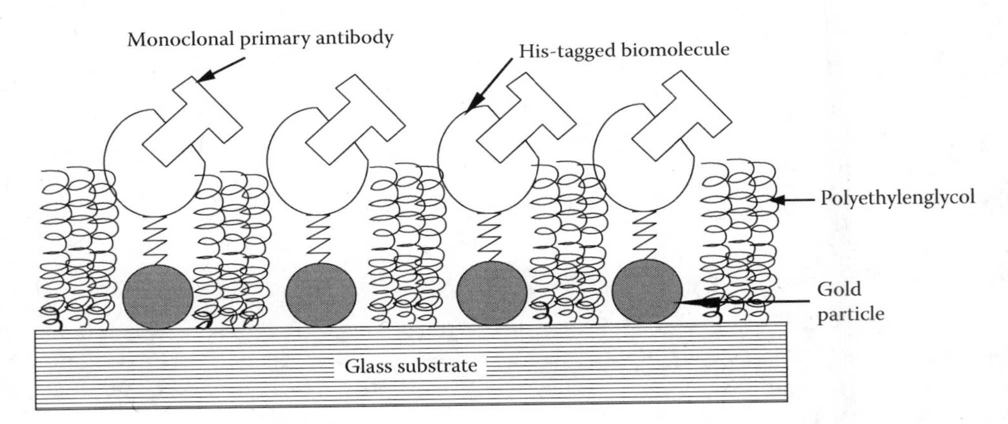

FIGURE 4.19 Scheme of the gold nanopattern and biomolecule presentation at a surface. The biomolecules used in this study are N-cadherin, L1, and agrin and are all expressed with a histidine (His) tag. The linker is mono-NTA-thiol (NTA-nitrilotriacetic acid). Diameter of Au, 6 nm; lateral spacing of 30, 96, or 160 nm. (Adapted from Wolfram et al., *Biointerfaces*, 2, 44–48, 2007.)

A novel approach for the coating and functionalization of substrates such as silicon and titanium for cell culture and tissue engineering is coating the substrate with an ultrathin film (30 ± 5 nm) of reactive star-shaped polyethyleneglycol prepolymers (star PEG). In order to allow specific cell adhesion, the films are modified with linear RGD peptides (gRGDsc) in different concentrations. Cell adhesion, spreading, and survival are observed for up to thirty days on linear RGD peptide (gRGDsc)–modified coatings, whereas no cell adhesion could be detected on unmodified star PEG.

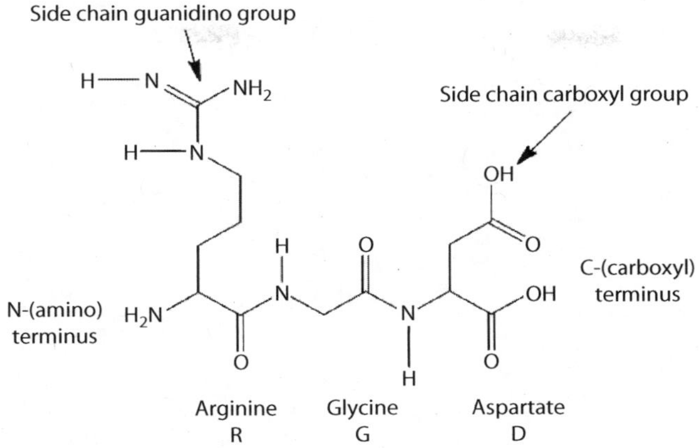

FIGURE 4.20 The molecular formula of the RGD sequence.

Figure 4.21 shows schematically the coating with star PEG (left) and the coating with both star PEG and a linear RGD peptide (right).

Among the many emerging materials, self-assembling peptides show promise for use in surface modification. An interesting class in this category is that of ionic-complementary peptides that contain sequences derived from a fragment of a Z-DNA binding protein in yeast. They have alternating hydrophobic and hydrophilic residues, as shown in Figure 4.22. They can self-assemble into nanofibrils rich in β-sheets that are very stable. These peptides have been studied for applications such as drug delivery, tissue scaffolding, and cell patterning. Depending on the desired application, the sequence can incorporate amino acids having certain functions. For example, for protein immobilization, residues containing COOH and NH_2 are incorporated.

4.8 ATTACHMENT OF PROTEINS TO SURFACES

The immobilization of proteins onto solid supports has become increasingly important, particularly in the generation of high-density protein microarrays for functional proteomics, including high-throughput analysis of protein-protein, protein–small molecule, and protein–nucleic acid interactions. However, the technical challenges associated with the generation of protein microarrays have prevented their widespread application. In addition, protein immobilization is important in other applications, such as drug screening, diagnostics, and biosensing involving fundamental studies in single-molecule enzymology.

The protein immobilization methodology is also important in developing biocatalysts for industrial processes where there is a need to conserve or recycle commercially valuable enzymes. Early methods for protein immobilization relied on nonspecific reactions, such as the attachment of amino groups on the protein surface with aldehyde-functionalized solid supports. Such methods not only require a pure sample of each protein to avoid coimmobilization of protein impurities, but also lead to a population of heterogeneously orientated proteins.

Noncovalent protein immobilization using biotin-avidin, anti-GST antibody, or His6-Ni interactions has also been explored.

Recent efforts have focused on site-specific covalent methods of protein immobilization. An impending, and often overlooked, issue in protein microarray technologies lies in the successful development of robust strategies that allow efficient immobilization of proteins onto glass surfaces, while maintaining their native biological functions. This calls for the development of novel protein immobilization methods, which ensure that proteins are uniformly and stably attached to the solid surface.

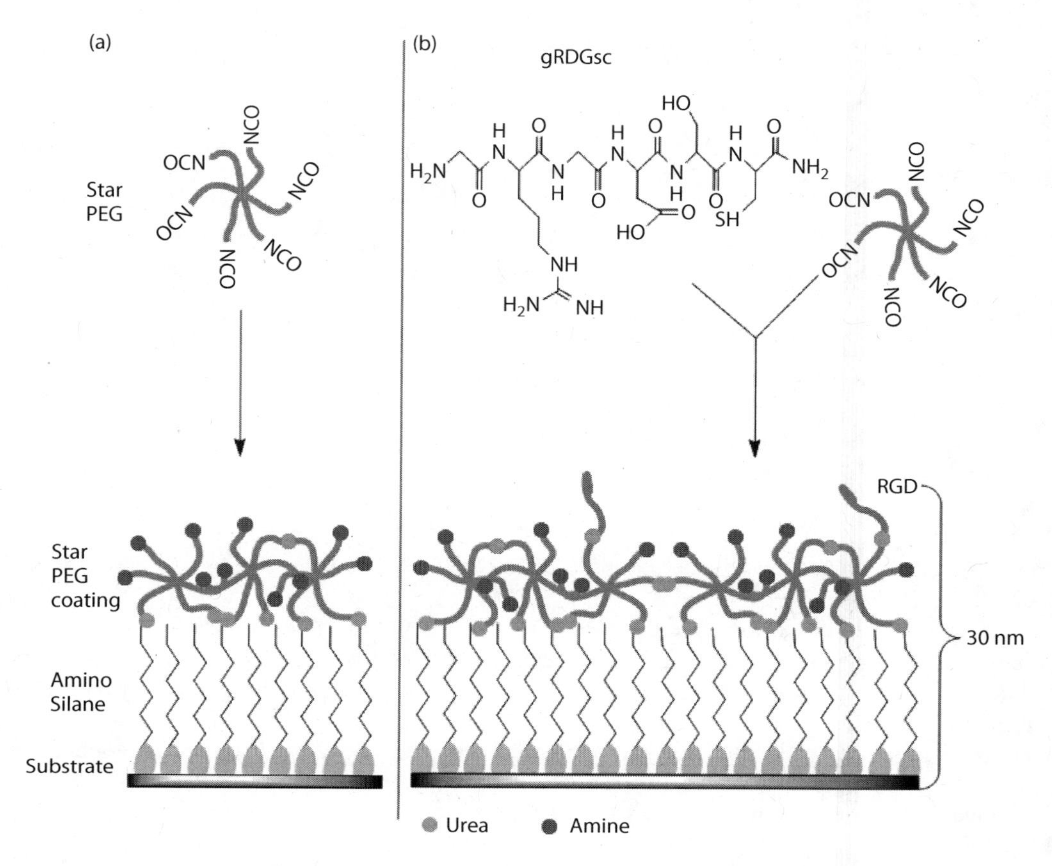

FIGURE 4.21 Scheme of the preparation of an unmodified surface with RGD (a) and RGD-modified (b) star PEG layers for increased cell adhesion. (Reproduced from Groll et al., *J. Biomed. Mater. Res.*, 74A, 607–17, 2005. With permission.)

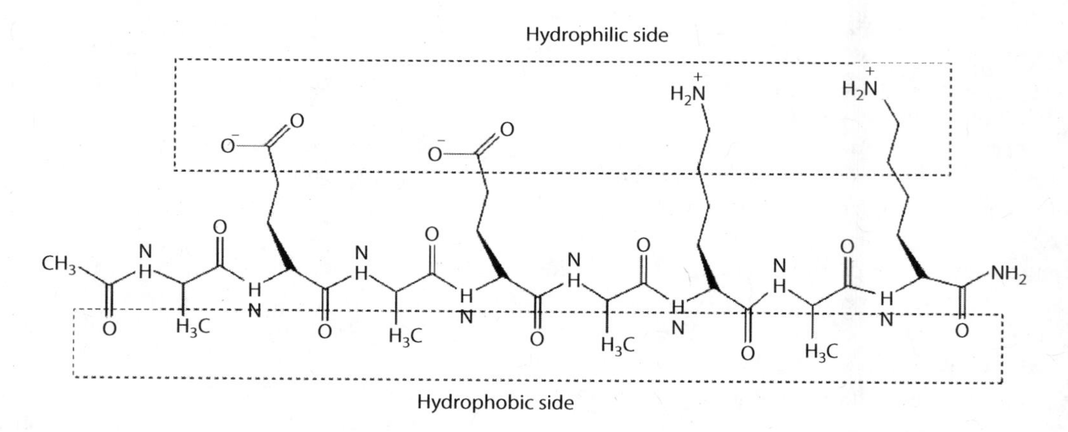

FIGURE 4.22 Schematic diagram of EAK16-II structure. The upper part shows the hydrophilic side of the peptide, and the lower part shows the hydrophobic side of the peptide. It has been demonstrated that ionic-complementary peptides can be used to modify both hydrophobic and hydrophilic surfaces. (Adapted from Yang et al., *PLoS ONE*, 12, e1325, 2007.)

It is suggested that a "universal" set of tools should be developed that would allow attachment of any biomolecule to any surface. This idea is discussed for the case of attaching proteins to surfaces or nanoparticles, as shown in Figure 4.23. These tools would allow control of the final orientation of the protein, its distance from the surface, and its affinity to the surface. However, these criteria, which are difficult to attain, would require accelerated research.

The important six criteria for immobilizing protein into nanoparticles are shown in Figure 4.23, along with protein-immobilized nanoparticles.

In this example, the proteins would cover the nanoparticle surface in three dimensions. Proteins could still have some rotational freedom around the axis connecting them to the nanoparticle while fulfilling the criteria. These criteria can be extended to the interaction of any biosensing molecule with the solid-state components of any biosensing device.

There are different methods for immobilizing a protein molecule to a surface. Three of the commonly used methods are described and compared. All the methods involve coupling of a free amine group in the protein molecule to a reactive functional group bound to the surface:

1. Activation of surface amines with glutaraldehyde (GA) to form an aldehyde-functionalized surface
2. Formation of an anhydride monolayer from a carboxylic acid
3. Formation of an active ester from a carboxylic acid

Figure 4.24 shows the reaction corresponding to the conversion of a carboxylic acid group into an ester. The ester reacts with an amino group ($R'NH_2$) to yield an amide (-NH-CO-), linking the protein to a SAM on the surface.

In this method, carboxylic acid–terminated SAMs on the substrate are activated with TFFA and converted to surface anhydride species that will further react with a free amino group (NH_2) on the protein.

This method involves the reaction of amino-terminated SAMs with glutaraldehyde. This molecule has two aldehyde groups separated by a short alkyl chain. One of the aldehyde groups will react with the amino-functionalized surface and form an imino group (Figures 4.25 and 4.26).

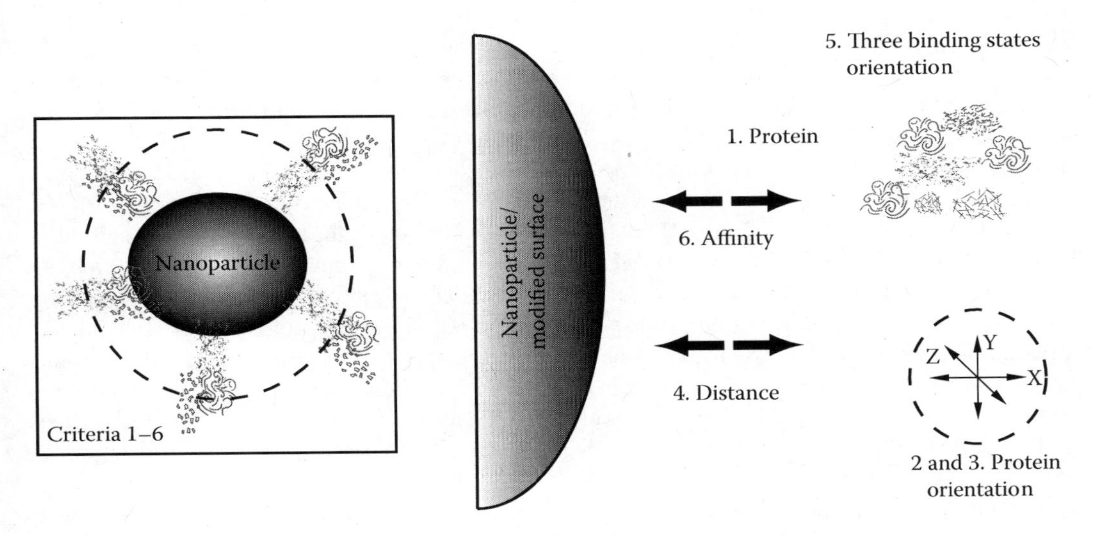

FIGURE 4.23 Schematic representation of the six criteria for a universal tool set that would allow controlled attachment of any protein to any nanoparticle or surface. (Reproduced from Medintz, *Nature Mater.*, 5, 842, 2006. With permission.)

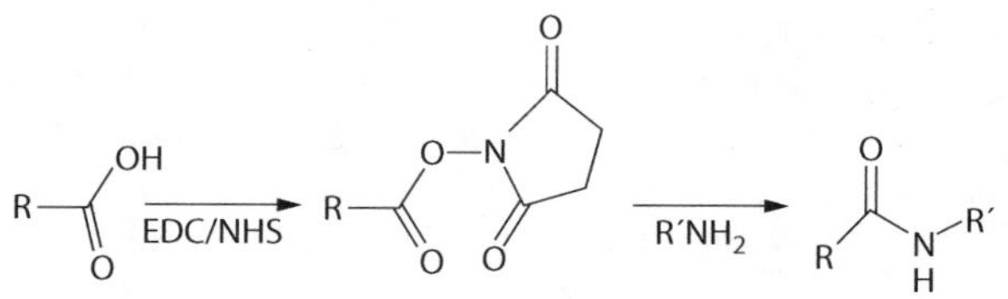

FIGURE 4.24 Activation of a carboxylic acid–terminated monolayer using EDC and NHS, followed by reaction with an amine to yield an amide linkage. EDC, 1-ethyl-3,3'-dimethyl carbodiimide; NHS, N-hydroxysuccinimide. (Adapted from Ducker et al., *Biointerphases*, 3, 59–65, 2008.)

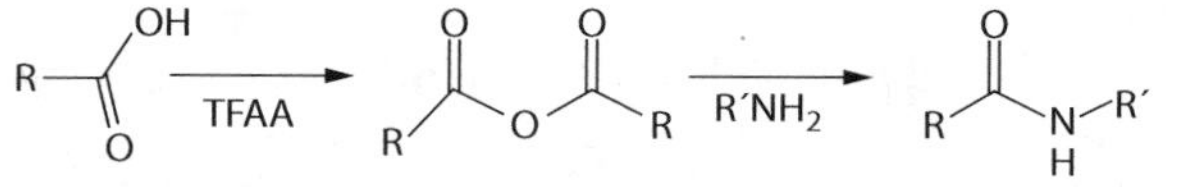

FIGURE 4.25 Activation of a carboxylic acid–terminated monolayer using trifluoroacetic anhydride (TFAA), followed by reaction with an amine to yield an amide linkage. (Adapted from Ducker et al., *Biointerphases*, 3, 59–65, 2008.)

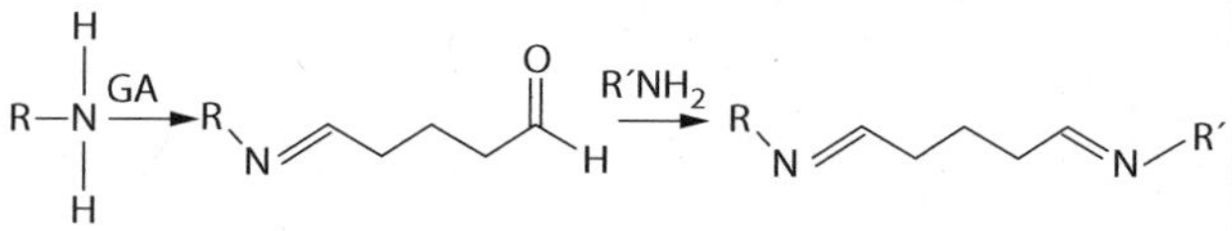

FIGURE 4.26 Reaction of glutaraldehyde (GA) with an amine-terminated monolayer to yield an aldehyde-functionalized surface that may subsequently react with an amine to form an imine linkage. (Adapted from Ducker et al., *Biointerphases*, 3, 59–65, 2008.)

4.9 SURFACE MODIFICATION OF BIOMATERIALS FOR TISSUE ENGINEERING APPLICATIONS

The biomaterial used for tissue engineering applications has to be biocompatible and biodegradable and should meet biological needs for the cell proliferation involved in new tissue formation. The specific interaction between biomaterials and cells depends on many parameters, principally the surface energy. It is also important to improve the surface area of the biomaterials that first come in contact with the biological environment. The interaction between cells and extracellular components is important for cell anchorage and signal transduction. Figure 4.27 shows the biological events occurring at the interface between the surface and biomolecules.

The events depicted in Figure 4.27 include water and biomolecule adsorption and cell adhesion to the top surface layer of the material. When a biomaterial is implanted, proteins from blood or serum will adsorb rapidly on the surface. If proteins are adsorbed in the right configuration, they will stimulate a positive cell response that will help tissue regeneration. It has been demonstrated that the interaction of scaffold with cells is determined by properties such as the presence and size of pores, the texture, and properties of the surface (free energy, ionic interaction, etc.). Biomaterials have to match the topography of the tissue as well. To overcome the hydrophobicity and the chemical inertness of some polymers and to create a surface roughness, oxygen plasma treatment can be used as well.

The entrapment method illustrated in Figure 4.28 is promising for improving the hydrophilicity of a biomaterial. In this method, the surface is treated with water-soluble polymers such as chitosan, gelatin, heparin, etc., that will induce surface swelling. The surface modified by swelling shows an enhanced proliferation of cells.

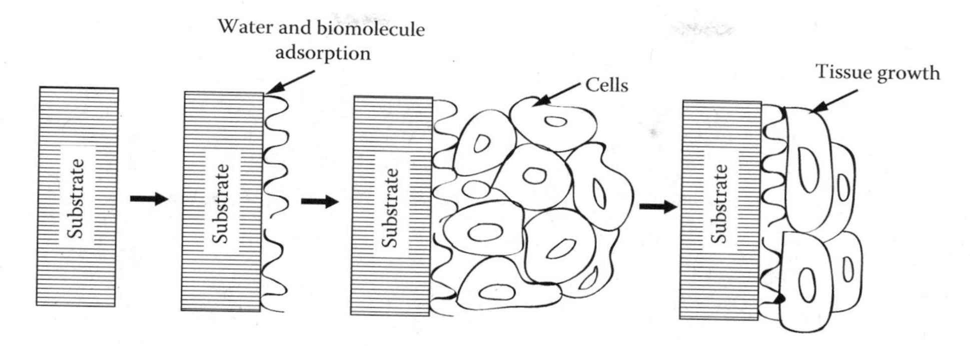

FIGURE 4.27 The sequences of events when biomaterials are implanted in the host. (Adapted from Jiao and Cui, *Biomed. Mater.*, 2, R24–37, 2007.)

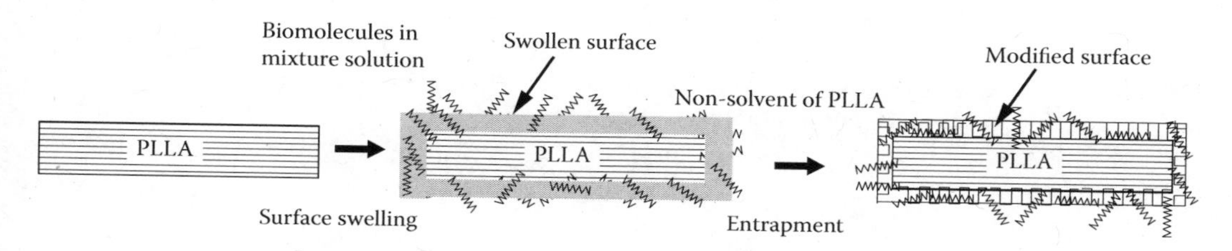

FIGURE 4.28 The modification of poly-L-lactide (PLLA) by the entrapment method. (Adapted from Jiao and Cui, *Biomed. Mater.*, 2, R24–37, 2007.)

A functional scaffold for tissue engineering should be able to mimic the structure and biological function of the extracellular matrix proteins and, at the same time, maintain the normal differentiated state of the cells. One of the most studied forms of tissue-engineered scaffolds is polycaprolactone fibers in the form of nonwoven membranes obtained by electrospinning. Although polycaprolactone has good biocompatibility properties, because of its hydrophobicity, it has to be functionalized to improve the cytocompatibility. To promote cellular growth, molecules such as collagen, gelatin, and chitosan are immobilized onto the surfaces of polymeric scaffolds. Sometimes, before the adsorption of these molecules, the surface is treated with 1,6-hexamethylene diamine (HMD) and the amino group is used to promote further functionalization, as shown in Figure 4.29. The amino groups introduced on the surface of polycaprolactone improve the hydrophilic properties of the surface and provide the sites necessary to further immobilize collagen, etc. The potential of these materials for tissue and cell culture is assessed with different cell lines. Collagen I, a RGD-containing protein, was found to be the best support for the attachment of cells.

Poor hydrophilic properties and the lack of natural recognition sites on the surface of polyesters have limited their applicability in the tissue engineering field. Among the methods used for surface modification of polyesters are surface morphological modification, surface chemical group/charge modification, etc.

Many applications in tissue engineering and drug delivery require the rapid transport and exchange of materials. Microfluidic hydrogels have been proposed as scaffolds for the transport of materials.[31] The main application of hydrogels in microfluidic channels is the sustaining the metabolism of embedded cells. They contain interconnected channels that have widths of 10 to 1,000 mm. The large size of these channels compared to the submillimeter size of pores greatly reduces the fluidic resistance of the gels. As a result, microfluidic gels require a much smaller driving pressure difference than bulk gels. Figure 4.30 shows the fabrication of microfluidic gels.

Chitosan is a pH-responsive polymer (aminopolysaccharide) that can be assembled at readily addressable sites in microfluidic channels from their aqueous environment. At low pH, chitosan is

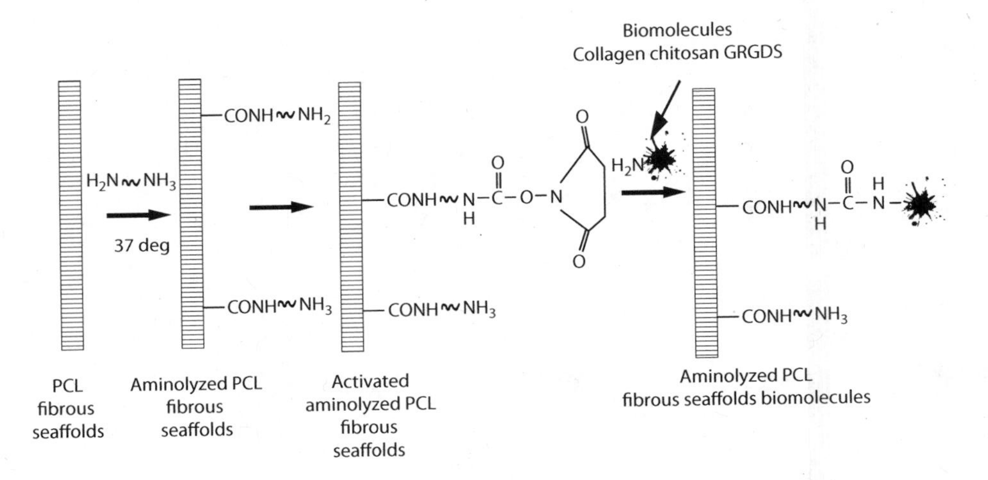

FIGURE 4.29 Chemical pathway for the immobilization of different biomolecules, such as collagen on the surface of the electrospun PCL fibrous scaffold. (Adapted from Mattanavee et al., *Appl. Mater. Interfaces*, 1, 1076–85, 2009.)

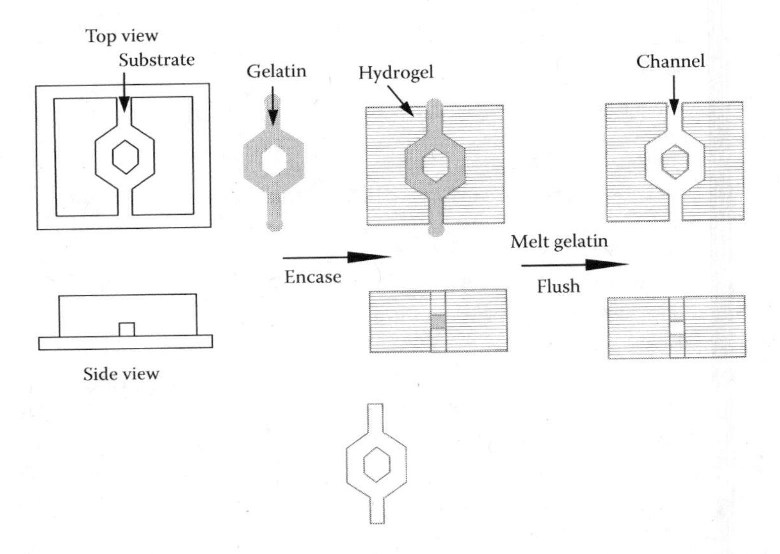

FIGURE 4.30 Schematic diagram of fabrication of microfluidic gels. Sealing a PDMS stamp to a substrate (glass or preoxidized PDMS) created a microfluidic network. Sequential introduction of Pluronic (1 to 6% in PBS) and liquid gelatin into the channels, and gelation at 4°C for 15 min and at 23°C for 2 h, yielded a gelatin mesh that easily separated from the channels. Encapsulation of the mesh in a liquid hydrogel precursor (type I collagen, fibrinogen, Matrigel), polymerization of the precursor, and flushing at 37°C yielded a hydrogel with open microchannels. (Reproduced from Golden and Tien, *Lab Chip*, 7, 20–25, 2007. With permission.)

soluble in aqueous solution while at neutral to high pH, the amine group becomes deprotonated and uncharged, making chitosan insoluble as shown in Figure 4.32.

Due to the high local pH near the electrode, chitosan will assemble from its aqueous solution onto a negatively biased electrode surface, as shown in the figure. In addition, the primary amine group of chitosan is nucleophilic at the neutral state, allowing various amine group reactive chemistries to be used for covalent conjugation of biomolecules onto chitosan. These reactions can also be carried out in a closed aqueous environment, avoiding drying or contamination.

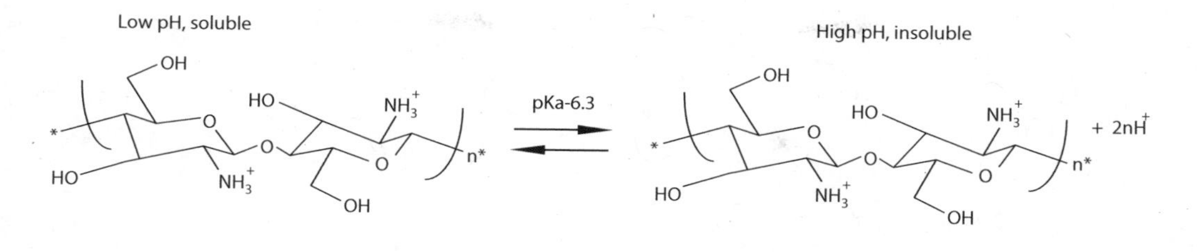

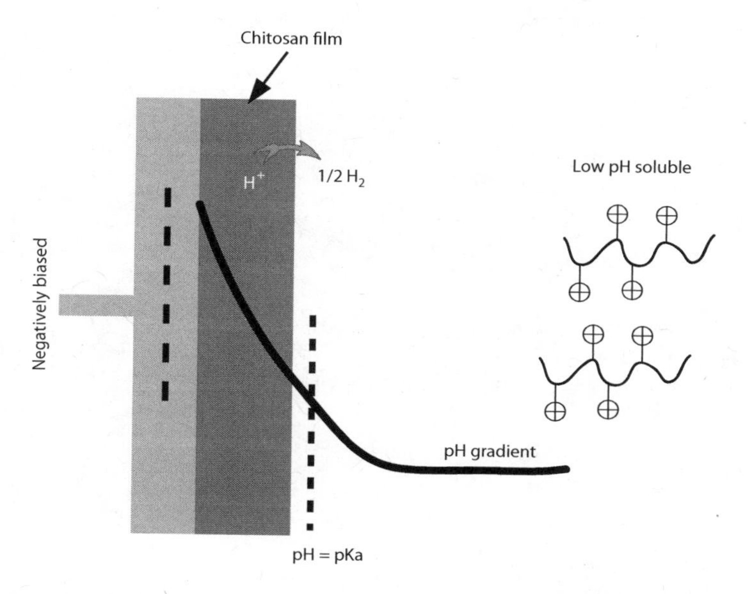

FIGURE 4.31 Image of microfluidic gel: Overlaid phase contrast and fluorescence images of a hexagonal network in collagen perfused with fluorescence microspheres. (Reproduced from Golden and Tien, *Lab Chip*, 7, 20–25, 2007. With permission.)

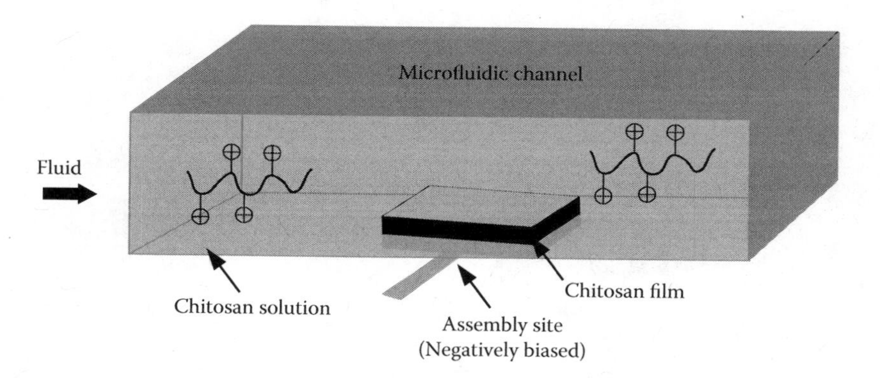

FIGURE 4.32 (a) pH-dependent protonation/deprotonation of the chitosan molecule. (b) Schematic view of chitosan deposition. (c) Schematic view of chitosan deposition in a microfluidic channel. (Reproduced from Park et al., *Lab Chip*, 6, 1313–20, 2006. With permission.)

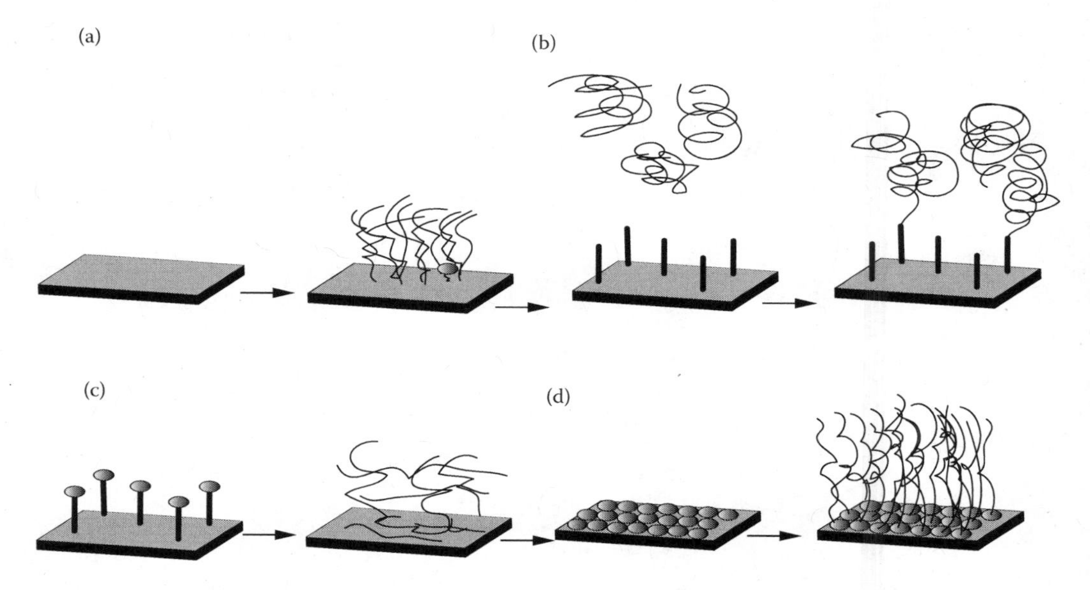

FIGURE 4.33 Schematic of surface modification methods using polymers: (a) EB irradiation, (b) grafting to, and (c) grafting from approaches (MBAAm, N,N-methylenebisacrylamide), and (d) surface-initiated living radical polymerization (atom transfer radical polymerization (ATRP)). (Adapted from Nagase et al., *J. R. Soc. Interface*, 6, S293–309, 2009.)

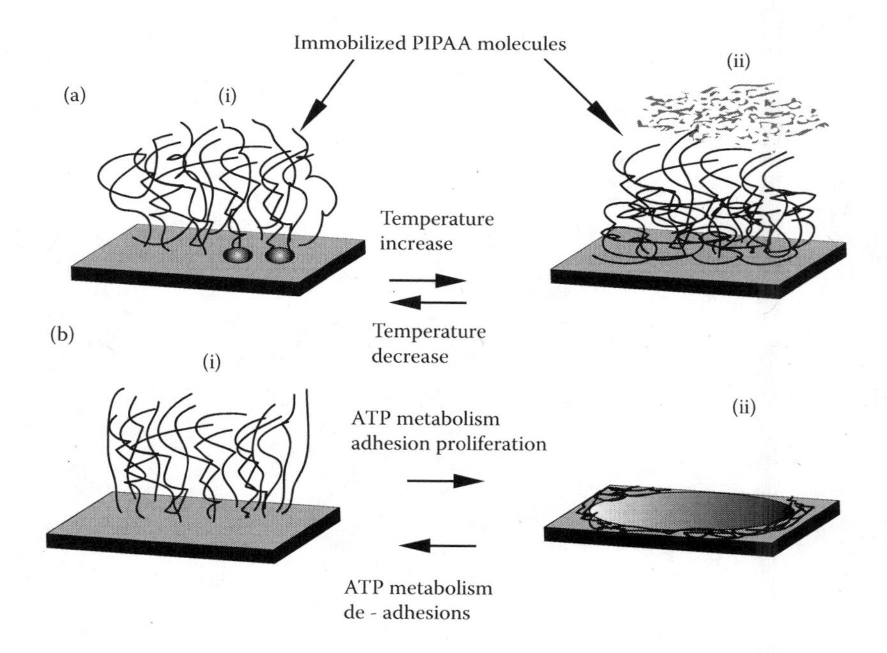

FIGURE 4.34 Temperature-responsive culture dish. (a) Temperature-dependent wettability changes for PIPAAm-grafted surfaces at 10 and 37°C: (i) hydrophilic surface and (ii) hydrophobic surface. (b) Schematic for the interactions of temperature-responsive surfaces with cells: (i) passive adhesion and (ii) active adhesion. (Adapted from Nagase et al., *J. R. Soc. Interface*, 6, S293–309, 2009.)

4.10 TEMPERATURE-RESPONSIVE INTELLIGENT INTERFACES

Some polymers, such as poly-N-isopropylacrylamide and its derivatives, exhibit a reversible temperature-dependent phase transition in solution. The phase transitions result in a change in the hydrophilic or hydrophobic properties of the polymer, with consequences upon the cell adhesion or detachment properties.

The advantage of these materials is the detaching of the adsorbed cells as the temperature is lowered because of the change in the hydrophilic properties of the polymer. The temperature-dependent hydrophilic or hydrophobic alteration of the polymer surface has found important application in chromatography and cell sheet culture substrates. Improved cell culture substrates have been prepared by using these novel intelligent surfaces. In the experiment shown in Figure 4.33, using EB irradiation, a monomer is polymerized and covalently immobilized onto a tissue culture polystyrene (TCPS) dish surface.

At 37°C, when the material is hydrophobic, cells adhere and proliferate on the material. The surface becomes hydrophilic as the temperature is lowered to 20°C and the detachment of cells is observed.

The concept is demonstrated in Figure 4.34, where the changes in the hydrophilic or hydrophobic properties (wettability) with the temperature are illustrated.

Cultured cells grown on the surface of a TCPS dish can be recovered by lowering the temperature. Because recovered cell sheets retain their structure and functions, transplantable tissues, composed exclusively of cells, can be engineered.

REFERENCES

1. Love, J.C., Estroff, L.A., Kriebel, J. K., Nuzzo, R.G., Whitesides, G.M. 2005. Self-assembled monolayers of thiolates on metals as a form of nanotechnology. *Chem. Rev.* 105: 1103-69.
2. Wang, X., Liu, L.-H., Ramström, O., Yan, M. 2009. Engineering nanomaterial surfaces for biomedical applications. *Exp. Biol. Med.* 234: 1128-39.
3. Kühn, G., Weidner, St., Decker, R., Ghode, A., Friedrich, J. 1999. Selective surface functionalization of polyolefins by plasma treatment followed by chemical reduction. *Surface Coating Technol.* 116-19: 796-801.
4. Makamba, H., Kim, J. H., Lim, K., Park, N., Hahn, J.H. 2003. Surface modification of poly (dimethylsiloxane) microchannels. *Electrophoresis* 24: 3607-19.
5. Klages, C.-P., Berger, C., Eichler, M., Thomas, M. 2007. Microplasma-based treatment of inner surfaces in microfluidic devices. *Contrib. Plasma Phys.* 47: 49-56.
6. Bodas, D., Khan-Malek, C. 2007. Hydrophilization and hydrophobic recovery of PDMS by oxygen plasma and chemical treatment – An SEM investigation. *Sensors Actuators B* 123: 368-73.
7. Dhayal, M., Jeong, H.G., Choi, J.S. 2005. Use of plasma polymerisation process for fabrication of bio-MEMS for micro-fluidic devices. *Appl. Surf. Sci.* 252: 1710-15.
8. Bouaidat, S., Winther-Jensen, B., Flygenring Christensen, S., Jonsmann, J. 2004. Plasma-polymerized coatings for bio-MEMS applications. *Sensors Actuators A* 110: 390-94.
9. Toh, A.G.G., Wang, Z.F., Ng, S.H. 2008. Fabrication of embedded microvalve on PMMA microfluidic devices through surface functionalization. Paper presented at DTIP of MEMS&MOEMS 2008, Nice.
10. Siow, K.S., Britcher, L., Kumar, S., Griesser, H.J. 2006. Plasma method for the generation of chemically reactive surfaces for biomolecule immobilization and cell colonization. A review. *Plasma Process. Polym.* 3: 392-418.
11. Ulman, A. 1996. Formation and structure of self-assembled monolayers. *Chem. Rev.* 96: 1533-54.
12. Nuzzo, R.G., Allara, D.L. 1983. Adsorption of bifunctional organic disulfides on gold surfaces. *J. Am. Chem. Soc.* 105: 4481- 83.
13. Prime, K.L., Whitesides, G.M. 1993. Adsorption of proteins onto surfaces contained end-attached oligo(ethylene oxide): a model system using self-assembled monolayers. *J. Am. Chem. Soc.* 115: 10714.
14. Chen, J., Tan, F., Ju, H. 2006. Gold nanoparticles doped three-dimensional sol-gel matrix for amperometric human chorionic gonadotrophin immunosensor. *Electroanalysis* 18: 1696-702.
15. Decher, G., Hong, J.D., Schmitt, J. 1992. Buildup of ultrathin multilayer films by a self-assembly process: III. Consecutively alternating adsorption of anionic and cationic polyelectrolytes on charged surfaces. *Thin Solid Films* 210: 831-35.

16. Tang, Z., Wang, Y., Podsiadlo, P., Kotov, N.A. Biomedical applications of layer-by-layer assembly: from biomimetics to tissue engineering. *Adv. Mater.* 18: 3203-24.
17. Davies, F., Higson, S.P.J. 2005. Structured thin films as functional components within biosensors. *Biosensors Bioelect.* 21: 1-20.
18. Koh, W.-G., Pishko, M. 2005. Immobilization of multi-enzyme microreactors inside microfluidic devices. *Sensors Actuators B* 106: 335-42.
19. Löfas, S., Johnsson, B. 1990. A novel hydrogel matrix on gold surfaces in surface Plasmon resonance sensors for fast and efficient covalent immobilization of ligands. *J. Chem. Soc. Chem. Commun.* 1526-28.
20. Lee, S., Pérez-Luna, V. 2005. Dextran-gold nanoparticle hybrid material for biomolecule immobilization and detection. *Anal. Chem.* 77: 7204-11.
21. Menard-Basquin, C., Kügler, R., Matzuzawa, N.N., Yasuda, A. 2005. Gold nanoparticle-assisted oligonucleotide immobilization for improved DNA detection. IEEE Proc. *Nanobiotechnol.* 152: 97-103.
22. Kim, H. M., Jin, S.M., Lee, S. K., Kim, M.-G. and Shin, Y.-B. 2009. Detection of biomolecular binding through enhancement of localized surface plasmon resonance (LSPR) by gold nanoparticles. *Sensors* 9: 2334-2344.
23. Fujiwara, K., Watarai, H., Itoh, H., Nakahama, E., Ogawa, N. 2006. Measurement of antibody binding to protein immobilized on gold nanoparticles by localized surface Plasmon spectroscopy. *Anal. Bioanal. Chem.* 386: 639-644.
24. Wolfram, T., Belz, F., Schoen, T. and Spatz, J.P. 2007. Site-specific presentation of single recombinant proteins in defined nanoarrays. *Biointerfaces* 2: 44-48.
25. Juergen Groll, J. Fiedler, E. Engelhard, T. Ameringer, S. Tugulu, H.-A. Klok, R.E. Brenner, M. Moeller. 2005. A novel star PEG-derived surface coating for specific cell adhesion , *J. Biomed. Mater. Res. Part A*, 74A: 607-617.
26. Yang, H., Fung, S.-Y., Pritzker, M. and Chen, P. 2007. Modification of Hydrophilic and hydrophobic surfaces using an ionic-complementary peptide. PLoS ONE December 2007, Issue 12, e1325.
27. Medintz, I. 2006. Universal tools for biomolecular attachment to surfaces, *Nature Materials* 5: 842. (Copyright: Nature Publishing Group).
28. Ducker, R.E., Montague, M.T., and Leggetta, G.J. 2008. A comparative investigation of methods for protein immobilization on self-assembled monolayers using glutaraldehyde, carbodiimide, and anhydride reagents. *Biointerphases* 3: 59-65.
29. Jiao, Y.-P. and Cui, F.-Z. 2007. Surface modification of polyester biomaterials for tissue engineering. *Biomed. Mater.* 2: R24-R37.
30. Mattanavee, W., Suwantong, O., Puthong, S., Bunaprasert, T., Hoven, V., and Supaphol, P. 2009. Immobilization of biomolecules on the surface of electrospun polycaprolactone fibrous scaffolds for tissue engineering, *Appl. Mater. Interfaces* 1: 1076-1085.
31. Golden, A.P. and Tien, J. 2007. Fabrication of microfluidic hydrogel using molded gelation as a sacrificial element. *Lab Chip* 7: 720-725.
32. Park, J. J., Luo, X., Yi, H., Valentine, T.M., Payne, G. F., Bentley, W.E., Ghodssi, R., Rubloff G.W. 2006. Chitosan-mediated in situ biomolecule assembly in completely packaged microfluidic devices. *Lab Chip* 6: 1313-1320.
33. Nagase, K., Kobayashi, J., and Okana, T. 2009. Temperature-responsive intelligent interfaces for biomolecular separation and cell sheet engineering. *J.R. Soc. Interface* 6: S293-S309.

REVIEW QUESTIONS

1. How can the biocompatibility of a device be improved?
2. Why is glass a hydrophilic substrate and how can it be made hydrophobic?
3. How can a surface be modified in order to resist protein adhesion?
4. What kind of surface would allow liquids to flow through the microchannels?
5. Explain the mechanism of plasma functionalization.
6. What are self-assembled monolayers (SAMs) and how can they be prepared?
7. How can hydroxylated surfaces such as glass be functionalized?
8. Explain how LB and LBL films are fabricated.
9. What are hydrogels and how are they used in BioMEMS applications?

5 Methods of Study and Characterization of Surface-Modified Substrates

Surface modification techniques were presented in Chapter 4. There exist many methods for surface modification, especially for biological and chemical applications. It is important to characterize these modified surfaces so that they become suitable for biological applications. In this chapter, different measures to characterize the surfaces and different experimental methods to evaluate these measures will be presented. Even though there exist a vast literature of methods for characterization in this area, this chapter will present only a few popular ones.

5.1 CONTACT ANGLE

5.1.1 Introduction to Contact Angle and Surface Science Principles

Contact angle describes the shape of a liquid droplet on a flat homogeneous solid surface when the liquid does not spread. It is one of the most sensitive surface analysis techniques, and the depth on the surface analyzed with this technique is between 3 and 20 Å. The shape of the droplet is determined by wettability of the surface, determined by the balance of the forces (γ_L, γ_{LS}, and γ_S) at the interfaces of the three phases, which are in contact as shown in Figure 5.1. It shows how strong the attraction of molecules within the droplet is, compared to attraction or repulsion toward the molecules of the underlying substrate (cohesion vs. adhesion).

In the case of a highly wettable surface, called a hydrophilic surface, the liquid is very strongly attracted to the solid surface. As a result, the droplet will completely spread out on the solid surface, and consequently, the contact angle will be less than 90°, usually in the range of 0 to 30°. If the solid surface is hydrophobic, the contact angle will be larger than 90°. The contact angle close to 0° indicates a highly hydrophilic surface, while the angle close to 180° indicates a highly hydrophobic surface. Thus, the contact angle becomes the important parameter to determine the wettability.

The relationship between the contact angle and the surface free energy is defined by Young's equation:

$$\gamma_s = \gamma_{SL} + \gamma_L \cos\theta$$

where γ_s is the solid surface free energy, γ_L is the liquid surface free energy, and γ_{sL} is the solid-liquid interfacial free energy. It is also possible to determine the cleanliness of a surface by measuring the contact angle. When a surface is contaminated, the contact angle is larger, due to the hydrophobicity, while for a clean surface, the wettability and adhesiveness improve and the contact angle decreases, as shown in Figure 5.2.

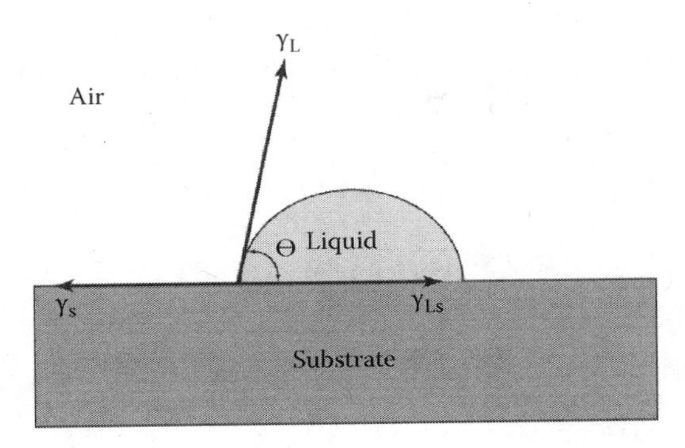

FIGURE 5.1 The contact angle θ is the angle between the tangent line to the droplet at the interface and the solid surface.

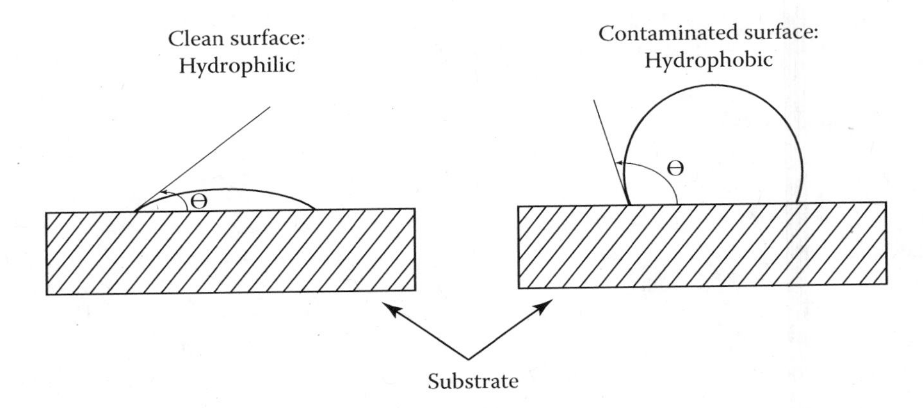

FIGURE 5.2 Influence of surface contamination on contact angle This method is currently used to determine wafer cleanliness in the semiconductor industry.

5.1.2 Contact Angle Measurement

There are several commonly used techniques that can be employed to measure the contact angle. An instrument generally called a *contact angle goniometer* (Figure 5.3) is used to measure the static, advancing, and receding contact angles, and also surface tension. With a goniometer, the contact angle is measured visually. The droplet is deposited by a syringe onto the sample surface and a high-resolution camera captures the image. This can later be analyzed, by either eye (with a protractor) or using image analysis software. By taking pictures as the droplet grows, a set of data can be acquired to obtain a good average. Multiple droplets can be deposited in various locations on the sample to determine heterogeneity.

In order to make the measurement more precise, modern contact angle systems (Figure 5.4) adopt precision optics and charge-couple device (CCD). There are four methods to measure contact angles: static or sessile drop method, Wilhemly plate method, captive air bubble method, and capillary rise method. The contact angle changes when the volume of the drop is changed. In the *static* or *sessile method*, fluid droplets are added successively until a plateau (advancing contact angle) is reached. The receding contact angle is obtained by retracting equivalent fluid volumes from the droplet (Figure 5.5). In the *captive air bubble method*, schematically shown in Figure 5.6, the solid surface is immersed in the fluid and the contact angle is measured between an air bubble of defined volume and the solid.

FIGURE 5.3 The first contact angle goniometer designed by Dr. Zisman of the NRL and manufactured by Ramé-Hart Instrument Co., New Jersey. (Courtesy of Ramé-Hart Instrument Co.)

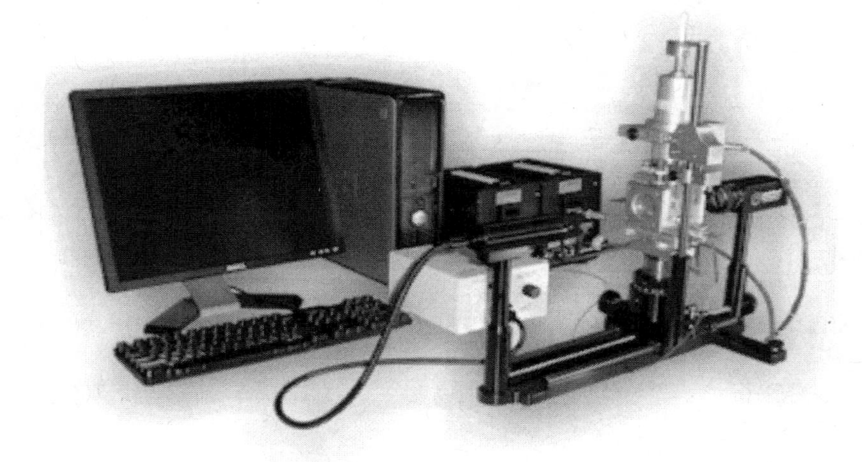

FIGURE 5.4 Modern goniometer (Model M200 Ramé-Hart Instrument Co.) for contact angle measurements. (Courtesy of Ramé-Hart Instrument Co.)

In the advancing contact angle mode, the volume of the drop increased, as shown in Figure 5.5a. The fluid volume is retracted in the receding contact angle mode, as shown in Figure 5.5b.

For double-sided samples, the best method for measuring the contact angle is the *Wilhemly plate method*, shown in Figure 5.7. This method allows temperature control of the wetting liquid.

The Wilhelmy plate is cleaned thoroughly, attached to a balance through a thin metal wire, and the force (F) on the plate due to wetting is measured using a tensiometer. The surface tension (σ) is calculated using the *Wilhelmy equation* as

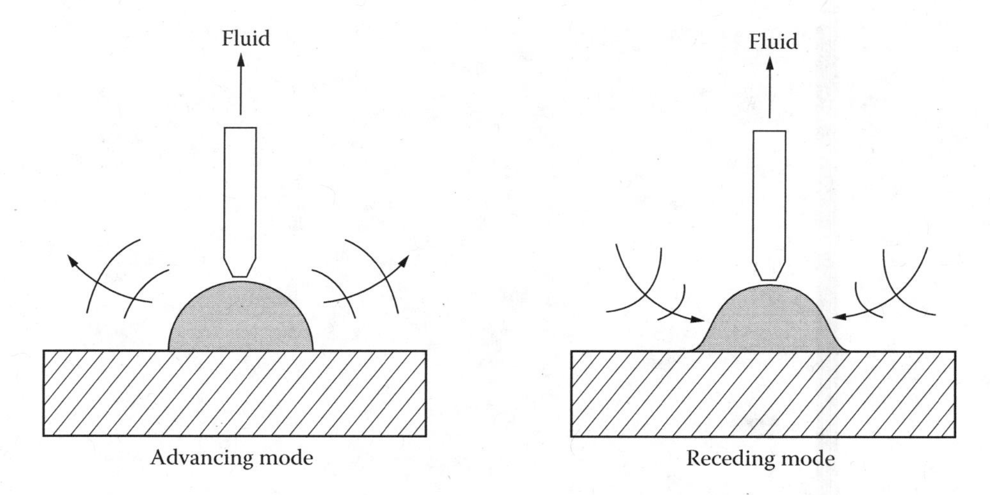

FIGURE 5.5 The contact angle changes when the droplet volume increases or when the droplet is retracted.

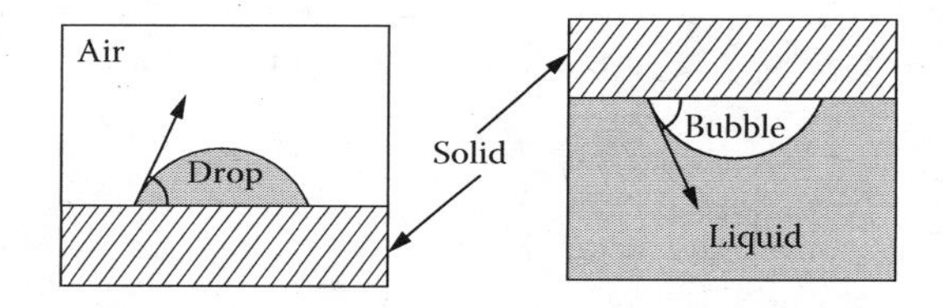

FIGURE 5.6 Static contact angle measured in a sessile drop or captive bubble configuration.

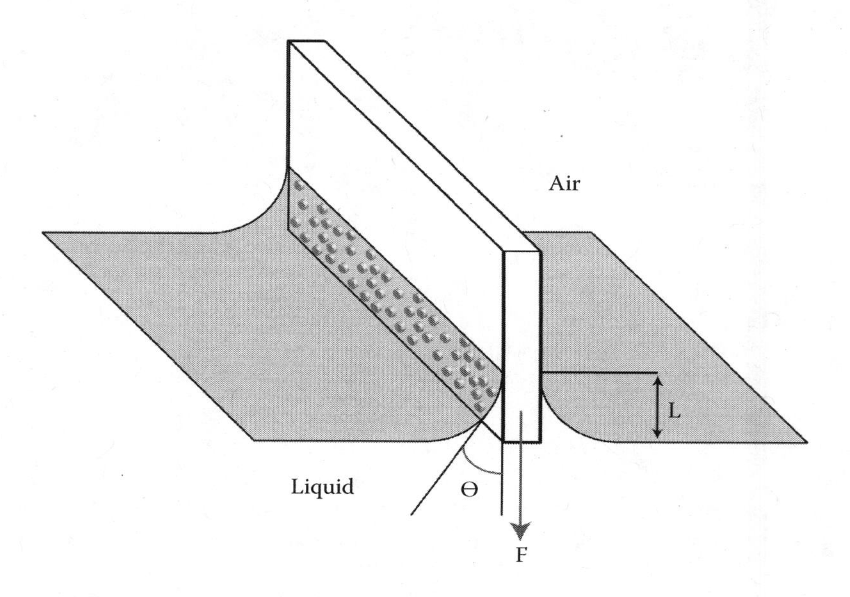

FIGURE 5.7 Contact angle measurements on double-sided samples.

$$\sigma = \frac{F}{2L.\cos\theta}$$

where L is the wetted length of the Wilhelmy plate and θ is the contact angle between the liquid phase and the plate. A systematic comparison of the different methods for contact angle measurement is reported in Krishnan et al.[1]

5.1.3 Evaluation of Hydrophobicity of the Modified Surfaces by Contact Angle Measurements: Case Studies

The adhesion between biomolecules, such as proteins and solid substrates, is important for biosensor applications because it determines the performance and reliability of sensing.

5.1.3.1 Sensitivity of Contact Angle to Surface Treatment

Various surface treatment methods have been used to modify silicon-based surfaces in order to improve the adsorption of a protein (for example, streptavidin) for sensing purposes.[2] Static contact angles were measured using a goniometer. In this study standard bare glass was the reference. But the contact angle was improved further by O_2 plasma treatment of the silicon surface (Figure 5.8).

The surface treatment experiment includes thermal oxidation, soaking in DI water, soaking in phosphate-buffered saline (PBS), silanization, and final coating with protein. The most hydrophobic surface of the highest contact angle corresponds to silica obtained by thermal oxidation at 1,100°C when the hydrophilic OH groups were driven off. When silica was heated in DI water, the silica surface was strongly hydroxylated and the corresponding contact angle decreased significantly. These results show clearly the sensitivity of the contact angle to small changes in the surface structure. Hence, one can modify the surface property by an appropriate treatment. The essence is to activate the OH groups on a surface.

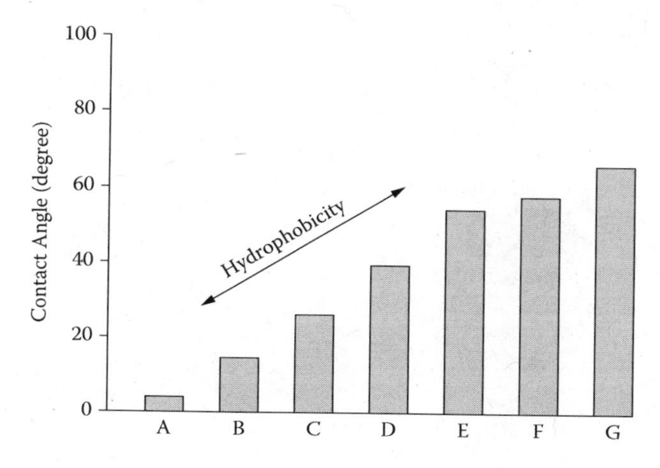

FIGURE 5.8 The water contact angles on various surfaces. Average values and standard deviations were obtained by several separate measurements of each substrate. Bare glass (C) was evaluated after the cleaning condition and O_2 plasma treatment (A). The 3-(trimethoxysilyl propylmethacrylate) (MPTES)–coated surface (E) was prepared by the silanization process after the treatment of O_2 plasma. The glycidylmethacrylate (GMA) surface (D) on the MPTES-treated surface was examined after the UV exposure. The various polyethyleneglycol (PEG) surfaces, including PEG thin film (B) and two kinds of PEG microstructures (F, G), were prepared on the GMA surface through the UV irradiation. (Adapted from Song et al., *Chem. Eng.*, 25, 1467–72, 2008.)

5.1.3.2 Contact Angle Measurements of Surfaces Functionalized with Polyethyleneglycol (PEG)

Polyethyleneglycol (PEG) and glycidylmethacrylate can be used to modify the contact angle and hydrophobicity for the selective immobilization of proteins and cells.[3]

The fabricated PEG microstructures could be easily manipulated, and they provide a biological barrier for the prevention of nonspecific binding of proteins and cells. In addition, the exposed glycidylmethacrylate region provided an efficient binding region of biomolecules with long-term stability of biomolecule patterns.

5.1.3.3 Study of Surface Wettability of Polypyrrole for Microfluidics Applications

Research in the field of BioMEMS and microfluidics has demonstrated that the topography of a surface may determine the wetting characteristics of a solid-liquid interface. The surface wettability of a conducting polymer (polypyrrole) changed when the topology of the surface was modified by applying low voltages. Contact angle measurements showed that with increasing voltages, the contact angle increases, making the surface less hydrophilic, as seen in Figure 5.9. The transition from hydrophilic to hydrophobic was found to occur at an applied potential of +1.0 V.[4]

5.1.3.4 Wetting Properties of an Open-Channel Microfluidic System

It is also possible to confine the liquid to a selective pathway by appropriately assigning the wettability of the surface. The fluid can be made to follow the hydrophilic surface, assisted by the hydrophobic surfaces as shown in Figure 5.10.

An open-channel surface-directed flow cell without sidewalls was fabricated. It consists of two planar parallel surfaces separated by spacers, as shown in Figure 5.10. In this figure, the liquid is confined to a hydrophilic pathway on the bottom surface, and it does not propagate into the surrounding hydrophobic areas. When the liquid moves into the gap by capillary forces, it moves along the pathways patterned with hydrophilic surfaces. The fluid is confined to the hydrophilic pathways by making the surrounding surfaces hydrophobic.

DI water is found to be more hydrophilic in the receding mode than in the advancing mode, as shown in Figure 5.11. The surface-directed flow system is fabricated by patterning a hydrophobic fluorinated polymer deposited by plasma polymerization. The areas of untreated glass substrate are hydrophilic pathways.

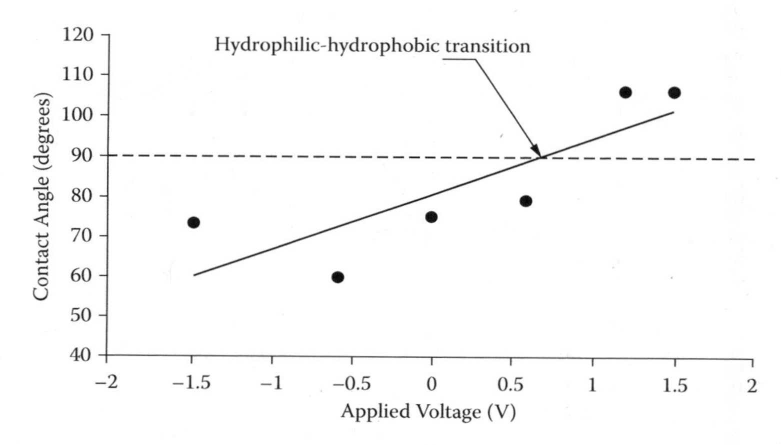

FIGURE 5.9 Effect of applied voltage on the contact angle of a polypyrrole film. (Adapted from The and Lu, "Topography and Wettability Control in Biocompatible Polymer for bioMEMS Applications," paper presented at Proceedings of the 3rd IEEE International Conference on Nano/Micro Engineered and Molecular Systems, Sanya, China, January 6–9, 2008.)

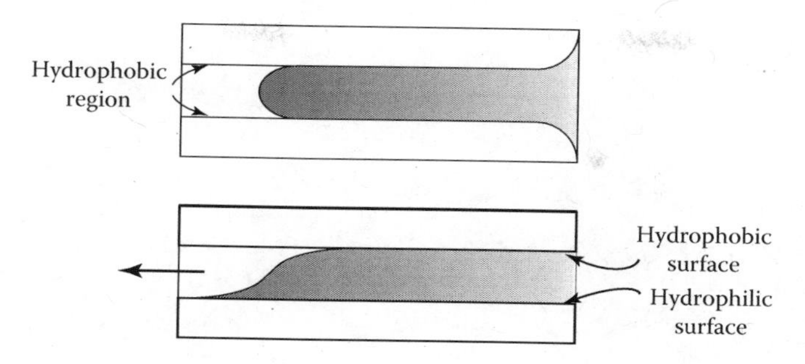

FIGURE 5.10 Capillarity-driven surface-directed flow system. (Adapted from Bouaidat et al., *Lab Chip*, 5, 827–36, 2005.)

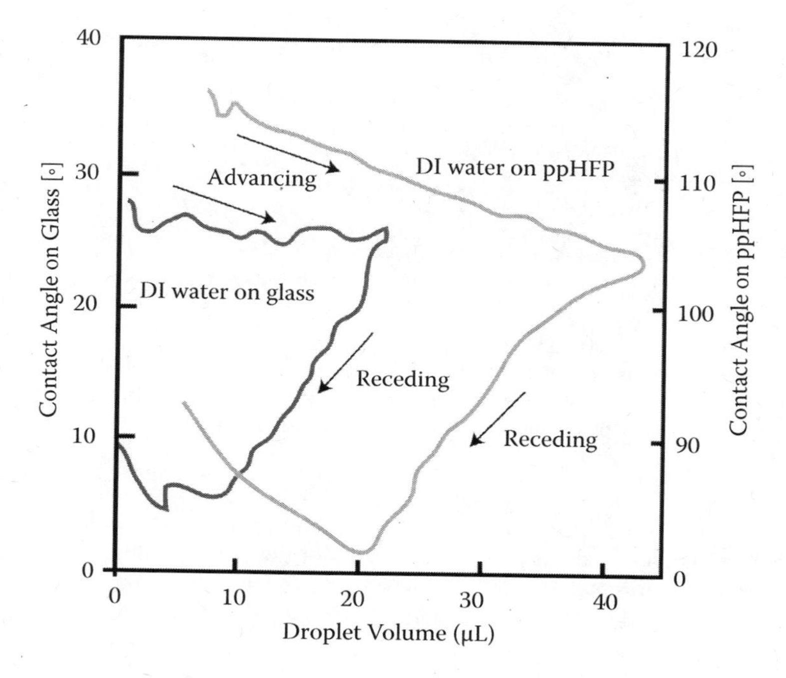

FIGURE 5.11 Contact angle of DI water as a function of droplet volume on hexafluoropropene (ppHFP) and borosilicate glass (BSG) in advancing and receding modes.

It is found that a fluorescent-labeled protein (fluorescein isothiocyanate–labeled bovine serum albumin) selectively adsorbs more onto the hydrophilic pathways on the bottom substrate and less onto the hydrophobic areas.

5.1.3.5 Contact Angle Analysis of the Interfacial Tension

Microcantilevers can also be used to investigate interaction of molecules with surfaces and binding of ligands to a variety of receptors, such as enzymes, antibodies, DNA, and membrane-bound proteins. It has been shown that the free energy released by the ligand-receptor binding reaction is proportional to the variation of solid-liquid interfacial tension as follows:

$$\Delta_r G^0 = -\frac{\Delta\gamma}{\Gamma_{LR}} - RT \ln K$$

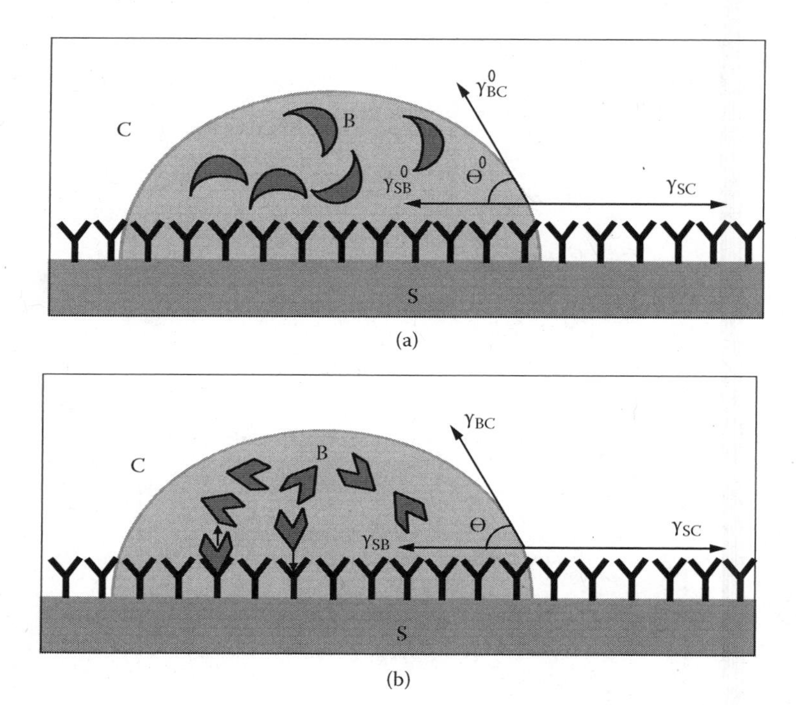

FIGURE 5.12 Schematic of sessile drop contact angle systems at equilibrium formed by a drop of ligand solution (phase B), a receptor-functionalized surface (phase S), and a surrounding phase (phase C); the contact angle is indicated by θ. (a) The unspecific case, where ligands and receptors do not bind. (b) The case where ligands and receptors bind. (Adapted from Bergese et al., *Langmuir*, 25, 4271–73, 2009.)

where $\Delta_r G^0$ is the change of the standard molar Gibbs free energy of the reaction in solution, $\Delta\gamma$ is the variation of the solid-liquid interfacial tension upon binding, Γ_{LR} is the surface concentration of the ligand-receptor complexes, and K is the surface equilibrium constant, that is, the surface binding affinity. The interfacial tension upon binding can be measured by contact angle analysis. According to this equation, a binding reaction gives a specific contribution to the solid-solution interfacial tension, as shown in Figure 5.12.

5.2 ATOMIC FORCE MICROSCOPY (AFM)

5.2.1 Basic Concepts of AFM and Instrumentation

The name of this method comes from the interactions (forces) between a sample and a probe during imaging or measuring these forces. AFM has proven to be a very important method for studying surfaces of biological material. Proteins and various biomaterials, including cells, can be imaged with high spatial nanometer resolution. Specific biointeractions between cell and cell, antigen and antibody, etc., can be investigated and quantified through force measurements at the nano-Newton scale. The force is experienced by the AFM tip.

The basic AFM setup is illustrated in Figure 5.13. The core piece of an instrument is a microcantilever, with a sharp tip at its end called a ***probe***. AFM microcantilevers are made from silicon or silicon nitride.

The forces between the tip and the sample surface are very small, usually in the order of less than 10^{-9} N. This interaction force between the tip and surface results in a deflection of the cantilever, which is elastic enough to undergo this deformation. The typical widths and lengths of the AFM cantilever are tens and hundreds of microns, while the thickness of the cantilever is around a few

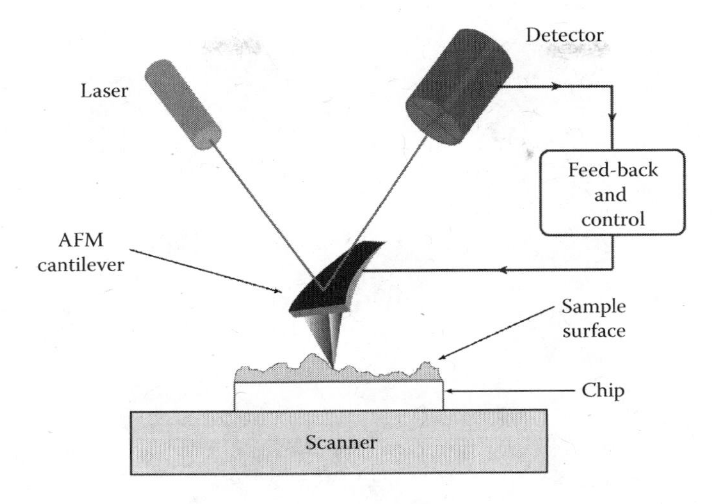

FIGURE 5.13 Schematic of atomic force microscope elements.

micrometers. One can vary the stiffness of the AFM probe by changing the geometry and dimensions of the cantilever. The cantilever can be rectangular or triangular. Depending upon the gap between the tip and surface, the probe could experience a negative or attractive force. As the tip is attracted to or repelled by the surface, the cantilever bends. This deflection is measured using an optical lever method in which a laser beam reflected from the surface of the cantilever is detected into a detector, as shown in Figure 5.13. A detector is an array of photodiodes sensitive to the position changes of light. Cantilever deflections of less than 0.1 nm can be detected using this method.

A feedback mechanism is employed in order to maintain a constant force between the tip and the sample by mounting the sample on a piezoelectric tube or a set of three piezoelectrical elements. A piezoelectric material elongates or contracts and produces force proportional to an applied voltage. The sample can be scanned in the x and y directions and moved in the z direction in order to compensate the cantilever deflection and maintain a constant force.

There are three modes of operation: contact mode, tapping mode, and jumping mode or force mapping mode. In the contact mode, the topography of the surface is mapped over an area as $S = f(x,y)$. A schematic of a modern AFM microscope for biological applications is shown in Figure 5.13.

Force-distance curves can be measured in the spectroscopy mode by attaching the tip of the cantilever to one end of the biomolecules to be analyzed, the other end being held by the substrate (Figure 5.14). When the tip is retracted, because of the force applied to the molecule, a bond or several bonds could be broken. The magnitude of the force applied to the molecule is given as $F = k.\delta$, where k is the spring constant of the cantilever and δ is the deflection of the cantilever. Forces in the pico-Newton regime can be measured by using special cantilevers, as will be seen in the case studies later.

Compared to the scanning electron microscope (SEM), AFM provides a three-dimensional surface profile. It does not need a metal coating, and it can be used to view biological samples because it works well in both ambient air (without vacuum) and a liquid environment. The only disadvantage is the size of the image, determined by the scanning window. Typically, an AFM can scan a maximum of 150 by 150 μm, while SEM can image considerably larger areas (mm by mm).

5.2.2 AFM Imaging of Biological Sample Surfaces

Many imaging studies in life sciences are performed in AFM using the *intermittent contact mode* (*tapping mode*), which has significant advantages over the contact mode. As the tip contacts intermittently the surface of the sample during scanning, the probability of damaging the sample or tip is

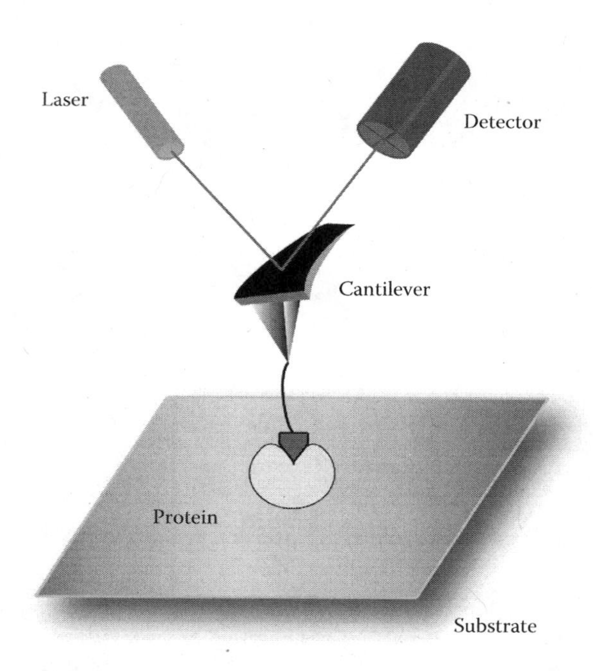

FIGURE 5.14 Setup of an AFM force spectroscopy experiment for quantifying biointeractions. (Adapted from Guthold et al., *Biomed. Microdevices*, 3, 9–18, 2001.)

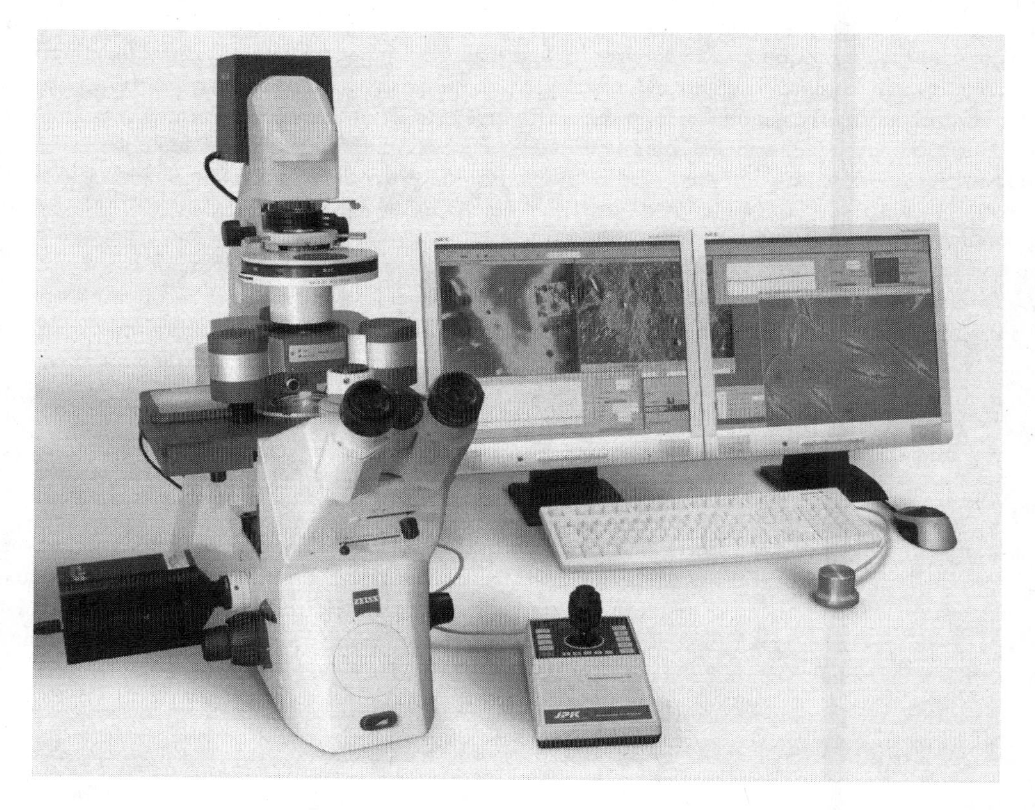

FIGURE 5.15 NanoWizard® II AFM setup mounted on a Zeiss Axiovert 200. (Courtesy of JPK Instruments AG, www.jpk.com.)

reduced. Unlike optical microscopy, AFM can directly visualize unstained biomolecules in physiological solutions at nanometer resolution (Figure 5.15).

AFM is a powerful tool for imaging macromolecules and various structures on a surface in solution. However, atomic force microscopes are limited by the speed at which they can successively record highly resolved images and require several minutes to capture images. A variable-controlled high-speed scanning method, as shown in Figure 5.16, was devised.[8] With this method nondistorted real images in liquid can be obtained.

Biomolecules must be kept and measured in a liquid environment under conditions that are close to those of the physiological environment. Under a physiological buffer solution, they retain their native structure and function. For this purpose, liquid cells offer a simple alternative for scanning a sample that is submerged in a liquid medium where both composition and temperature can mimic the natural environment.

The light path in a liquid cell is illustrated in Figure 5.17.

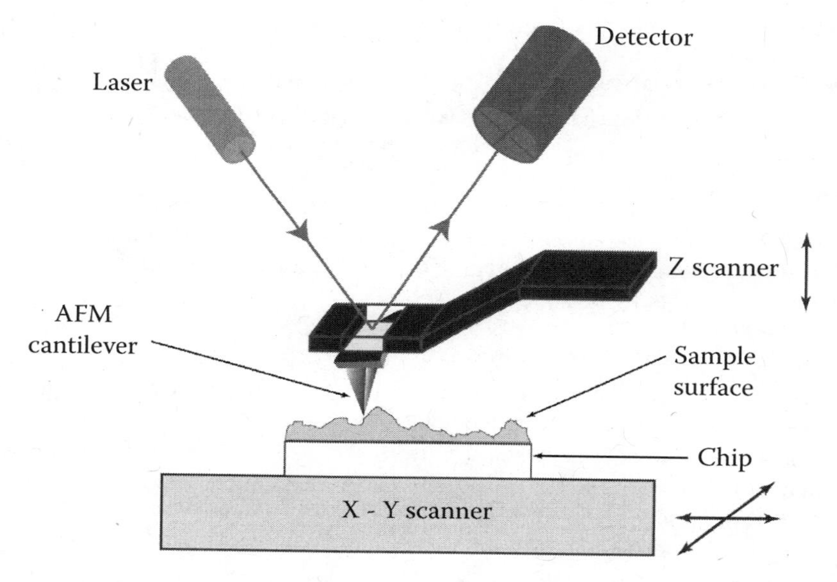

FIGURE 5.16 Schematics of a cantilever moving during measurement. (Adapted from Choi et al., *Nanotechnology*, 19, 445701, 2008.)

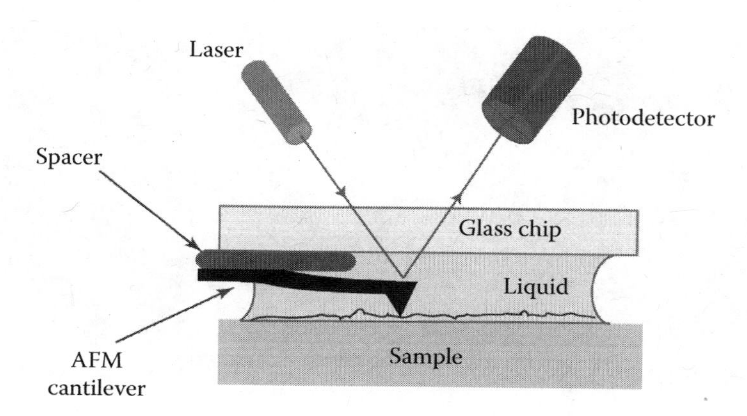

FIGURE 5.17 Light path in the liquid cell on Nano-R™ AFM microscope (Agilent, former Pacific nanotechnology).

The atomic force microscope (AFM) is now widely used for the investigation of structure and mechanical properties of biological materials. From these studies, information has been obtained for the development of biosensors. Sample preparation is a crucial step for successful biological AFM. In some cases, good results have been obtained using physical adsorption or chemical fixation in supports such as mica. Some case studies on biomolecules and biological entities are presented below.

5.2.2.1 *Ex Situ* and *In Situ* AFM Characterization of Phospholipid Layers Formed by Solution Spreading (Casting) on a Mica Substrate

The results of this study show how important it is to take the AFM measurements of phospholipids under conditions that are close to those of their physiological environment.[9] The surface structure of the casted DPPC layers (phospholipid), as shown in Figure 5.17a, appears to be relatively nonuniform, containing many defects. The surface evolved after hydration for 30 min from small island arrangements to a lamellar structure with expanded terraces. After immersion in HEPES buffer solution at room temperature, the typical DPPC multilayer structure is formed, as shown in Figure 5.18c.

5.2.2.2 Study of Bacterial Surfaces in Aqueous Solution

For these studies,[9] large entities such as bacteria have to be trapped in the pores of a polymer membrane having a pore size comparable to the cell size, as shown in Figure 5.19; otherwise, they would be detached by the AFM tip during scanning. The cells are sucked into the pores of the membrane and imaged by AFM. By using this method, single bacteria, and fungal spore cells, the growth and division of *Staphylococcus aureus* can be monitored by AFM. Structural changes have been revealed directly in growth medium by imaging the surface architecture of cells.

5.2.2.3 AFM Study of Native Polysomes of *Saccharomyces* in a Physiological Buffer Solution

Polysomes (or polyribosomes) are complexes of mRNA and ribosomes on which polypeptides are synthesized by transcription and translation.[10] For the visualization of polysomal fractions, mica is used as a substrate because when it is freshly cleaved, it has an atomically flat surface. In order to anchor biomolecules to the mica surface, divalent cations (Ni^{2+} as $NiCl_2$) are used. To study the native three-dimensional form, the translation was temporarily stopped. This substrate is submerged under the buffer solution, and the imaging conditions have to be optimized before the measurement.

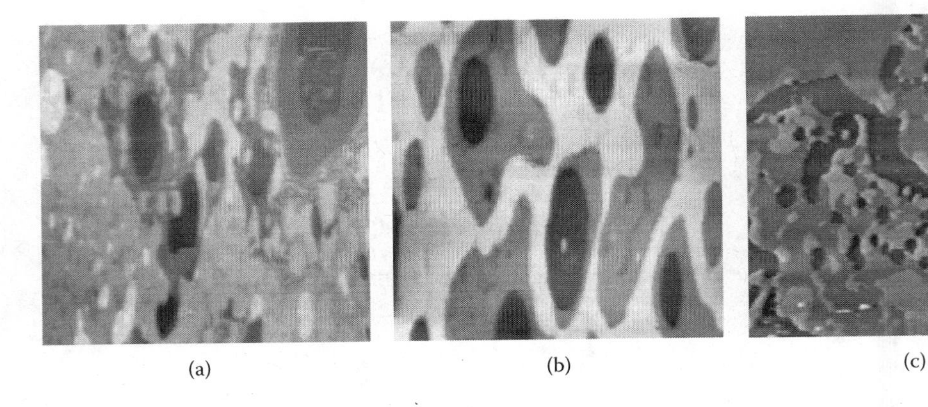

FIGURE 5.18 AFM images of the phospholipid DPPC after hydration levels: DPPC (a) cast on mica, (b) hydrated for 30 min, and (c) immersed in a buffer solution. (Reproduced from Spangenberg et al., *Phys. Stat. Sol. A*, 201(5), 857–60, 2004. Copyright Wiley-VCH Verlag GmbH & Co. KGaA. With permission.)

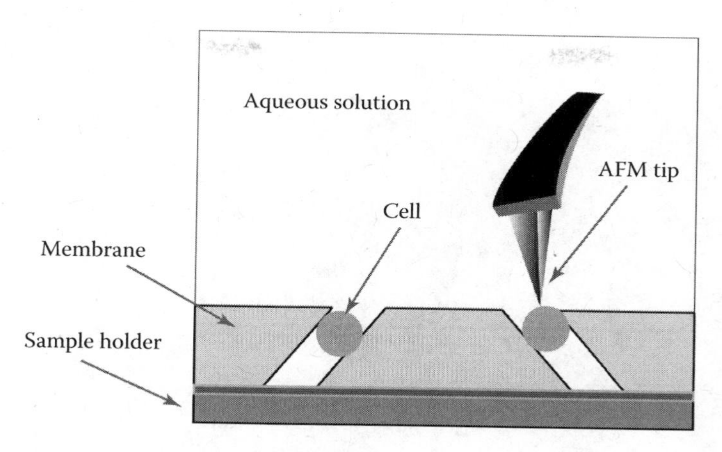

FIGURE 5.19 AFM image of an individual cell under physiological conditions. A cell suspension is sucked through pores in a polycarbonate membrane and imaged by an AFM tip. (Adapted from Dufrêne, *J. Bacteriol.*, 186(11), 3283–85, 2004.)

The AFM images seen in Figure 5.20 show the presence of ribosomes and subunits of ribosomes with ring shapes and linear structure. mRNA chains are visible between adjacent ribosomes in the close-up images in Figure 5.20b. The heights of different particles were determined from the AFM images.

5.2.2.4 Single DNA Molecule Stretching Experiments by Using Chemical Force Microscopy

The aim of these experiments[12] is to measure the elastic properties and to study the force-induced melting transition in single DNA molecules. In these experiments, DNA molecules are attached to the substrate surface at one end and to the cantilever tip at the other end, as shown in Figure 5.21.

When the surface is pulled away from the cantilever, its deflection is used to determine the force required to stretch the DNA molecule. The behavior of DNA under high forces during melting or breaking of base pairs has also been studied. It has been shown that at high forces of around 65 pN, dsDNA is stretched beyond its B-form length, resulting in an overstretching equilibrium transition corresponding to a force-induced melting process. At even higher forces, a new transition corresponding to the separation of strands occurs.

5.2.2.5 AFM Measurements of Competitive Binding Interactions between an Enzyme and Two Ligands

Competitive binding can be studied by AFM by measuring adhesion forces.[13] In this case, the enzyme called shikimate kinase is immobilized on a gold-coated tip, and one ligand (adenine, a mimic of the head group of ATP) is immobilized on a substrate, and a second one is free in the solution. ATP is dissolved in the buffer solution.

The adhesion force between the enzyme and the ATP mimic (adenine on a SAM) is measured for different concentrations of the added substrate ATP and also for different concentrations of enzymes, as shown in Figure 5.22. Experiments with competitive binding of other ligands in solution show that the observed adhesion forces arise predominantly from specific interactions between the immobilized enzyme and surface-bound adenine derivative.

The results showed that the adhesion force is reduced in the presence of ATP, meaning that the enzyme-adesine interaction is a combination of specific binding interactions and hydrophobic interactions. Other classes of enzymes could be tested against potential inhibitors in solution. Enzyme-ligand interactions are very important for the discovery of therapeutically useful enzyme inhibitors.

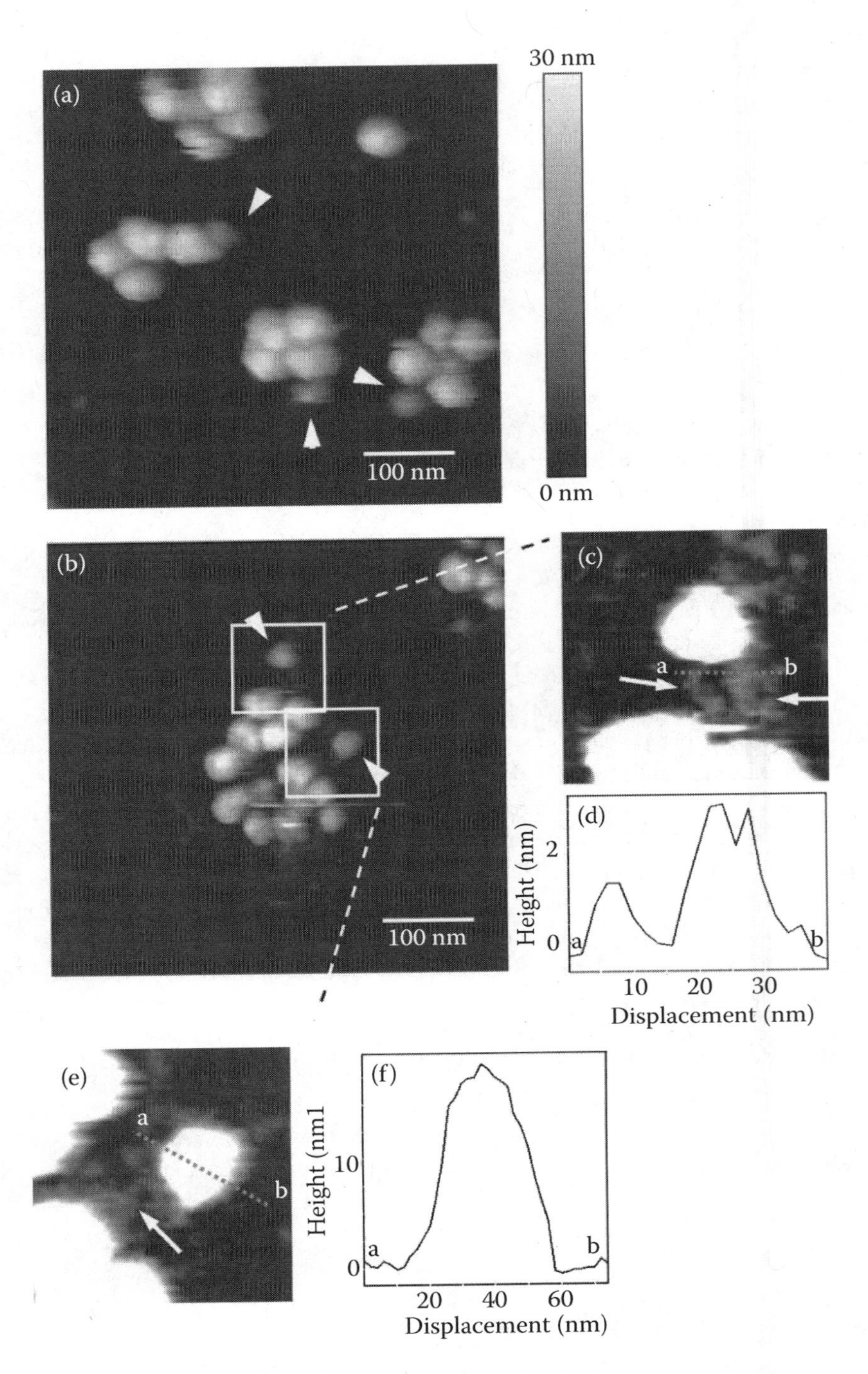

FIGURE 5.20 AFM images of polysomal fractions measured in liquid solution. (a) Images of polysomal fraction 5. (b) Images of fraction 6. (c) Close-up images of (b). (e) Cross-section of mRNA in (c). (f) Cross-section of small subparticles in (e). (Reprinted from Mikamo et al., *J. Struct. Biol.*, 15, 106–10, 2005. Copyright 2005. With permission from Elsevier.)

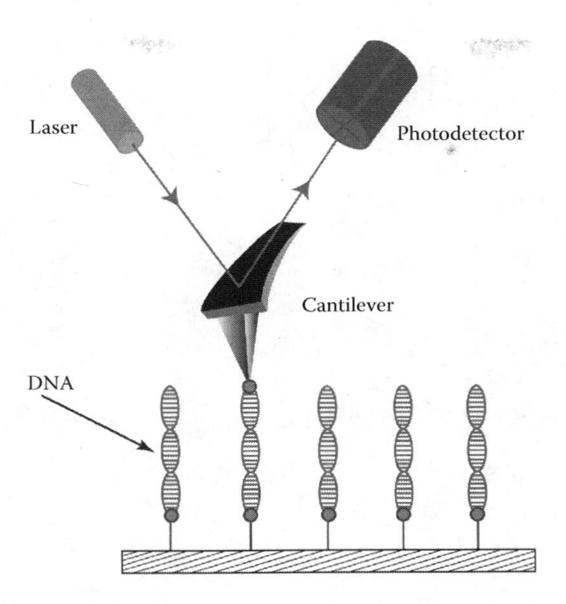

FIGURE 5.21 AFM measurement of force required to stretch the DNA molecules using AFM. (Adapted from Mikamo et al., *J. Struct. Biol.*, 15, 106–10, 2005.)

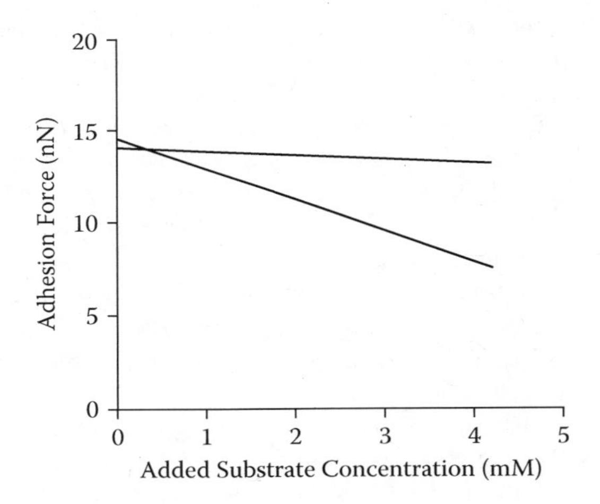

FIGURE 5.22 A graph showing the dependence of the adhesion force between an AFM tip derivatized with shikimate kinase and a self-assembled monolayer of an ATP mimic at various concentrations of added substrate ATP (lower curve) and shikimate (upper curve) in 100 mM ethanolamine buffer. (Reprinted from Fiorini et al., *Biophys. J.*, 80, 2471–76, 2001. Copyright 2005. With permission from Elsevier.)

5.2.2.6 Study of Antigen-Antibody Interactions by Molecular Recognition Force Microscopy (MRFM)

In this experiment,[13] the antibodies are attached to the tip of the cantilever and the antigen molecules (lysozyme) are immobilized on a mica surface in a liquid environment (Figure 5.23).

When a low concentration of lysozyme is added to the liquid system (until the black arrow above c), the antibody is bound only to the surface-bound lysozyme. At higher concentrations, the antibody will be bound to lysozyme in the solution.

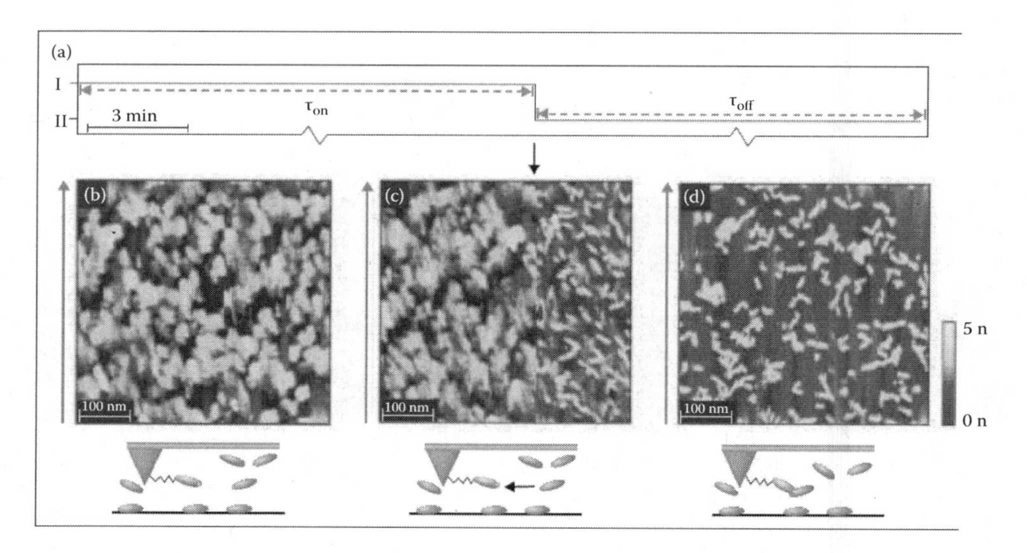

FIGURE 5.23 (See color insert.) Force recognition imaging of antibody-antigen interaction. (Reprinted from Allison et al., *Curr. Opin. Biotechnol.*, 13, 47–51, 2002; original figure in Raab et al., *Nat. Biotechnol.*, 17, 901–5, 1999. Copyright 2005, with permission from Elsevier.)

5.2.2.7 Study of Cancer Alterations of Single Living Cells by AFM

Cancer transformation introduces alterations in cell growth, morphology, and cell membrane. These differences may also induce changes in mechanical properties of the cell, leading to adhesive interactions that can be related to the higher invasive capacity of cells.

AFM can be used for the determination of cell stiffness and also for measuring the force between a ligand and a receptor. Cell stiffness is an important parameter describing the overall changes in the cell structure. The cell deformation (i.e., cell stiffness) is usually quantified by the Young's modulus, while the unbinding force determines the adhesive properties of a cell by characterization of unbinding forces between a pair of molecules.

The determination of cell elastic properties is determined on the basis of the *force curves*, representing the dependence between the cantilever deflection (converted into a loading force) and the relative sample position (displacement), as shown schematically in Figure 5.24a. In the case of measuring soft elements like cells, cantilever deflection is much smaller than for rigid samples, and the resulting force curve has a nonlinear character, as shown in Figure 5.24a.

The Young's modulus distribution as shown in Figure 5.24b is formed from values determined for all force curves recorded for a single cell. The calculated Young's modulus reflects only the elastic properties that originate from the layer, defined by the indentation depth.

The results suggest that the cell stiffness can be used as a marker and a diagnostic parameter for the disease.

5.3 X-RAY PHOTOELECTRON SPECTROSCOPY

5.3.1 Introduction

X-ray photoelectron spectroscopy (XPS), also called electron spectroscopy for chemical analysis (ESCA), is a powerful surface spectroscopic technique based on the interaction of monochromatic x-rays with the atoms in a specimen. Monochromatic x-rays used as a source are aluminum ($K\alpha$ = 1,486 eV) and magnesium ($K\alpha$ = 1,253 eV). The electron receives enough energy to leave the atom and escape from the surface of the sample with certain kinetic energy, as shown in Figure 5.25.

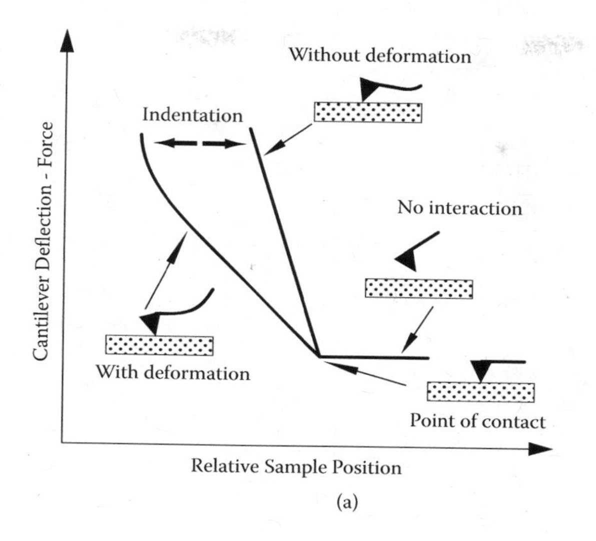

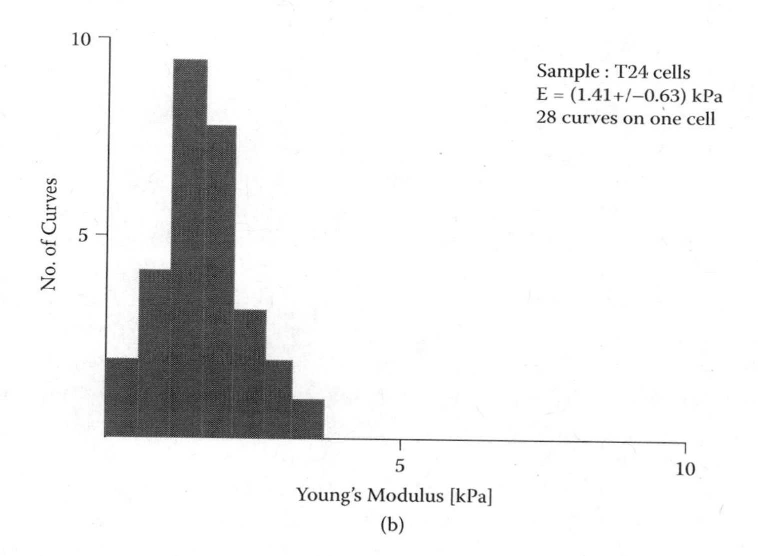

FIGURE 5.24 Alterations of single living cells by AFM. (a) Idea of cell stiffness determination. (b) The Gaussian distribution fitted to a histogram of Young's modulus determined for a single cell (T24 cells). (Reproduced from Lekka and Wiltowska-Zuber, *J. Physics Conference Ser.*, 146, 012023, 2009. With permission.)

The energy of this photoelectron provides information on the atom and its environment. As each chemical element has a unique binding energy, XPS recognizes and quantifies the chemical elements in the surface region of a solid and, at the same time, gives information on their binding states, oxidation numbers, and functionalities. The emitted electrons are analyzed, counted, and their binding energy in the atom of origin is determined. Inelastic interactions on the surface will contribute to the scattering background. The energy of the photoelectrons is related to the atomic and molecular environment from which they originated, while the number of electrons emitted is related to the concentration of the emitting atoms in the sample. In the recorded spectrum, each peak is characteristic of a given electron energy level of a given element, and its position is influenced by the chemical environment. Therefore, XPS provides an elemental analysis of the surface because the electrons interact

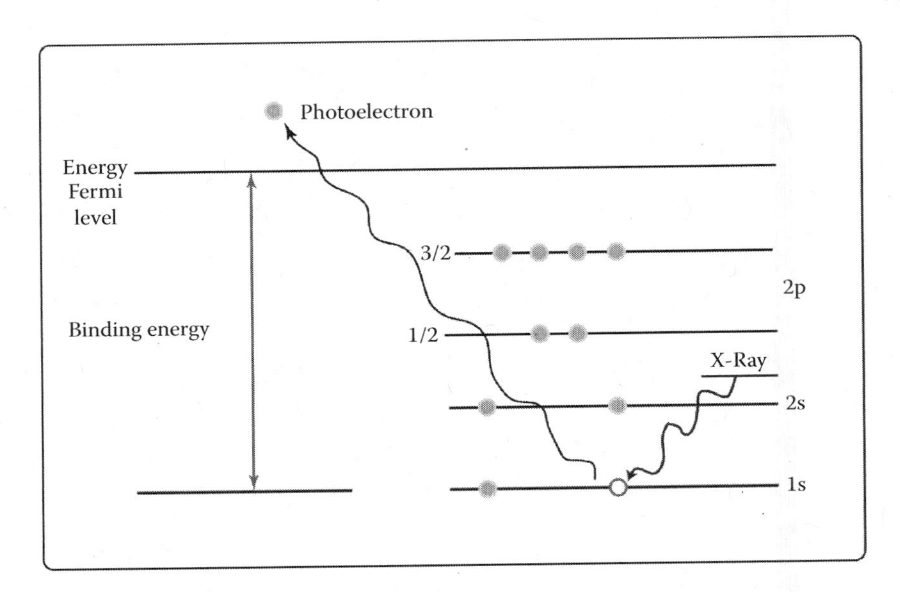

FIGURE 5.25 Interaction of x-rays with a sample.

strongly with materials, and only those emitted from the top few atomic layers can escape from the surface without energy loss. XPS is measured, to determine not only the elemental composition of the surface, but also its uniformity (the sampling volume extends from the surface to a depth of approximately 5 to 7 nm). XPS can be used for the analysis of all atoms of the periodic table. It is a highly sensitive method, and it can detect 0.1 atom% or less of a chemical element, except for hydrogen and helium.

Among the advantages of XPS are the high information content and the flexibility in addressing a wide variety of samples. Any sample that is vacuum stable can be analyzed by XPS. However, when the x-ray exposure is too long, certain materials, such as polymers and biological samples, may undergo some chemical changes due to the secondary electrons in the XPS source. X-rays will penetrate deeply into a sample, and stimulate electron emission throughout the specimen.

An XPS instrument has two main components: an x-ray source, preferably monochromatic, and an electron energy analyzer, usually a spherical sector analyzer, as shown in Figure 5.26. The measurements must be made in ultra-high vacuum to allow the photoelectrons to travel from the surface of the sample to the detector without striking a gas atom. In addition, if a clean surface is prepared for analysis, it would become contaminated if it were not under vacuum.

A schematic diagram of one such instrument is shown in Figure 5.26. Photoelectrons from the sample, which is above a magnetic immersion lens, are passed through a set of electrostatic lenses, which transfer an image of the analysis area on the sample to the entrance slit of the electron analyzer. The concentric hemispherical analyzer is the heart of the XSP instrument. The energy-filtered electrons are amplified and counted by a position-sensitive detector. Since the technique is surface sensitive, the whole instrument is kept under ultra-high vacuum.

Depth profiling of the near-surface region can be done using angle-resolved x-ray photoelectron spectra (Figure 5.27).

5.3.2 X-Ray Photoelectron Spectroscopy of Biologically Important Materials

Standard XPS has been used successfully to study a number of modified and unmodified inorganic materials and biomolecules involved in sensing experiments. Although this technique has the potential to provide valuable information, the use of x-ray photoelectron spectroscopy

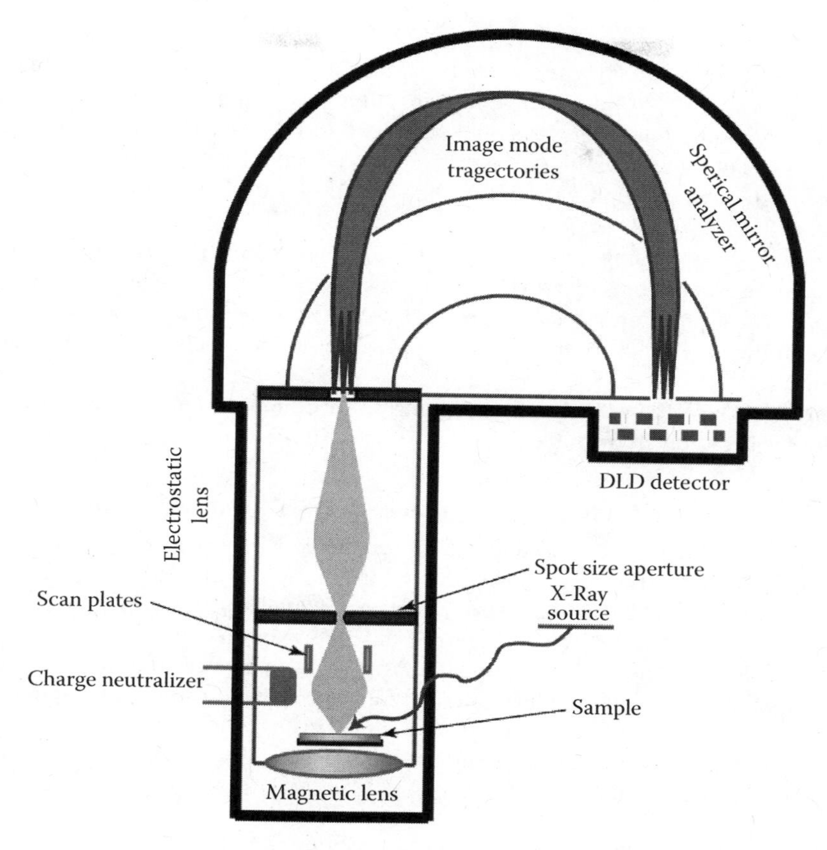

FIGURE 5.26 Schematic diagram of a Kratos Axis Ultra XPS instrument operating in imaging mode. DLD, delay line detector. (Courtesy of Kratos Analytical.)

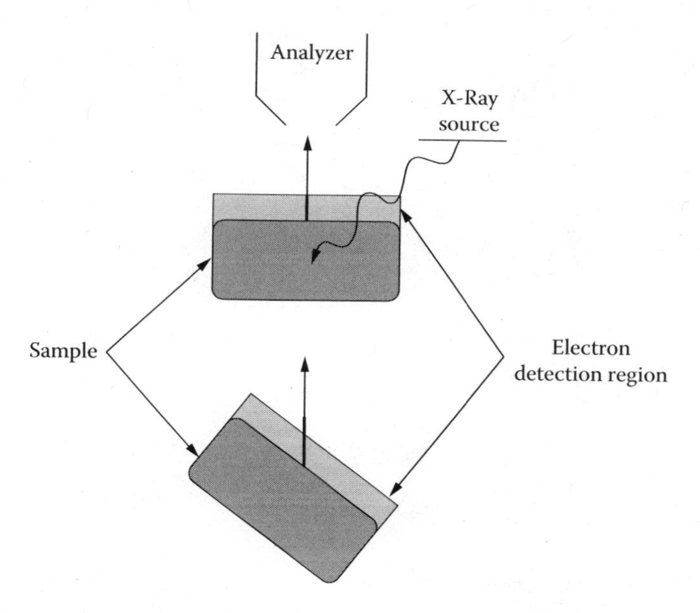

FIGURE 5.27 Depth profiling by rotating the sample.

in the characterization of biological materials has faced some limitations. One important challenge in using the XPS technique for biologically important molecules is the handling of the hydrated samples. Sometimes, the wet sample is freeze-dried and a 1 to 2 nm thick ice overlayer is left on the top. Subsequently, the sample is transferred quickly to a chilled temperature stage in the analysis chamber at high vacuum at approximately –100°C, and the ice is removed by sublimation.

However, deep freeze-drying may damage cell walls and cause structural changes, and for this reason, in some cases, high-pressure (in the mbar range) XPS is used. A high-intensity x-ray source is necessary, and a synchrotron radiation is used. However, there are many experimental difficulties associated with the application of this technique because of the effects of ultra-high vacuum and x-ray exposure.

The strength of the XPS technique in providing surface information on biomolecules and biological entities is showed here with the help of three case studies.

5.3.2.1 Peptide Nucleic Acids on Gold Surfaces as DNA Affinity Biosensors

Peptide nucleic acid (PNA) is a well-studied DNA analogue where the sugar-phosphate backbone is replaced with a peptide-like polyamide structure. Unlike DNA, peptide nucleic acids form self-assembled monolayers (SAMs) and bind strongly to complementary DNA, making them good candidates for biosensing applications. The results of the XPS experiments are shown in Figure 5.28. The increase of the normalized N1s peak from 2.6 to 3.2 times after the hybridization event is shown in Figure 5.28. The same result is obtained for the P2p that originates from the DNA phosphate group.

The results of this study encourage the use of PNA-based biosensors for detection of nucleic acids in complex biological samples.

5.3.2.2 Application of XPS to Probing Enzyme-Polymer Interactions at Biosensor Interfaces

In this study, electropolymerized conducting polypyrrole and nonconducting poly-o-phenylenediamine polymers are used to entrap glucose oxidase for fabricating enzyme electrodes for a glucose biosensor.[16] The relative proportions of enzyme and polymer in the surface layer are estimated using the

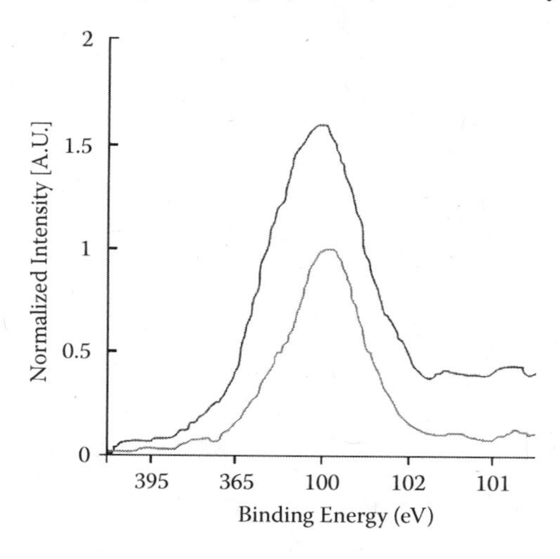

FIGURE 5.28 XPS experiments showing the interaction of immobilized ssPNA with complementary ssDNA. High-resolution XPS spectrum for N1s core level normalized to Au4f corresponding to the immobilized ssPNA. (Reproduced from Briones et al., *J. Mol. Catal. A Chem.*, 228, 131–36, 2005. With permission.)

enzyme-polymer interactions measured by XPS. The characteristic amidic peak in the C1s spectrum of the enzyme molecule at 288 eV (carbonyl) and 286 eV (N-C-CO-) was used to estimate the surface composition of the polypyrrole-enzyme composite by deconvolution. The results demonstrated high sensitivity of this method for determining the amount of the enzyme at a biosensor interface.

5.3.2.3 Detection of Adsorbed Protein Films at Interfaces

Proteins contain carbon, oxygen, and nitrogen and only small amounts of sulfur, phosphorus, and metals. Their detection using XPS is based on surface nitrogen detection, and the detection limit may be as low as 10 ng/cm^2. The identification of proteins adsorbed from mixture also involves using other methods, such as radiolabeling in conjunction with XPS.[17,18]

5.4 CONFOCAL FLUORESCENCE MICROSCOPY

5.4.1 INTRODUCTION

A *fluorescence microscope* is used to study the properties of organic or inorganic substances using the phenomena of *fluorescence* and *phosphorescence*. A component of interest in the specimen is specifically labeled with a fluorescent molecule called a *fluorophore*, for example, fluoresceine. The specimen is illuminated with light of a specific wavelength, which is absorbed by the fluorophores, causing them to emit longer wavelengths of light from their electronically excited states. The magnified image in a fluorescence microscope is produced by the longer-wavelength light emitted by the fluorescent species. The difference between the fluorescence emission and absorption (excitation) wavelengths is called the *Stokes shift*, and if it is sufficiently large, the exciting and fluorescence signals can be separated by filters so that only the fluorescence light will reach the detector. Typical components of a fluorescence microscope are the light source (xenon arc lamp or mercury vapor lamp), the excitation filter, the dichromatic mirror or beam splitter, and the emission filter. The filters and mirror are chosen to match the spectral excitation and emission characteristics of the selected fluorophores. In this manner, a single fluorophore of specific color is imaged at a time. In a conventional wide-field microscope, the entire specimen is bathed in light from a mercury or xenon source, and the image can be viewed directly by eye or projected onto an image capture device or photographic film. In a *confocal microscope*, shown in Figure 5.29, the image is formed differently. Illumination is achieved by scanning across the specimen with one or more focused beams of light, from a laser or arc discharge source. This point of illumination is brought to focus in the specimen by the objective lens, and laterally scanned using a scanning device with computer control. The sequences of points of light from the specimen tagged with fluorophores are detected by a low noise detector, such as avalanche photodiodes, through a pinhole or a slit, and the output is built into an image and displayed by the computer. The most important feature of a confocal microscope is the capability of isolating and collecting a plane of focus from within a sample, thus eliminating the out-of-focus haze normally seen with a fluorescent sample.

Confocal optics improved dramatically the spatial resolution in the vertical direction. It is able to produce sharp images of thick samples at various depths by taking images point by point and reconstructing them with a computer. This technique makes it possible to visualize features in living cells and tissues and even single molecules.

Biological laser scanning confocal microscopy relies heavily on fluorescence as an imaging mode, due to the high degree of sensitivity afforded by the technique and the ability to specifically target structural components in living cells and tissues. Using the confocal detection principle, femtoliter-sized sample volumes can be analyzed. Many fluorescent probes use synthetic aromatic organic compounds designed to bind with biological macromolecules, for example, a protein or nucleic acid. Fluorescent dyes are also useful in monitoring cellular integrity, membrane fluidity, protein trafficking, and enzymatic activity. In addition, fluorescent probes have been widely applied for imaging the genetic material within a cell in genetic mapping and chromosome analysis in the

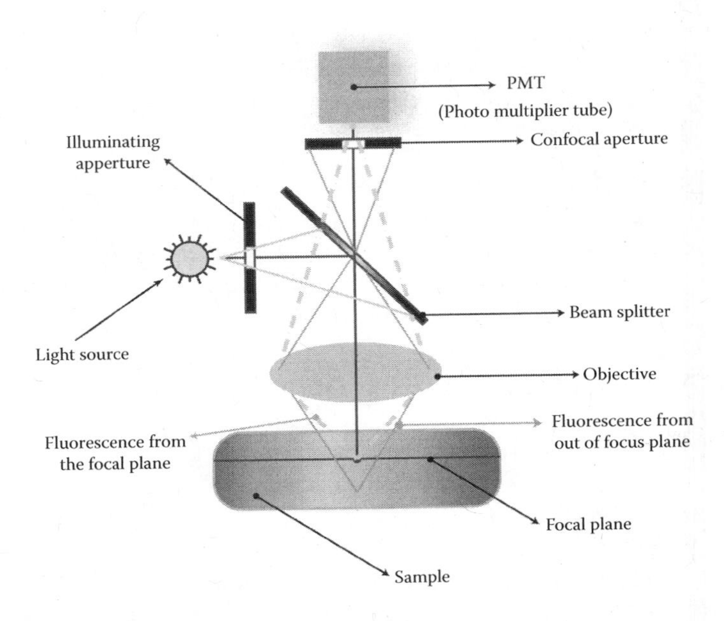

FIGURE 5.29 The components of a confocal microscope.

field of molecular genetics. The broad range of applications available to laser scanning confocal microscopy includes studies in neuroanatomy and neurophysiology, as well as morphological studies of a wide spectrum of cells and tissues.

Other applications include stem cell research, DNA hybridization, membrane and ion probes, and bioluminescent proteins. Confocal microscopy is a widely used tool for fluorescent imaging of biological objects. The advantage of fluorescence for microscopy is the possibility to analyze location and expression of many target molecules at the same time.

In practice, the confocal principle is combined with a scanning system utilizing a laser light source that builds up an image by scanning a laser light across the sample in the x and y directions (Figure 5.30). The signal from the photomultiplier tube is converted to a digital form that contains information on the position of the laser in the image and the intensity of light coming from the sample. A computer is used to store the intensity value of each point from the detector, and present these on a high-resolution video monitor to display the image. The image is displayed on the computer screen as a shaded gray that can be suitably colored later. To collect a series of images, the computer then shifts the focus by a fixed amount, and the object is scanned again to produce

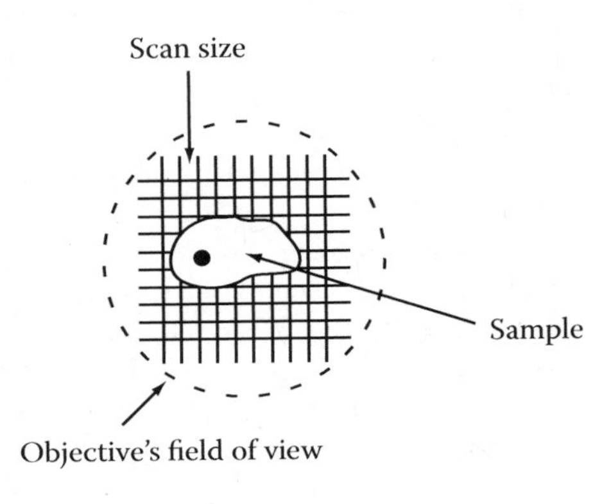

FIGURE 5.30 Scanning of a sample in a confocal microscope.

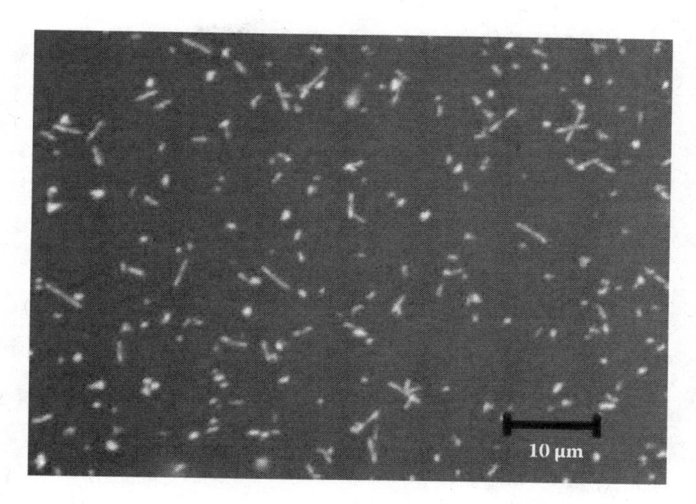

FIGURE 5.31 Confocal microscopy image of the labeled CNT attached to the labeled antibodies. (Reproduced from Teker et al., *Nanobiotechnology*, 1, 171–82, 2005. With permission.)

the next image, at the different z position. This image is stored and the process repeated to build the three-dimensional data set.

5.4.2 Biological Confocal Microscopy: Case Studies

Applications include the use of confocal microscopy for live cell imaging. This is often done by making use of a green fluorescent protein (GFP) or other fluorochromes attached to cellular components, such as proteins, genes, and cell structures for imaging their position or movements. Confocal microscopy is also used to analyze pH gradients and membrane potentials, using specific fluorescent dyes, and to measure intracellular changes in ion concentrations of calcium, sodium, magnesium, zinc, and potassium.

5.4.2.1 Bioconjugated Carbon Nanotubes for Biosensor Applications

The noncovalent biological functionalization of single-wall carbon nanotubes (SWNTs) with antibodies and the change in their properties have been investigated.[19] The carbon nanotubes are labeled with a dye that emits in green when illuminated with an argon laser at 488 nm excitation line. The antibody (polyclonal goat anti-mouse IgG) is labeled as well by using a dye that emits in red with a HeNe laser (543 nm excitation line). The two labeled solutions, CNT and the antibody, are mixed and left to interact before the confocal microscopy analysis. Figure 5.31 shows a confocal microscopy image of green-dye-labeled CNTs, coated with the labeled antibodies. The image shows a high degree of attachment of the antibodies to the carbon nanotubes.

Antibody-functionalized CNT has vast potential for applications in drug and gene delivery systems.

5.5 ATTENUATED TOTAL REFLECTION (INTERNAL REFLECTION) INFRARED SPECTROSCOPY

5.5.1 Introduction: ATR-FTIR Basics

Infrared spectroscopy (Fourier transform infrared spectroscopy (FTIR)) is one of the most important spectroscopic techniques used for the characterization of organic functional groups. It is based

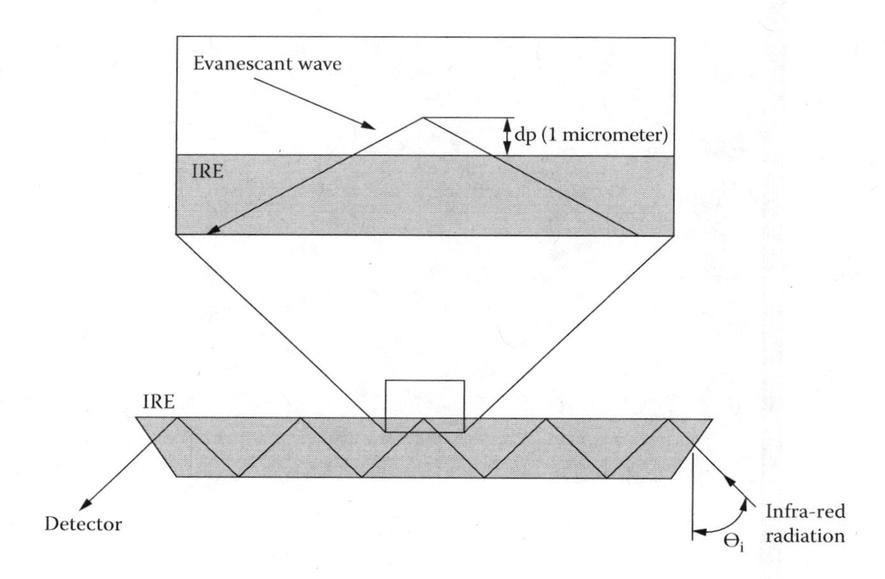

FIGURE 5.32 Optical path in a high refraction index ATR crystal.

on transitions between the vibrational states of the molecules corresponding to wavenumbers in the range of 200 to 4,000 cm^{-1}. Each functional group has characteristic stretching and bending bands at well-known wavenumbers, and their positions depend on the neighboring groups. Attenuated total reflection (ATR) is a surface technique that allows the recording of the infrared spectrum of a thin film at the interface between an optically dense crystal called the ATR crystal or internal reflection element (IRE) and the air. The technique was developed by Harrick[20] and Fahrenfort,[21] and since then, the ATR method has found a variety of applications. The optical path in an ATR crystal is shown in Figure 5.32. The IR light enters the crystal and penetrates a distance d_p beyond the reflective interface into the rarer medium. The beam propagates down the length of the crystal by multiple internal reflections from the opposing flat surfaces. Upon the interaction with the sample, the energy of the propagating wave is attenuated, giving rise to a reflection spectrum with the same bands as in the absorption (or transmission) spectrum of the sample. The crystal is in a holder in a FTIR accessory that has two mirrors that direct the beam into the crystal and then back into the optical path of the FTIR spectrometer.

In ATR, the beam of the spectrometer is directed into a prism at an angle that exceeds the critical angle, so that the total reflections occur. When a sample is placed in optical contact with the prism at the point at which a total reflection occurs, the sample absorbs IR energy at wavelengths equivalent to those in a transmission experiment. The total reflection generates an evanescent wave that extends beyond the surface of the crystal into the sample. The intensity of the wave decays exponentially with the distance from the surface, in the order of less than 1.5 μm, meaning that the thickness of the sample medium no longer has an effect on the measured reflection.

Solid samples must be mechanically pressed into the crystal to achieve a good optical contact. When liquids are analyzed by ATR, the optical contact is achieved readily. The sensitivity of the method depends on several parameters, such as the number of reflections allowed by the geometry of the crystal, the refractive indices of both the crystal and the sample, and the wavelength of the radiation. The depth to which the radiation penetrates the sample is proportional to the wavelength; that is, as the wavelength increases, the penetration increases as well. The penetration depth is the distance from the interface where the electric field decreases to 1/e of its initial value.

The penetration depth (d_p) and the number of reflections (N) through the ATR element are calculated as follows:

$$d_p = \frac{\lambda / n}{2\pi[\sin^2 \theta_i - (n_s / n)^2]^{1/2}}$$

and

$$N = \frac{L}{t} \cos \theta_i$$

where λ is the wavelength, n is the refractive index of the ATR crystal, n_s is the refractive index of the sample, θ_I is the incident angle, L is the length of the crystal, and t is its thickness.

ATR spectroscopy is particularly suitable to measure substances with very strong infrared absorption, such as water in thin films, coatings, and aqueous solutions or hydrated biomolecules.

5.5.2 Applications of ATR-FTIR Spectroscopy to Biomolecules and Biomedical Samples: Case Studies

5.5.2.1 Hydration Studies of Surface Adsorbed Layers of Adenosine-5'-Phosphoric Acid and Cytidine-5'-Phosphoric Acid by Freeze-Drying ATR-FTIR Spectroscopy

A new methodology, suitable for spectral measurements on polar, biologically important molecules, has been developed.[22–24] Hydrated nucleotides are important model compounds for ATP, and their infrared spectroscopic study provides important information on their structure in a water environment. Surface deuteration and protonation studies of the hydrated samples have permitted the assignment of the main infrared bands and evidenced the strong dependence of a general spectral pattern of these molecules on the relative humidity.

The A_{1680}/A_{1725} absorbance ratio is found to be a measure of degree of protonation. Interestingly, while the transmission spectrum is not sensitive to the hydration degree, in the ATR spectrum, the absorbance ratio increases with the relative humidity and reaches a maximum at around 66% relative humidity. It is demonstrated that the surface adsorbed water molecules are involved in the proton transfer process. Because of the strong ionic character of the TlBrI (KRS-5) ATR crystal, the water adsorption is dissociative. In a subsequent work, it was demonstrated that the protonating agent is the hydronium ion (H_3O^+) formed on the ATR crystal surface. It is shown that the phosphate group is the main hydration site in the mononucleotide molecules. The freeze-drying technique developed in these works may be useful for the study of membrane-associated phenomena and models of real-life biological systems.

5.5.2.2 Study of the Interaction of Local Anesthetics with Phospholipid Model Membranes

The ATR technique is very suitable for studying the molecular structure of biomembranes and their interaction with pharmacologically active substances such as local anesthetics. In this work, the mechanism of local anesthetic action is studied *in situ* by using the lipid bilayer model membrane formed of a dipalmitoyl phosphatidic acid monolayer attached to the zinc selenide or germanium ATR crystal with the polar head group and a 1-palmitoleoyl-sn-glycero-3-phosphocholine (POPC) monolayer with the polar head group toward the aqueous phase.[25] The local anesthetics (procain, oxybuprocaine, falicaine, and cinchocaine) investigated in this work are dissolved in a phosphate-buffered solution, and the ATR spectra are recorded with both parallel and perpendicular polarized light using cells. The position and the integrated optical density values of the carbonyl stretching band, $\nu_{C=O}$, corresponding to fatty acid esters are evaluated.

The results show that the four local anesthetics tend to adsorb in multilayers to a lipid bilayer membrane and point to an unspecific mechanism of local anesthesia. It is suggested that the

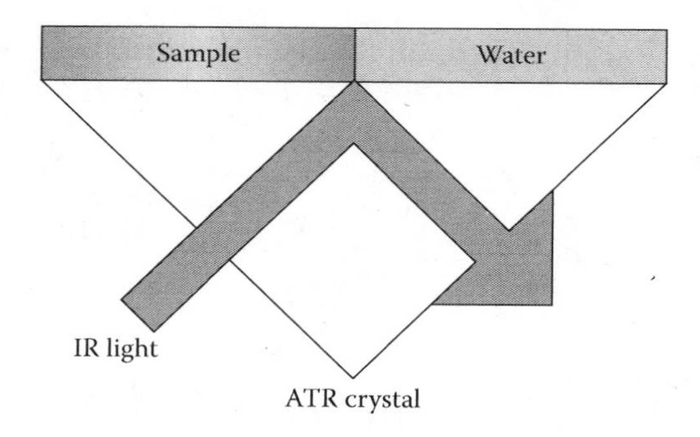

FIGURE 5.33 ATR study of the polymer-water interface.

anesthetics seal the membrane and block the transmembrane signal transfer. The results of this work show the applicability of ATR spectroscopy for *in situ* studies of drug-membrane interactions.

5.5.2.3 Assessment of Synthetic and Biologic Membrane Permeability by Using ATR-FTIR Spectroscopy

In this work, the membrane permeation phenomena are investigated by using ATR-FTIR spectroscopy.[26] The diffusion coefficient of a permeant within the stratum corneum is determined by the inherent nature and arrangement of structured lipids (ceramides ~ 50%, cholesterol ~ 25%, cholesterol sulfate ~ 5%, and free fatty acids ~ 15%) in the intracellular domains of the stratum corneum.

The membrane (a silastic membrane with a thickness of 127 μm) is placed on the top of the ATR crystal, and the molecule of interest (here acetonitrile in polyethyleneglycol) is in a cell. Spectra are recorded at defined time intervals, and the concentration of acetonitrile at the membrane-crystal interface is determined by monitoring the absorbance corresponding to the carbon-nitrogen stretching band ν_{CN} in acetonitril or p-cyanophenol (~2,230 cm^{-1}) and that of the Si-O band (~3,000 cm^{-1}) associated with the membrane. The potential of the ATR technique to yield information about the state of the stratum corneum lipids and the changes that the penetrants bring about is demonstrated.

5.5.2.4 ATR Measurement of the Physiological Concentration of Glucose in Blood by Using a Laser Source

The normal concentration of glucose in human blood is relatively low (typically 90 to 120 mg/dl), and water has a high absorption in infrared; therefore, the use of a CO_2 laser improves considerably the resolution of the measurement. The authors of this work describe the quantitative measurements of glucose in blood by using a powerful laser source in combination with the ATR technique.[27] Calibration experiments are conducted with pig blood samples containing known glucose concentrations, and a liquid cell in contact with an ATR plate is used. A calibration curve is constructed using blood samples with known glucose concentrations and used to calculate the concentration. The results are then compared to independent measurements made with a standard laboratory glucose analyzer. This work was important in the development of glucose sensors for diabetic patients.

5.5.2.5 Application of ATR-FTIR Spectroscopic Imaging in Pharmaceutical Research

It is shown that FTIR spectroscopic imaging in the ATR mode is a powerful technique for the characterization of biomedical materials because it relies on the characteristic absorbance of the important molecular vibrations in the sample.[28] This technique is used to gather quantitative information on the spatial distribution of chemical components in pharmaceutical tablets in contact with water as a function of time. A heated single-reflection ATR accessory with a zinc selenide (ZnSe) crystal is used, and the polymer-drug sample is sandwiched between the crystal and a cover glass with a

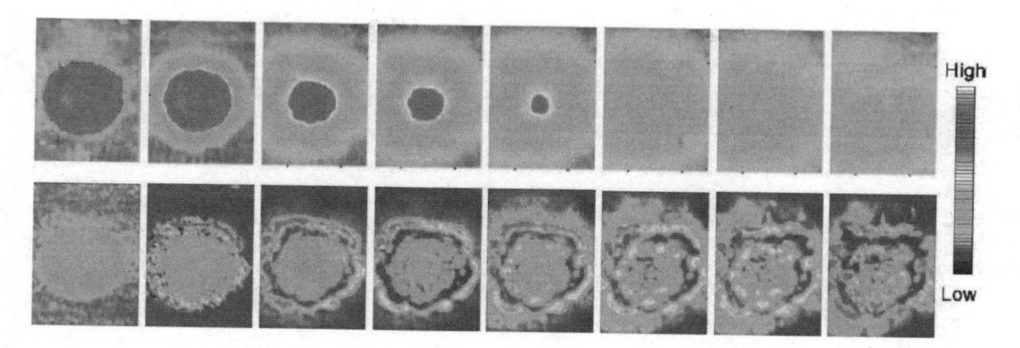

FIGURE 5.34 (See color insert.) ATR-FTIR images of the PEG/ibuprofen formulation as a function of time.

thin spacer. One of the systems studied by using this approach was sodium benzoate incorporated in a polyethyleneglycol matrix.

Sodium benzoate is used here as a model substance because it is frequently used as an antimicrobial preservative and it is very soluble in water. The spatial distribution of both PEG and sodium benzoate is studied with the ATR technique as a function of time. The results showed that the polymer and the drug dissolve immediately, and indicate that the drug is not released through diffusion through the polymer (Figure 5.34).

Figure 5.34 shows the results of the study of possible crystallization of ibuprofen upon dissolution as a function of time. The integrated absorbances of the characteristic bands of PEG and ibuprofen are mapped; the last images on the right were acquired 40 min after addition of water.

This work demonstrates the potential of ATR-FTIR imaging for the study of dissolution and drug release under various conditions.

5.6 MECHANICAL METHODS: USE OF MICRO- AND NANOCANTILEVERS FOR CHARACTERIZATION OF SURFACES

Molecular adsorption on a solid surface is driven by surface Gibbs free energy reduction, which leads to a change in surface stress. The mechanical response of the solid to stress can be exploited to detect the chemical interaction occurring at the surface. This can be done by using microcantilever (MC) beams as effective mechanical transducers of biochemical reactions. When operated in static mode, the microcantilever is functionalized on one face with a receptor molecule that can selectively bind to the ligand species. The specific receptor-ligand binding changes the surface stress on that face, and the MC bends as shown in Figure 5.35, providing the mechanical transduction of the reaction. This phenomenon has found a breakthrough application in cantilever biosensors.

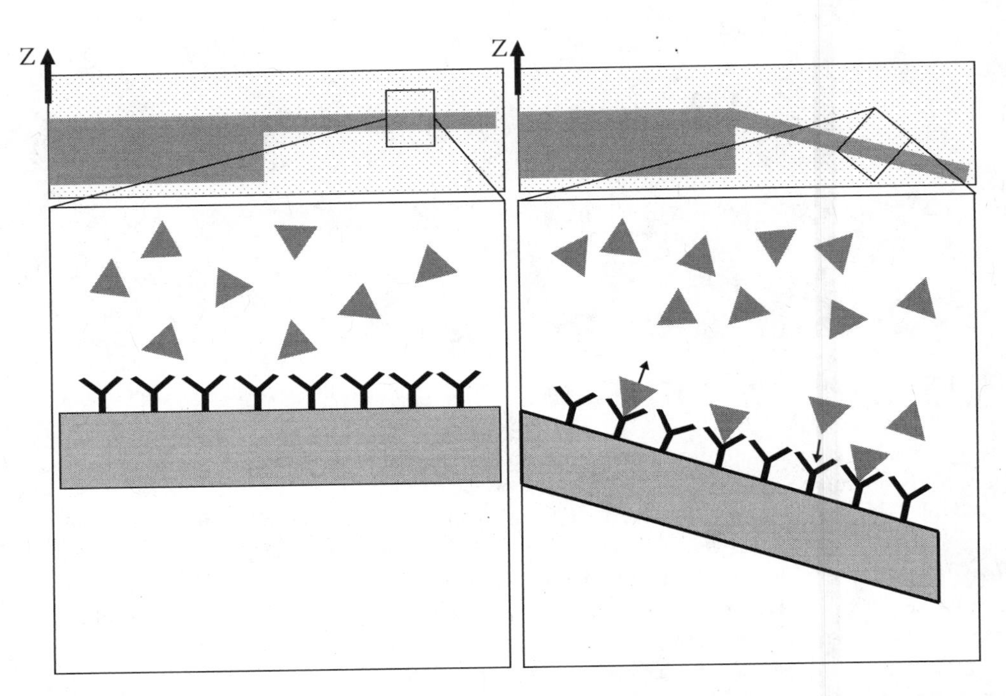

FIGURE 5.35 Binding of ligand molecules to receptor molecules confined to the top surface of the microcantilever changes the surface stress on that face, and there is a downard bending. (Adapted from Bergese et al., *J. Colloid Interface Sci.*, 316, 1017–22, 2007.)

REFERENCES

1. Krishnan, A., Liu, Y.-H., Cha, P., Woodward, R., Allara, D., Vogler, E. A. 2005. An evaluation of methods for contact angle measurements. *Colloids Surfaces B Biointerfaces* 43:95–98.
2. Bhusnan, B., Tokachichu, D. R., Keener, M. T., Lee, S. C. 2005. Morphology and adhesion of biomolecules on silicon based surfaces. *Acta Biomater.* 1:327–41.
3. Song, H.-M., Lee, C.-S., Korean, J. 2008. Simple fabrication of functionalized surface with polyethylene glycol microstructure and glycidyl methacrylate moiety for the selective immobilization of proteins and cells. *Chem. Eng.* 25:1467–72.
4. The, K. S., Lu, Y.-W. 2008. Topography and wettability control in biocompatible polymer for bioMEMS applications, Paper presented at Proceedings of the 3rd IEEE International Conference on Nano/Micro Engineered and Molecular Systems, Sanya, China, January 6–9, 2008.
5. Bouaidat, S., Hansen, O., Bruus, H., Berendsen, C., Bau-Madsen, N. K., Thomsen, P., Wolff, A., Jonsmann, J. 2005. Surface-directed capillary system: Theory, experiments and applications. *Lab Chip* 5:827–36.
6. Bergese, P., Oliviero, G., Colombo, I., Depero, L. E. 2009. Molecular recognition by contact angle: Proof of concept with DNA hybridization. *Langmuir* 25:4271–73.
7. Guthold, M., Superfine, R., Taylor II, R. M. 2001. The rules are changing force measurements on single molecules and how they relate to bulk reaction kinetics and energies. *Biomed. Microdevices* 3:9–18.
8. Choi, I., Kim, Y., Kim, J. H. Yang, Y. Lee, J., Lee, S., Surin Hong, S., Jongheop Yi, J. 2008. Fast image scanning method in liquid-AFM without image distortion. *Nanotechnology* 19:445701.
9. Spangenberg, T., de Mello, N. F., Creczynski-Pasa, T. B., Pasa, A. A., Niehus, H. 2004. AFM *in-situ* characterization of supported phospholipid layers formed by solution spreading. *Phys. Stat. Sol. A* 201(5):857–60.
10. Dufrêne, Y. F., 2004. Refining our perception of bacterial surfaces with the atomic force microscope. *J. Bacteriol.* 186(11):3283–85.
11. Mikamo, E., Tanaka, C., Kanno, T., Akiyama, H., Jung, G., Tanaka, H., Kawai, T. 2005. Native polysomes of *Saccharomyces cerevisiae* in liquid solution. *J. Struct. Biol.* 15:106–10.

12. Williams, M. C., Pant, K., Rouzina, I., Karpel, R. L. 2004. Single molecule force spectroscopy studies of DNA denaturation by T4 gene 32 protein. *Spectroscopy* 18:203–11.
13. Fiorini, M., McKendry, R., Cooper, M. A., Rayment, T., Christopher Abell, C. 2001. Chemical force microscopy with active enzymes. *Biophys. J.* 80:2471–76.
14. Allison, D. P., Hinterdorfer, P., Han, W. 2002. Biomolecular force measurements and the atomic force microscope. *Curr. Opin. Biotechnol.* 13:47–51.
15. Lekka, M., Wiltowska-Zuber, J. 2009. Biomedical applications of AFM. *J. Physics Conference Ser.* 146:012023.
16. Briones, C., Mateo-Marti, E., Gomez-Navarro, C., Román, E., Martin-Gago, J. A. 2005. Structural and functional characterization of self-assembled monolayers of peptide nucleic acids and its interaction with complementary DNA. *J. Mol. Catal. A Chem.* 228:131–36.
17. Griffith, A., Glidle, A., Cooper, J. M. 1996. Probing enzyme polymer biosensors using x-ray photoelectron spectroscopy: Determination of glucose oxidase in electropolymerized films. *Biosensors Bioelect.* 11(6/7):625–31.
18. McArthur, S. L. 2006. Applications of XPS in bioengineering. *Surf. Interface Anal.* 38:1380–85.
19. Teker, K., Sindeshmukh, R., Sivakumar, K., Lu, S., Wickstrom, E., Wang, H. N., Vo-Dinh, T., Panchapakesan, B. 2005. Applications of carbon nanotubes for cancer research. *Nanobiotechnology* 1:171–82.
20. Tao, S. L., Lubeley, M. W., Desai, T. A. 2003. Bioadhesive poly(methyl methacrylate PMMA) microdevices for controlled drug delivery. J. *Controlled Release* 88:215–28.
21. Harrick, N. J. 1967. *Internal reflection spectroscopy*. New York: Interscience.
22. Fahrenfort, J. 1961. Attenuated total reflection: A new principle for the production of useful infrared spectra of organic compounds. *Spectrochim. Acta* 17:689.
23. Ashrit, P. V., Badilescu, S., Girouard, F. E., Truong, V.-V. 1989. Water absorption studies in thin films by the IR attenuated total reflection method. *Appl. Optics* 28(3):420–22.
24. Badilescu, S., Sandorfy, C. 1987. Freeze-drying Fourier transform infrared attenuated total reflection spectroscopy of surface adsorbed layers. Hydration of adenosine-5'-phosphoric acid. *Appl. Spectrosc.* 41:10–15.
25. Badilescu, S., Sandorfy, C. 1986. Freeze-drying Fourier transform infrared attenuated total reflection spectroscopy of surface adsorbed layers. Water-assisted proton transfer in cytidine-5'-phosphoric acid. *Can. J. Chem.* 64:2404–8.
26. Schöpflin, M., Fringeli, U. P., Perlia, X. 1987. A study of the interaction of local anesthetics with phospholipid model membranes by infrared ATR spectroscopy. *J. Am. Chem. Soc.* 109:2375–80.
27. Watkinson, A. C., Hadgraft, J., Walters, K. A., Brain, K. R. 1994. Measurement of diffusional parameters in membranes using ATR-FTIR spectroscopy. *Int. J. Cosmetic Sci.* 16:199–210.
28. Mendelsohn, Y., Clermont, A. C. 1990. Blood glucose measurement by multiple attenuated total reflection and infrared absorption spectroscopy. *IEEE Trans. Biomed. Eng.* 17:458–65.
29. Kazarian, S. G., Chan, K. L. A. 2006. Applications of ATR-FTIR spectroscopic imaging to biomedical samples. *Biochim. Biophys. Acta* 1758:858–67.
30. Bergese, P., Oliviero, G., Alessandri, I., Depero, L. E. 2007. Thermodynamics of mechanical transduction of surface confined receptor/ligand reactions. *J. Colloid Interface Sci.* 316:1017–22.

REVIEW QUESTIONS

1. Give a definition of the contact angle and explain how the contact angle is related to the intra- and intermolecular forces.
2. How can the contact angle be measured experimentally?
3. What is the spatial resolution of the AFM technique?
4. What are the main parts of an AFM instrument?
5. How can biomolecules be measured with AFM?
6. What is chemical force microscopy and how can this technique be used for the study of DNA?
7. What is molecular recognition force microscopy (MRFM)?
8. What is x-ray photoelectron spectroscopy and how this technique can be used to determine the elemental chemical composition of the surface of a material?

9. How can biological materials be handled for XPS measurements?
10. What are the most important applications of confocal microscopy?
11. What is a fluorescent microscope and how does it work?
12. What is the detection limit of the confocal microscopy method?
13. Give an example of application of confocal microscopy to the study of biological materials.
14. What are the applications of ATR-FTIR spectroscopy to biomolecules and biomedical samples?

6 Biosensing Fundamentals

6.1 BIOSENSORS

6.1.1 INTRODUCTION

There is a growing interest in the use of biosensors in various applications, including medical, environmental, toxicological, defense monitoring, etc. A *biosensor* is a compact sensing device that incorporates a specific *biological* or *biomimetic element* connected to, or integrated within, a physicochemical *transducer* or a transducing microsystem. When the specific biological element, such as a biomolecule, a biological entity, recognizes a *target*, that is, a specific *analyte*, the result of this recognition process, namely, a *biochemical signal*, is converted by a transducer into a quantifiable electrical signal, such as current or voltage or other forms of signals, such as optical, mechanical, electrochemical, piezoelectric, etc. A *bioreceptor* is a biological molecular species or a biological system that adopts a biochemical mechanism for analyte recognition. The magnitude of the electrical signal is proportional to the concentration of a specific analyte to which the biological element binds. Biosensors operate through a *biorecognition process* due to the affinity between the biological entity and the analyte.

One can say that biosensors are the technological counterpart of our sense organs, combining the recognition with a biological recognition element by a transducer. Another commonly cited definition is: "A biosensor is a sensing device in which a biologically recognizable entity is coupled to a transducer, to quantify the development of some complex biological parameter." The IUPAC definition seems to be the most general: "A device that uses specific biochemical reactions mediated by isolated enzymes, immunosystems, tissues, organelles or whole cells to detect chemical compounds usually by electrical, thermal or optical signals."[1] With the advent of present-day technologies, there are other efficient mechanical signals, such as deflection, rotation, and twist, that are widely used. There are other terms sometimes used for biosensors, such as *immunosensors*, *optrodes*, *glucometers*, *biochips*, etc. In many cases, an enzyme is used for interactions, based on the enzyme-substrate pair, while interaction pairs such as the antibody-antigen and nucleic acids–complementary sequences are also used for biosensing. Other commonly used biological materials in biosensors are tissues, ion channels, microorganisms, cell receptors, and biomimetic components. The schematic representation of a biosensor operation is shown in Figure 6.1.

The uniqueness of a biosensor is defined by the way the two components, namely, the analyte and the *bioreceptor*, are allowed to interact in a sensor (Figure 6.2).

In many cases, this combination enables the detection of analytes without using reagents. For example, the glucose concentration in blood samples can be measured directly by a biosensor that was made specifically for glucose measurement, by simply dipping the sensor in the sample. This is in contrast to the commonly performed assays, in which many sample preparation steps are necessary, and each step may require a reagent to treat the sample. The main advantages of biosensing are simplicity and speed of measurement that requires no specialized laboratory skills. As shown in Figure 6.3, there is direct spatial contact between the recognition element and the signal transducer equipped with an electronic amplifier, leading to a compact functional unit.

The combination of highly specific biological reactions with appropriate transduction elements provides biosensors with other advantages, such as high sensitivity, specificity, short response time, and real-time monitoring measurement capability. The nature of biosensors allows miniaturization

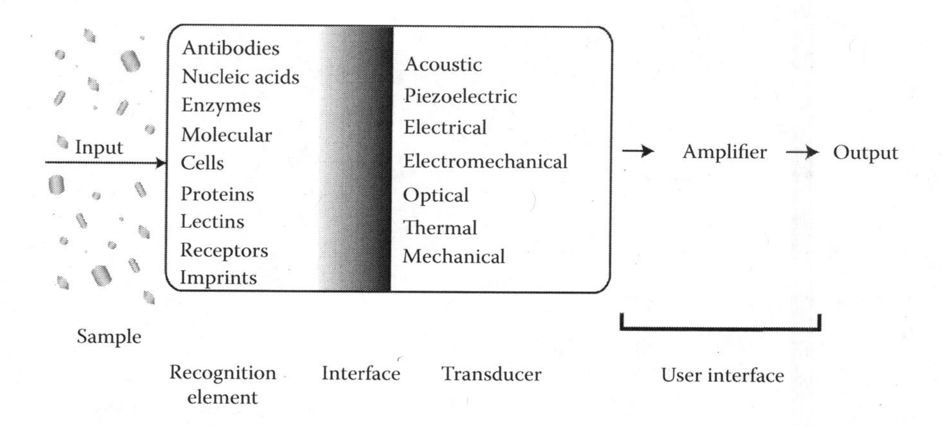

FIGURE 6.1 Conceptual illustration of an immunosensor. (Adapted from Chambers et al., *Curr. Issues Mol. Biol.*, 10, 1–12, 2008.)

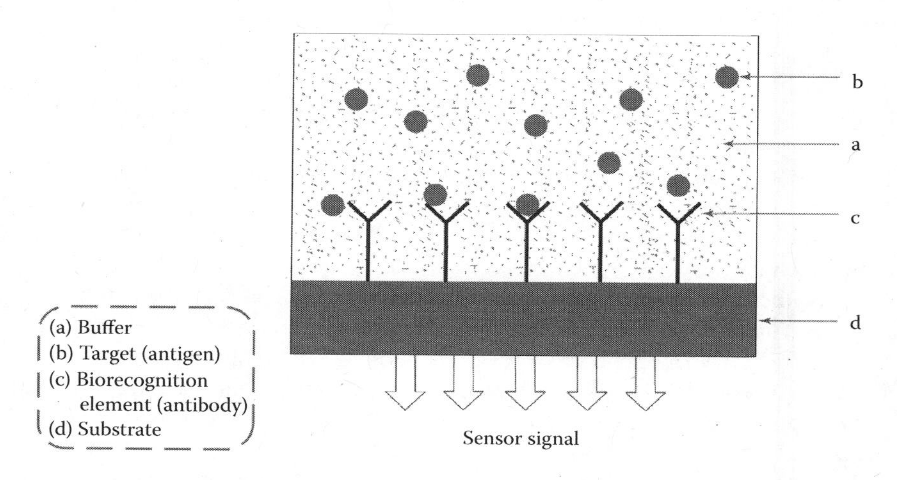

FIGURE 6.2 Configuration of a biosensor showing biorecognition: (a) buffer solution, (b) target (antigen), (c) biorecognition element, and (d) substrate.

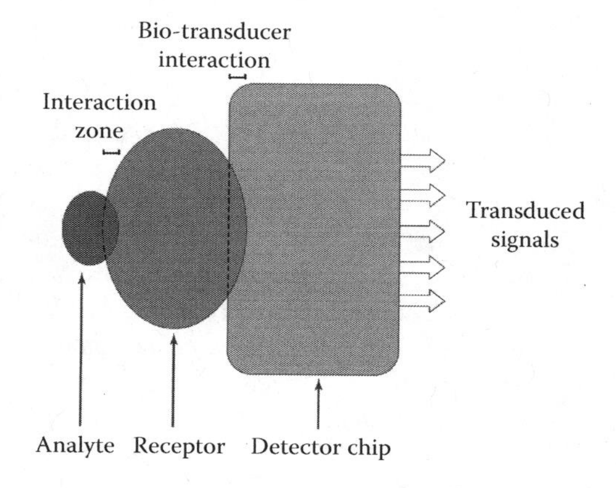

FIGURE 6.3 Schematic diagram of a biosensor.

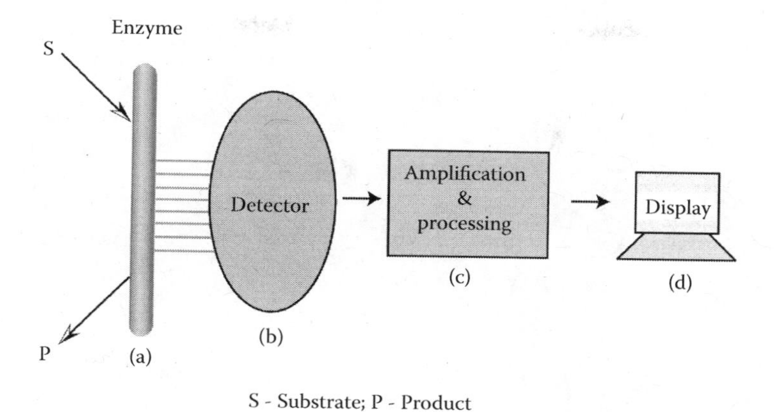

FIGURE 6.4 Schematic representation of a biosensor. The enzyme (a) interacts with the substrate and converts it to product. This interaction is "sensed" by the transducer (b), which converts it to an electrical signal. The output from the transducer is amplified (c), processed (d), and displayed (e).

and integration to achieve portable and handheld instruments, making biosensors an ideal choice for on-site measurements. Figure 6.4 shows schematically the components of a typical biosensor.

The first biosensor was described by Clark and Lyons in 1962[3] to monitor glucose levels, in which a Clark oxygen electrode was combined with the enzyme glucose oxidase (GOD) entrapped in a dialysis membrane. During the enzymatic oxidation of glucose, the coreactant oxygen was consumed and was monitored amperometrically. Alternatively, the production of hydrogen peroxide during the enzyme reaction could also be measured. Clark and Lyons's biosensor is shown in Figure 6.5.

Since then, many works have been reported, especially on detection using enzyme electrodes. Enzymes were immobilized with different procedures, and reaction substrates or products were detected by different methods. In 1969, Guilbault and Montalvo[4] reported the first enzyme biosensor based on potentiometry to detect urea, where urease was immobilized on an ammonium-selective liquid membrane electrode. The first commercial glucose analyzer based on amperometric detection of hydrogen peroxide was built in 1973 by Yellow Springs Instrument Company (see Figure 6.6). Liedberg et al. were the first to use surface plasmon resonance to monitor antibody-antigen reactions in real time.[6]

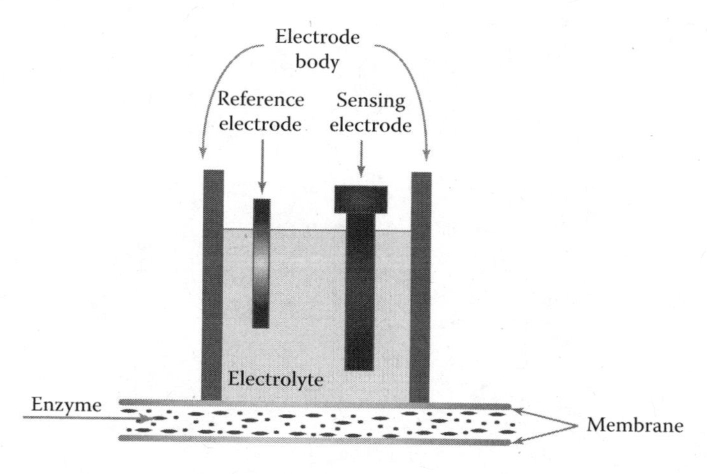

FIGURE 6.5 Clark and Lyons enzyme electrode. (Adapted from Clark Jr. and Lyons, *Ann. NY Acad. Sci.*, 102, 29–45, 1962.)

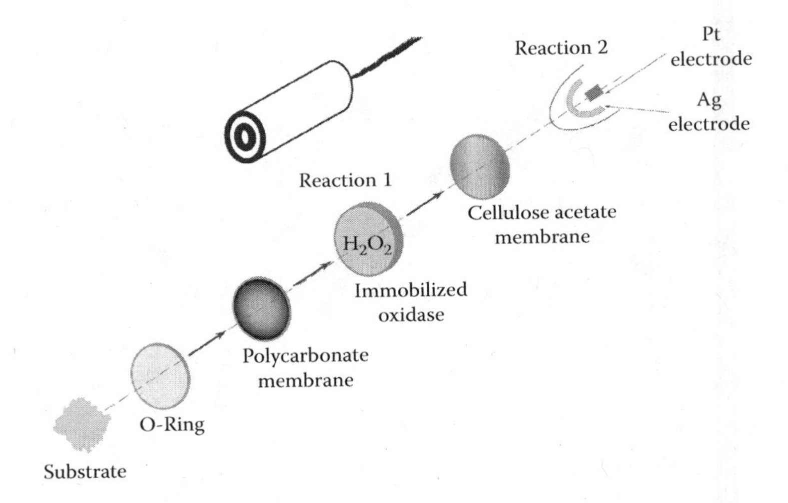

FIGURE 6.6 Schematic diagram of a glucose sensor developed by Yellow Springs Instruments, Inc. (Adapted from Magner, *The Analyst*, 123, 1967-70, 1998.)

The biological components may be immobilized physically or chemically onto the transducer. Physical immobilization of a biomolecule on the sensing element can be realized through physical adsorption, encapsulation, and entrapment.

The sensitivity, reliability, and usefulness of biosensors are determined by the stability of the immobilization techniques. Figure 6.7 shows the different techniques of immobilization.

Although chemical links allow more stable devices, there is a possibility of denaturation, which is the deformation of tertiary structures of biomolecules during the chemical reactions at different

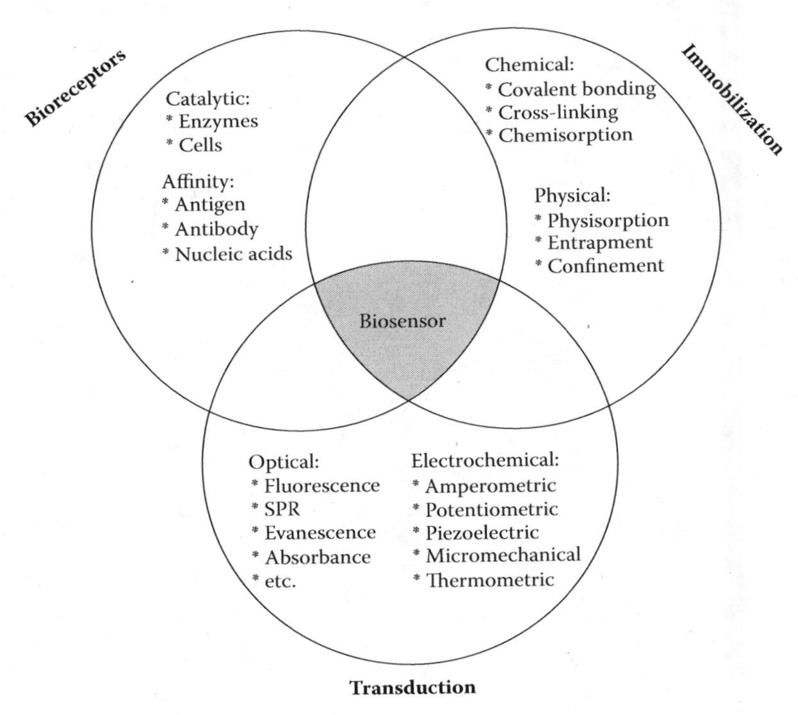

FIGURE 6.7 Map of bioreceptors, immobilization methods, and transducers.

steps. It is important to select an immobilization method that would allow the biomolecules to retain their selectivity and binding ability. Bifunctional linkers or long-chain spacers are used to attenuate the strong effect of the surfaces. In the case of nonpolar substrates, physical adsorption is based on hydrophobic forces and van der Waals forces, while for hydrophilic surfaces it is based on hydrogen bonding interactions and ionic forces.

In the porous entrapment method, an encapsulation matrix is bound around the biomolecule, and it helps to bind into the surface of the sensor. The biological component is embedded within a membrane by coprecipitation or polymerization, or trapped between several layers of a semipermeable membrane or gel. The different methods of functionalization of the sensor surface have been discussed in detail in Chapter 4. The key part of a biosensor is the transducer, which makes use of physical change that accompanies the reaction. For example, a change of mass or change of stress can be sensed. Biosensors are often classified by the method of signal transduction, as shown in Figure 6.8.

The different types of biosensors will be discussed in detail later in this chapter.

The important performance characteristics of a sensor are working range, sensitivity, limit of detection, selectivity and specificity, response time, and stability, and they will be discussed as part of the case studies.

6.1.2 Classification: Case Studies

As emphasized in the previous section, to design an efficient biosensor, one has to match the appropriate biological and transduction components to produce a relevant signal. Amplification and processing of the signal change provides the opportunity for quantification of the analyte (ligand)-receptor binding. A variety of biological species, such as enzymes, antibodies, nucleic acids, living cells, etc., and transducers may be considered for the design of a biosensor. If an appropriate operating format can be devised, virtually any biological recognition element can be interfaced to any transducer. *Molecular biosensors* can be classified upon the biological species as molecular, cellular, and tissue sensors. Biological recognition utilizes three basic mechanisms: biocatalytic, bioaffinity, and microbe-based systems. Depending on the principle of the

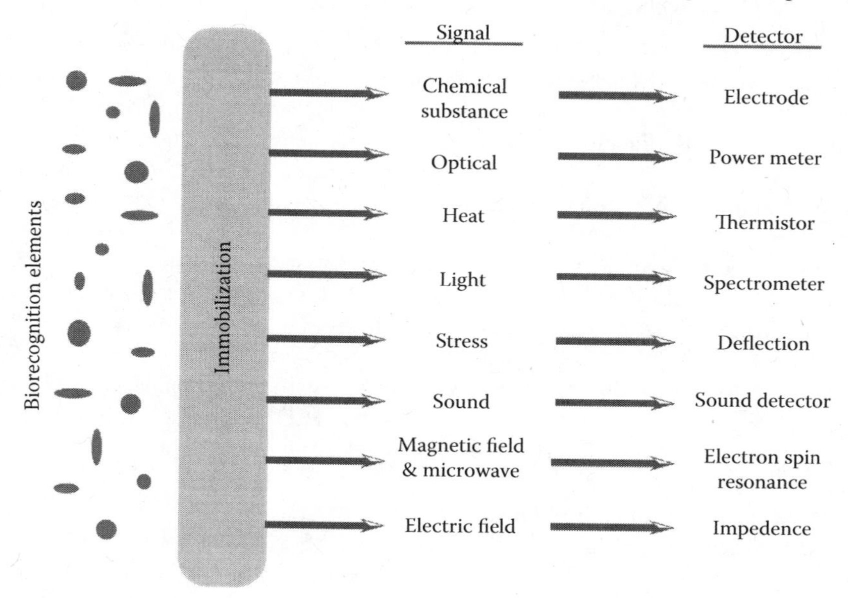

FIGURE 6.8 Different types of signals and corresponding transducers.

detection, there are various types of biosensors: electrochemical, optical, piezoelectric, thermometric, mechanical, and magnetic. Among the *electrochemical biosensors*, the *amperometric* sensors monitor the current on the working electrode, resulting from oxidation or reduction of an electroactive species at a constant potential. The current generated by the redox reaction of the analyte at the sensing electrode is directly proportional to the concentration of analyte at the electrode surface. Amperometric transducers that constitute a highly developed area in biosensing are the most described in the literature, with a few of them commercialized successfully. In addition to amperometric biosensors, there are also *conductometric* and *potentiometric* electrochemical biosensors. Conductometric sensors measure the electrical conductance (or resistance) of the solution, while for the potentiometric devices the measured parameter is the oxidation or reduction potential corresponding to an electrochemical reaction. Other types of electrochemical detection techniques are those measuring impedance and field effect with transistor technology. The majority of electrochemical biosensors use enzymes or a combination of enzymes for the determination of analytes, such as glucose, lactate, sucrose, acetylcholine, etc. The enzymes are used because of their biocatalytic activity. These sensors usually work with three electrodes: a reference electrode (Ag/AgCl), an auxiliary (counter) electrode, and a working electrode (Pt or carbon), which serves as the transducing element. An example of an amperometric biosensor is a *glucose biosensor*, which is the most commercially successful sensor. Several glucose biosensors will be described and discussed below.

6.1.2.1 Enzyme-Based Biosensors

Enzyme-based biosensors are molecular biosensors relying on specificity and selectivity of an enzymatic protein toward a target substrate. Blood components such as glucose, urea, etc., can be measured with various enzyme-based biosensors. Some biosensors use an enzyme system; for example, the device that allows the quantification of acetylcholine needs the presence of three enzymes: acetylcholinesterase, choline oxidase, and horseradish peroxidase. The catalytic reactions release measurable products that can be detected by different methods. Of all the enzyme-based biosensors, because of the medical needs, the most widely studied is the glucose biosensor. A few of these sensors and their integration into microfluidic systems are described below.

6.1.2.1.1 Enzyme-Based Amperometric Glucose Biosensor

The first glucose biosensor was developed by Leland Clark in the 1950s and was commercialized by Yellow Springs Instruments.[5] Diabetes is a severe disorder of carbohydrate metabolism that results in a high concentration of glucose in the blood. If the glucose level is not controlled, because of the possible complications, diabetes can become a life-threatening condition. The enzyme *glucose oxidase* (GOD) is the *recognition element* that catalyzes the oxidation of glucose to gluconolactone. GOD is a dimeric protein containing one flavin adenine dinucleotide (FAD) per monomer as cofactor (Figure 6.9).

It is trapped in the proximity of the Pt electrode, and its activity depends on the surrounding oxygen concentration. GOD is reduced by the glucose and, as a result of the redox reaction, glucolactone and hydrogen peroxide are formed, as shown below.

$$\text{Glucose} + \text{GOD}\ (\text{FAD}^+) \rightarrow \text{glucolactone} + \text{GOD}\ (\text{FADH}_2)$$
$$\text{FADH}_2 + \text{O}_2 \rightarrow \text{FAD}^+ + \text{H}_2\text{O}_2$$
$$\text{H}_2\text{O}_2 \rightarrow 2\text{H}^+ + \text{O}_2 + 2\text{e}^-$$

where $FAD^+/FADH_2$ is the cofactor of GOD.

The subunit structure of the glucose oxidase showing FAD is given in Figure 6.10.

Glucose concentration is proportional to the concentration of hydrogen peroxide produced through the reaction, as well as to the consumption of oxygen. The amperometric reading (current vs. glucose concentration) shows a linear relationship up to a specific glucose concentration.

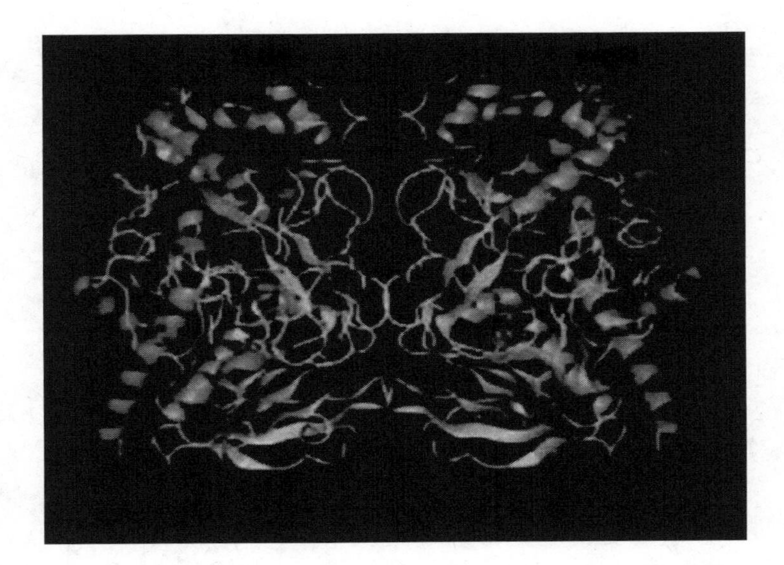

FIGURE 6.9 Topology of glucose oxidase. (Reproduced from Tatke et al. *Biopolymers*, 89, 582-94, 2008. With permission.)

Because the redox center of an enzyme is buried in a thick protein shell, the electron transfer between the enzyme and electrode is too slow (Figure 6.11).

To improve the response time of bioelectrodes, various electroactive compounds were used as *mediators*. Common mediators for electron transfer in glucose oxidase are ferrocene and its derivatives, benzoquinone, polyviologen, chloranil, and methylene blue. A mediator is a low molecular redox couple that reacts with the enzyme and regenerates it. It can be immobilized at the electrode surface, together with the enzyme, without using a membrane. The first glucose biosensor specifically designed for home use was released by MediSense, Inc. in 1987. Test strips are manufactured by screen printing of different layers on a polyvinylchloride strip. The sensor has two working electrodes and uses a ferrocene derivative as a mediator. A more modern screen-printed amperometric biosensor is shown in Figure 6.12.

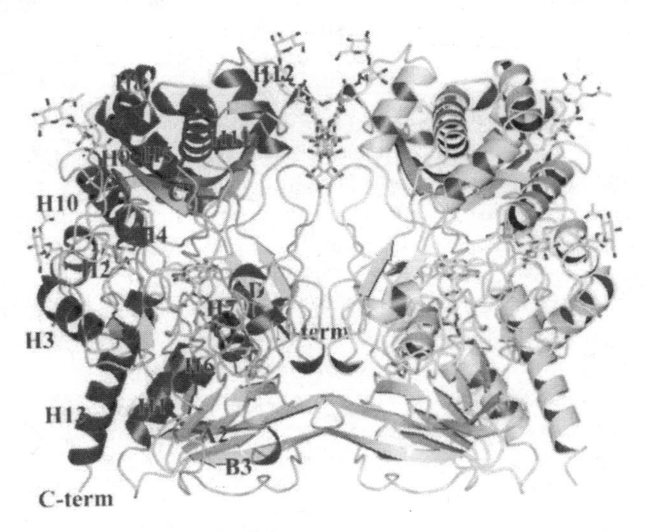

FIGURE 6.10 Subunit structure of the enzyme showing the FAD cofactor. (Reproduced from Tatke et al., *Biopolymers*, 89, 582-94, 2008. With permission.)

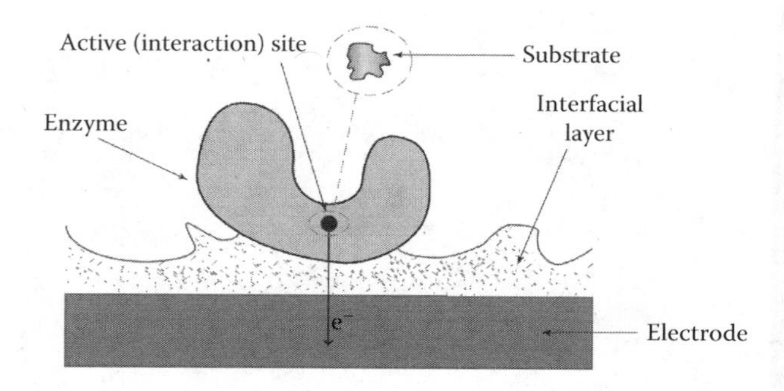

FIGURE 6.11 Electron transfer in a biocatalytic sensor.

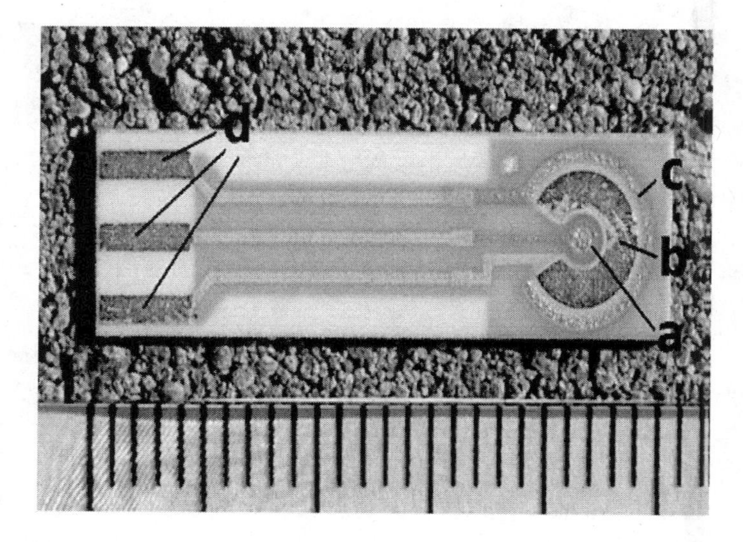

FIGURE 6.12 Three-electrode screen-printed sensor produced by BVT (Brno, Czech Rep.). The sensor body is made from ceramics. A gold working electrode (a) is surrounded by an Ag/AgCl reference electrode (b) and gold auxiliary electrode (c). Letter d means silver output contacts. The ruler in the bottom is in millimeter scale. (Reproduced from Pohanka and Skládal, *J. Appl. Biomed.*, 6, 57–64, 2008. With permission.)

Other commercial amperometric biosensors available are SIRE P201 (Chemel AB, Lund, Sweden), FreeStyle Freedom Blood Glucose Monitoring System, Precision Xtra (Abbot Diabetes Care, Alameda, California), etc. Graphite electrodes are also used instead of the platinum originally used by Clark. A disposable amperometric glucose biosensor consists of the following layers: metallic substrate, graphite layer, isolating layer, mediator-modified membrane, immobilized enzyme membrane, and a cellulose acetate membrane. The outer cellulose acetate layer acts as a barrier for interfering substances, such as ascorbic acid, uric acid, etc. Metal-dispersed carbon paste enzyme electrodes have also been used due to their high stability and good dynamic performance.[9] Because of the large market, there is a strong incentive to further develop blood glucose tests. It would be very important to develop biosensors for other analytes such as myoglobin and lactate dehydrogenase involved in myocardial infarction, but their concentration is usually too low to be detected by direct electrochemical methods. Whole cell biosensors having several enzymes have also been reported to be less expensive than the purified enzyme biosensors. The tissue materials with the whole cells can also be incorporated in the graphite paste.

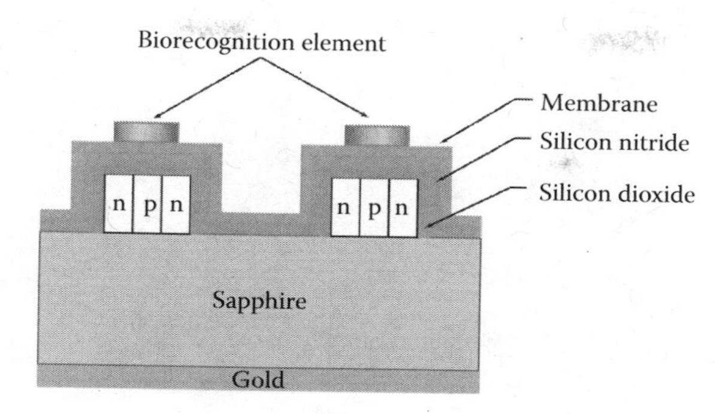

FIGURE 6.13 Schematic drawing of a field effect transistor (FET) (n-p-n type)–based biosensor. (Adapted from Pohanka and Skládal, *J. Appl. Biomed.*, 6, 57–64, 2008.)

6.1.2.1.2 *Enzyme-Based Potentiometric Biosensors Based on Ion-Selective Field Effect Transistors (ISFETs) for Glucose and Penicillin Determination*

Potentiometric biosensors used in the past had a pH glass electrode as a transducer, but now they use ion-sensitive field effect transistors (ISFETs), enzyme field effect transistors (ENFETs), and extended gate field transistors (EGFETs). ENFETs are based on biocatalytic reactions affecting the charge at the gate surface. The sensing performance of ENFET is determined by the immobilization method of enzymes, that is, the integration method of the enzyme and the ISFET device. The general configuration of the ENFET device based on a silicon substrate is shown in Figure 6.13.

In previous works, enzymes were immobilized on polymer films such as polyacrylamide hydrogels, polyvinylchloride, or polyurethane, yielding active sensing devices. However, the sensitivity of the device was quite low because of the diffusion barriers of substrates through polymer membranes. By the functionalization of the Al_2O_3 gate surface with an enzyme monolayer, the performance of the ENFET device can be improved considerably. The authors used 3-aminopropyltriethoxysilane to functionalize the Al_2O_3 gate of the ISFET, and then activated it with glutaric aldehyde and immersed the chip in the solution of the respective enzyme (glucose oxidase, urease, acetylcholine esterase, and α-chimotrypsin) to form the monolayer (Figure 6.14).

A nanoporous thin film is functionalized with a biocatalytic enzyme monolayer. The sensing functions of the resulting ENFET systems were characterized, and shorter response times and

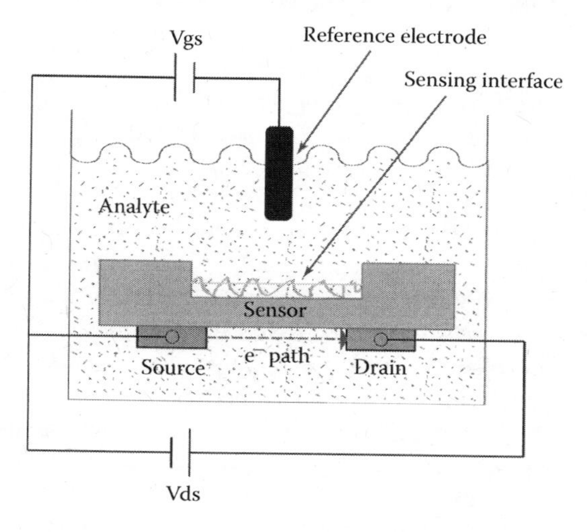

FIGURE 6.14 General configuration of the ENFET device. (Adapted from Kharitonov et al., *Sensors Actuators B*, 70, 222–31, 2000.)

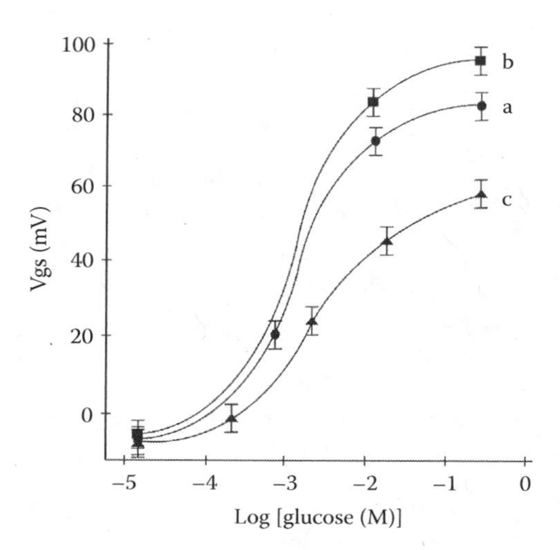

FIGURE 6.15 Calibration plots of the GOx-modified ENFET in the presence of different concentrations of glucose. (a)The freshly prepared ENFET. (b) After two days of continuous operation. (c) After five days of continuous operation. Data recorded in an electrolyte solution composed of 0.1 M NaCl. (Reproduced from Kharitonov et al., *Sensors Actuators B*, 70, 222–31, 2000. With permission.)

increased concentration ranges in the sensing of substrates have been obtained. The enzymes are attached to the gate surface through a covalent bond via amino groups of lysine residues. The calibration plot of the glucose sensor (Figure 6.15) showed a linear response in a range of concentration extending from 3.5×10^{-4} M to 9.0×10^{-3} M.

Because the upper detection limit is much higher than that of previously described ENFETs, this device is adequate for *in vivo* usage. The novel approach to fabricate ENFETs allowed the sensing of urea, glucose, acetylcholine, and N-acetyl-L-tyrosine ethyl ester, respectively.

6.1.2.1.3 Enzyme-Based Glucose Sensing and Automatic Insulin Injection Using a Microfluidic Chip

Recently, an integrated microfluidics device has been developed for the continuous sensing of glucose levels and injection of insulin when the glucose concentration is too high. Figure 6.16 shows a schematic representation of the integrated microfluidic system glucose sensing and insulin injection (GSII) biochip.[11]

The chip is composed of two modules: the upper module and the sensing module. The upper module is a polydimethylsilaxane (PDMS)-based control module that consists of microvalves, micropumps, and microchannels, while the sensing module is composed of an electrochemical glucose sensor.

In order to immobilize the enzyme GOD, electropolymerized pyrrole was used to entrap the enzyme on the surface of the Pt working electrode. The assembled chip is shown in Figure 6.17 (photograph), and the calibration curve showing the output current for various concentrations of glucose is given in Figure 6.18. The results showed that glucose concentration in the range of 1.61 to 30 mM could be detected, and also a precise amount of insulin was injected in real time.

Optical biosensors detect a change in optical properties, such as reflectance, light scattering, absorbance, polarization state, phase shift, interference, fluorescence, evanescence, etc. They can be designed to respond to ultraviolet, visible, or infrared radiation, or to the production of chemiluminiscence, and can be easily adapted to devices containing optical fibers. Reagents can be immobilized on a fiber or fiber bundle, and their optical properties will be changed by their interaction with the analyte. These devices are known as *optrodes***.** The response time of an optical biosensor is shorter than that of an electrochemical device, and the signal is free of electrical interferences. The

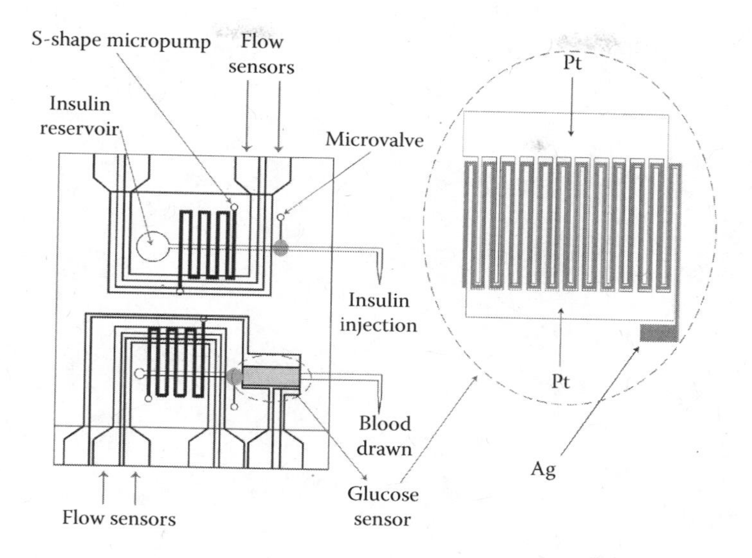

FIGURE 6.16 Glucose sensing and insulin injection biochip. (Reproduced from Huang et al., *Sensors Actuators B*, 122, 461–68, 2007. With permission.)

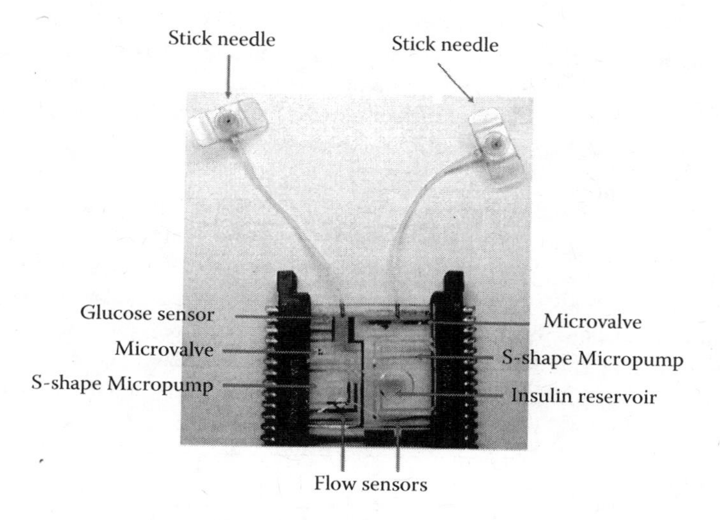

FIGURE 6.17 Photograph of the MEMS-based microfluidic chips (5 × 6 cm). (Reproduced from Huang et al., *Sensors Actuators B*, 122, 461–68, 2007. With permission.)

most used optical sensors are based on fluorescence, surface plasmon resonance, and interferometry, and are often coupled with fiber optics. The detection may be direct or indirect through labeled probes. Presently, a substantial proportion of optical sensors are direct and do not involve labeling. Several reviews focus on optical biosensors.[12–14] *Fluorescence* is by far the most used optical method because of its high sensitivity and accuracy compared to other optical methods. A powerful fluorescent tag is the green fluorescent protein (GFP), a dye added in some cases to study living cells (indirect biosensing). There are other types of fluorescence biosensing, such as *fluorescence energy transfer* (FRET), when two different fluorophobes are used, and *molecular beacons*, where the electronic transfer occurs between a fluorescent molecule and a fluorescent quencher. Case studies where the fluorescence detection mode is used will be described in Section 6.1.2.3.

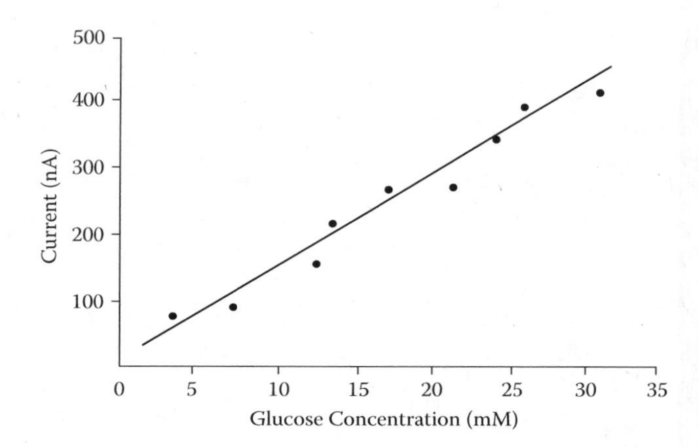

FIGURE 6.18 Sensor output current for different concentrations of glucose. (Adapted from Huang et al., *Sensors Actuators B*, 122, 461–68, 2007.)

6.1.2.1.4 Enzyme-Based Noninvasive Optical Absorbance Measurement System for Detecting Human Body Metabolites

A complete lab-on-a-chip (LOC) that uses submicroliter droplets as reaction chambers was fabricated and characterized.[15] To actuate microdroplets, the electrowetting technique is used because it can be easily integrated with the optical detection. This is an alternative to continuous flow devices, as it manipulates the liquid as unit-sized discrete microdroplets. Figure 6.19 shows the schematic of a device where the sample and the reagent droplets are transported and mixed. This figure shows the droplet sandwiched between two ITO electrodes, as well as the green LED and photodiode necessary for the optical detection.

A *colorimetric* enzyme-based method is used for the detection of glucose. A violet-colored compound called quinoneimine is formed by the reaction of hydrogen peroxide (product of the enzymatic oxidation of glucose) and 4-amino antipyrine (4-AAP) and a second reagent, N-ethyl-N-sulfopropyl-*m*-toluidine (TOPS). The violet compound has a strong absorption band at 545 nm, and its concentration, proportional to that of glucose, can be calculated from the absorbance value of this band by using the LED-photodiode setup. The glucose assay consists of dispersing droplets of the reagent and glucose sample, merging the droplets, mixing at an actuation voltage of 50 V, and

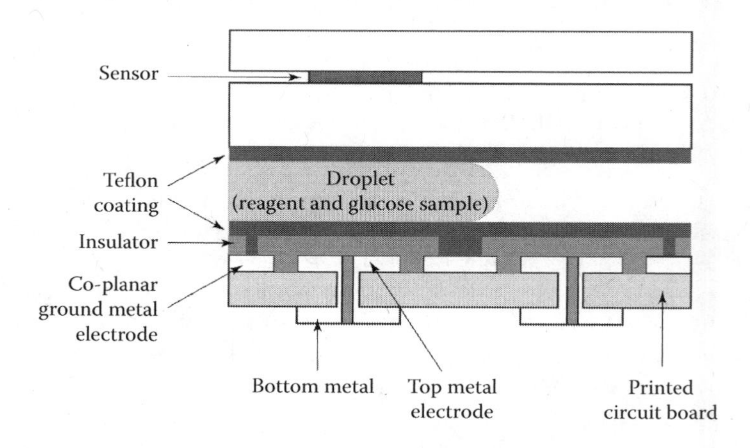

FIGURE 6.19 Electrowetting chip integrated on a printed circuit board. Sensor is shown on the top. (Adapted from Luan et al., *IEEE Sensors J.*, 8, 628–35, 2008.)

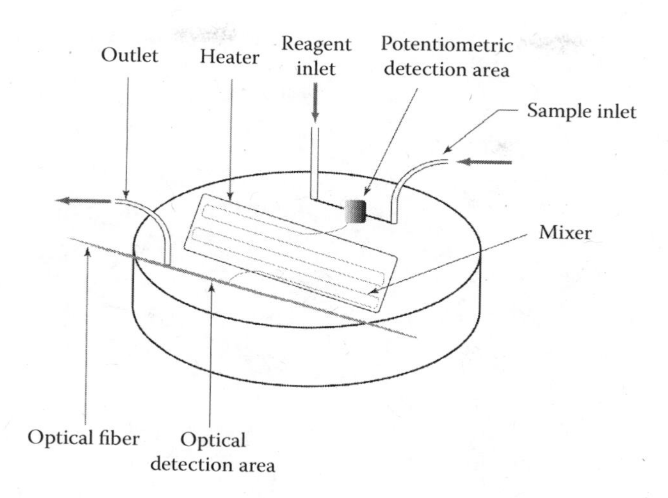

FIGURE 6.20 Optical flow cell with microreactor. (Adapted from Grabowska et al., *Microelect. Eng.*, 84, 1741–43, 2007.)

measuring the absorbance. The whole process is carried out in less than 40 sec. This device uses enzymes in solution, while most of the biosensors use immobilized enzymes.

6.1.2.1.5 Enzyme-Based Microfluidic System with Dual Detection of Urea and Creatinine

In order to make the results more reliable, the system described here[16] uses two independent detection methods: electrochemical and optical (Figure 6.20).

The concentrations of creatinine and urea are important in everyday clinical analysis for estimation of kidney function. The microsystem-microreactor is made of PDMS with a Y-microchannel that has a mixing area where the reagent is mixed with the sample. Both analytes form a colored complex with appropriate reagents. In the optical detection area there is a flow cell in which the absorbance measurements are made, while the opposite side is dedicated to potentiometric detection. Both test curves showed a good linearity in the region of clinical interest.

6.1.2.1.6 Microfluidic System Based on a Fabry-Perot Filter for Determining Protein Concentration in Human Biological Fluids Using a Photodetector

The microsystem described in this chapter is composed of one wafer containing the microlaboratory and another one the detection system.[18] The whole system, which includes the microchannel layout and the detection system, is depicted in Figure 6.21.

The Fabry-Perot filter is used in this work as a wavelength selecting element. Figure 6.21 shows the Fabry-Perot filter that consists of two highly reflective silver mirrors and a cavity medium (SiO_2). The protein concentration (albumin) is proportional to the absorbance at 628 nm due to the colored complex formed by the protein with bromcresol green.

A large category of optical biosensors is based on *refractive index detection*. The most important technique of optical sensing, which today concentrates the majority of research activities, is surface plasmon resonance (SPR). *Interferometer-based biosensors* are highly sensitive devices that detect changes in the refractive index.[19] The most important, the Mach-Zehnder interferometer, is shown in Figure 6.22a.

A change in refractive index at the surface of a sensor arm results in a change of optical phase on the sensing arm, and a subsequent change in the light intensity measured at the photodetector. Interferometric biosensors constitute one of the most sensitive integrated optic alternatives, compared to other optical biosensors (i.e., plasmonic biosensors) for label-free detection. In these sensors the guided light interacts with the analyte through its evanescent field. Mach-Zehnder and

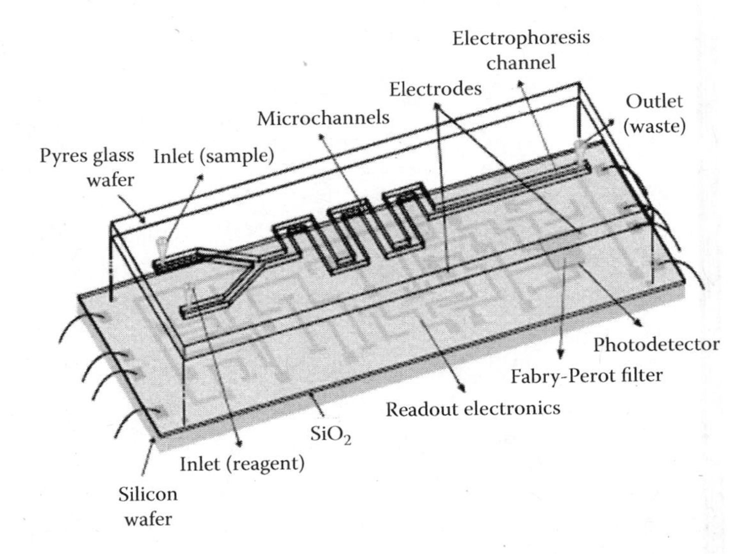

FIGURE 6.21 Micro total analysis system (μTAS) developed for clinical applications. (Reproduced from Minas et al., "High-Selectivity Detection in Microfluidic Systems for Clinical Diagnostics," in *EUROSENSORS XIV: The 14th European Conference on Solid-State Transducers*, Copenhagen, Denmark, August 27–30, 2000, pp. 873–79. With permission.)

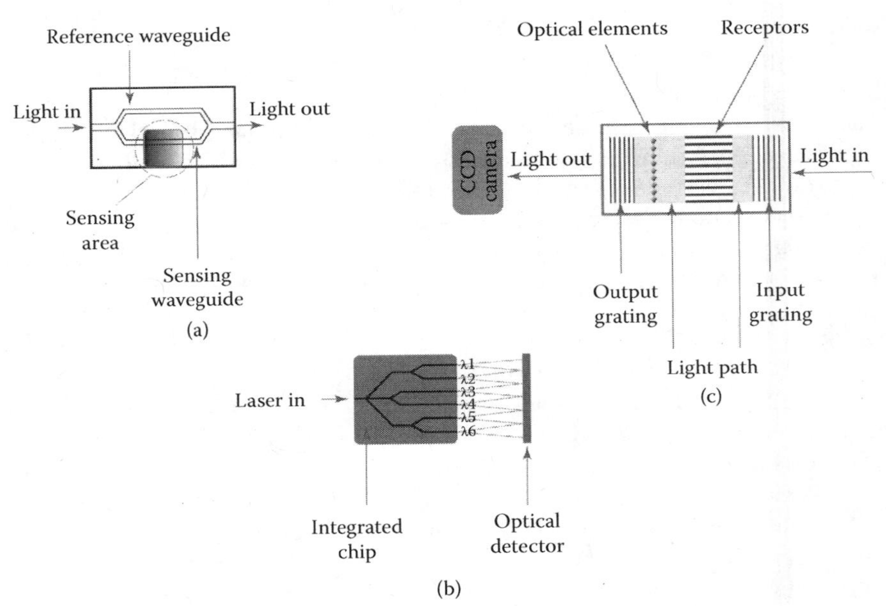

FIGURE 6.22 (a) Scheme of the Mach-Zehnder interferometer sensor. (b) Multichannel Young's interferometer sensor. (c) Hartman interferometer sensor.

Young interferometers are composed of an incident waveguide that is split into two single-mode waveguide branches, in which one of them contains a sensing window, as seen in Figure 6.22.

6.1.2.2 Nucleic-Acid-Based Biosensors

Nucleic-acid-based biosensors rely on nucleic acid binding events. In spite of the high accuracy, nucleic-acid-based biosensors are less developed than those based on antibody-antigen interactions.

The main reason is the difficulty to isolate nucleic acids from cells that require extensive sample preparation before amplification, hybridization, and detection. However, in recent years, there has been a considerable increase in the use of nucleic acids (DNA or RNA) as a tool in recognition and monitoring of many compounds of analytical interest, such as pollutants, toxic substances, antitumor drugs, pathogenic nucleic acid sequences, etc., due to the high stability and huge variability of nucleic acid sequences. Nucleic acid layers combined with electrochemical transducers produce a new kind of affinity biosensors for analytes of interest. Various types of detectors have been used in conjunction with the nucleic acids.

6.1.2.2.1 *Portable Lab-on-a-Chip Microsystems Based on Integrated Mach-Zehnder Interferometer (MZI) for Sensing Based on Nucleic Acids*

In these sensors, the guided light intensity interacts with the analyte through its evanescent field.

MZI biosensors have been used in gaz detection, as well as in the detection of different pollutants. In the present work, a single-stranded DNA was immobilized on the Si_3N_4 sensor surface by using silane chemistry. The MZI output signal (in V) is measured, and the phase change after immobilization is calculated. Finally, hybridization with the complementary DNA brings about a new phase change. The detection limit of this method was established at 10 nM.

6.1.2.2.2 *DNA-Based Biosensor for the Detection of Genetically Modified Organisms*

Genetically modified organisms (GMOs) are those whose genome has been modified by the introduction of an exogenous gene. In this work,[20] the hybridization of a nucleic acid (DNA) is associated with an electrochemical sensor. DNA is immobilized on the screen-printed electrode transducer, as shown in Figure 6.23, while the target of complementary oligonucleotides is in the solution.

Several electrochemical methods, such as cyclic voltammetry and chronopotentiometry, were used to detect the hybridization reaction. The multistep procedure of modifying a gold working electrode is presented in Figure 6.24.

To quantify the electrochemical signal, an enzyme-amplified detection scheme was used.

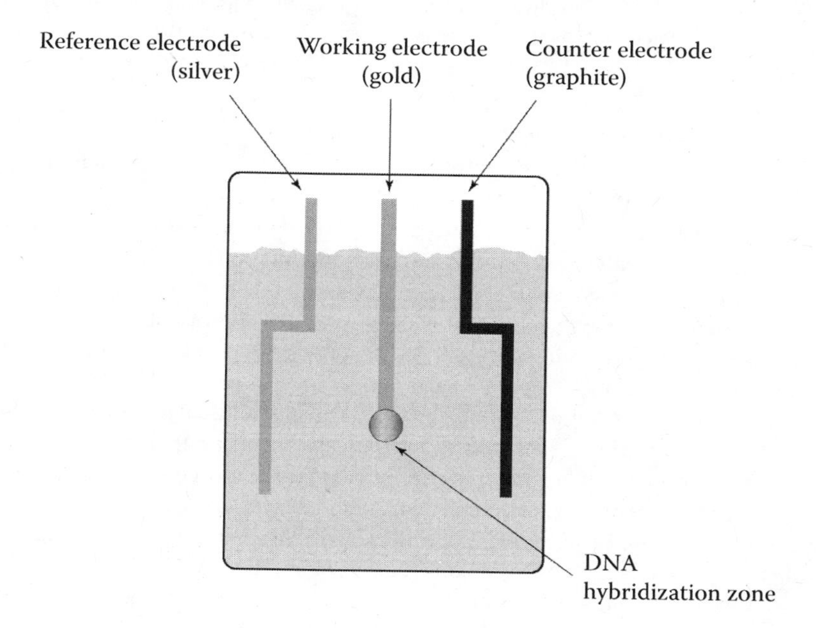

FIGURE 6.23 DNA hybridization on screen-printed electrodes. (Adapted from Nica et al., *Analele Univ. Bucuresti Chimie*, I-II, 85–94, 2004.)

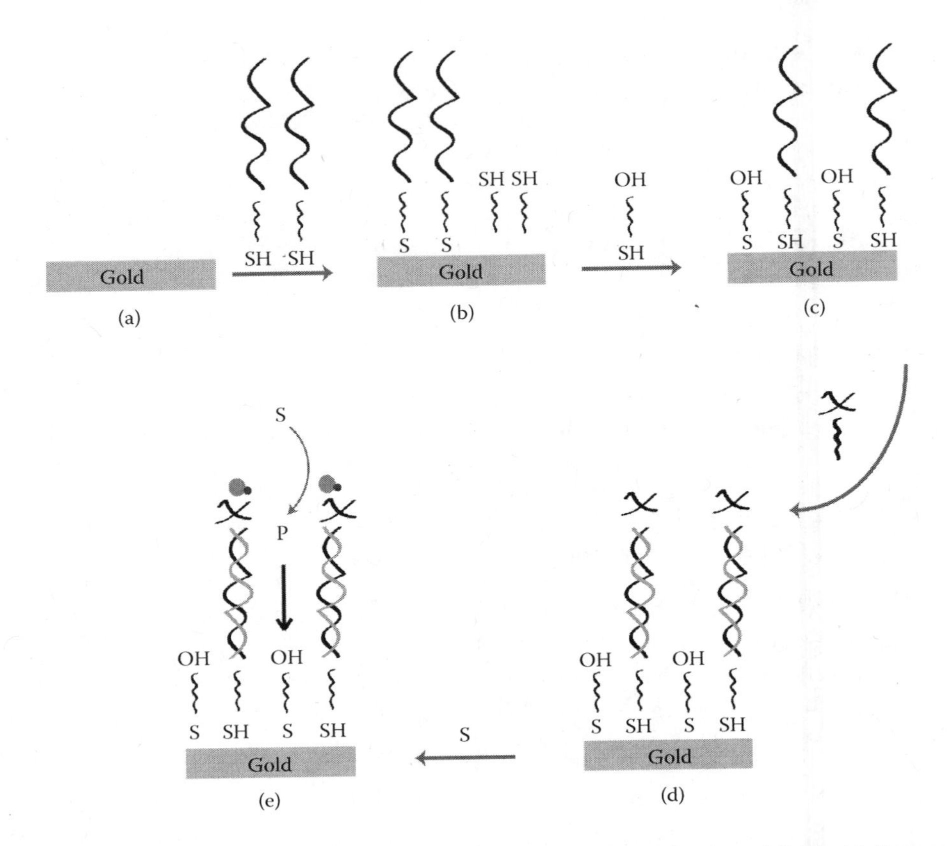

FIGURE 6.24 Procedure for the preparation of electrodes and hybridization of DNA (a) Bare gold (b) Chemisorption of 5'-mercaptohexyl-DNA on gold (c) Chemisorption of 6-mercapto-1-hexanol with the formation of a mixed monolayer (d) Hybridization with 5'-biotinylated complementary target (e) Incubation of enzyme-labeled DNA modified electrode with enzymatic substrate (S) with the formation of a precipitate. (Adapted from Nica et al., *Analele Univ. Bucuresti Chimie*, I-II, 85–94, 2004.)

6.1.2.3 Antibody-Based Biosensors

Antibody-based biosensors take advantage of specific antibody-antigen interactions, as shown in Figure 6.25.

Antibody-antigen interactions are reversible, and the strength of the binding is measured by the *affinity factor*, which represents the ease of association and dissociation.

Due to their target specificity, antibodies are considered the best molecular recognition units. Antibody-based biosensors are also called *immunosensors*. The use of antibodies for detection purposes increased significantly following the work of Köhler and Milstein[21] on the technology of monoclonal antibodies. Large quantities of monoclonal antibodies can be produced by this method using cell clones. Both polyclonal (pAb) and monoclonal (mAb) antibodies are widely used as diagnostic tools in clinical and research areas.[22] Antibodies and antigens are generally large, structurally complex proteins, and their interaction with the biosensor substrate may lead to a loss of specificity and biological activity. This may happen even when monoclonal antibodies are used, which are more specific. For these reasons, antibodies have been engineered to develop robust and stable formats that will facilitate the preparation of antibody fragments for specific biosensor applications.[23] The immobilization methods that are used to attach antibodies or antibody fragments to the transducer surface are physical adsorption, covalent coupling, and more elaborated methods, such as antibody fragment tag or antibody fragment fusion. Because the method of immobilization is a critical step in biosensing, it has to be optimized in order to have an oriented immobilization, as schematically depicted in Figure 6.26.

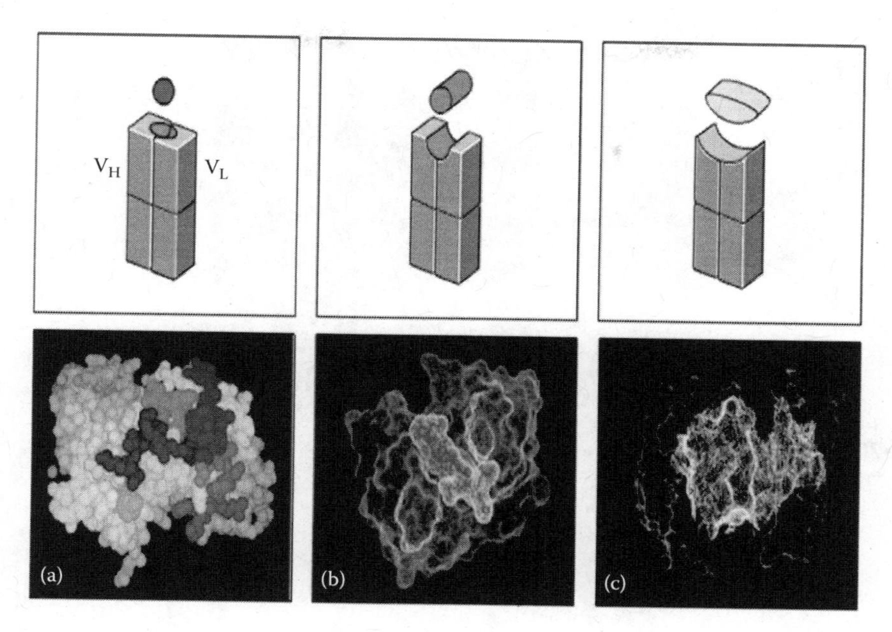

FIGURE 6.25 Antibody-antigen interaction. (a), (b) and (c) show the interaction for different shapes.

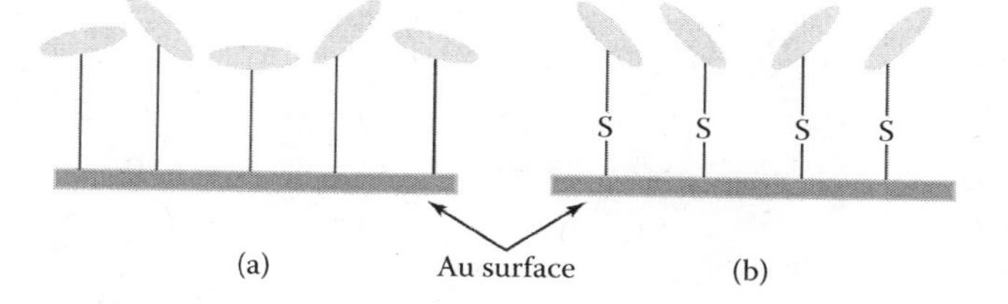

FIGURE 6.26 Antibody-antigen immobilization through random coupling (a) and directional coupling (b).

Oriented immobilization of antibodies onto the gold-coated biosensor surface is often realized by using self-assembled monolayers (SAMs) with thiol (SH) groups. Other methods to orient antibodies or antibody fragments make use of the high affinity between biotin and streptavidin. Antibodies are first biotinylated, and then coated onto the transducer surface, and orientation is achieved using the interaction with streptavidin.

The major advantage of antibody-based biosensors is that the target does not have to be purified before detection. Optical and electrochemical signal transduction techniques have been developed; the most used ones are enzyme fluorescence based. Some case studies on antibody-based biosensors are discussed below.

6.1.2.3.1 Antibody-Based Microfluidic Chip on Silicon Platform Integrated with a Spectrometer-on-Chip for Fluorescence-Based Biosensing

Experiments carried out with different silicon substrates showed that polished silicon is well suited for the immoblilization of fluorescent antibodies.[24] The microfluidic chip (see Figure 6.27) consisted of a detection channel and a rinsing channel with a mixing zone, and the channels were fabricated by anisotropic etching.

The schematic setup used for the experiment is shown in Figure 6.28, and a photo of the actual instrument is shown in Figure 6.29.

The antigen (antigen sheep IgG) was first adsorbed on the surface of the channel, then rinsed with phosphate-buffered saline (PBS) and the blocker (bovine serum albumin (BSA)), and finally,

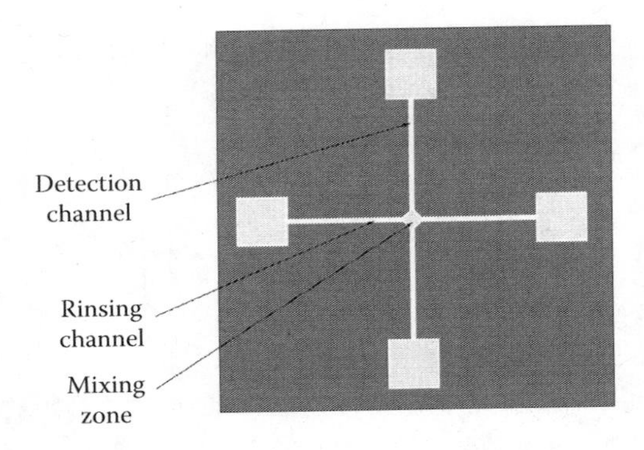

FIGURE 6.27 Schematic of the microfluidic channel. (Adapted from Chandrasekaran et al., *Sensors*, 7, 1901–15, 2007.)

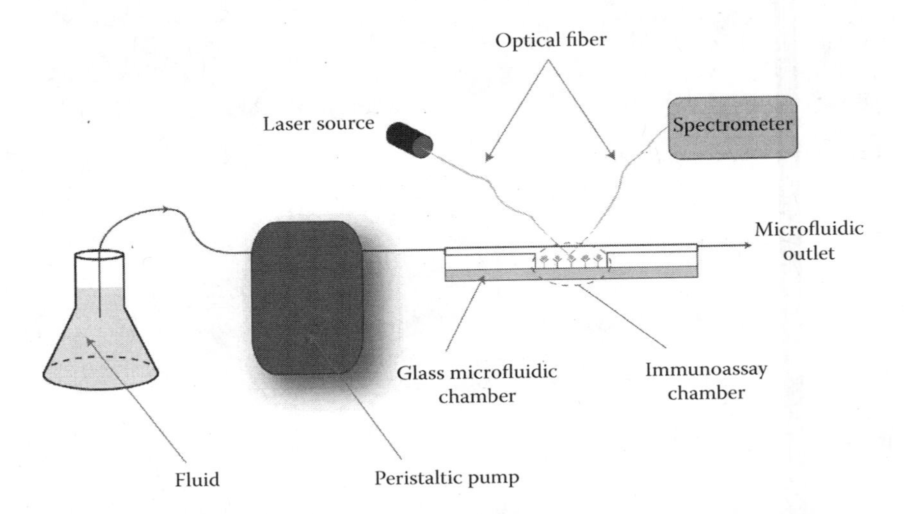

FIGURE 6.28 Biooptical fluorescence detection setup. (Adapted from Chandrasekaran et al., *Sensors*, 7, 1901–15, 2007.)

the antibody was introduced and fluorescence measurements were carried out for different concentrations of the antibody.

Among the advantages of this method are the inexpensive packaging techniques with Pyrex glass and polycarbonate materials. The immobilization of the antigen molecules onto the surface of silicon microfluidic channels has been carried out without using specialized surface treatment techniques. Later, a method for the amplification of fluorescence produced by the fluorophobes within the microfluidic channels was devised by the same group. Gold was used as a reflector material, and the schematic corresponding to the integration of reflectors into the channel surfaces.[25]

6.1.2.3.2 Gold Nanoparticle Ring and Hole Structures for Sensing Proteins and Antigen-Antibody Interactions

A new experimentally simple nanosphere lithography method was used to fabricate gold ring and nanohole structures.[26] The method is based on the simultaneous self-assembly of polystyrene microspheres and gold colloids in multilayers, by a vertical deposition method. Dissolution of polystyrene microspheres resulted in the formation of a monolayer, where holes are surrounded by gold nanoparticles (Figure 6.30a).

FIGURE 6.29 Experimental setup of the antibody-based biosensor built in the Optical Microsystem Laboratory, Concordia University. (Adapted from Chandrasekaran et al., *Sensors*, 7, 1901–15, 2007.)

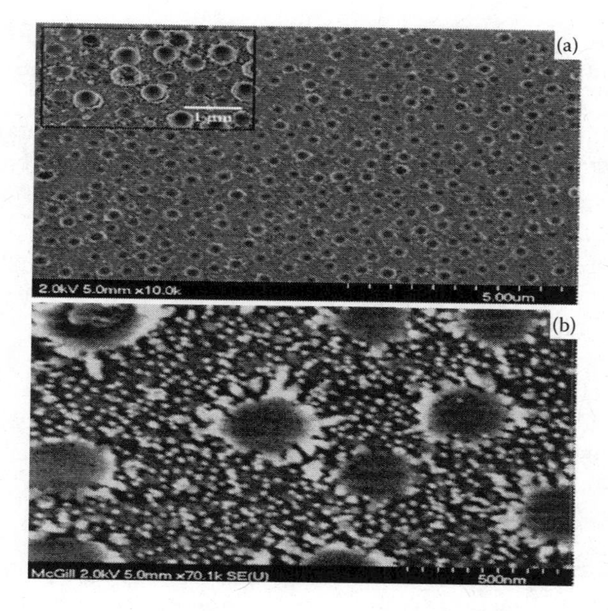

FIGURE 6.30 (a) SEM image of a nanohole array prepared with 530 nm of PS and 20 nm of Au. Inset: Enlarged image of a region where PS microspheres were not completely removed by dissolution. (b) High-resolution image of the nanohole structures. (Adapted from Fida et al., *Plasmonics*, 4, 201–7, 2009.)

The detection method used in this work is an optical method based on surface plasmon resonance (SPR) in gold nanoparticles (localized surface plasmon resonance (LSPR)). The method is based on the sensitivity of the LSPR band of gold to the refractive index of the environment. For an excellent review, see Homola.[27]

The sensitivity of the sensor platform depends on the size of the holes and their density. Furthermore, sensing experiments have shown a high sensitivity of the hole structure toward

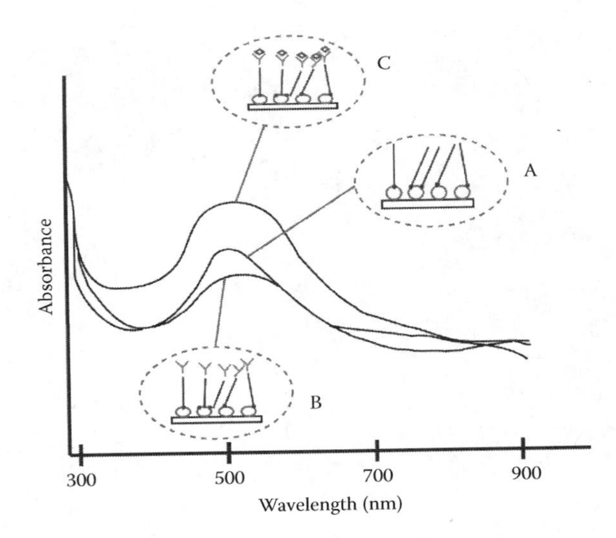

FIGURE 6.31 Spectra corresponding to the antigen-antibody interaction: (A) spectrum of the functionalized substrate, (B) spectrum of the antibody incubated on the functionalized gold substrate, and (C) spectrum of the antibody-antigen pair on the functionalized gold substrate. (Reproduced from Fida et al., *Plasmonics*, 4, 201–7, 2009. With permission.)

fibrinogen, amyloid-derived diffusible ligands, and a plant protein (AT5G07010.1). The position and shape of the localized surface plasmon resonance band changed significantly as a result of the antigen-antibody recognition event (Figure 6.31).

The figure shows that after the adsorption of the protein (the concentration of the protein was 4.3×10^{-7} M), the protein-antibody interaction brought about a significant broadening of the band toward the longer wavelengths. The broadening of the band and the presence of the new band are indicative of a strong interaction between the antigen and antibody. Due to the nonuniformity of the sensor platform in terms of sites of adsorption (nanoparticles, small clusters, holes), the LSPR band of gold and functionalized gold is broad. However, the adsorption of antibody results in a shift of 10 nm, and next, the adsorption of the protein brings about a new change—the strong broadening of the band toward longer wavelengths with the 610 nm shoulder.

It is concluded that the antigen-antibody recognition event resulted in significant changes in the position and shape of the LSPR band, showing good potential for the ring and hole structures to be further used to study these interactions.

6.1.2.3.3 Antibody-Based Fluorescence Biosensor Based on OLED Technology

In this work, a fluorescence detector was coupled to a microfluidic device.[28] The light source in this device is an organic light-emitting diode (OLED) fabricated on ITO-coated glass substrates and then encapsulated by a glass cover to prevent exposure to moisture. Donkey anti-sheep IgG conjugates tagged to Alexafluor 488 are used as a sample. Based on the fluorescence response, the authors have found that the minimum volume detected in the chamber was 0.012 μl of the sample that contained 2.4 ng of donkey anti-sheep IgG conjugates.

6.1.2.4 Microbial Biosensors

This category of biosensors relies on microorganisms as the biological recognition element, and they generally involve the measurement of microbial respiration. Microbial biosensors interfaced with optical or electrochemical detectors are less expensive than the enzyme-based approaches, but they involve long assay times because of the cellular diffusion characteristics. Microbial biosensors are particularly applicable for environmental monitoring.

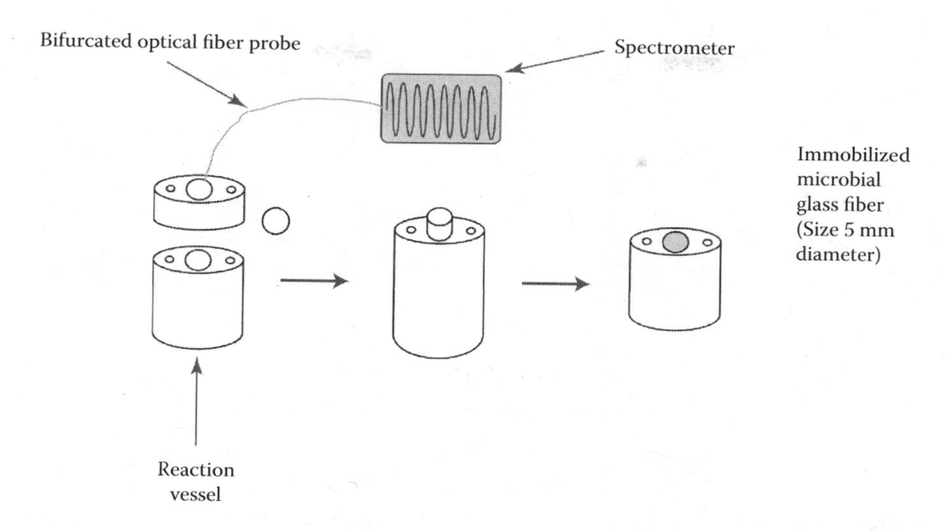

FIGURE 6.32 Operating system of the microbial optical biosensor. (Adapted from Kumar et al., *Biosensors Bioelect.*, 21, 2100–5, 2006.)

6.1.2.4.1 *Optical Microbial Biosensor for Detection of Organophosphorus Pesticides*

The authors[29] developed an optical biosensor for the detection of methyl parathion pesticide, a highly active organophosphorus insecticide that is toxic to mammals. The biosensor contains a disposable microbial membrane (the whole cells of *Flavobacterium* sp. were immobilized in a small disk of glass fiber filter paper), and the concentration of methyl parathion is determined optically by measuring the absorbance of *p*-nitrophenol that results by the hydrolysis of the pesticide. The schematic diagram of the microbial biosensor is shown in Figure 6.32.

The results were compared with those obtained previously by using an enzyme-based biosensor, and a lower detection limit (0.3 μM) was found from the linear range of the calibration plot.

6.1.2.4.2 *Amperometric Microbial Biosensor for Monitoring the Microbial Corrosion of Metallic Materials in Industrial Systems*

In this work,[30] microbes are isolated from a microbiologically corroded metal surface and immobilized on a porous acetylcellulose membrane. The oxygen consumption was measured to estimate the respiratory activity of the microbe.

The microbial biosensor assembled, as shown in Figure 6.33, was placed into the measuring cell containing a sulfuric acid solution of known concentration. Sulfuric acid is the most corrosive acid involved in the microbial corrosion, and it results from the various sulfur contents oxidized by bacteria such as *Thiobacillus* sp. It was found that the response of the microbial biosensor was stable and reproducible. A linear relationship between the response and the concentration of sulfuric acid was found under various conditions of pH and temperature. The biosensor can be used for monitoring the microbial corrosion.[31]

6.2 IMMUNOASSAYS

6.2.1 INTRODUCTION

An *immunoassay* is a biochemical test used to detect and quantify a specific substance called analyte in a blood or body fluid sample, using an *immunological reaction*, that is, the specific binding of an antibody to its antigen. An *antibody* is a protein (immunoglobulin) produced by B-lymphocytes (immune cells), in response to stimulation by an antigen. Antibodies are found in the blood of

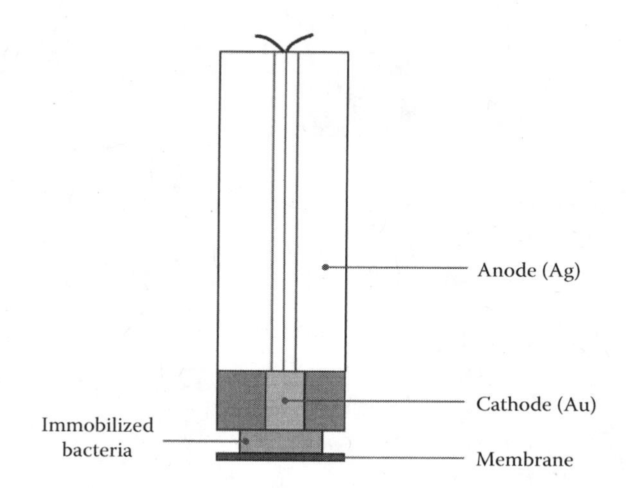

FIGURE 6.33 *Pseudomonas*-based amperometric microbial biosensor species. (Reproduced from Dubey and Upadhyay, *Biosensors Bioelect.*, 16, 995–1000, 2001. With permission.)

vertebrates, and are used by the immune system to identify and neutralize foreign organisms such as bacteria and viruses. They are typically made of basic structural units, each with two large heavy chains and two small light chains (see Figure 6.34).

Generally, *monoclonal antibodies* are used because they bind to one site of a particular molecule, providing a more specific test. Immunoassays detect the formation of antibody-antigen complexes through an *indicator reaction*. An *antigen* is a substance that prompts the generation of antibodies and causes an immune response. Immunoassay takes advantage of the ability of antibodies to bind selectively to the specific physical structure of the target analyte present in a sample. Working like a key and lock, the binding sites on an antibody attach noncovalently to their corresponding target analyte, also known as the antigen. Because the binding is based on the antigen's physical shape rather than its chemical properties, antibodies do not respond to substances that have dissimilar structures. Compound-specific immunoassay kits have been developed to detect only the target analyte and its metabolites. In the environmental field, class-specific kits have demonstrated the most use. Immunoassay is the method of choice for measuring analytes normally present at very low concentrations (in the range of pM) that cannot be determined accurately by any other test.

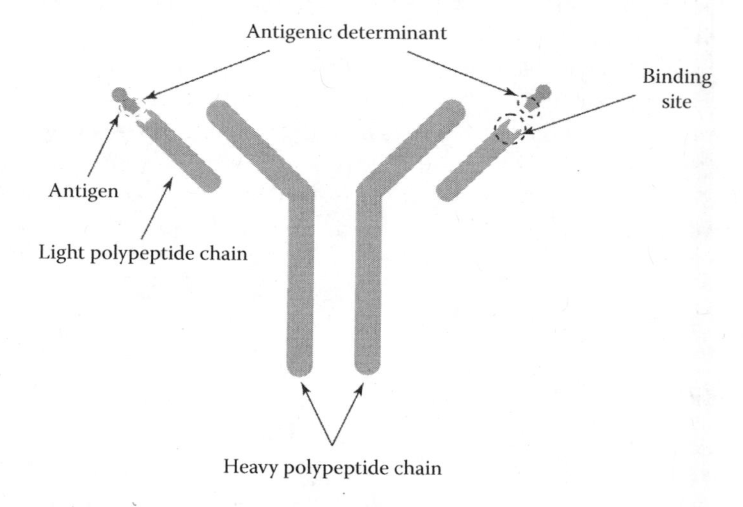

FIGURE 6.34 Antibody's structure and its interaction with an antigen.

Common uses of immunoassays include the measurement of many clinically important analytes, such as drugs, hormones, specific proteins, tumor markers, lipoproteins, viruses, bacteria, etc., present in biological samples.

Immunoassays have an important role in the diagnosis of many infectious diseases. When trying to detect the presence of an infection, the concentration of antibody specific to that particular pathogen is measured. The response of the sample should be, of course, compared to standards of known concentration. Detecting the quantity of antibody or antigen can be achieved by a variety of methods. One of the most common is to label either the antigen or the antibody. Immunoassays with labeled reagents are usually divided into homogenous and heterogeneous. *Heterogeneous immunoassay* refers to the adsorption of an antigen onto an antibody attached to a polymer or glass substrate, and it requires an additional step to remove the unbound antibody or antigen from the site using a solid phase reagent.

Immunoassays can be competitive or noncompetitive. In a *competitive immunoassay*, the antigen in the unknown sample competes with a labeled antigen to bind with antibodies, and the amount of labeled antigen bound to the antibody site is then measured. The response is inversely proportional to the concentration of antigen in the sample because the greater the response, the less antigen in the unknown sample that was available to compete with the labeled antigen. In *noncompetitive immunoassays*, also referred to as the sandwich assays, the antigen in the unknown sample is bound to the antibody site, and then a labeled antibody is bound to the antigen. Unlike the competitive method, the results of the noncompetitive method are directly proportional to the concentration of the antigen because a labeled antibody will bind only if the antigen is present in the unknown sample. More accurate analytical results are obtained by using the sandwich immunoassay because an antigen may form a complex with two antibodies and the free ones can be easily separated from the complex. *Homogeneous assays* are typically faster and easier to perform. In this format, the immune complex is formed between the unimmobilized antigen and antibody. However, the sensitivity and selectivity of this test are low because of the one-to-one antigen-antibody interaction, and an additional separation step of the unreacted antigens and antibodies may be required.

6.2.2 Enzyme-Linked Immunosorbent Assay (ELISA)

Enzyme-linked immunosorbent assay (ELISA; also called enzyme immunoassay (EIA)) is a biochemical technique used in immunology to detect the presence of an antibody or an antigen in a sample. ELISA has been used as a diagnostic tool in medicine and plant pathology, as well as a quality control method in biotechnology. In simple terms, in ELISA, an unknown amount of antigen is affixed to a surface, and then a specific antibody is washed over it so that it can bind to the antigen. The antibody is linked to an enzyme, and in the final step, a substance is added that the enzyme will convert to a detectable signal. Sensitive colorimetric reactions are used to identify the concentration of the analyte.

The presence of the analyte is determined by comparing the color developed by the sample of an unknown concentration with the color formed by a reference containing the analyte at a known concentration. The concentration of the analyte is determined by the intensity of color in the sample, in comparison with the color/concentration values on a chart. It can be measured more accurately with a photometer or spectrophotometer, and the measurement can be compared to a reference value. In the case of *fluorescence ELISA*, when light of an appropriate wavelength is shone upon the sample, antigen-antibody complexes will fluoresce, and the amount of antigen in the sample is inferred through the intensity of the fluorescence. ELISA has also found applications in the food industry for detecting potential food allergens, such as milk, peanuts, walnuts, almonds, and eggs. Prior to the development of the EIA/ELISA, immunoassays were conducted using radioactively labeled antigens or antibodies by a technique called *radioimmunoassay*. In this case, it is the radioactivity that provides the signal that indicates whether a specific antigen or antibody is present in the sample. Radioimmunoassay was first described in a paper by Rosalyn Sussman Yalow and

Solomon Berson published in 1960.[32,33] Since its introduction, radioimmunoassay (RIA) has made it possible to measure analytes such as hormones and drugs at extremely low concentrations, thus opening a new avenue of sensitive assays. In spite of its success, the use of an isotopically tagged antigen or antibody led to problems associated with waste disposal, short half-life, and radiolysis of the labeled marker. In order to avoid these drawbacks, while retaining the specificity of an immunoassay, various alternatives to RIA have been developed, such as enzyme immunoassay (EIA), fluorescent immunoassay, particle immunoassay, spin immunoassay, chemiluminescent immunoassay, and bioluminescent immunoassay. Among these, the enzyme immunoassay methods described previously are the most useful techniques, since they are as sensitive as RIA and the required laboratory equipment is relatively inexpensive and simple. Recently, chemi- and bioluminescence have had an impact on biochemical analysis, medicine, and clinical diagnosis. *Chemiluminiscence* is the phenomenon observed when the excited product of a chemical process reverts to its ground state with emission of light. A large number of molecules are capable of exhibiting chemiluminiscence, but in most cases, the chemiluminiscence is too weak to be used for analytical purposes. When certain enzymes such as peroxidase react with appropriate substrates, the change in color can be used as a signal. However, the signal has to be associated with the presence of an antibody or antigen, which means that the enzyme has to be linked to an appropriate antibody. The linking process was independently developed by Stratis Avrameas and G. B. Pierce.[34,35] The immunosorbent is prepared by attaching an antibody or antigen to the surface of the container. The ELISA test, or the enzyme immunoassay (EIA), was the first screening test commonly employed for HIV diagnosis. In an ELISA test, a person's serum is diluted four-hundred-fold and applied to a plate to which HIV antigens have been attached. If antibodies to HIV are present in the serum, they may bind to these HIV antigens. The plate is then washed to remove all the other components of the serum, and a specially prepared "secondary antibody"—an antibody that binds to human antibodies—is then applied to the plate, followed by another wash. This secondary antibody is chemically linked in advance to an enzyme. The enzymes generally used in immunoassays are alkaline phosphatase (ALP) or galactosidase (which catalyze a hydrolysis reaction) and horseradish peroxidase or glucose oxidase (which catalyzes an electron transfer reaction). The plate will contain enzyme in proportion to the amount of secondary antibody bound to the plate. When a substrate for the enzyme is applied, the catalysis by the enzyme leads to a change in color or fluorescence. Detection becomes possible when a second antibody is added, which is prepared from the serum of an animal injected previously with human antibody. The human antibody in this case serves as an antigen, and the animal thus produces an antibody against the human antibody. Once isolated, the second antibody can be chemically linked to a system that can produce a detectable signal. In ELISAs, the antigen-antibody complex is exposed to the second antibody, which binds to the antibody portion of the complex, creating a sandwich type structure. The signaling system consists of an enzyme attached to the second antibody. As has been previously mentioned, the enzyme converts it to a colored substance that can be measured when the appropriate chemical is added (Figure 6.35).

This test quantifies how much enzyme is present by the intensity of the color produced. The presence of more enzyme means that more secondary antibody needs to be attached. The amount of secondary antibody is determined by the amount of the target that is available. As the first antibody binds to antigen, the more antigen that is accessible, the higher the amount of first antibody that will be retained. The measure of the absorbance at a wavelength, therefore, reflects the amount of the antigen initially present. The technology is used widely for environmental field analysis, as the antibodies can be highly specific to a target compound or group of compounds. Immunoassay kits are relatively simple to use. ELISA is used in the field of the environment because of its speed, sensitivity, selectivity, long shelf life, and simplicity of use. Antibodies for ELISA have been developed specifically to bind to a selected environmental contaminant(s). The selective response is used to confirm the presence of the contaminant(s) in samples. During the first step of the immunoassay, the walls of a test tube may be coated with the antibodies, or the antibodies may be introduced into the test tube coated with magnetic or latex particles. With either method, the quantity of antibodies

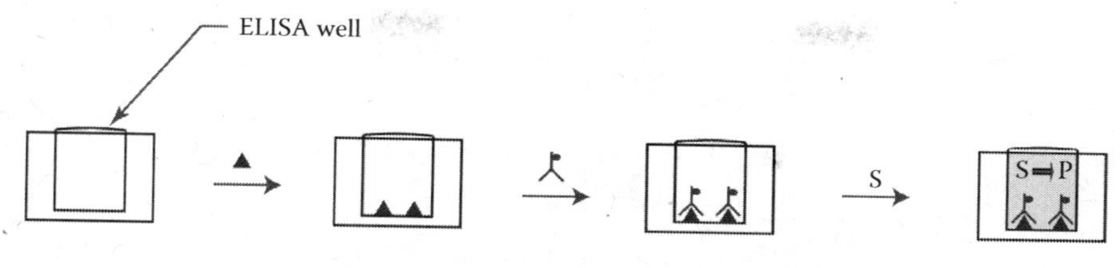

Antigen (▲) is added to walls or a microtitration plate

An enzyme-linked Antibody (⅄) is added. The antibody binds to antigen.

A colorless substrate (S) is added. The substrate is converted to a colored product (P) by the action or the enzyme-lined antibody.

FIGURE 6.35 Sequences of ELISA immunoassay.

and their binding sites are known. Some of the contaminant, or the antigen, is combined with an enzyme that will react with an appropriate colorimetric reagent to produce a color change that does not interfere with the antigen's ability to bind with the antibodies. The enzyme labels the antigen and allows for the detection of the antigen's presence. The solution containing the labeled antigen is called the *enzyme conjugate*. When the colorimetric reagent, or *chromogen*, is added to the solution, it reacts with the enzyme on the labeled antigen and forms a colored compound. During the analytical procedure, a known amount of sample and enzyme conjugate are introduced into the test tube that contains the antibodies, and the target analyte present in the sample will compete with the labeled antigen in the enzyme conjugate for a limited number of antibody binding sites. According to the law of mass action, if more analyte is present in the sample, the more enzyme conjugate of the analyte that will be displaced from the binding sites. The amount of bound conjugate determines the intensity of the color and is inversely proportional to the amount of analyte in the sample. The original concentration of the analyte can be determined by measuring the amount of enzyme conjugate bound to the antibody. For example, the immunoassay system for the measurement of mercury shows a direct relationship between the concentration and the absorbance.

In most clinical and surveillance laboratories of the developing world, this process relies on serological techniques aimed at the detection of pathogen-specific antibodies. These are frequently measured with ELISA. The success of these tests is due to their high sensitivity and specificity, good reproducibility, and a high throughput capacity with affordable costs. The specificity of the ELISA test is shown in Figure 6.36.

6.2.3 Microfluidic Immunoassay Devices

Immunoassays are commonly performed on microtiter plates with typically 6, 24, 96, 384, or 1,536 sample wells. The immunoreaction itself is a rapid process. Due to many mixing and washing steps, and the long incubation times involved in the test, an ELISA test in a standard microtiter plate may take from several hours to one or two days. In addition, the test typically requires large amounts of quite expensive reagents and about 0.5 to 2 ml of sample per test. In order to shorten the test, several methods are available to integrate the immunosorbent assay system into a microfluidic device.[36–38] Relevant information related to microfluidic immunosensing is given in Liu and Wang.[9] In microfluidic immunoassays, the microchannel may serve as an immunoreactor chamber that leads to a drastic reduction of the amounts of reagents and samples. Fused silica capillaries have been used as immunosensors.[39–41] Fluorescence, surface plasmon resonance (SPR), and electrochemical detection techniques are implemented in microfluidic immunoassays. Surface plasmon fluorescence spectroscopy (SPFS) and total internal reflection fluorescence microscopy (TIRFM)

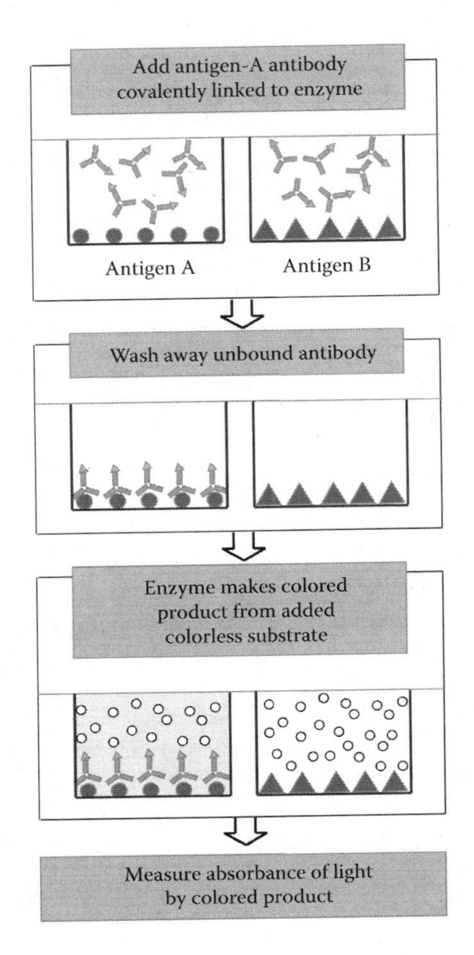

FIGURE 6.36 Specificity of the ELISA test.

are used as well. In the latter, the laser is internally reflected on the surface-fluid interface, and it is the evanescent field that excites the fluorescent molecules.[42] Two case studies are presented below.

6.2.3.1 A Compact-Disk-Like Microfluidic Platform for Enzyme-Linked Immunosorbent Assay

As has been mentioned previously, due to the series of steps involved and long incubation times, conventional ELISA involves a tedious protocol due to inefficient mass transport of the reagents from bulk solution to solid surface. The test requires hours of incubation for diffusion-limited reactions on the surface, and bulky instruments for optical detection. Integrated *microfluidic platforms* provide advantages such as high specific surface area, low reagent consumptions, and short diffusion length, and can significantly reduce the assay time and sample or reagent consumptions, and also enhance the reaction efficiency. An integrated microfluidic device on a compact disk (CD) has been developed to automatically perform ELISA for rat IgG from hybridoma cell culture.[42] The design of the CD-like ELISA chip is shown in Figure 6.37, with twenty-four sets of ELISA.

The microfluidic platform combined several microfluidic functions, such as capillary valving, centrifugal pumping, and flow sequencing. The rotation speed of the disk is computer controlled, and the flow sequence of several different solutions involved in the ELISA process is controlled by centrifugal and capillary forces. The whole ELISA process is carried out automatically after preloading the reagents into corresponding reservoirs. The antigen solution is released into the

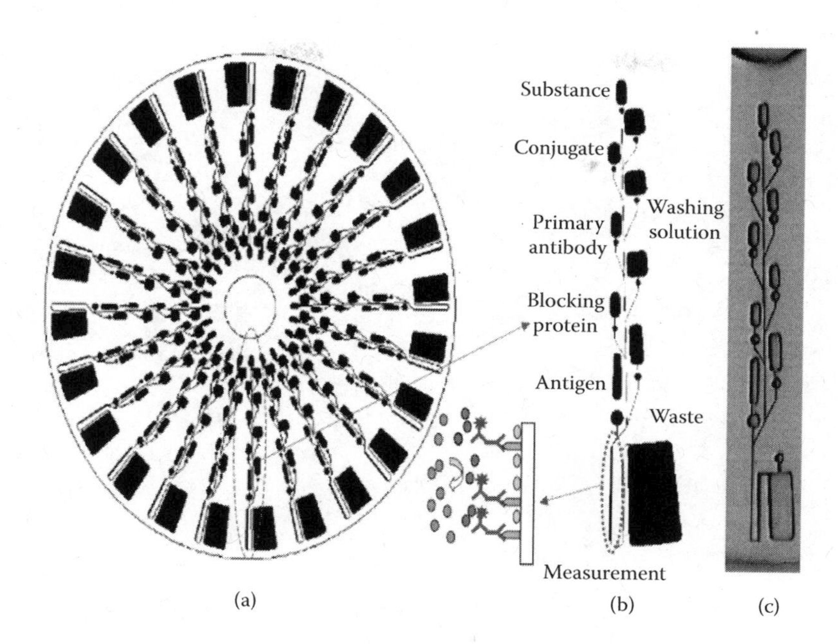

FIGURE 6.37 Schematics of (a) a CD ELISA design with twenty-four sets of assays. (b) A single assay (1, waste; 2, detection; 3, first antibody; 4, 6, 8, 10, washing; 5, blocking protein; 7, antigen/sample; 9, second antibody; 11, substrate). (c) Photo of a single assay. (Reproduced from Lai et al., *Anal. Chem.*, 76, 1832–37, 2004. With permission.)

measurement site at a low rotation speed, and after incubation, the unbounded antigen is washed out to the waste reservoir. The blocking protein, washing solution, primary antibody, conjugate solution, washing solution, and finally, substrate are delivered sequentially to the measurement site. Within the same detection range, similar results can be found by controlling the rotation speed in the microchip and microtiter plate. The detection limit was found to be 5 mg/l (31 nM) of the rat IgG (MW ~ 160,000). The results showed that both the consumption of reagents (antigen and antibody) and the assay time are reduced to about one-tenth of that required by the microtiter plate.

6.2.3.2 Portable Low-Cost Immunoassay for Resource-Poor Settings

An integrated approach to a portable and low-cost immunoassay for resource-poor settings has been described by Sia et al.[44] Interestingly, in this miniaturized immunoassay (Figure 6.38), antibodies are conjugated to gold colloids, instead of enzyme-conjugated secondary antibodies.

A solution of silver nitrate is added, and silver ions are reduced to silver atoms by hydroquinone. An InGaAlP a red semiconductor laser diode is used as a light source and the transmission of the silver film is measured optically. This immunoassay was developed in a microfluidic format as well, and the results for anti-HIV-1 antibodies in the sera of HIV-1-infected patients were compared to those obtained in the microwell format. The incubation times were found to be 10 min in the microfluidic device compared to 1 to 3 h in the microwells. The advantage of using a silver film for detection is the possibility of using a wide variety of laser diodes, as the silver film blocks the light over a broad region of wavelengths. The use of this integrated immunoassay is recommended for applications in resource-poor settings such as developing countries.

6.3 COMPARISON BETWEEN BIOSENSORS AND ELISA IMMUNOASSAYS

Biosensors that monitor antibody-antigen interactions are also termed *immunosensors*. They are based on the principles of ELISA, with either antibody or antigen immobilized on the sensor

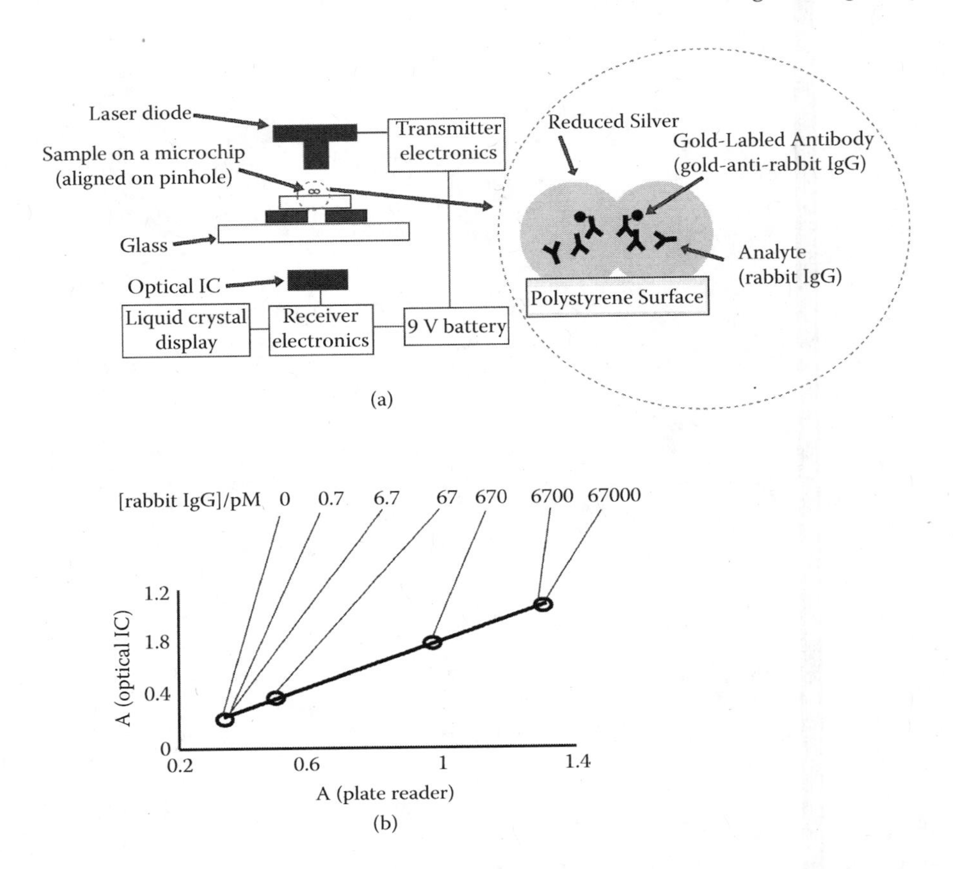

FIGURE 6.38 Schematics of the POCKET immunoassay device. (a) Red light from the laser diode passed through the silver-coated microwell containing the sample to the optical IC. A pinhole was used to block stray light that did not pass through the sample. The laser diode and the optical IC were driven by the same circuit, which also had an integrated liquid crystal display that showed the measured transmittance value. (b) An immunoassay using silver reduction was performed on a ninety-six-well plate that detected rabbit IgG. Optical micrographs of the silver films on microwells are shown for each sample. (Adapted from Sia et al., *Angew. Chem. Int. Ed.*, 43, 498–502, 2004.)

surface. The sensitivity and specificity of these particular biosensors are determined by the same characteristics as in ELISA, that is, the affinity and specificity of the binding.

As has been shown, a biosensor is an analytical device composed of a biological recognition element in intimate contact with a signal transducer that relates the concentration of a target analyte to a measurable response. A biosensor is a chemical sensor in which the detection of the analyte is done using a biologically based mechanism. They are considered to be located in the center of a continuum of analytical technologies that ranges from chemical sensors such as a pH electrode to complex multistep bioanalytical assays such as ELISA. However, there are two characteristics that distinguish biosensors from ELISA and other immunoassays:

- Biosensors are able to detect and measure the analyte directly, in real time and in a single step (for example, unlabeled antibody), whereas for most immunoassays the measurement of the antibody-antigen binding is a complex multistep process. Regardless of the detection system employed, all ELISAs require the use of an appropriate enzyme label and a matching substrate that is suitable for the detection system being used.

- In biosensors, the immobilized biological recognition element may be regenerated and used for several assays, whereas immunoassays are typically based on irreversible binding and are used only once.

Other authors consider that there are direct and indirect biosensors: In the *direct biosensor* the antibody-antigen interaction is detected in real time (true biosensor). The *indirect biosensor* (immunoassay) needs a preliminary biological reaction (reagents are used only once), and the product of this reaction is detected. The combination of highly specific biological reactions with an appropriate transducer gives biosensors other advantages, such as high sensitivity, specificity of the biological element, and short response time.

Biosensors make use of the scientific knowledge and clinical practice accumulated during the last century, when immunoassays were the only analytical tool for the detection and measurement of biologically important analytes. Nowadays, combining advances in bio- and nanotechnology, biosensors are moving out of the laboratory and more and more into commercial applications. They are exploring the benefits of miniaturization and large-scale fabrication to ensure their commercial success.

REFERENCES

1. Nagel, B., Dellweg, H., Gierasch, L.M. 1992. Glossary for chemists of terms used in biotechnology (IUPAC Recommendations). *Pure & Appl. Chem.* 64:143-168.
2. Chambers, J. P., Arulanandam, B. P., Matta, L. L., Weis, A., Valdes, J. J. 2008. Biosensor recognition elements. *Curr. Issues Mol. Biol.* 10:1–12.
3. Clark Jr., L. C., Lyons, C. 1962. Electrode systems for continuous monitoring in cardiovascular surgery. *Ann. NY Acad. Sci.* 102:29–45.
4. 4. Guilbault, G.G., Montalvo, J. 1969. Ureea-specific enzyme electrode. *J. Am. Chem. Soc.* 91: 2164.
5. Magner, E. 1998. Trends in electrochemical biosensors. *Analyst* 123:1967–70.
6. Liedberg, B., Nylander, C., Lundstrm, I. 1983. *Sensors Actuators* 4:299–304.
7. Tatke, S.S., Loong, C.K., D'Souza, N., Schoephoerster, R.T., Prabhakaran, M. 2007. Large scale motions in a biosensor protein glucose oxidase: A combined approach by QENS, normal mode analysis, and molecular dynamics studies. *Biopolymers* 89: 1-13.
8. Pohanka, M., Skládal, P. 2008. Electrochemical biosensors—Principles and applications. *J. Appl. Biomed.* 6:57–64.
9. Liu, J., Wang, J. 2001. A novel improved design for the first-generation glucose biosensor. *Food Technol. Biotechnol.* 39:55–58.
10. Kharitonov, A. B., Zayats, M., Lichtenstein, A., Katz, E., Willner, I. 2000. Enzyme monolayer-functionalized field-effect transistors for biosensor applications. *Sensors Actuators B* 70:222–31.
11. Huang, C.-J, Chen, Y.-H., Wang, C.-H., Chou, T.-C., Lee, G.-B. 2007. Integrated microfluidic system for automatic glucose sensing and insulin injection. *Sensors Actuators B* 122:461–68.
12. Velasco-Garcia, M. N. 2008. Optical biosensors for probing at the cellular level: A review of recent progress and future prospects. *Semin. Cell Dev. Biol.* doi:10.1016/j.semcdb.2009.01.013.
13. Ligler, F. S. 2009. Perspective on optical biosensors and integrated sensor systems. *Anal. Chem.* 81:519–26.
14. Erickson, D., Li, D. 2004. Integrated microfluidic devices. *Anal. Chem. Acta* 507:11–26.
15. Srinivasan, V., Pamula, V., Pollack, M., Fair, R. 2003. A digital microfluidic biosensor for multianalyte detection. In The Sixteenth Annual International Conference on Micro Electro Mechanical Systems, Kyoto, 2003, MEMS-03, pp. 327–30.
16. Luan, L., Evans, R. D., Jokerst, N. M., Fair, R. B. 2008. Integrated optical sensor in a digital microfluidic platform. *IEEE Sensors J.* 8:628–35.
17. Grabowska, I., Sajnoga, M., Juchniewicz, M., Chudy, M., Dybko, A., Brzozka, Z. 2007. Microfluidic system with electrochemical and optical detection. *Microelect. Eng.* 84:1741–43.
18. Minas, G., Martins, J. S., Correia, J. H. 2000. High-selectivity detection in microfluidic systems for clinical diagnostics. In *EUROSENSORS XIV: The 14th European Conference on Solid-State Transducers*, Copenhagen, Denmark, August 27–30, 2000, pp. 873–79.
19. Schneider, B. H., Edwards, G., Hartman, N. F. 1997. Hartman interferometer: Versatile integrated quantification of nucleic acids, proteins, and pathogens. *Clin. Chem.* 43:1757–63.

20. Nica, A. G., Mascini, M., Ciucu, A. A. 2004. DNA-based biosensor for detection of genetically-modified organisms. *Analele Univ. Bucuresti Chimie* I-II:85–94.
21. Köhler, G., Milstein, C. 1975. Continuous cultures of fused cell secreting antibodies of predefined specificity. *Nature* 256:495–97.
22. Jayasena, S. D. 1999. Aptamers: An emerging class of molecules that rival antibodies in diagnostics. *Clin. Chem.* 45:1628–50.
23. Saerens, D., Huang, L., Bonroy, K., Muyldermans, S. 2008. Antibody fragments as probe in biosensor development. *Sensors* 8:4669–86.
24. Chandrasekaran, A., Acharya, A., You, J. L., Soo, K. Y., Packirisamy, M., Stiharu, I., Darveau, A. 2007. Hybrid integrated silicon microfluidic platform for fluorescence based biodetection. *Sensors* 7:1901–15.
25. Chandrasekaran, A., Packirisamy, M. 2008. Enhanced fluorescence-based bio-detection through selective integration of reflectors in microfluidic lab-on-a-chip. *Sensor Rev.* 28:33–38.
26. Fida, F., Varin, L., Badilescu, S., Kahrizi, M., Truong, V.-V. 2009. Gold nanoparticle ring and hole structures for sensing proteins and antigen-antibody interactions. *Plasmonics* 4:201–7.
27. Homola, J. 2003. Present and future of surface plasmon resonance biosensors. *Anal. Bioanal. Chem.* 377:528–39.
28. Azam, S., Boukadoum, M., Izquierdo, R., Acharya, A., Packirisamy, M. 2008. Integrated multifunctional fluorescence biosensor based on OLED technology. In *NEWCAS-TAISA 2008 Joint 6th International IEEE Northeast Workshop on Circuits and Systems*, pp. 173–76, Montreal (Canada).
29. Kumar, J., Jha, S. K., D'Souza, S. F. 2006. Optical microbial biosensor for detection of methyl parathion pesticide using *Flavobacterium* sp. Whole cells adsorbed on glass fiber filters as disposable biocomponent. *Biosensors Bioelect.* 21:2100–5.
30. Dubey, R. S., Upadhyay, S. N. 2001. Microbial corrosion monitoring by an amperometric microbial biosensor developed using whole cell of *Pseudomonas* sp. *Biosensors Bioelect.* 16:995–1000.
31. Sharpe, M. 2003. It's a bug's life: Biosensors for environmental monitoring. *J. Environ. Monit.* 5:109N–13N.
32. Yallow, R. S., Berson, S. 1959. Assay of plasma insulin in human subjects by immunological methods. *Nature* 184:1648–49.
33. Yalow, R. S., Berson, S. A., 1960. Immunoassay of endogeneous plasma insulin in man. *J. Clin. Invest.* 39:1157–75.
34. Avrameas, S. 1968. Detection d'anticorps et antigens a l'aide d'enzymes. *Bull. Soc. Chim.* 50:1169–78.
35. Nakane, P. K., Pierce, G. B. 1966. Enzyme-labelled antibodies. Preparation and applications for the localization of antigens. *J. Histochem. Cytochem.* 18:9–20.
36. Rossier, J. S., Girault, H. H. 2001. Enzyme linked immunosorbent assay on a microchip with electrochemical detection. *Lab Chip* 1:153–57.
37. Eteshola, E., Leckband, D. 2001. Development and characterization of an ELISA assay in PDMS microfluidic channels. *Sensors Actuators B* 72:129–33.
38. Sato, K., Tokeshi, M., Odake, T., Kimure, H., Ool, T., Nakao, M., Kitamori, T. 2000. Integration of an immunosorbent assay system: Analysis of secretory human immunoglobulin A on polystyrene beads in a microchip. *Anal. Chem.* 72:1144–47.
39. Henares, G. T., Mizutani, F., Hisamoto, H. 2008. Current development in microfluidic immunosensing chip. *Anal. Acta* 611:17–30.
40. Narang, U., Gauger, F. S., Ligler, F. S. 1997. Capillary-based displacement flow immunosensor. *Anal. Chem.* 69:1961–64.
41. Salim, M., O'Sullivan, B., McArthur, S. L., Wright, P. C. 2007. Characterization of fibrinogen adsorption onto glass microcapillary surfaces by ELISA. *Lab Chip* 7:64–70.
42. Moran-Mirabal, J. M., Edel, J. B., Meyer, G. D., Throckmorton, D., Singh, A. K., Craighead, H. G. 2005. Micrometer-sized supported lipid bilayer arrays for bacterial toxin binding studies through total internal reflection fluorescence microscopy. *Biophys. J.* 89:296–305.
43. Lai, S., Wang, S., Luo, J., Lee, J., Yang, S.-T., Madou, M. J. 2004. Design of a compact disk-like microfluidic platform for enzyme-linked immunosorbent assay. *Anal. Chem.* 76:1832–37.
44. Sia, S. K., Linder, V., Parviz, B. A., Siegel, A., Whitesides, G. M. 2004. An integrated approach to a portable and low-cost immunoassay for resource-poor settings. *Angew. Chem. Int. Ed.* 43:498–502.

REVIEW QUESTIONS

1. Define an antibody, an antigen, and an immunological reaction.
2. What is an immunoassay?
3. How is the antibody-antigen complex formed? Why is this interaction highly specific?
4. Give three examples of clinically important analytes that can be detected by immunoassays.
5. What is a homogeneous and a heterogeneous immunoassay?
6. What is an ELISA test? How is it commonly performed? Describe the principal steps involved into the analytical procedure.
7. What kind of enzymes are generally used in an ELISA test?
8. Explain what an enzyme conjugate is and its role in the ELISA test.
9. What are the advantages of an integrated microfluidic platform over the classical ELISA immunoassay performed in a microtiter plate?
10. What is a biosensor and how does it work? Describe briefly the principal components of a biosensor.
11. Describe briefly Clark and Lyons's glucose biosensor.
12. How are biomolecules immobilized on the surface of a sensor?
13. How are biosensors classified upon the recognition elements?
14. How are biosensors classified upon the detection systems?
15. Give an example of an enzyme-based amperometric biosensor.
16. How can an integrated microfluidic system be used for automatic glucose sensing and insulin injection?
17. Describe the principle of an optical biosensor and give an example.
18. What is the most used optical biosensing method today?
19. How can a DNA-based biosensor be used for detection of genetically modified organisms?
20. What is an immunosensor and how does it work? Define the affinity factor.
21. Give an example of an antibody-based biosensor.
22. What is a microbial biosensor and how is it used for environmental monitoring?
23. What are the characteristics that distinguish biosensors from ELISA and other immunoassays?

7 Fabrication of BioMEMS Devices

7.1 BASIC MICROFABRICATION PROCESSES

7.1.1 Introduction

Microfabrication initially used integrated circuit (IC) manufacturing technologies to fabricate objects with well-controlled features that range in size from micrometers to millimeters. Microelectromechanical systems (MEMS) have been primarily based on a silicon (single-crystal and polycrystalline silicon materials) platform, as they were developed historically with well-established manufacturing approaches that were used to make integrated circuits. However, these methods are quite expensive and limited in use, as they were optimized only for a limited number of materials, especially silicon and other semiconductor materials, and for required material properties. Traditional silicon micromachining techniques have proved to be the methods of choice for fabricating electronic and mechanical devices at the micrometer and millimeter scales, and have offered a number of possibilities for the design of devices capable of sensing, actuation, and interacting with biological systems. In recent times, other materials and fabrication technologies have been developed, as researchers are trying to reduce feature sizes beyond the optical diffraction limit to cater for expanding applications to many fields. Advances in microfabrication technology are providing new opportunities for many applications in biology and medicine. They include medical devices and biosensors used in diagnostics, devices used in therapeutics, and various analytical tools for molecular biological sciences. The diversity of biomedical applications encouraged the MEMS community to develop new materials that are more biologically friendly and new technologies that are different from traditional silicon-based methods. Polymers are gaining more and more importance in biological and biomedical applications, as they offer a wide range of tunable material properties with application potential in a cost-effective way. Microfluidic device fabrication employs many of the traditional MEMS fabrication methods to build embedded channels in materials such as silicon, glass, and polymers. This chapter focuses on the diverse and available micro- and nanofabrication techniques and their applications to the fabrication of various devices. In the first sections, the schemes of traditional microfabrication methods for fabricating structures are presented and discussed.

Microfabrication consists of many processes, such as (1) pattern transfer and lithography to define features and geometry on substrate and thin-film layers, (2) additive processes for obtaining thin films on substrates, (3) a micromachining process to machine or make features on substrates and thin films, and (4) bonding of chips. Physical vapor deposition (PVD), chemical vapor deposition (CVD), low-pressure chemical vapor deposition (LPCVD), plasma-enhanced chemical vapor deposition (PECVD), spin coating, and oxidation form part of the additive processes. The processes, namely, bulk micromachining, surface micromachining, wet etching, dry etching, gas phase etching, reactive ion etching, LIGA, and laser micromachining, are part of the micromachining processes. There are many types of bonding, such as fusion bonding, anodic bonding, low-temperature glass bonding, and adhesive bonding. The next sections will provide an overview of some microfabrication methods in addition to other nanofabrication techniques. For detailed information, readers should read a dedicated microfabrication book.

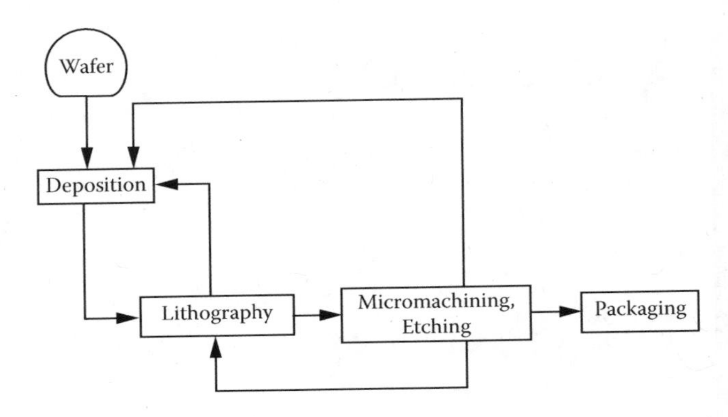

FIGURE 7.1 Sequence of processes for building a MEMS device.

Devices are generally built by a sequence of processes. Different manufacturing processes have been traditionally used in the fabrication of microelectromechanical systems (MEMS): *bulk micromachining* of silicon or other crystalline materials; *surface micromachining* of silicon and other materials; *LIGA*, a lithographic process employing electroforming and micromolding used for creation of high-aspect-ratio, two-dimensional metal and ceramic structures; and *laser micromachining*. In many cases, a combination of bulk and surface micromachining is utilized in the fabrication of the desired system. The basic processes used to build a device include thin-film deposition, photolithography, etching, and substrate bonding,[1,2] as shown in Figure 7.1.

7.1.2 Thin-Film Deposition

A variety of techniques are used for the deposition of thin films of different materials on a substrate. Thin films can also be used as functional or masking or sacrificial layers in microfabrication and micromachining. Thin films are deposited, either by a physical evaporation process such as thermal evaporation or by a chemical-reaction-driven process or by an oxidation process. The most important deposition methods include *thermal oxidation*, *chemical vapor deposition* (CVD), *physical vapor deposition* (PVD), and *electrodeposition*.

By heating silicon at high temperatures in an oxygen atmosphere, silicon oxidizes and forms a thin layer of silicon dioxide (SiO_2) on the surface. In the case of wet oxidation, the oxide growth can be accelerated due to the presence of humidity. It is called dry oxidation and humidity is not present. In microstructures, silicon dioxide is used as etch masks or for electrical isolation.

Inorganic materials such as silicon compounds and polysilicon are usually deposited by *low-pressure CVD* (LPCVD) at higher temperatures or *plasma-enhanced CVD* (PECVD) at lower temperatures. As shown in Figure 7.2, the CVD process involves a reaction between gases and the

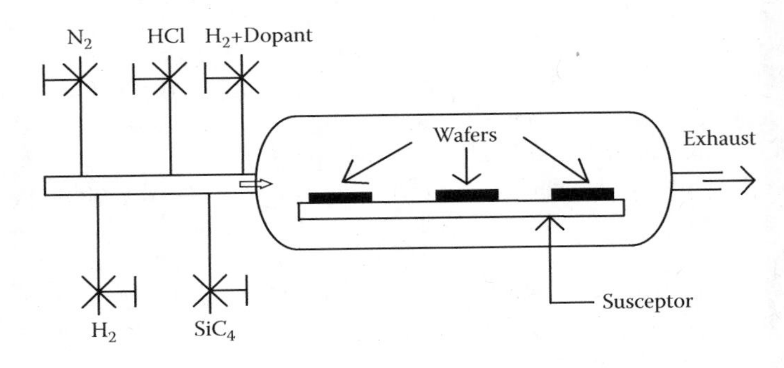

FIGURE 7.2 The chemical vapor deposition process.

substrate, in a chamber under vacuum and high temperature, producing a solid film on the substrate. The deposited layer is the product of chemical reaction between the gases and substrate. Other reactants are carried to the outlet by the flow of gases. For example, silicon nitride results from the reaction between dichlorosilane and ammonia. Amorphous and polycrystalline silicon thin films that are usually produced by LPCVD or PECVD are important structural materials in many microsystems. Typically, the deposition is done at low pressures (around 300 mT) and at moderate temperatures (SiO_2 at 450°C, polysilicon at 580 to 650°C, Si_xN_y at 800°C). After the gas is injected into the chamber, the molecules will react directly with the substrate surface and form a thin layer of the desired material. If the reaction happens under high temperatures and low pressure, it is called LPCVD. If the plasma is enhanced with additional radio frequency (RF) power, then reaction can happen at lower temperatures, as in the case of PECVD. Due to higher temperatures, the thin films deposited in the LPCVD process have more residual stress than the ones deposited by the PECVD process.

Metals and some metal oxides can be deposited by physical vapor deposition, based on thermal evaporation or sputtering carried out under high-vacuum conditions. Metal films can be deposited by *vacuum evaporation*, *sputtering*, *CVD*, and *plating*, and are mostly used for interconnections, ohmic contacts, and forming optical surfaces.

Physical vapor deposition under vacuum is used to deposit single-element conductors, resistors, and dielectrics. In this system, the target material is vaporized through heating in vacuum and the vapor is allowed to deposit on substrate. Resistive and electron beam heating are the two most common heat sources for physical vapor deposition.

Thermal evaporation, shown in Figures 7.3 and 7.4, is suitable for the deposition of metals with low melting temperatures. A hybrid technique that combines magnetron sputtering and pulsed laser ablation to deposit metal, ceramic, and diamond-like materials is also possible.[3]

In the case of the sputtering method, metals, compound materials, and refractive metals can be deposited by sputtering a cathode target with positive ions from an inert gas discharge. Under high

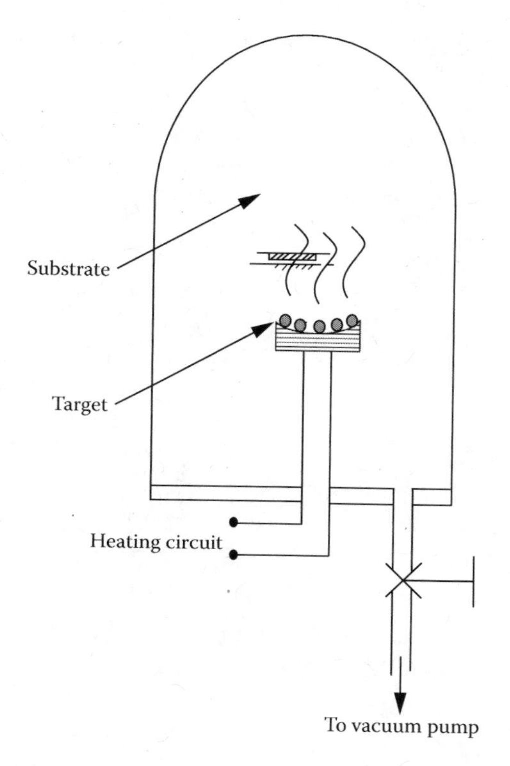

FIGURE 7.3 Schematic diagram of the thermal evaporation system.

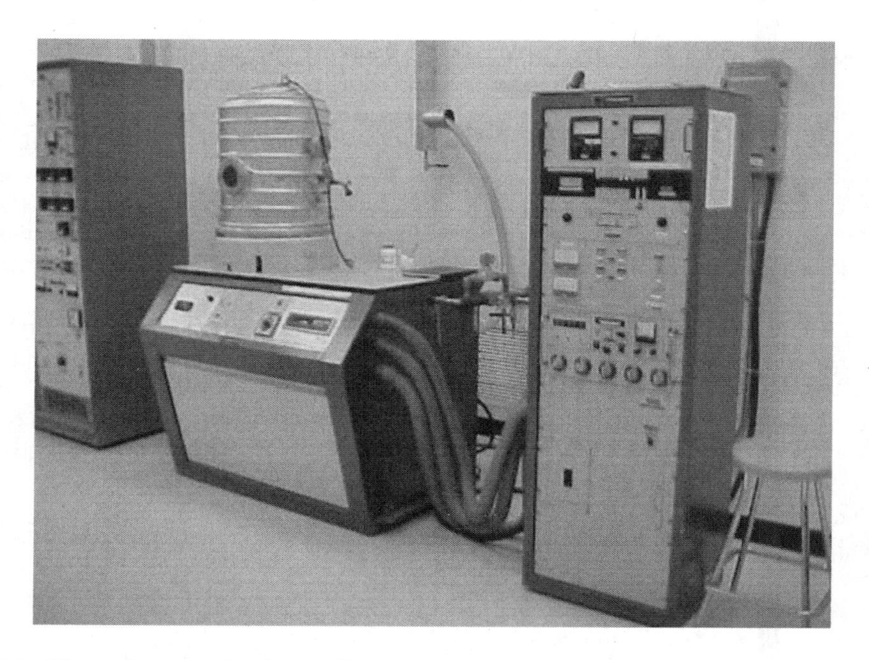

FIGURE 7.4 Thermal evaporation system. (Courtesy Edwards.)

vacuum, plasma of inert gases such as argon is produced in a chamber under high electric field. When the gas is ionized in the plasma state, the positive argon ions are attracted by the target connected to the cathode. When these gas ions bombard the target, the target atoms are released. The released target atoms are attracted by the anode, resulting in a thin-film coating on the substrate (Figure 7.5). Introduction of noninert gases into the ambient atmosphere during sputtering will result in reactive sputtering.

In the spin-casting process, the material is dissolved in a volatile liquid solvent, as shown in Figure 7.6. The solution is poured onto a wafer, which is rotated at a high speed. The solvent evaporates as the liquid spreads over the surface of the wafer, leaving behind the thin film of the solid material. Spin casting is useful for depositing photoresists, as well as some inorganic glasses. Spin

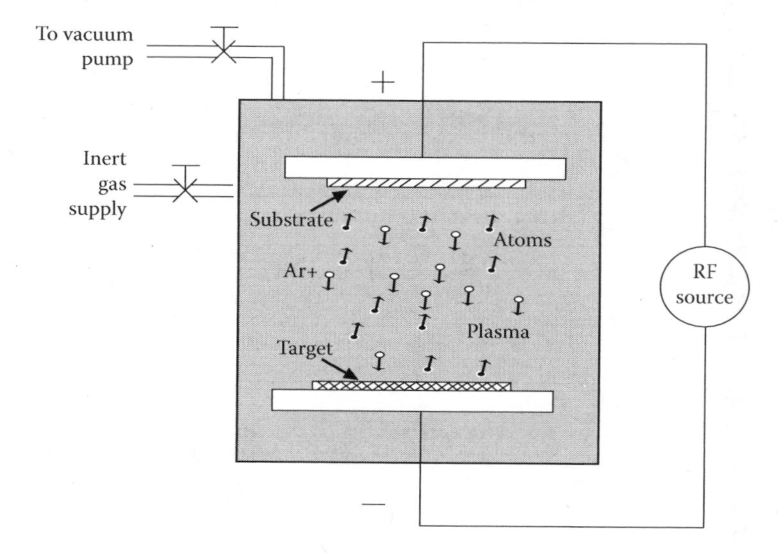

FIGURE 7.5 Schematic of the sputtering deposition process.

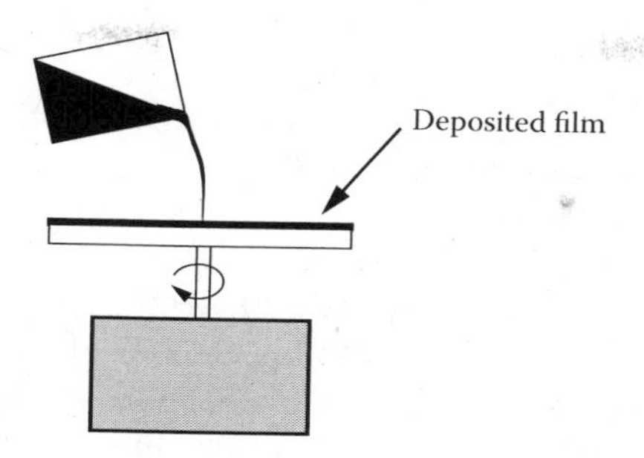

FIGURE 7.6 Schematic of the spin-coating process.

casting is a simple method to fabricate films having a smooth surface. The thickness of the film can be varied by changing the rotational speed and the viscosity of the liquid.

The main disadvantage of spin-coat materials is their high residual stress and the susceptibility to shrinkage when the solvent is removed. Spin-coated films are also less dense and more susceptible to chemical attack than materials deposited by other techniques.

The *dip coating* technique, shown in Figure 7.7, can be described as a process where the substrate to be coated is immersed in a liquid and withdrawn with a well-defined withdrawal speed under controlled temperature and atmospheric conditions. The coating thickness is mainly defined by withdrawal speed, solid content, and viscosity of the liquid. The dip-coating technique is used to produce ceramic and glass materials by applying a sol-gel process. In general, the sol-gel process involves the transition of a system from a liquid sol (colloidal solution) into a solid gel phase. Thin-film coatings, ceramic fibers, microporous inorganic membranes, glasses, and porous aerogel materials can be fabricated using this technique. The starting material, called a precursor, used in the preparation of the sol, is usually an inorganic metal salt or metal organic compound such as metal alkoxide. In a typical sol-gel process, the precursor is subjected to a series of hydrolysis and polymerization reactions to form a colloidal suspension, or a sol. Thin films of silicon oxide, titanium dioxide, indium tin oxide (ITO), etc., can be deposited by this method.[4–9]

FIGURE 7.7 Dip-coating apparatus.

7.1.3 Photolithography

Photolithography is one of the most well-established techniques for microfabrication used to transfer a geometric pattern onto a material (silicon, glass, etc.) by selective exposure to light. The lithographic process proceeds as shown in Figure 7.8. The pattern is first drawn with a computer program and transferred onto a photomask. The *photomask* is usually a glass plate with an opaque material in the desired pattern. The photosensitive material (photoresist) is deposited onto a substrate such as silicon or glass by spin coating. After the deposition in the order of a few microns thickness, the substrate is heated to 60 to 100°C in order to improve the adhesion and remove the trace of solvents. This process step is called *soft baking*. The photoresists are thermoset polymer resins that cross-link under exposure to ultraviolet light. In the next step, called *aligning*, the substrate and the mask are then brought in contact in a machine called the aligner and the photoresist is exposed to a UV source. Depending on the desired resolution, UV, EUV (extreme UV), x-ray, or e-beam sources are used. Depending upon the type of photoresist, the exposed part of the photoresist under the transparent part of the mask or the unexposed part becomes soluble in the developing solution, and it is subsequently

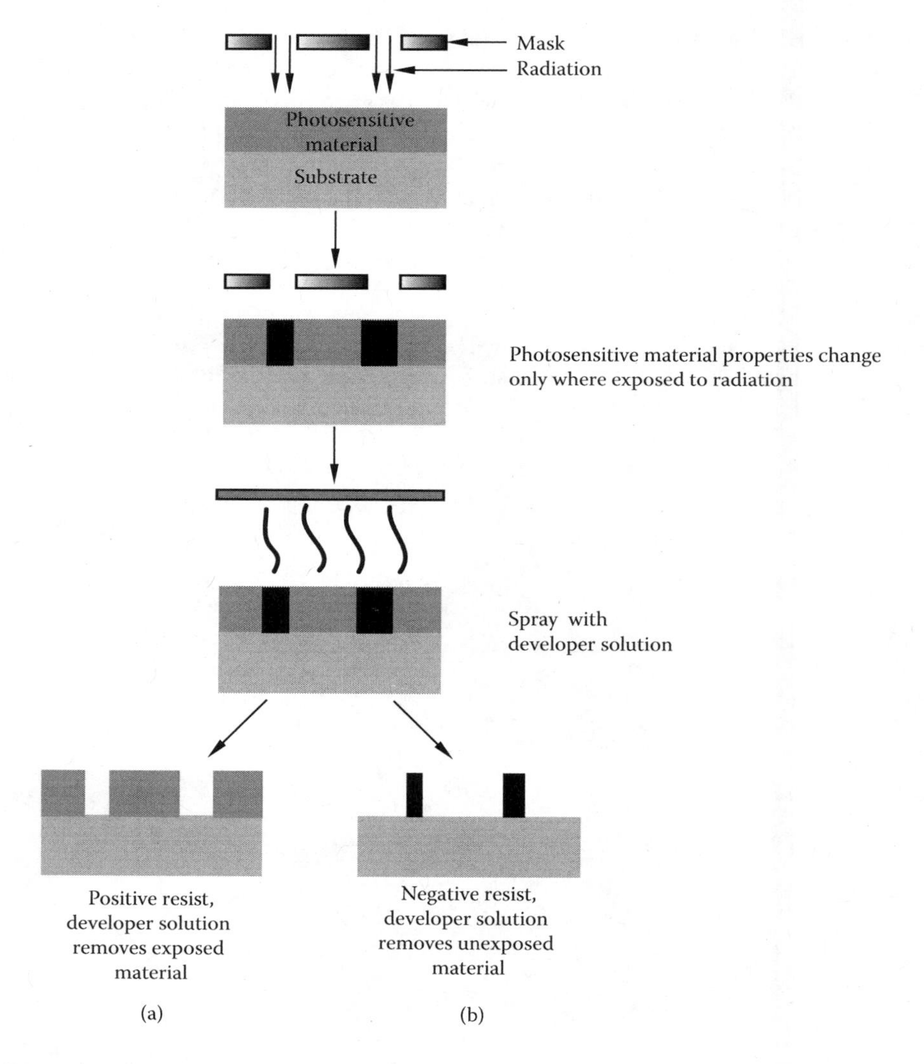

FIGURE 7.8 Process of photolithography and pattern definition in (a) positive and (b) negative resists.

washed off. In the next step, called developing, for a positive photoresist, the exposed area becomes more soluble in the developing solution and can be easily removed. In the case of a negative photoresist, the unexposed part is removed in subsequent developing. A negative photoresist, such as SU8 when exposed to UV light, will become cross-linked and can be developed with propylenglycol monoether acetate, ethyl acetate, etc. One of the popular negative photoresist materials is called SU8 because of its monomer that contains eight epoxy groups. SU8 has proved to be a very stable material both thermally and mechanically because of the strong cross-linking that takes place during exposure to UV light. The sections of the photoresist not covered by the opaque regions of the photomask are exposed by irradiation with UV light. Generally, shorter wavelengths are recommended for higher resolution. For nanoscale lithography, x-ray, electron beam, or EUV is used.

The patterning of positive and negative resists is shown in Figure 7.8. The photolithography sequence is completed by a hard-baking or postbaking step to improve the adhesion. In summary, the photolithography process involves photomask making, spin coating of photoresist, soft baking, UV exposure, and developing and postbaking steps.

7.1.4 Etching

Following the hard bake, the transfer of the desired pattern from the photoresist to the underlying film or wafer is done by a process known as *selective etching*. Etching is the process of removing selectively the unwanted regions of a film or substrate. Etching is used not only to fabricate patterns on mask layers, but also to remove the surface damage induced by the previous process steps and to fabricate two- and three-dimensional structures. Semiconductors, metals, and insulators can all be etched with the appropriate etchants. The various etching techniques can be divided into two categories: *wet etching*, carried out with liquid chemicals, and *dry etching*, using gas phase chemistry. Both types of etching can lead to isotropic or anisotropic etching, but wet etchants are generally considered isotropic in reactivity and have a better material selectivity than the dry etchants. However, monocrystalline substrates such as crystalline silicon can be etched anisotropically by using etchants that attack silicon along preferred crystallographic directions, such as potassium hydroxide (KOH) and tetramethyl ammonium hydroxide (TMAH). Anisotropic etching of silicon (100) wafer results in a pyramid-shaped cavity, as shown in Figure 7.9a, due to the variation of etch rate with crystallographic designs. But, in the case of isotropic etching, the etch rate is independent of crystallographic direction, resulting in a cavity shape, as shown in Figure 7.9b. Silicon dioxide and silicon nitride are used as masking materials for anisotropic and isotropic etching of silicon.

Dry etching, which uses gases for etching, is suitable for making very small features in thin films, and it is used when a high resolution is necessary. There are many classes of dry etching techniques called high-pressure plasma etching, reactive ion etching (RIE), deep reactive ion etching (DRIE), and ion milling. In *high-pressure plasma etching*, the reactive species that are created react with the material to be etched, and as a result, the material is dissolved at the surface. In the RIE technique, the material to be etched is introduced in a reactor where a plasma is formed by using an RF power source (Figure 7.10). The high-energy ions may either hit the material and remove atoms

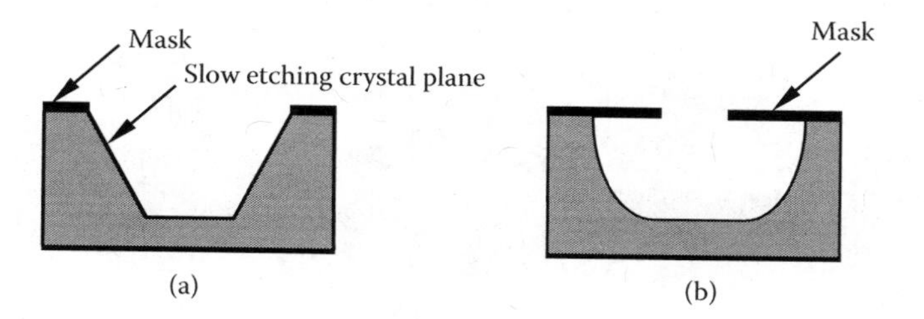

FIGURE 7.9 Profile of (a) isotropic and (b) anisotropic wet etching using a photomask.

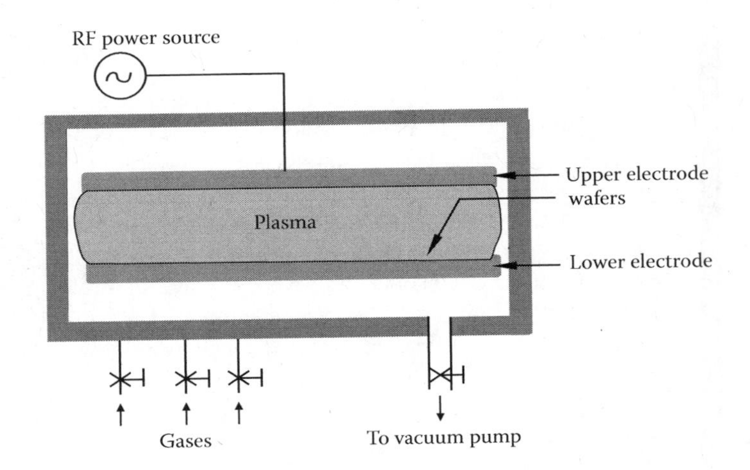

FIGURE 7.10 Reactive ion etching system.

from the surface or react at the surface of the material. Because there are many more collisions with the horizontal surfaces than with the walls, the etching rate will be higher.

Deep reactive ion etching (DRIE), also called deep trench etching, is a more recent development using an additional passivation step in addition to the etching step. In this method, an etched structure is coated with a layer of polymer (a fluoropolymer is used in commercial DRIE silicon etchers) and etched subsequently by ion bombardment. In DRIE, an inductively coupled plasma (ICP) is unit attached to a RIE system to increase the plasma energy, and it is possible to etch deep features and cavities with very high aspect ratios in the order of one hundred. As DRIE is highly anisotropic in nature, it is possible to obtain complex geometries with vertical sidewalls. Due to the highly anisotropic nature of etching, the ions are directed toward the bottom of the pit, and the sidewalls are left intact, leading to vertical walls. Compared to wet etching, the cost of the dry etching is considerably higher. The process is preferred for the patterning of thin films.

7.1.5 Substrate Bonding

Bonding is used to fabricate complex systems, principally from silicon and glass substrates. There are several bonding techniques used in fabricating micromechanical devices. One of the mostly used methods to join two materials together is *fusion bonding*. The technique is used to bond surfaces that were previously cleaned and rendered hydrophilic by hydroxylation. The surfaces to be bonded have to be smooth and flat. By pressing the substrates together, they are bounded by the surface attraction of two hydrophilic surfaces. Silicon-silicon, silicon-glass, and glass-glass bondings can be realized by this method, as shown in Figure 7.11. The device is usually annealed at high temperatures to increase the bond strength.

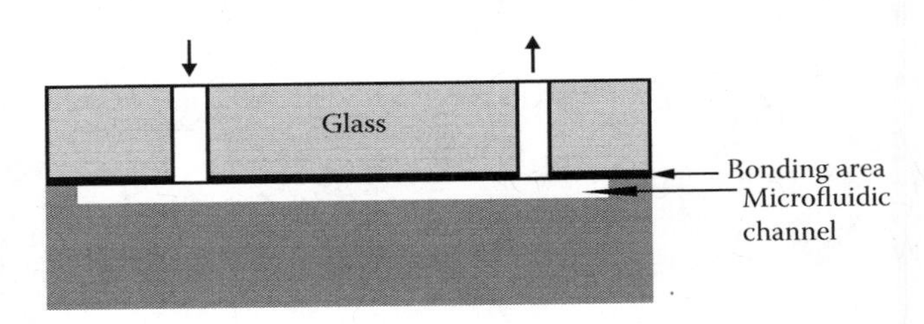

FIGURE 7.11 Silicon-glass bonding to form flow channels. (From Schmidt, Proceedings of the IEEE, 86, 1575–1585, 1998.)

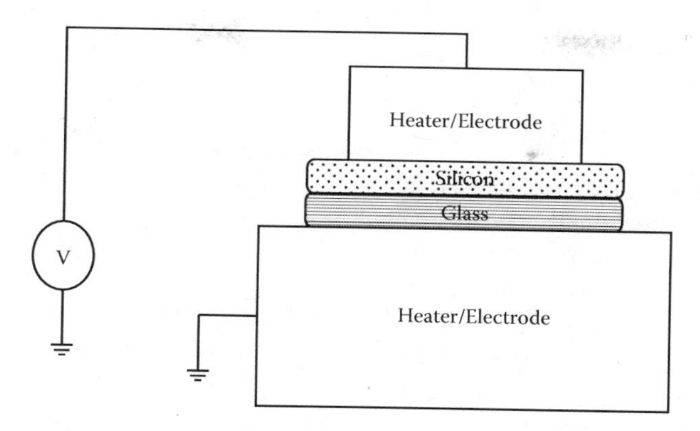

FIGURE 7.12 Anodic bonding apparatus. (From Schmidt, Proceedings of the IEEE, 86, 1575–1585, 1998.)

Another substrate joining technique used for device fabrication is the *anodic (electrostatic) bonding*. Figure 7.12 shows an anodic bonding setup used for silicon-to-glass bonding.

This technique allows lower bonding temperatures (300 to 400°C). The glass-silicon sandwich is first heated to around 200°C, and a high voltage of 200 to 1,000 V is applied. The glass is connected to the cathode and the silicon to the anode. At high temperatures, there is a migration of Na^+ ions from the glass toward the cathode, and a strong electrostatic force at the interface facilitates the formation of covalent silicon-oxygen bonds that promote the fusion. When using this technique, it is important to use glass with a thermal coefficient of expansion close to that of silicon.

Wafer-to-wafer bonding can also be carried out by using intermediate layers such as Au thin films (*eutectic bonding*), *glass-frit bonding*, *solder bonding*, etc. Other bonding methods, such as thermal pressure bonding, solvent bonding, ultrasonic welding, etc., are also used for joining polymer parts. The bonded of materials can be characterized by several methods. Nondestructive techniques that can be used to monitor the bonding process include infrared transmission, ultrasonic, and x-ray topography.

An adhesive bonding method using SU8 negative photoresist as an intermediate layer has been reported for microfluidic applications.[11] This method enables bonding at a low temperature (below 200°C), and the planarity of the wafers is not as critical as in the case of other bonding methods. Due to this low temperature, the stress induced by the bonding process is low. The adhesive bonding technique was used for the fabrication of microfluidic devices.

Figure 7.13 shows a cross section through a microfluidic device fabricated in glass using adhesive bonding. The depth of the channel is 20 μm. The SU8 photoresist is used as an adhesive to bond the top glass plate to silicon.

The techniques described in the previous sections are used by both the semiconductor and MEMS industries. The MEMS-specific technologies, called micromachining, are briefly discussed in the following section.

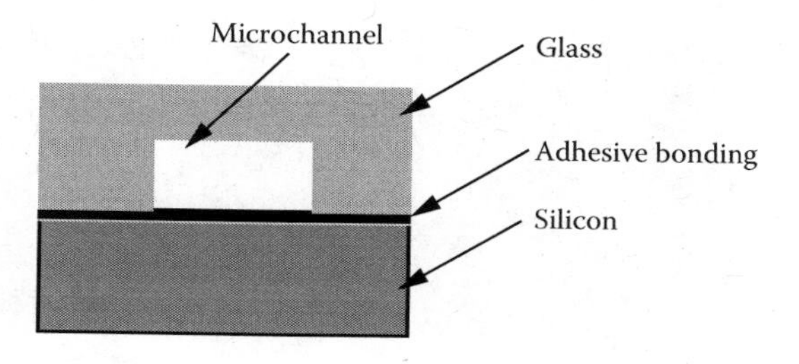

FIGURE 7.13 Microchannel fabricated by photoresist adhesive bonding.

7.2 MICROMACHINING

7.2.1 Bulk Micromachining

Bulk micromachining techniques are used to remove significant amounts of materials, such as silicon glass, GaAs, etc., from substrates. This is done to create various components of movable micromachined structures, such as beams, membranes, plates, cavities, chambers, and channels.[1,12] Depending upon the chemical used, it could be wet etching using liquid chemicals or dry etching using gaseous chemicals. In the case of etching of (100) silicon with liquids such as KOH or TMAH, the etch is stopped by bounded (111) planes, as seen in Figure 7.14. The (111) planes act as etch stops as the etch rate in the <111> direction is negligible compare to the etch rate in the <100> direction. One could obtain many useful structures using this crystallographic-dependent anisotropic nature of wet etching. Some examples include V-grooves for the alignment of optical fibers, micronozzles for microrockets, microjest and microchannels for ink jet printers, etc.

An anisotropic wet etch on a silicon wafer creates a cavity with a trapezoidal cross section. The bottom of the cavity is a (100) plane, and the sides are (111) planes. One can also adopt the wet bulk micromachining for making freestanding structures such as cantilevers, shown in Figure 7.15.

Gases are used for etching in the case of the dry bulk micromaching process. In the case of the bulk micromachining etching process, it starts with the transport of reactants, through a boundary layer to the surface to be etched. Then, the reaction between reactants and the film(s) to be etched takes place at the surface, and after, the reaction is followed by the transport of reaction products from the surface (Figure 7.16). The etch rate is high (4 to 20 μm/min), and the process can be controlled through any of these steps, but it is easier to control the reaction rate by changing the temperature and the concentration of the etchant. Wet etching is easy to implement; it is a low-cost process that shows a good selectivity for most materials. *Isotropic etching* is preferred for large geometries when the sidewall slope is not important, while *anisotropic etching* is best for making small gaps and vertical sidewalls. But the process is typically more costly.

In the case of reactive ion etching, free radicals of electrically neutral species that have incomplete outer shells, e.g., CF_3·, react with the film to be etched and form volatile by-products. The process starts with the mass transport of reactive species from the gas stream to the reaction surface, where the reaction takes place, followed by mass transport of reaction products back to the gas stream. Oxygen is added to CF_4 plasma to increase the amount of reactive F species (O reacts with CF_3 and CF_2 and hence reduces the recombination rate of F). The ionic species are accelerated toward each electrode by an alternating electric field. The ionic species, such as Cl_2^+, CF_4^+, and CF_3^+ (or Ar^+ in a purely physical sputter etch), strike the wafer surface and remove the material to be etched.

When anisotropicity is not important, the SF_6/O_2 mixture can be used as an etchant of high selectivity. Chlorine-based chemistry (Cl_2, HCl, $SiCl_4$, BCl_3) results in anisotropic and selective etching, but the etch rate is lower than in the CF_4-based process. The etching rate can be increased by ion bombardment, and the anisotropicity enhanced by adding a small amount of O_2. Bromine-based etching chemistry (HBr, Br_2) is similar to that of chlorine-based etchants.

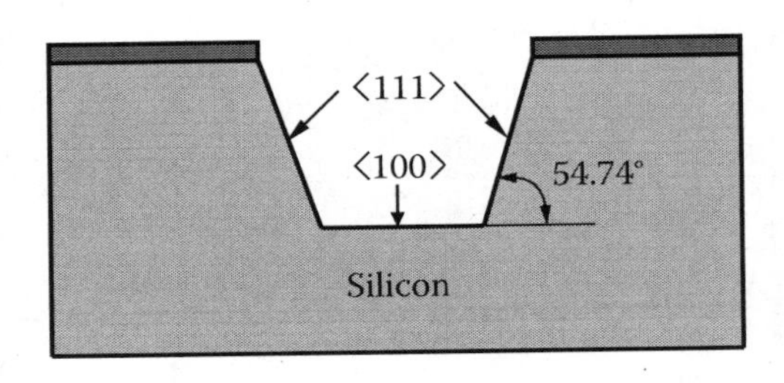

FIGURE 7.14 Anisotropic etching of silicon.

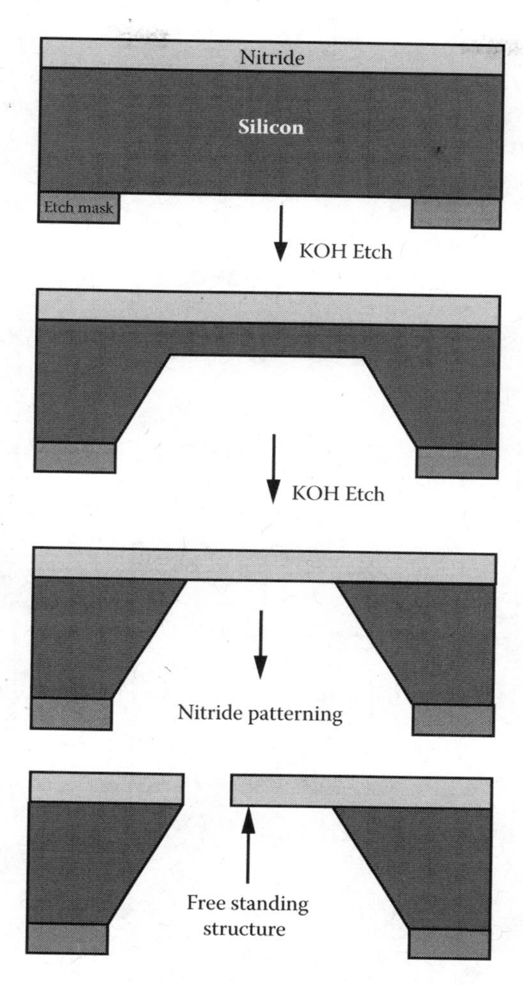

FIGURE 7.15 Structure fabricated with bulk micromachining. (From Madou, *Fundamentals of Microfabrication: The Science of Miniaturization*, CRC Press, Boca Raton, FL, 2001.)

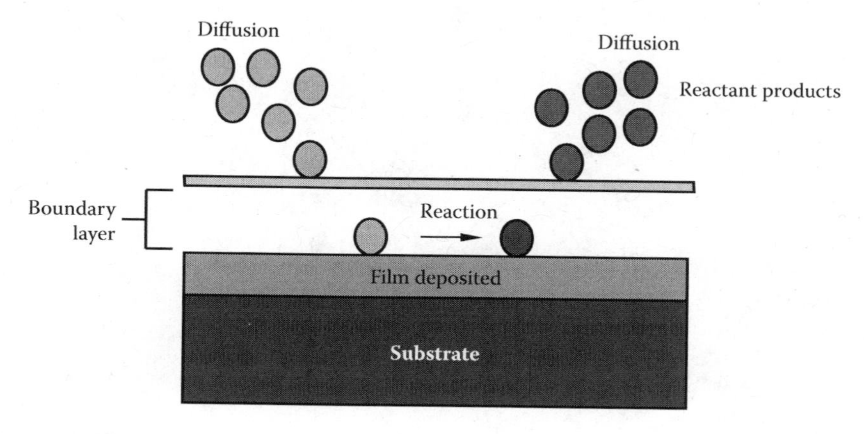

FIGURE 7.16 Illustration of wet etching.

Similarly, room temperature gas phase etching of silicon and polysilicon is possible with XeF_2 gas (xenon difluoride). XeF_2 gas has extreme selectivity for silicon and polysilicon and high selectivity for silicon dioxide, silicon nitride, and photoresist.[13] As a result, many materials, such as metals, silicon dioxide, silicon nitride, and photoresist, can be used as a hard mask for etching silicon or polysilicon. The simple xenon difluoride gas phase etch setup is shown in Figure 7.17a.

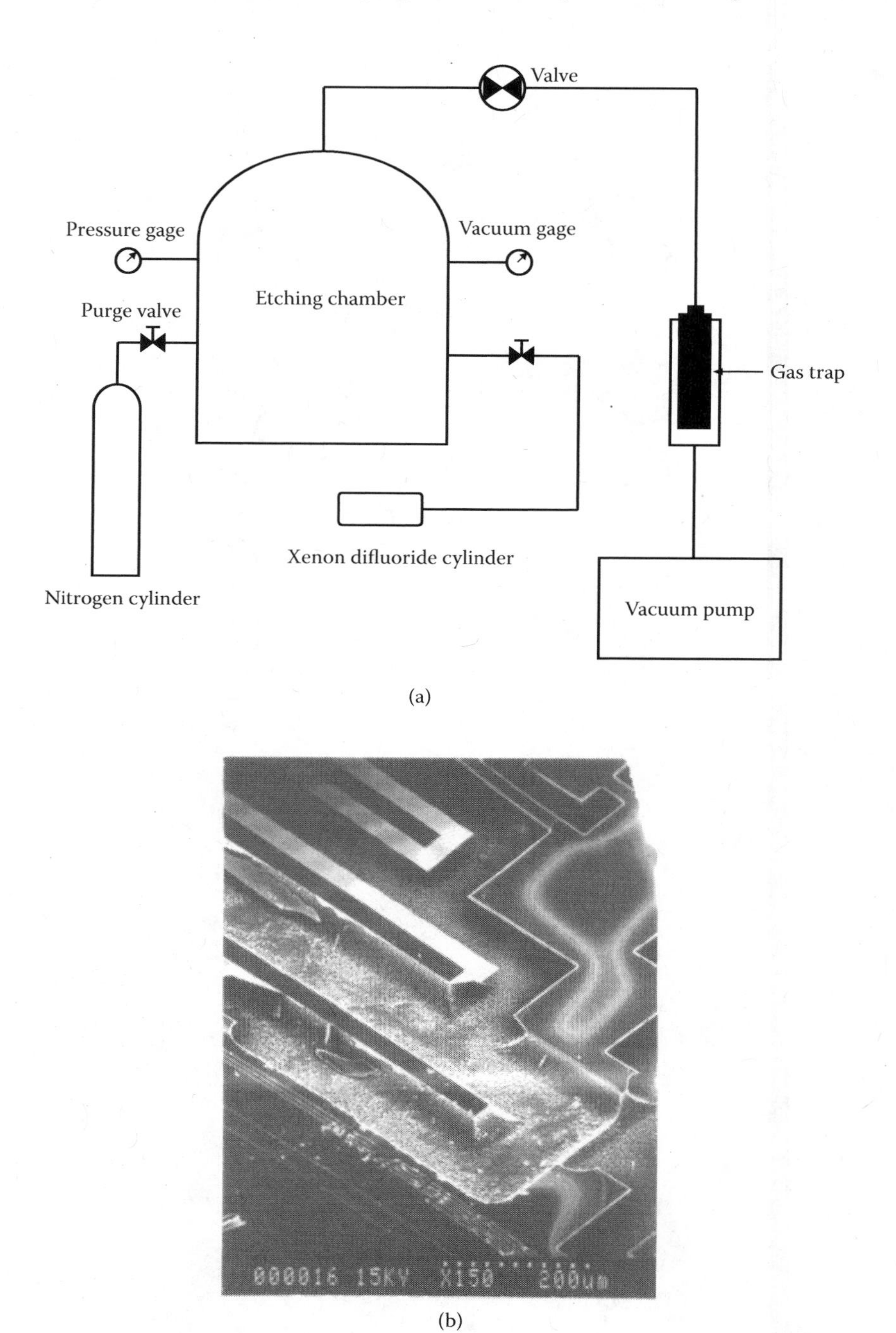

FIGURE 7.17 (a) Schematic of the etching setup. (b) SEM image of the etched silicon surface.

Xenon difluoride is a solid white crystalline with a sublimation pressure of about 4 Torr at room temperature. Pulse etching of XeF_2 is more efficient than continuous etching. In the case a pulse etching, pressure in the chamber is reduced below 4 Torr and the gas valve is opened; as result, XeF_2 sublimates in the chamber. The XeF_2 vapor etches silicon as follows:

$$Si + 2XeF_2 \rightarrow SiF_4 + 2Xe$$

Once the etching is completed for a minute or pulse duration, the purge valve is opened to purge the gases out. One could obtain an etch rate around 3 to 5 microns per minute, depending upon the mask size and gas concentration.

The etched silicon channel and structures are shown in Figure 7.17b. The overhanging oxide mask is seen in this figure. XeF_2 can be use for both bulk micromachining and surface micromachining. Even though the etch rate is higher for XeF_2 etching, the surface roughness is high compared to wet etching. One can use a hybrid micromachining technique in order to improve this surface roughness. Following the gas phase etching, a wet anisotropic etching with tetramethyl ammonium hydroxide is carried out in hybrid micromachining in order to reduce the surface roughness.[13]

7.2.2 Surface Micromachining

Surface micromachining is a microfabrication technique used to create movable microstructures on a silicon substrate by selective etching of thin-film layers called *sacrificial layers*. The removal of the sacrificial layer results in movable structures such as membranes and beams. Figure 7.18 illustrates the surface micromachining process. The common combinations of sacrificial and structural layers are given below.

Surface micromachining is a versatile technology, as the patterning of the structural and sacrificial layers is accomplished by chemical etching processes that are insensitive to the crystalline structure of the films. In addition, by using multiple layers of structural and sacrificial materials, surface micromachining enables the fabrication of integrated multilevel structures. Different combinations of structural-sacrificial materials can be used, suiting the fabrication processes, as shown in Figure 7.19. Some of the surface micromachined structures using XeF_2 etching are given in Figure 7.20.

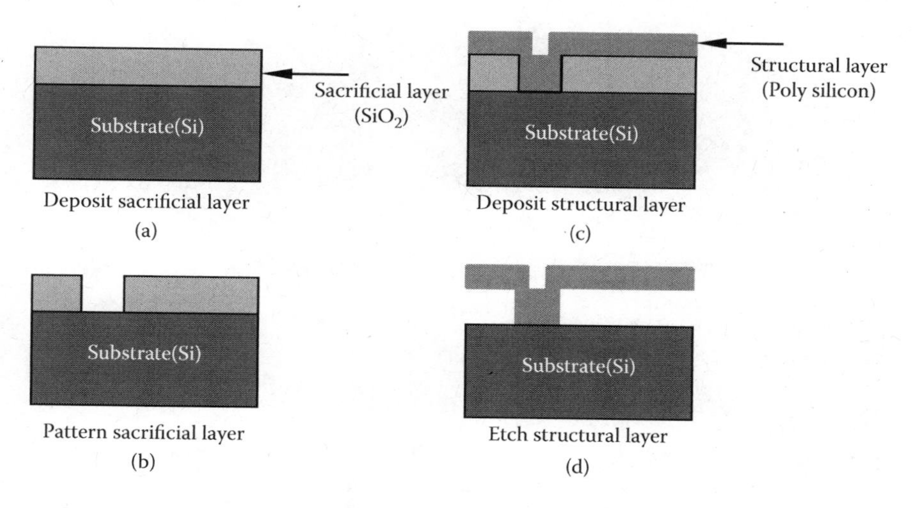

FIGURE 7.18 Basic surface machining fabrication process: (a) Deposition of the sacrificial layer, (b) patterning of the sacrificial layer, (c) deposition and patterning of the structural layer, and (d) selective etching of the sacrificial layer.

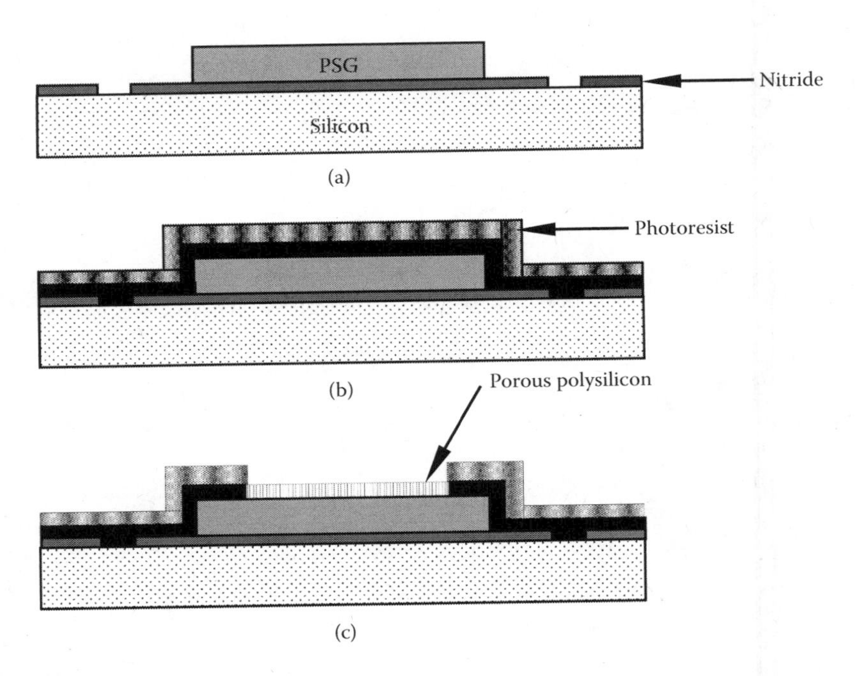

FIGURE 7.19 Surface micromachining of a freestanding porous polysilicon shell. (a) Deposit nitride layer and the sacrificial PSG layer. Pattern sacrificial layer. (b) Deposit photoresist for patterning the structural layer of polysilicon. (c) Pattern porous polysilicon layer. The PSG layer can be dry etched for releasing the polystructure. (Adapted from He et al., *J. Micromech. Syst.*, 16, 463–73, 2007.)

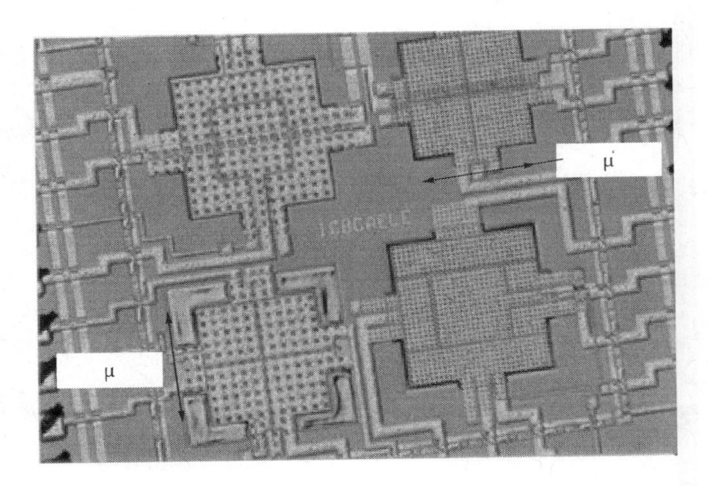

FIGURE 7.20 Photographs of plate structures surface micromachined from the Mitel 1.5 μm process using XeF_2 etching. Structures are made out of metal and oxide layers.

7.2.3 High-Aspect-Ratio Micromachining (LIGA Process)

The LIGA technology is used largely in the fabrication of biomedical devices (Figure 7.21). This German acronym means lithography, electroplating, and molding. The process can be used for manufacturing high-aspect-ratio three-dimensional microstructures in a wide variety of materials, such as metals, polymers, ceramics, and glasses. It relies on the use of high-energy x-rays from a synchrotron to induce damage in the molecules of a sensitive resist. After exposure to the radiation,

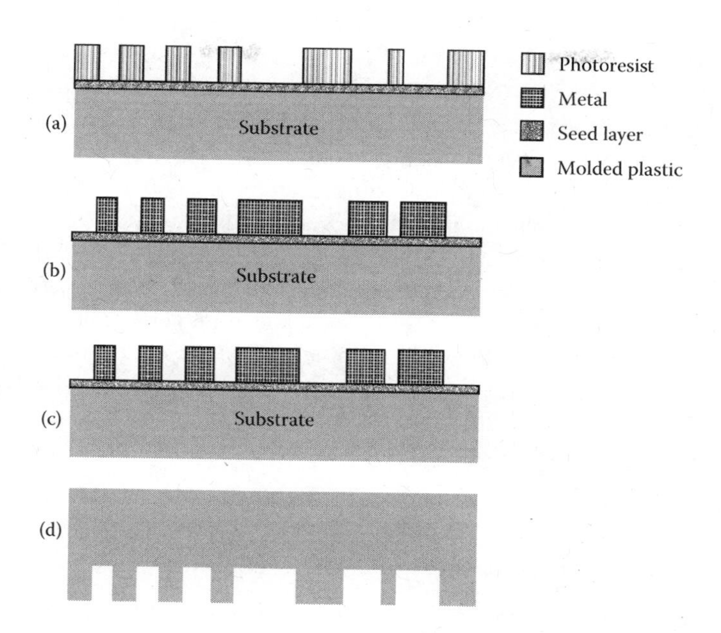

FIGURE 7.21 Outline of the micromolding process using LIGA technology: (a) photoresist patterning, (b) electroplating of metal, (c) resist removal, and (d) molded plastic components.

these regions may be selectively etched by a chemical solvent, and the remaining structure may be replicated by a number of electroplating or casting processes.

As shown in Figure 7.21, high-intensity and low-divergence x-rays are used as the exposure source for lithography. The x-rays are usually produced by a synchrotron radiation source. Polymethylmethacrylate (PMMA) is used as the x-ray resist. The thicknesses of several hundreds of microns and aspect ratios (depth-to-width ratio) of more than one hundred have been achieved by using LIGA technology. A characteristic x-ray wavelength of 0.2 nm allows the transfer of a pattern from a high-contrast x-ray mask to a resist layer with a thickness of up to 1,000 μm, so that a resist relief may be generated with an extremely high ratio. The openings in the patterned resist can be preferentially plated with metal, yielding a highly accurate complementary replica of the original resist pattern. The mold is then dissolved away to leave behind plated structures with sidewalls that are vertical and smooth. It is also possible to use the plated metal structures as an injection mold for plastic resins. After curing, the metallic mold is removed, leaving behind microreplicas of the original pattern. By combining LIGA with the use of a sacrificial layer, it is also possible to realize freestanding micromechanical components.

An important drawback of the LIGA process is the need for a short-wavelength collimated x-ray source such as a synchrotron. LIGA-like processes using conventional exposure sources have also been developed. Photoresists with high transparency and high viscosity can be used to achieve a single-coating mold thickness in the range of 15 to 500 μm.

7.3 SOFT MICROMACHINING

7.3.1 Introduction

The fabrication of microstructures and micropatterns for BioMEMS devices with soft materials such as polymers and gels is sometimes called soft lithography. The term, a collective name for several techniques, was coined in the 1990s by Professor Whitesides and his group at Harvard University.[15] They have developed many of the unconventional methods and unconventional materials for microfabrication that will be discussed later in this chapter. Soft polymeric materials possess many attractive properties, such as high toughness, recyclability, excellent biocompatibility, and biodegradability. Polymers are very desirable materials in the microfabrication of devices in contact with biological materials. Due to the variety of polymers, they offer a broad spectrum of physical and chemical properties for the design of biocompatible microsystems. Soft materials have many attributes that make them ideally suited for defining microfluidic, optical, and nanoelectromechanical structures with low-cost replication processes, such as molding and templating. One of the widely used materials for microfabrication of devices is polydimethylsiloxane (PDMS). PDMS is a silicone elastomer with properties that make it attractive for biomedical applications. It is thermally stable, permeable to gases with good optical and mechanical properties, in addition to being biocompatible, and it can be implanted *in vivo*. In addition, PDMS is not hygroscopic; that is, it does not swell with water and it shows little autofluorescence. Due to its very good sealing and bonding properties, PDMS has become the material of choice for microfluidics. It is also a cheap and commercially available material and has good elastic properties, allowing the fabrication of nonplanar structures.

7.3.2 Molding and Hot Embossing

Two of the commonly used fabrication techniques for polymers are molding and embossing, which are well-known techniques in the macroworld of polymers. The micromolding process as well as other soft lithography methods, starts with the fabrication of a *master* (a mold), made with a hard micromachining technique. Silicon masters, because of their brittleness, can only be used for casting and embossing. Polymers (SU8 and other epoxy resins) have been recently used for the fabrication of the master, but their lifetime is quite limited. The geometrical accuracy of the master structure will determine the quality of the structures that are fabricated with it. In order to have low frictional forces in the demolding step, the roughness of the master structure should be below 100 nm. After making the master mold, a prepolymer solution (a mixture of a PDMS linear polymer and a curing or cross-linking agent) is poured into the master and heated for several hours, until the mixture becomes cured and can be peeled off. The micromolding process, which is also called microcasting, is illustrated in Figure 7.22. The advantage of the micromolding process is the possibility to reuse the same master many times.

Hot embossing is another microfabrication technique that is used in nanofabrication. The process is illustrated in Figure 7.23.

A quartz stamp made of quartz, silicon, or other metal, having the negative image of the desired structure, is brought in contact with a polymer (polycarbonate, PMMA) heated in vacuum at a temperature slightly above its glass transition temperature. At the same time, a uniformly distributed pressure is applied on the heated stamp, pressing it into the polymer substrate (the pressure is around 500 N cm^{-2} in the case of PMMA). When the chip is cooled below the glass transition temperature, the polymer chip embossing the stamp's features can be released in a process called demolding. The advantages of the hot embossing process are simplicity, short overall cycle time (5 to 7 min), variety of available suitable materials,[17–19] and availability of commercial equipment. The accuracy of the replication in the hot embossing process enables the fabrication of structures in the few tens of nanometers range.[18] Hot embossing is a successful replication technique used for the fabrication of microfluidic chips. The structural resolution that can be obtained with hot embossing is shown in Figure 7.24. The structure shown in this figure corresponds to a two-dimensional channel array where the width of a microchannel is less than 1 μm.

The basic replication process has been developed into many techniques, namely, micro contact printing, micro transfer molding, micromolding in capillaries, solvent-assisted micromolding, etc.

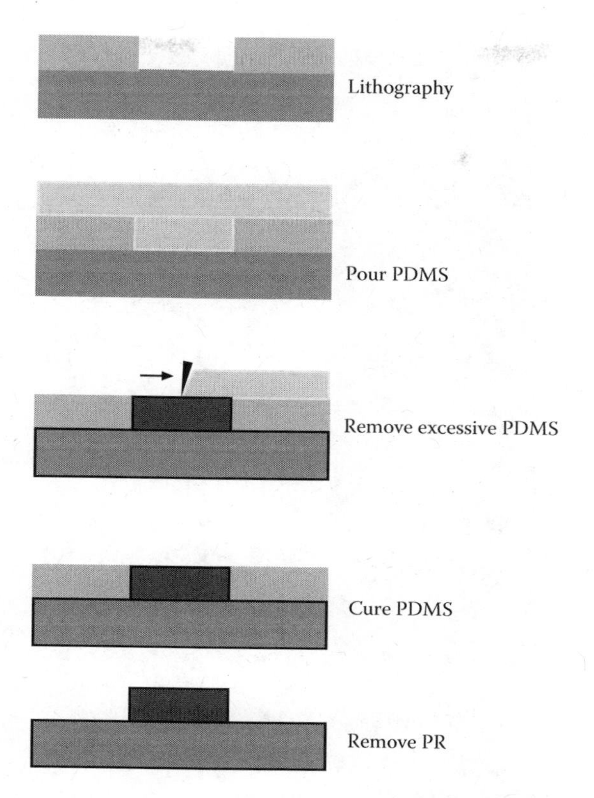

FIGURE 7.22 Schematic diagram of a process for patterning PDMS with fine spatial features. (From Liu et al. The 226th Am. Chem. Soc. National Meeting, New York, Sept. 7–11, 2003.)

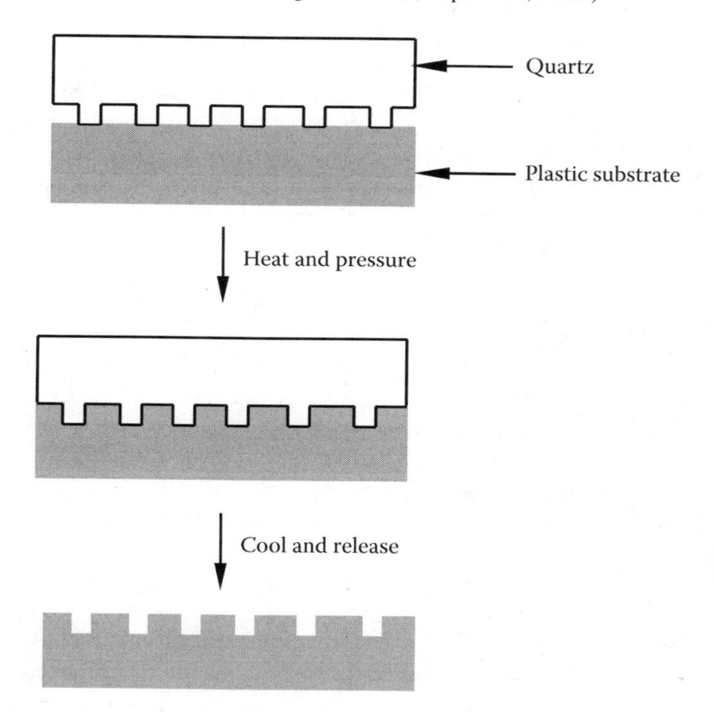

FIGURE 7.23 Schematic of the hot embossing process.

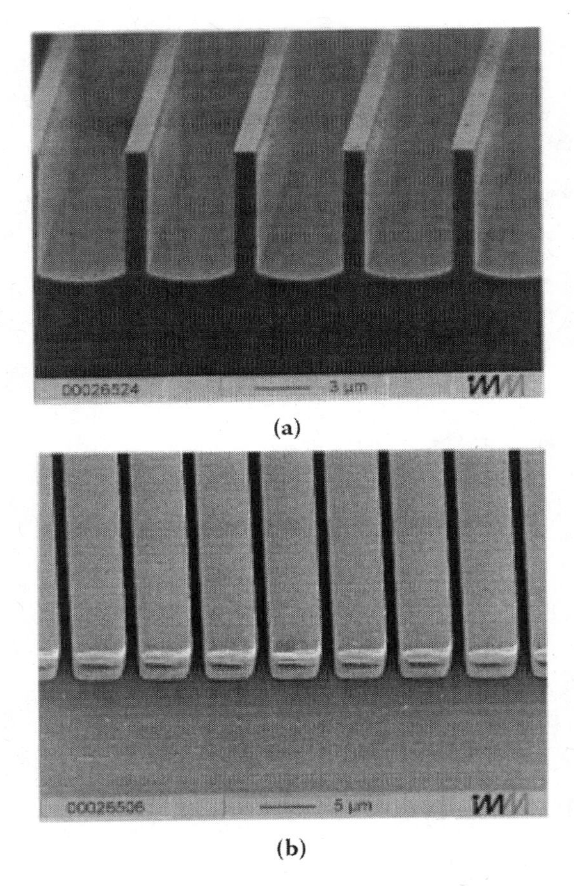

FIGURE 7.24 Hot embossed structures on PDMS. (a) Embossing tool fabricated with an advanced silicon etch (b) Resulting channel array structure. (Reproduced from Becker and Gärtner, *Electrophoresis*, 21, 12–26, 2000. With permission.)

7.3.3 Micro Contact Printing (μCP)

Micro contact printing (μCP), shown in Figure 7.25, is a convenient technique to create patterned self-assembled monolayers (SAMs) of alkanethiolates on a metal surface. The pattern is formed by "printing" with an elastomeric "stamp" by using the alkanethiol as an "ink" to form the patterned SAMs on a gold film. The structure and preparation of SAMs have been discussed in detail in Chapter 4. Alkanethiolates adsorb spontaneously from solution on a gold surface, forming a self-assembled monolayer.

Self-assembly is a spontaneous organization of molecules into a stable structure. The stamp is fabricated by molding PDMS using a master prepared by soft lithographic techniques.

After the PDMS stamp is fabricated, it is exposed to a solution of alkanethiol in anhydrous ethanol and dried. The inked stamp is then applied to a gold surface for 10 to 20 sec in order to transfer the thiols to the metal surface and create patterns of SAMs on the surface of gold. After the transfer, the stamp is peeled away from the gold surface.

The micro contact method[20,21] is experimentally simple and can be conducted in any chemical laboratory. The stamps can be used many times without deterioration. The technique may be useful in large-area patterning.

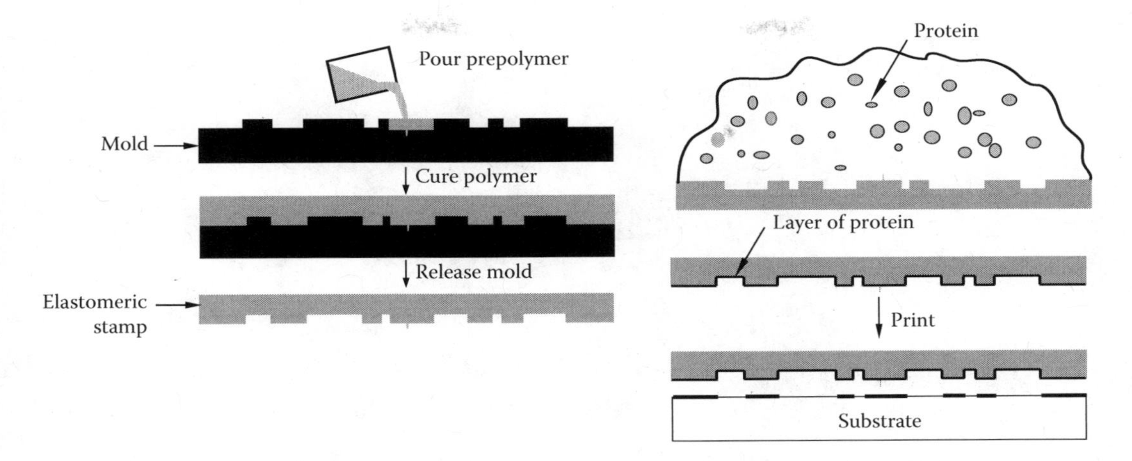

FIGURE 7.25 Schematic of the micro contact printing technique.

By adding hydrochloric acid solution to the etching solution, the CuCl precipitate (bright dots in the SEM image) is dissolved. Various microstructures with features in the nanometer range can be generated by using micro contact printing followed by wet etching (Figure 7.26).

7.3.4 Micro Transfer Molding (μTM)

A PDMS stamp is filled with a prepolymer or ceramic precursor and placed on a substrate. The material is cured and the stamp is removed. The technique generates features as small as 250 nm and is able to generate multilayer systems. The technique can fabricate isolated and interconnected structures over a large area in the range of 3 cm^2. Micro transfer molding has the capacity to fabricate a variety of structures by using many materials, including sol-gel silicon-based materials prepared by using a variety of organo-alkoxysilanes as precursors. Figure 7.27 shows the main steps of the micro transfer molding process. The sol-gel material is obtained by acid-catalyzed hydrolysis and polycondensation of organic precursors such as methyltrimethoxysilane and phenyltrimethoxysilane at room temperature. The different functional groups provide specific properties to the sol-gel material. Polymeric pillars and microwells can be obtained on large surface areas with very good periodicity.

For example, the microwell structures with a fine definition of the sidewalls of these microstructures are shown in Figure 7.28.

The method is able to generate complex topologies and accepts nonplanar surfaces as substrates.

7.3.5 Micromolding in Capillaries

Micromolding in capillaries is a soft lithographic technique based on a network of microchannels created between a PDMS mold with relief features and a support. The channels are spontaneously filled with a low-viscosity liquid prepolymer by capillary action, as shown in Figure 7.29. After the solidification of the polymer and the removal of the PDMS mold, polymeric structures are formed on the support.[24] Narrow trenches of 30 nm have been created by this MIMIC technique, as shown in Figure 7.30.

The metal lines are formed by physical vapor deposition of the metal onto the PS lift-off of the polystyrene.

It has been shown that the cross-linked PDMS stamp tends to swell in solvents such as chloroform, toluene, and tetrahydrofuran. In the fabrication of molecularly imprinted polymer filaments, dimethylformamide was used successfully in acrylic acid–based and polyurethane-based imprinting systems.

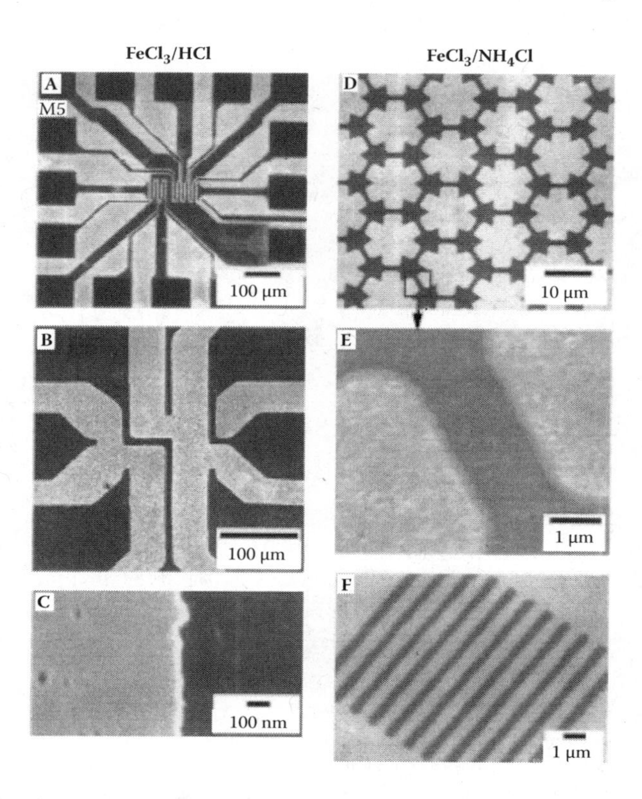

FIGURE 7.26 SEMs of copper films that were patterned with SAMs of hexadecanethiolate and etched in solutions containing $FeCl_3/HCl$ (A–C) or $FeCl_3/NH_4Cl$ (D–F) for 20 sec. The bright regions correspond to copper covered by the SAM; the dark regions are Si/SiO_2 where copper has dissolved. (Reproduced from Xia et al., *Chem. Mater.*, 8, 601–3, 1996. With permission.)

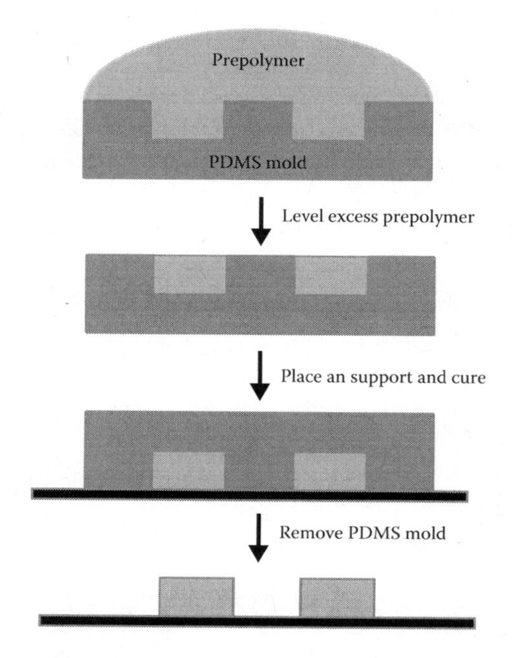

FIGURE 7.27 Micro transfer molding. (Adapted from Meenakshi et al.,. *J. Chem. Ed.*, 84, 1795–98, 2007.)

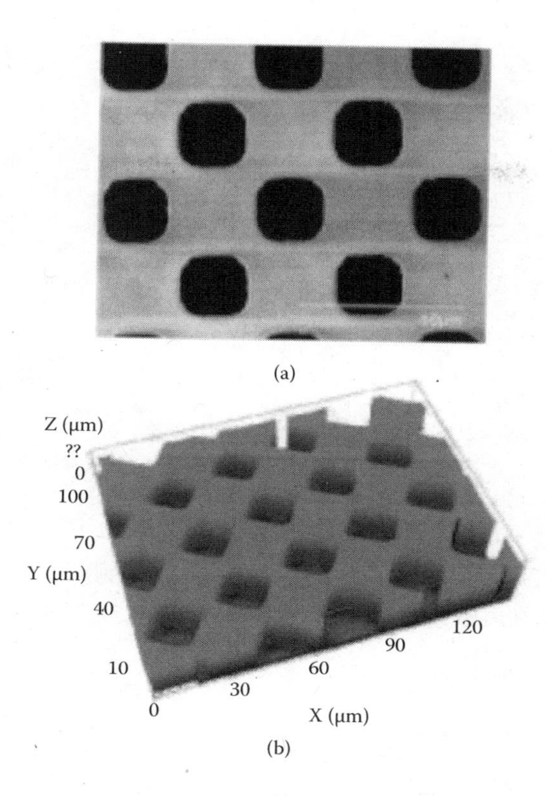

FIGURE 7.28 (a) SEM image and (b) confocal image of 7 μm deep, 4 μm wide microwells defined on a thin sol-gel layer using the μTM process. (Reprinted from Fernández-Sánchez et al., *Chem. Mater.*, 20, 2662–68, 2008.)

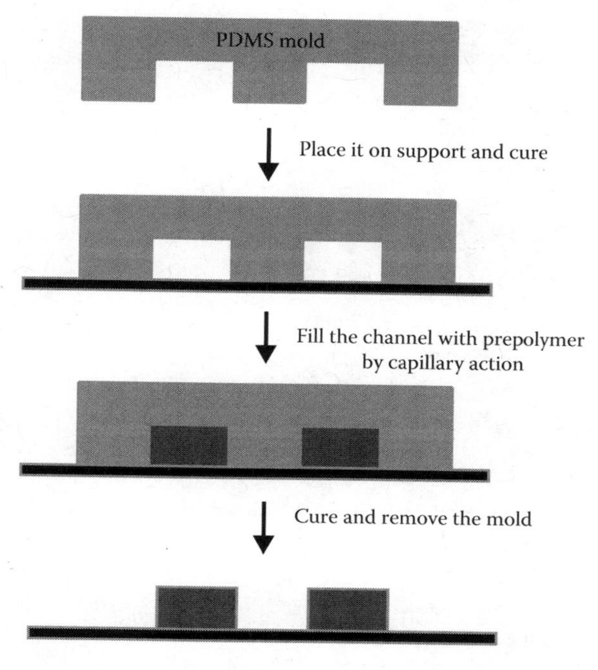

FIGURE 7.29 Micromolding in capillaries (MIMIC).

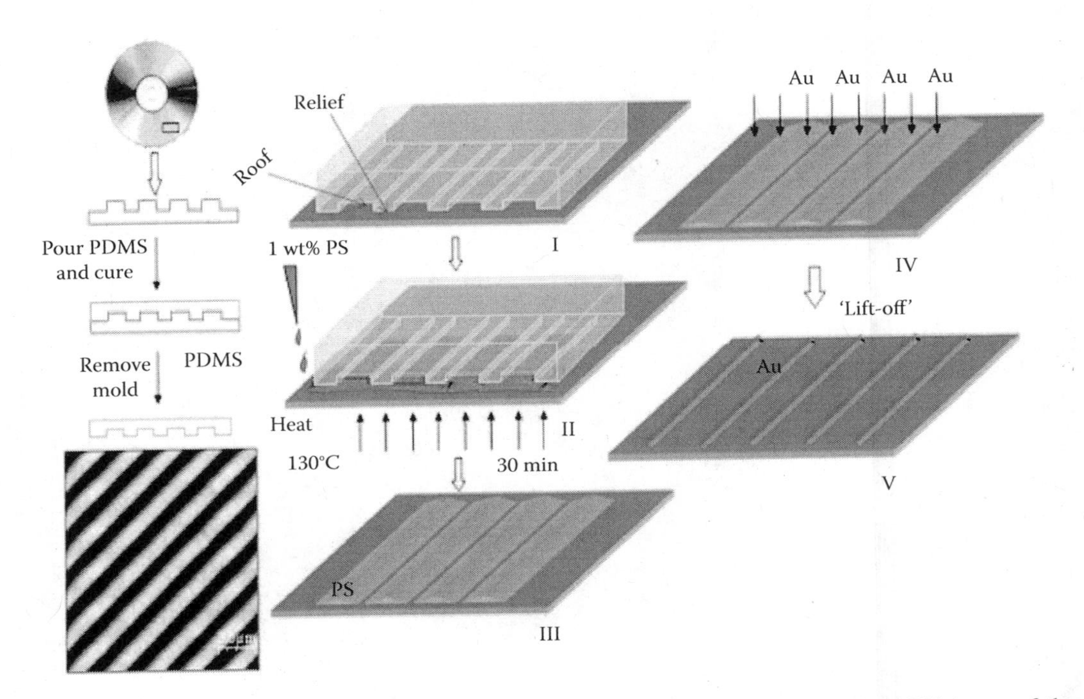

FIGURE 7.30 (See color insert.) Preparation of the PDMS stamp from a CD and the AFM image of the stamp; molding of polystyrene using the PDMS stamp (steps I to III) and the metal deposition (steps IV and V). (Reprinted from Radha et al., *Appl. Materials & Interfaces* 1, 257-260, 2009.)

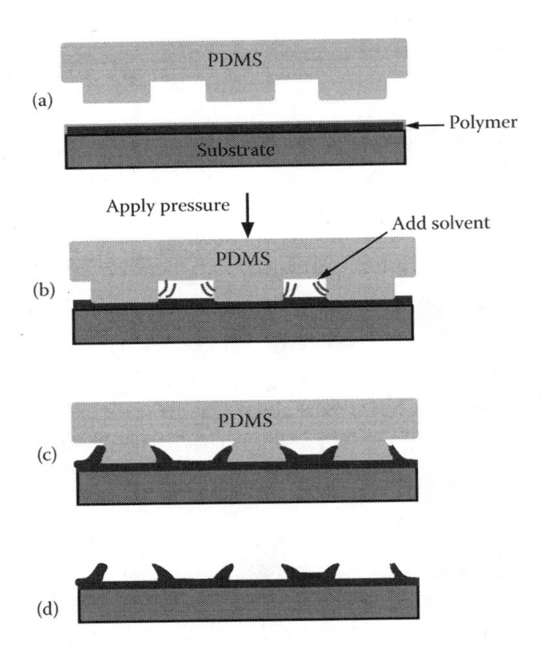

FIGURE 7.31 Schematic illustration of solvent-assisted capillary force lithography. (a) PDMS mold with the desired topography is swollen by the solvent. (b) The swollen mold is pressed on a thin polymer film spin coated on a substrate. (c) The solvent diffuses into the thin polymer film due to the capillary force. (d) The mold is removed. (Adapted from Yu et al., *Polymer*, 46, 11099–103, 2005.)

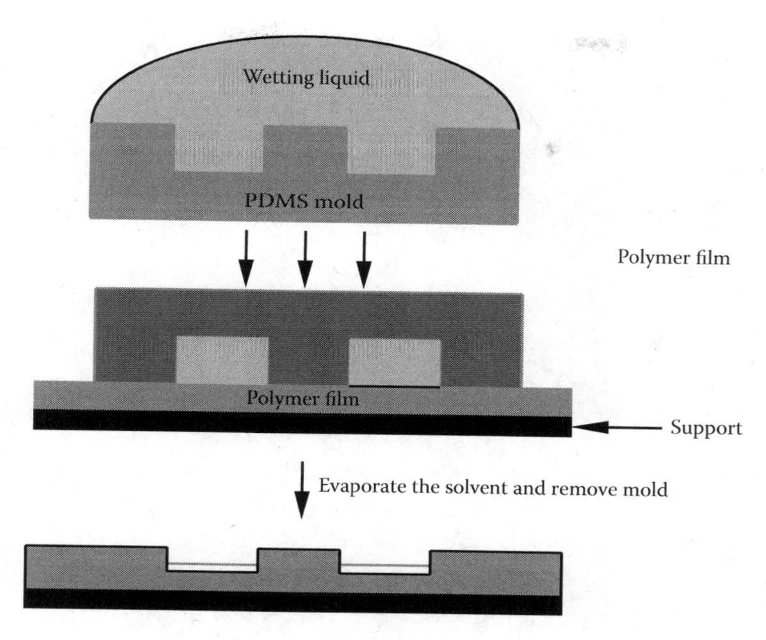

FIGURE 7.32 Solvent-assisted micromolding (SAMIM).

While PDMS films are usually patterned by micromolding techniques, recently, a method for patterning PDMS using a combination of wet and dry etching was developed. The concept of this method is to dry etch the PDMS from 10 μm thickness to a thickness of 2 to 3 μm, producing vertical sidewalls, and then wet etch to remove the remaining material, producing a smooth substrate. To obtain a good quality pattern, Al was used as a hard mask for reactive ion etching of PDMS with a CF_4:O_2 mixture. At the end of the process PDMS was wet etched before etching aluminum, as shown in Figure 7.33.

7.4 MICROFABRICATION TECHNIQUES FOR BIODEGRADABLE POLYMERS

Most of the implantable and drug delivery devices are based on a special class of biodegradable polymers. They have the advantage of natural degradation in tissues over a period of time, usually several months, and are biocompatible. Biodegradable polymers are used for the fabrication of three-dimensional scaffolds for tissue engineering. Some of the most important biodegradable polymers with a wide range of medical applications are poly-D-lactic acid (PDLA), poly-ε-caprolactone (PCL), polyvinylalcohol, etc. While some of the replication techniques discussed in the previous sections have been proved useful for the microfabrication of biodegradable polymers, new fabrication processes, such as pressure-assisted microsyringe method and stereolithography, have also been developed.

A *pressure-assisted microsyringe method*[27] was used to fabricate a polylactic-co-glycolic acid (PLGA) scaffold with microscale porosity. The deposition system, illustrated in Figure 7.34, is based on a stainless steel syringe that contains the solution of the polymer in a volatile solvent. The solution is squeezed through a glass capillary needle with compressed air. The syringe is mounted on the z-axis of a micropositioning system, while a glass substrate on a x-y horizontal stage moves relative to the syringe. The dimensions of the structures deposited with the syringe depend on the pressure applied to the syringe, viscosity of the solution, and dimensions of the syringe tip. The system, including valves, pressure regulators, sensors, and position controllers, is interfaced with a computerized controlling system that allows the fabrication of a wide range of patterns.

After the first layer has been deposited, subsequent layers are deposited by moving the syringe.

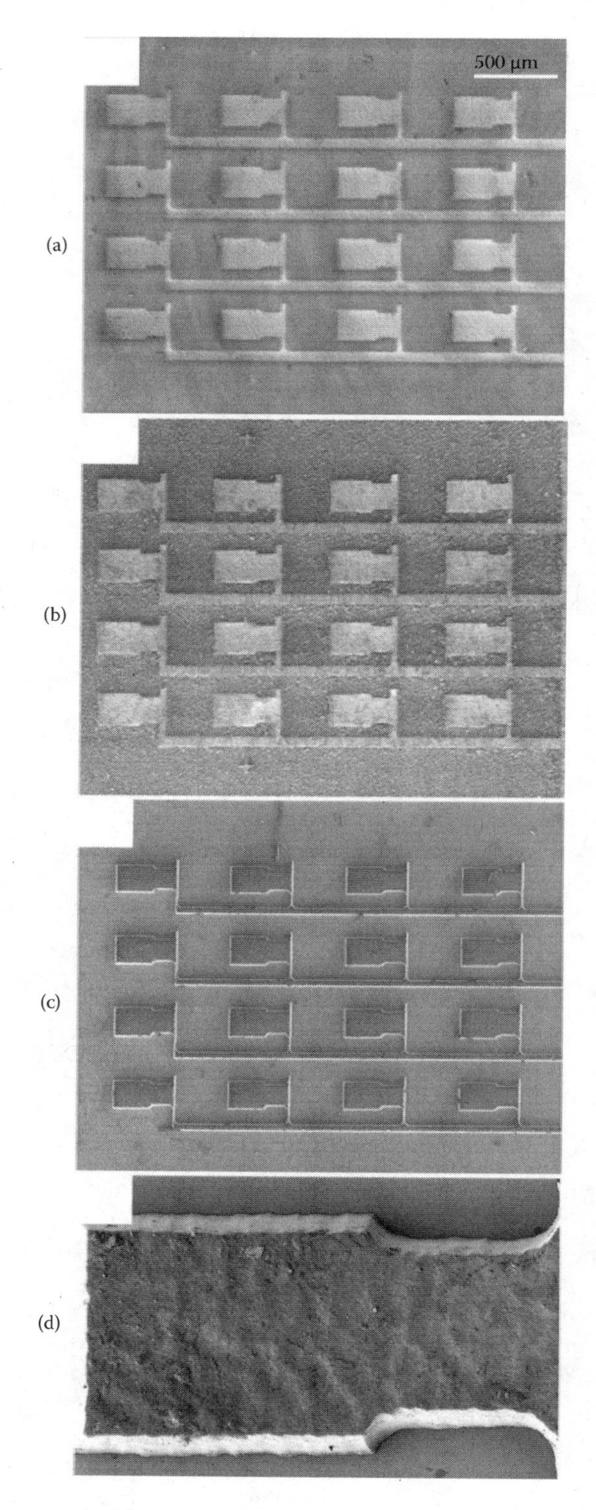

FIGURE 7.33 (a) SEM micrograph of an Al mask on PDMS. (b) Sample after dry etching. (c) Sample after wet etching. (d) Close-up on one of the PDMS structures. (Reproduced from Balakrisnan et al., *J. Micromech. Microeng.*, 19, 047002, 2009. With permission.)

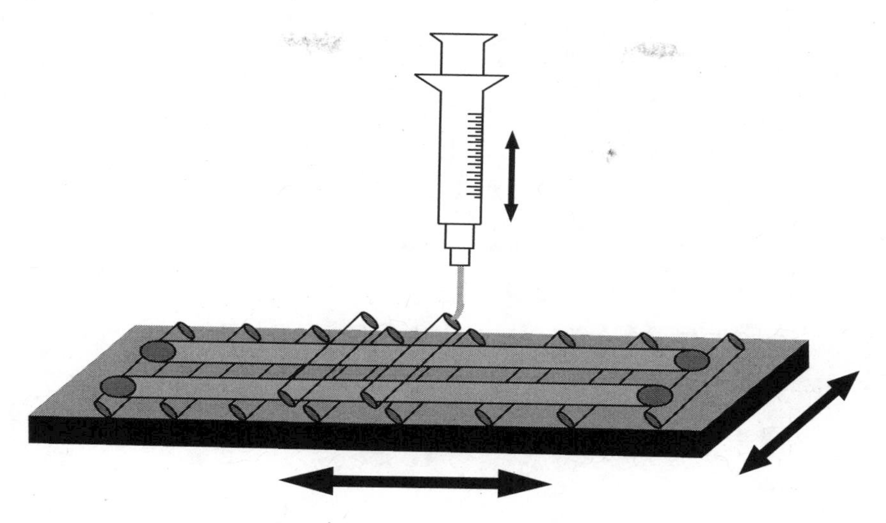

FIGURE 7.34 Schematic setup of a pressure-assisted microsyringe method. (Adapted from Chen and Lu, in *Handbook of Biodegradable Polymeric Materials and Their Applications*, ed. Mallapragada and Narasimhan, Vol. 1, 2005, pp. 1–17.)

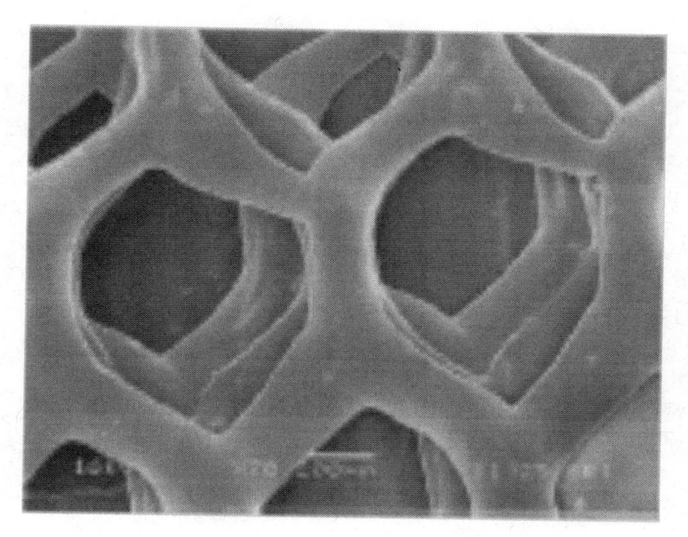

FIGURE 7.35 Three-dimensional PLGA scaffold made by the microsyringe method. (Reproduced from Vozzi et al., *Biomaterials*, 24, 2533–40, 2003. With permission.)

Three-dimensional scaffolds can be made by stacking two-dimensional layers (Figures 7.34 and 7.35). These direct deposition methods are automated and have a dynamic control of the scaffold. The advantage of the technique is its simplicity and the need for inexpensive facilities. But only a few polymers are suitable for this method, and the geometries that can be produced are also limited.

Laser stereolithography[28,29] is a method that allows real three-dimensional microfabrication in a way similar to three-dimensional printing, but working in a liquid environment. The technique is also called three-dimensional printing, photosolidification, solid free-form fabrication, and solid imaging. Laser stereolithography is based on the photoinitiated polymerization or cross-linking in a specified volume. The essence of the process involves the generation, upon irradiation with a UV laser, of active centers such as radicals and ions that interact with the monomer molecules and initiate the polymerization process in the presence of a photoinitiator, as shown in Figure 7.36. The

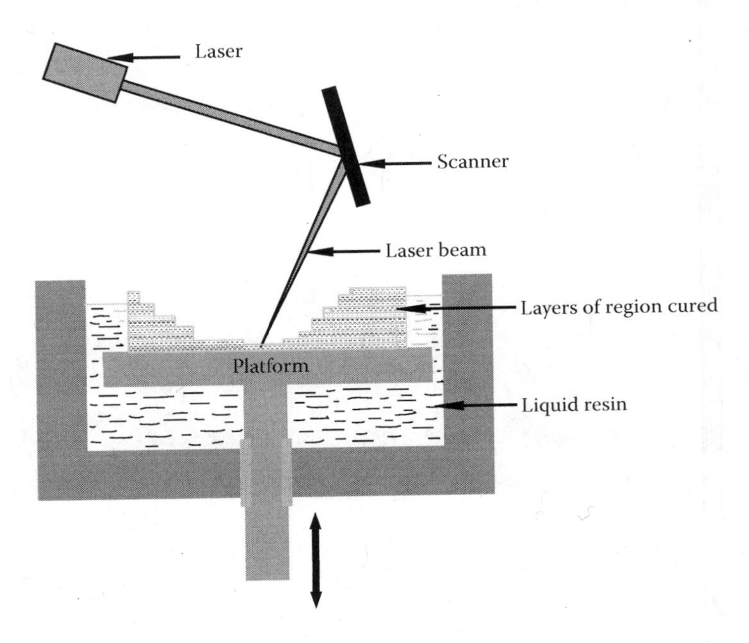

FIGURE 7.36 Schematic of stereolithography.

polymer cures and solidifies only at the focal point of the laser. A three-dimensional solid model designed with CAD software is numerically sliced into a series of two-dimensional layers with an equal thickness. The code generated from each sliced two-dimensional file is then executed to control a motorized x-y-z platform immersed in a liquid photopolymer. The liquid monomer or polymer, for example, diethyl fumarate, is selectively exposed to the focused laser light, which moves in x-y directions. After the first layer is formed, the platform moves downward and a new layer of polymer is solidified according to the design. This layer-by-layer microfabrication enables the creation of complex internal features such as passageways and curved surfaces. Furthermore, by using different proteins and microparticles containing polymer solutions for each layer, a precise spatial distribution of biochemical microenvironments can be created.

A cellular structure produced by stereolithography is shown in Figure 7.37. Square pores were formed by solidifying vertical walls with a 400 μm interline spacing.

7.5 NANOFABRICATION METHODS

7.5.1 Laser Processing, Ablation, and Deposition

Machining by femtosecond pulsed laser ablation is a versatile and convenient process for creating precision shapes in a wide variety of materials. For example, this technique is used to create microchannels in glass and polymer surfaces for microfluidic device fabrication. Due to the high-resolution capability, *laser ablation* is one of the most efficient physical methods for micro- and, more recently, nanofabrication. When focused on the surface of a solid target, the pulsed-laser radiation is absorbed, leading to heating, melting, and finally ablation of the target material. Focusing of femtosecond (fsec) laser pulses into a target leads to the generation of a microplasma. Highly localized laser effects can be induced while minimizing thermal and mechanical damage to the surrounding material, for example, the optical pulses produced by Ti. In the case of *laser micromachining*, Al_2O_3 regenerative amplifiers are typically about 150 fsec in duration and can easily produce peak laser power densities greater than 1,012 W/cm^2 when focused onto a material surface. The short-pulse length ensures that the laser energy is confined within and ionizes only relatively a thin surface layer of absorbing material.

FIGURE 7.37 Cellular type structure produced by laser stereolithography. Fluorescein microparticles were embedded in PEG-DMA walls. 400 × 400 μm pore size. (From Lu, Y. and Chen, S.C. 2004. *Advanced Drug Delivery Reviews,* 56: 1621-1633.)

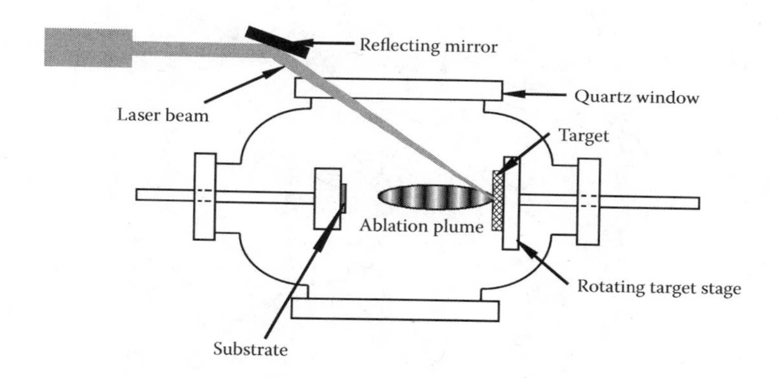

FIGURE 7.38 Schematic of a vacuum chamber.

When the target is ablated in vacuum, as shown in Figure 7.38, nanoclusters of the target material can be deposited on a substrate placed at some distance from the target, leading to the formation of a thin nanostructured film. The properties of the nanostructures can be controlled by the pulse duration and the wavelength of the radiation.

Femtosecond laser ablation is opening up new perspectives in ultraprecise processing of organic materials and in synthesis of thin films with tailored structures and compositions.

Machining by femtosecond pulsed laser ablation is a versatile and convenient process for creating precision shapes in a wide variety of materials while minimizing thermal or mechanical damage on the remaining surface. The process is well suited for the surface structuring, drilling, and cutting operations. It is also effective for machining solid metal, glass, quartz, and polymer materials used in the fabrication of many biomedical devices. A typical structure created for tissue growth studies is shown in Figure 7.39.

7.5.2 High-Precision Milling

High-precision machining with diamond tools provides a cost-effective manufacturing platform for many applications. For materials that cannot be directly diamond machined, such as steel and glasses, a hybrid process utilizes diamond machining as a mask-making process, and then plasma

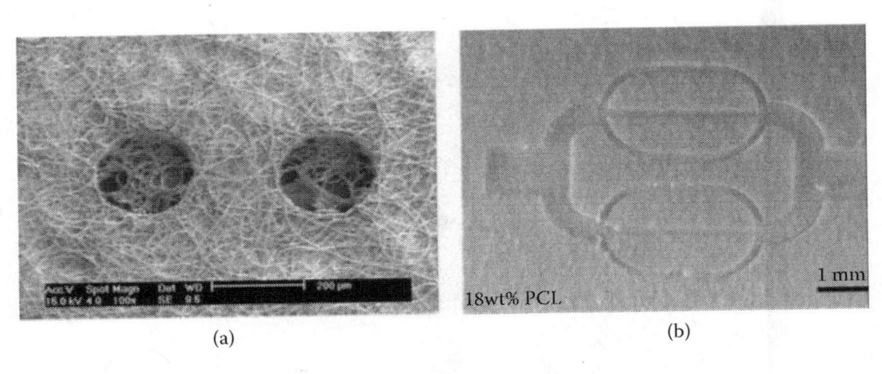

FIGURE 7.39 Structures machined in enzyme/substrate poly-ε-caprolactone nanofiber tissue scaffold by femtosecond laser ablation. (a) Cylindrical pockets with 200 μm diameter for cell growth studies. The pocket on the right side of the image contains cell cultures. (b) Branching channel structure for flow studies. (Reproduced from Yi et al., *Adv. Polymer Technol.*, 27, 188–98, 2008. With permission.)

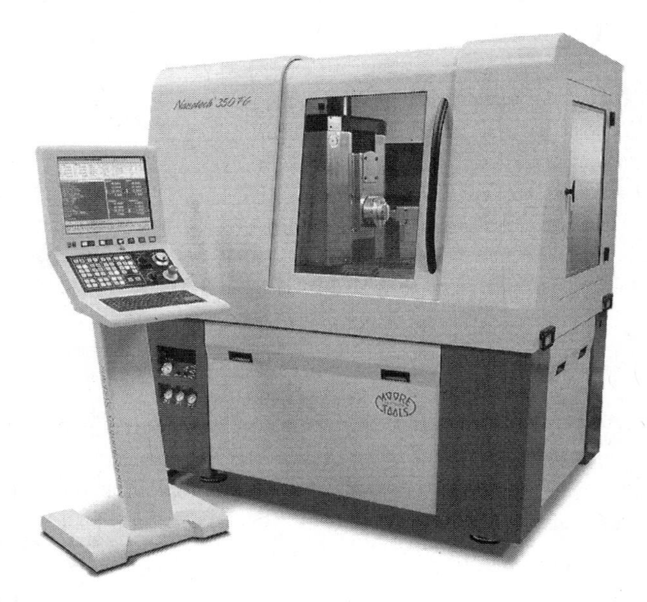

FIGURE 7.40 High-precision milling with 350FG ultraprecision milling machine. (Reproduced from Nanotech Systems. With permission.)

etching to the final dimensions is developed. Ultraprecision machining has the capability of producing the desired structures in a single setup, greatly reducing manufacturing errors (Figure 7.40).

Typical applications for this machine include machining of aspheric and toroidal surfaces, linear diffractive, microprismatic optical structures, and micromilling of lens arrays. As an example, a microsize diamond tool is controlled to contour a diffractive optical surface with lateral features that are 10 nm deep in the vertical direction.

7.5.3 Inductively Coupled Plasma (ICP) Reactive Ion Etching

A reactive ion etching (RIE) system with an inductively coupled plasma (ICP), called deep reactive ion etching (DRIE), has immense capability for batch fabrication of deep anisotropic microstructures in silicon and enables high-aspect-ratio deep silicon and SiO_2 etching.

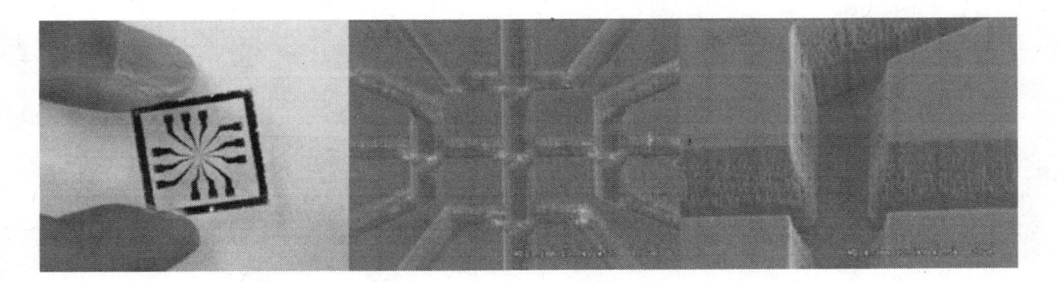

FIGURE 7.41 A microfluidic channel with electrodes fabricated by ICP reactive ion etching. (From Yi et al. *Adv. Polym. Technol.*, 27, 188–198, 2008.)

Microfluidic devices with deep channels are fabricated by this technique for manipulating micro- and nanoparticles and biomolecules. Near-vertical sidewalls and smooth etched surfaces are obtained by optimizing etching parameters. Microfluidic devices of various sizes on Pyrex glass and silicon can be fabricated. Microfluidic devices with smooth surfaces and sidewalls were fabricated for an electrokinetic flow (EKF) pattern study. Two types of EKF patterns—extensional and rotational flow under different biases—have been successfully demonstrated, with five-cross microfluidic devices shown in Figure 7.41.

7.5.4 Electron Beam Lithography

In recent years electron beam lithography (EBL) has become the principal nanofabrication technique for defining nanostructures. EBL is often combined with traditional photolithography for patterning of larger surrounding structures.

Electron beam lithography (often abbreviated as e-beam lithography), explained in Figure 7.42, is the technique of scanning an electron beam in a patterned fashion across a surface covered with a resist film, and of selectively removing either the exposed or the nonexposed regions of the resist. The purpose, as in photolithography, is to create very small structures in the resist that can subsequently be transferred to the substrate material, often by etching. It was developed for manufacturing integrated circuits, and is also used for creating nanostructures. The primary advantage of electron beam lithography is that it is one of the ways to exceed the diffraction limit of light and make features in the nanometer regime. This form of maskless lithography has found wide usage in photomask making used in photolithography and low-volume production of semiconductor components.

On the other hand, the key limitation of electron beam lithography is low throughput, as it takes a long time to expose an entire silicon wafer or glass substrate. A typical EBL system consists of the following parts: (1) an electron gun or electron source that supplies the electrons, (2) an electron column that focuses the electron beam, (3) a mechanical stage that positions the wafer under the electron beam, (4) a wafer handling system that automatically feeds wafers to the system and unloads them after processing, and (5) a computer system that controls the equipment.

7.5.5 Dip Pen Nanolithography

Dip pen nanolithography (DPN), shown in Figure 7.43, is an established method of nanofabrication in which materials are deposited onto a surface through a sharp probe tip. Molecules are transferred from the tip to the surface through a water meniscus, which forms in ambient conditions as the tip nears the surface. DPN enables controlled deposition of a variety of nanoscale materials onto many different substrates. The vehicle for deposition can include pyramidal scanning probe microscope tips, hollow tips, and even tips on thermally actuated cantilevers.

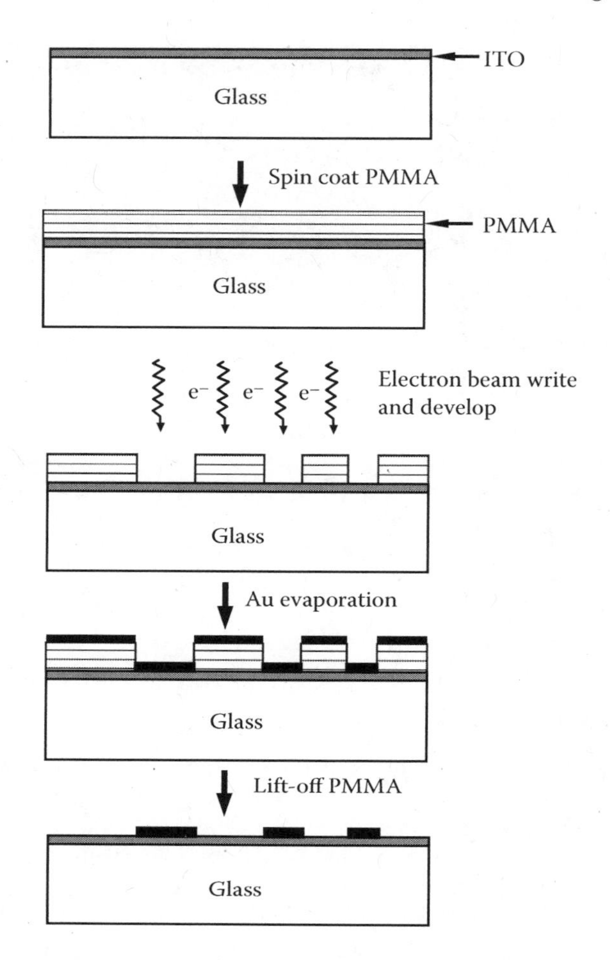

FIGURE 7.42 Electron beam lithography process.

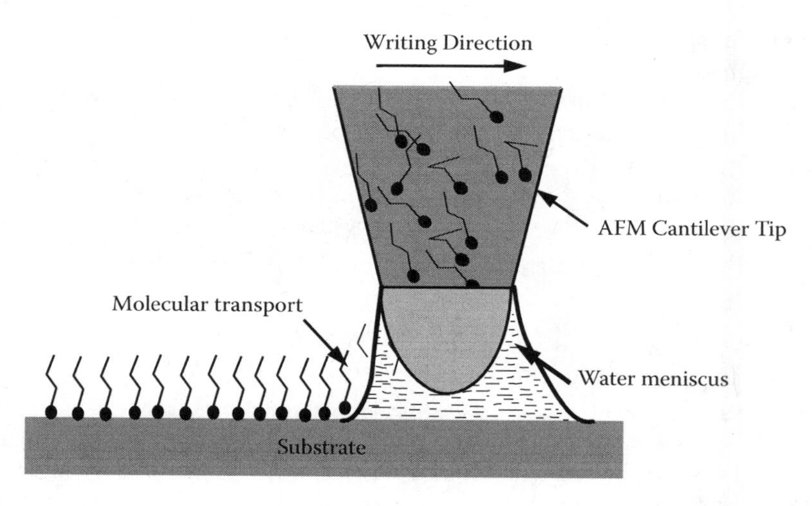

FIGURE 7.43 Dip pen nanolithography. (Adapted from Salaita et al., *Nat. Nanotechnol.*, 2, 145–55, 2007.)

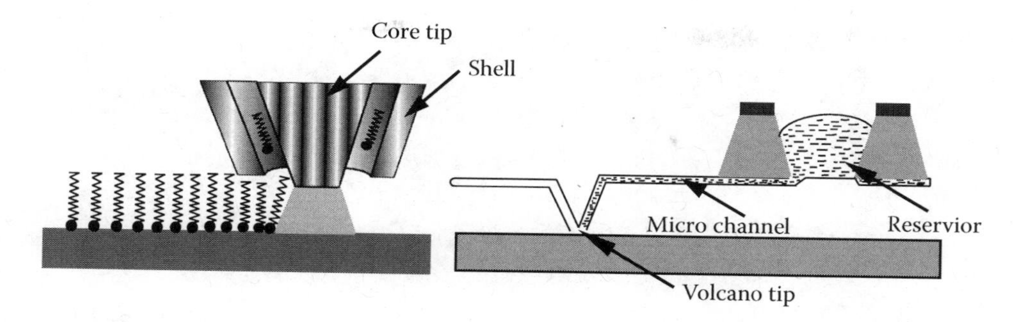

FIGURE 7.44 Microfluidic reservoir-integrated dip pen lithography. (Adapted from Salaita et al., *Nat. Nanotechnol.*, 2, 145–55, 2007.)

In array approach, it is possible to scale up the number of tips in the order of fifty thousand. The versatility of dip pen nanolithography offers a number of benefits over other techniques, making it a leading method of nanofabrication. This method could yield a resolution of features as small as 14 nm, and spatial resolution of 5 nm. Figure 7.44 shows the reservoir-integrated DPN system in which ink from a reservoir is fed to the tip through a set of microfluidic channels.

DPN has proved capable of patterning a variety of materials onto a variety of substrates. Because DPN works in ambient conditions, it offers a significant advantage for materials that are environmentally sensitive or incompatible with existing microfabrication techniques, such as biomolecules, conductive polymers, or ceramics.

7.5.6 Nanosphere Lithography (Colloid Lithography)

The top-down nanofabrication platforms are useful in surface patterning; however, cost and process considerations limit their applicability. Fabrication of periodic nanoscale structures over large areas using self-organizing systems is of great interest because of the simplicity and low cost. These techniques utilize *self-assembly*, which uses chemical or physical driving forces. Self-assembly techniques are bottom-up methods that allow simple control of the pattern size by using block copolymers or colloidal particles. Spherical colloids are commonly used as building blocks to create polymeric micro- and nanostructures because of the relative ease of forming long-range ordered structures. Monodispersed spherical particles can be easily close packed into hexagonal structures, as shown in Figure 7.45.

As shown in Figure 7.46, ordered structures can be obtained by using either mono- or multilayer templates, respectively. Gold-polystyrene (PS) composites are prepared by this technique, and after removing the polystyrene template by dissolution in ethanol in an ultrasound bath, an ordered gold structure is obtained (Figure 7.45).

In the case that a monolayer of polystyrene nanospheres or copolymers is used as a template, the patterning process begins by the deposition and formation of monolayers of hexagonally close-packed nanospheres using self-assembly techniques. Metal is then deposited through the monolayer mask, partially filling the gaps between the nanospheres. A lift-off process then removes the monolayer of nanospheres and a periodic array of triangularly shaped nanometer-scale particles consisting of the metallic material is left on the substrate.[35,36] The quality of the resulting array is influenced by a number of parameters, such as the solvent evaporation rate, colloid and polystyrene concentrations, and wetting characteristics of the substrate. The nanosphere layers can be used directly as a protective barrier against etching. Although large, defect-free areas have been obtained using nanosphere lithography, the range of types of patterns and pattern conditions inherently limits the process. Moreover, the etching process can significantly alter the surface chemistry by redepositing various by-products, which might not be suitable for clinical usage.[37]

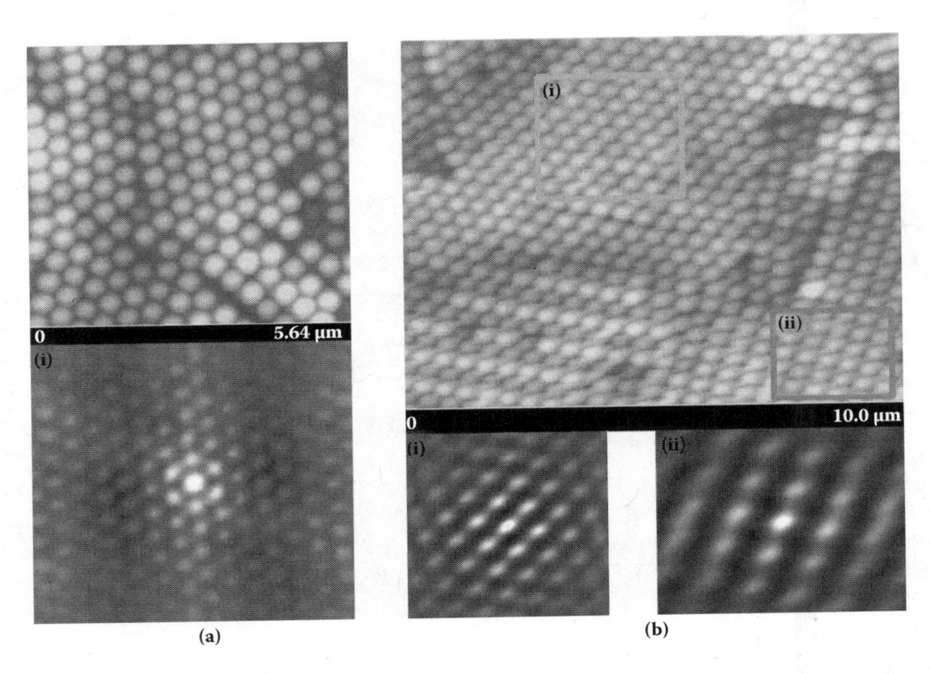

FIGURE 7.45 AFM images of a PS-Au composite film prepared by the following two methods: (a) monolayer film by the flow-controlled inclined deposition (FCID) method and (i) its corresponding autocorrelation image, and (b) multilayer film obtained by vertical deposition method and (i, ii) autocorrelation images of two parts of the AFM image shown in the light and dark boxes, respectively. (Reproduced from Djaoued et al., *Appl. Spectrosc.*, 61, 1202–10, 2007. With permission.)

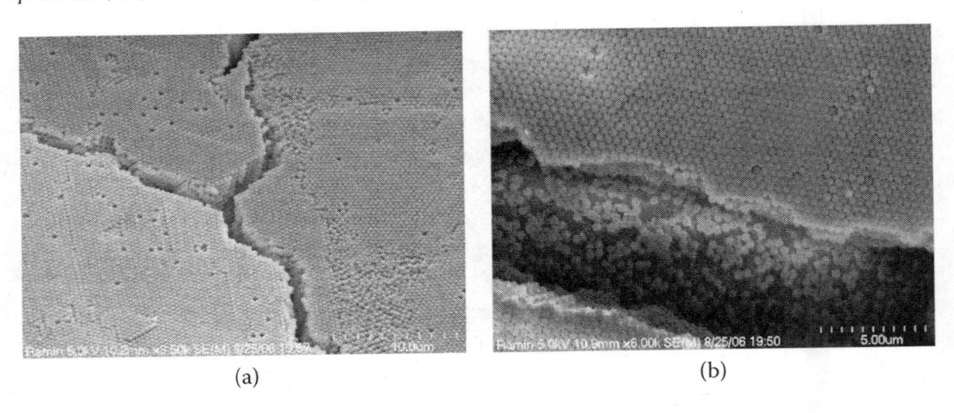

FIGURE 7.46 SEM image of the Au-PS composite multilayer film prepared by vertical deposition. (a) The formation of fissures and the multilayered structure of the composite film. (b) The magnified image of a fissure of 6 μm and the reduced number of PS microspheres at the bottom of the fissure. (Reproduced from Djaoued et al., *Appl. Spectrosc.*, 61, 1202–10, 2007. With permission.)

7.5.7 Surface Patterning by Microlenses

A different arrangement of microspheres on a substrate is possible using a self-assembly processes. If these microspheres are made of transparent materials, then they can also perform as microlenses. As explained earlier, two-dimensional lattices of microlenses are formed by self-assembly processes from colloidal suspensions of microspheres. Microspheres of transparent materials that are transparent for laser radiation, such as amorphous quartz (a-SiO_2), focus the incident laser radiation onto the substrate.

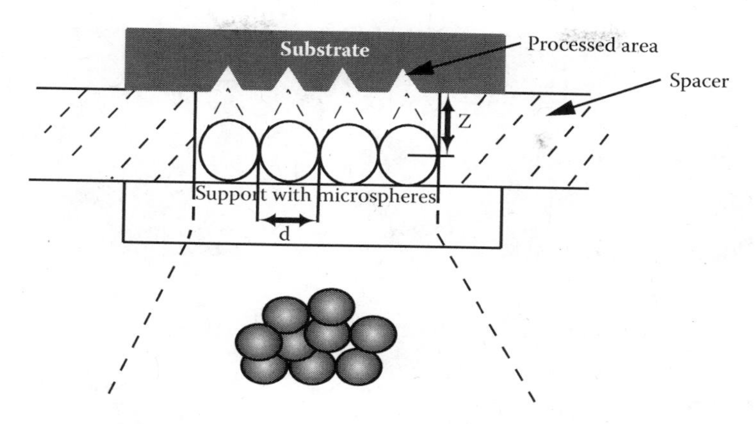

FIGURE 7.47 From a colloidal suspension of a-SiO_2 (quartz glass) microspheres, a two-dimensional hexagonal lattice is formed by self-assembly. Each sphere acts as a microlens that focuses the laser radiation onto the substrates. (Adapted from Bäuer et al., *Romanian Rep. Phys.*, 57, 936–52, 2005.)

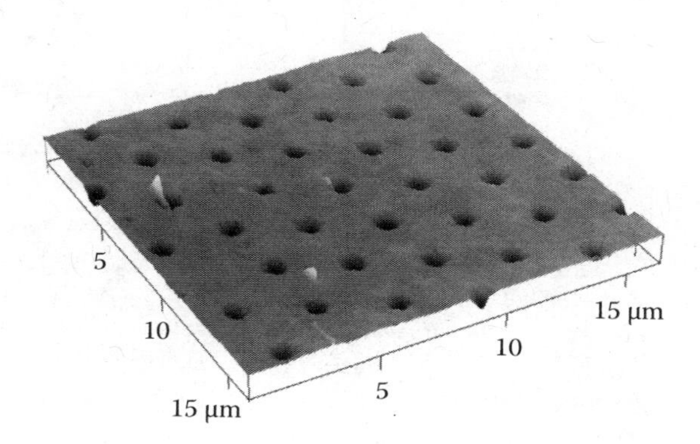

FIGURE 7.48 Holes produced by local ablation of polyimide (PI) by means of a-SiO_2 microspheres of 3 μm size and KrF-laser radiation. (Adapted from Bäuer et al., *Romanian Rep. Phys.*, 57, 936–52, 2005.)

This is schematically shown in Figure 7.47. The technique allows the patterning of thousands or millions of single submicron features on a substrate with a single or a few laser shots. Because of their thermal stability and optical properties, microspheres of a-SiO_2 on fused quartz supports are commonly utilized.

Surface patterning by laser ablation of a polyimide film resulted in holes in a hexagonal pattern, as shown in Figure 7.48.

7.5.8 Electrochemical Patterning

A nanolithography technique that generates topological patterns by electrochemical reduction of metal on a solid-state ion conducting membrane has been developed.

As shown in Figure 7.49, a voltage bias is applied for a short time period between two electrodes: one is a Ag-coated atomic force microscope (AFM) tip and the other is a sputtered Ag film bulk counterelectrode attached to the $RbAg_4I_5$ ion conductor. The Ag atoms are drawn from the counterelectrode to the tip and eventually deposited on the ion conductor in the vicinity of the probe. When a negative bias is applied to the tip, silver atoms at the counterelectrode are oxidized into positively charged ions, which then migrate to the tip electrode due to the applied electric field. The silver ions are electrochemically reduced into Ag atoms at the tip, resulting in nanosize Ag patterns. A pattern produced with this method is shown in Figure 7.50.

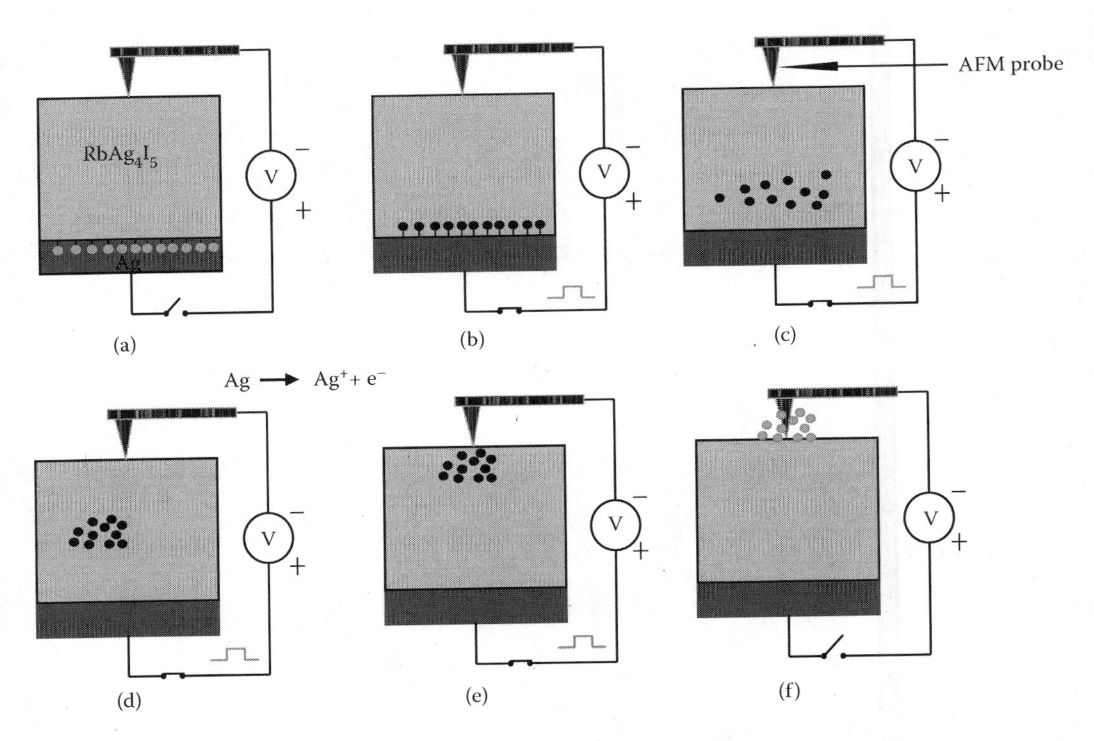

FIGURE 7.49 Simplified schematic diagram of nanopatterning of Ag on $RbAg_4I_5$, showing the time evolution of Ag atoms and ions under voltage bias. The migration of silver ions in the ion conductor under the electric field is shown in steps between (b) and (e). Oxidation and reduction processes are depicted in steps (b) and (f), respectively. (Adapted from Lee et al., *Appl. Phys. Lett.*, 85, 3552–54, 2004.)

7.5.9 Electric-Field-Assisted Nanopatterning

A new method for producing ordered arrays of metal particles on a reconstructed ceramic surface has been developed. The stepped surface is formed as a result of annealing an alumina surface, and it serves as a template for the production of arrays of metal nanoparticles. The authors demonstrated the preferential condensation of liquid metal on the crests in the form of an array of droplets due to the electric field that exists along the surface.

7.5.10 Large-Area Nanoscale Patterning

Hemispherical wells in silicon were generated with 100 nm silica spheres by laser-assisted embossing (Figure 7.51). By changing the size of the spheres in the mold, the energy density of the incident laser pulse, and the applied mechanical pressure between the mold and the surface, the size of the nanowells can be controlled. Thermal oxidation of the patterned silicon substrates can reduce the size of the wells by 10 to 20 nm. The nanowells patterned in silicon or silicon oxide can be functionalized with hydrophobic or hydrophilic silane molecules. These nanowells are ideal reaction vessels or "beakers" to grow isolated and monodisperse nanocrystals that are well ordered on a substrate. Because of their small volumes and chemical functionality, low concentrations of precursor materials can enable the growth of different types and sizes of nanoparticles.

Because isolated particles can be formed in well-defined locations, detailed studies of the properties of individual nanoparticles are possible with these wells.

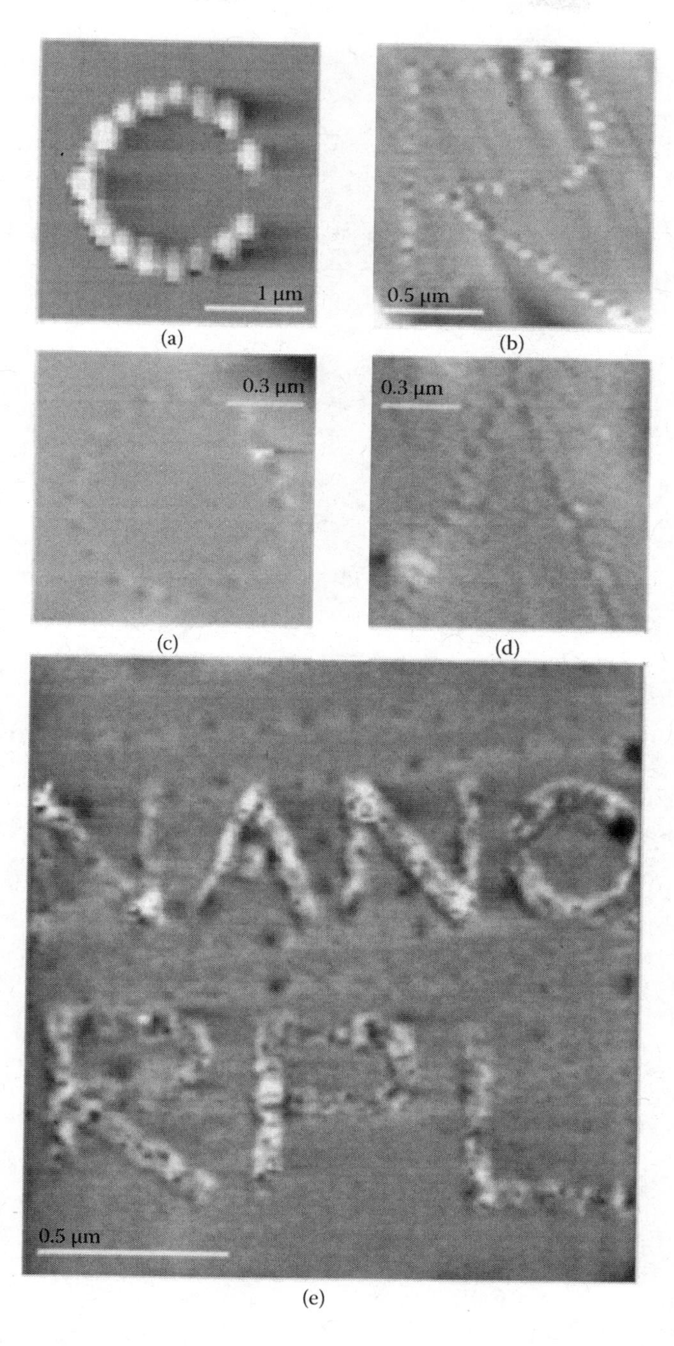

(a) (b) (c) (d) (e)

FIGURE 7.50 AFM contact mode deflection images of electrochemically induced Ag nanofeatures on $RbAg_4I_5$. (a) Positive features generated by 200 mV (5 ms pulses for each dot). (b) Positive features generated by 400 mV (1 ms pulses for each dot). (c) Negative features generated by 200 mV (1 ms pulses for each dot). (d) same as in (c). (e) Positive features generated by 200 mV pulses. (Reproduced from Lee et al., *Appl. Phys. Lett.*, 85, 3552–54, 2004. With permission.)

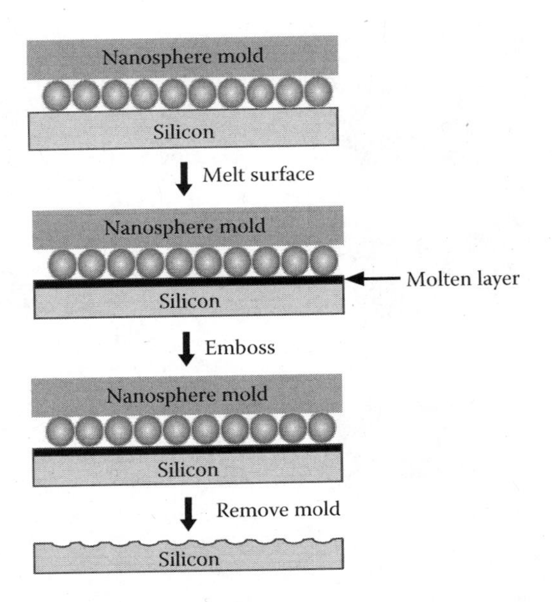

FIGURE 7.51 Scheme for generating nanowells. (Adapted from Barton and Odom, *Nano Lett.*, 4, 1525–28, 2004.)

7.5.11 Selective Molecular Assembly Patterning (SMAP)

This patterning method is based on selective self-assembly of alkane phosphates on metal oxide surfaces. The method is based on a combination of metal oxide surfaces (Figure 7.52) with the selective self-assembly of alkyl phosphates and poly-L-lysine–g–polyethyleneglycol (PLL-g-PEG) on distinct metal oxide surfaces. Alkyl phosphates self-assembled on TiO_2 renders the surface hydrophobic, while PLL-g-PEG renders negatively charged metal oxides (SiO_2) resistant to protein adsorption. The structured surface consists of a TiO_2 pattern surrounded by a matrix of SiO_2, and by sequentially adsorbing alkyl phosphate and PLL-g-PEG, large-scale, biologically relevant chemical patterns can be generated. Protein adsorption to patterned surfaces was studied by fluorescence microscopy, and the fluorescence signal was found to be localized on the hydrophobic protein-adhesive TiO_2 squares, as shown in Figure 7.53. This patterning approach can be applied in the area of microarray biosensor chips.

7.5.12 Site-Selective Assemblies of Gold Nanoparticles on an AFM Tip-Defined Silicon Template

Controlled arrays of gold nanoparticles on silicon can be fabricated by combining AFM nanolithography and self-assembly techniques.[51] SAMs terminated with amino and methyl groups can be adsorbed at different surface locations and then used to guide the assembly of gold nanoparticles. The different affinities of gold nanoparticles for amino (or thiol) groups and methyl groups create the basis for controlling their assembly on solid surfaces. The procedure is shown in Figure 7.54. After the adsorption of a monolayer of octadecyltrichlorosilane, this is locally degraded by applying a voltage pulse between the AFM tip and the substrate. After the removal of the degraded molecules, an amino-terminated surface is generated by adsorption of a monolayer of aminopropyltrimethoxysilane on the exposed silicon oxide. Gold nanoparticles are adsorbed to the silane amino groups through electrostatic interactions and can be positioned to give ordered nanoparticle arrays. Figure 7.55 shows the silicon oxide dots formed by the AFM nanodegradation technique.

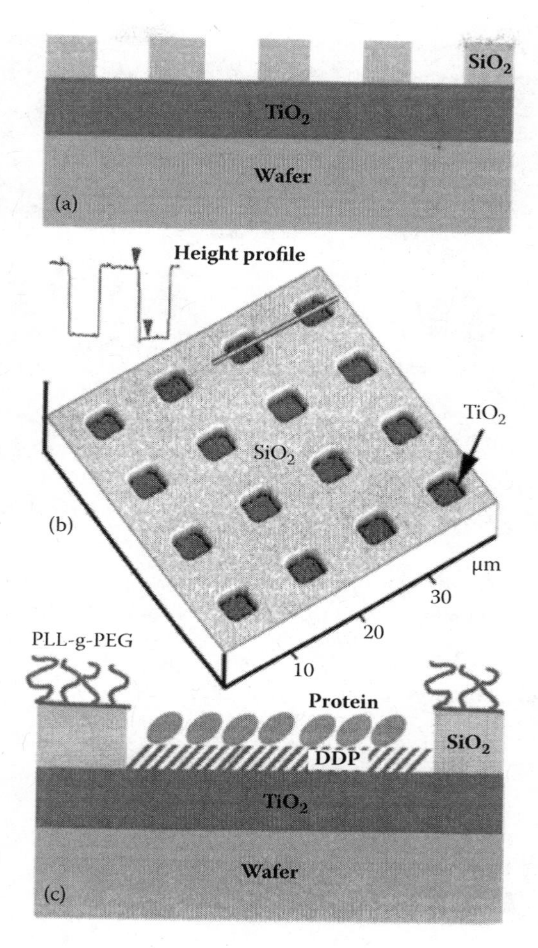

FIGURE 7.52 A schematic illustration of the SMAP methodology. (a) Sample exhibiting a material contrast, produced using common photolithographic techniques. TiO_2 squares within the SiO_2 matrix are shown. (b) Atomic force microscope image of the surface shown in (a). TiO_2 squares are located ~35 nm below the SiO_2 matrix (inset: a height profile of the surface). (c) Schematic view of the surface after the surface modification procedures: DDP on TiO_2, protein on DDP, and PLL-*g*-PEG on SiO_2, with the polylysine backbone lying flat on the surface and PEG chains extending away from it. (Adapted from Michel et al., *Langmuir*, 18, 3281–87, 2002.)

7.5.13 Highly Ordered Metal Oxide Nanopatterns Prepared by Template-Assisted Chemical Solution Deposition

The preparation of metal oxide patterns composed of ordered nanofeatures of various morphologies, i.e., perforations (craters), rings, canyons, wires, dots, or channels with typical lateral dimensions of less than 40 nm and thickness below 15 nm, is possible by a simple chemical solution deposition (CSD) of molecular inorganic precursors and commercial block copolymers. Self-assembly during evaporation and subsequent stabilization at 500°C leads to the formation of various nanopatterns.[43] Compared to other techniques for surface nanopatterning, this method has the advantage of being cheap, reproducible, and easy to scale up, and does not require specialized equipment. Critical parameters for controlling the nanostructure of the patterns are the dissolution conditions, substrate surface energy, temperature, and organic-to-inorganic ratio.

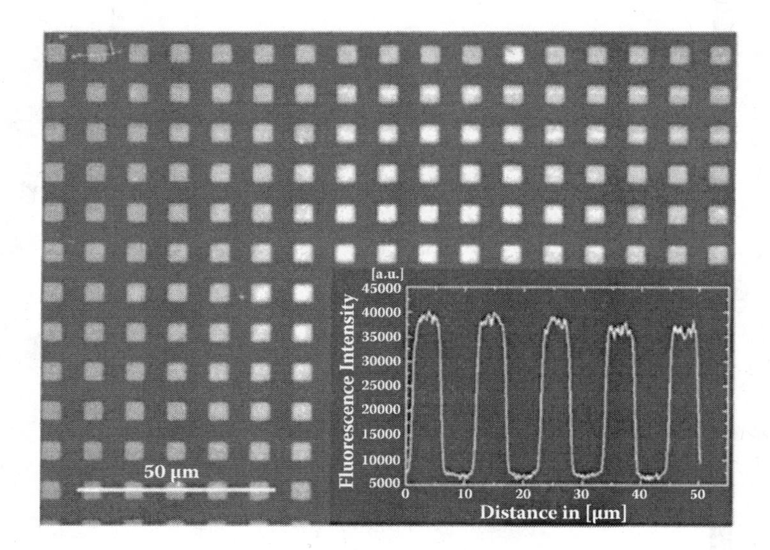

FIGURE 7.53 Fluorescence microscopy image on Oregon Green–labeled streptavidin subjected to the SMAP-treated, 5 × 5 μm² TiO_2 in SiO_2 substrate. Streptavidin adsorption can only be observed on the TiO_2/DDP spots, while the SiO_2/PLL-*g*-PEG remains protein resistant. The inset shows the local distribution of fluorescence of the Oregon Green–labeled streptavidin across the surface (in arbitrary units). A contrast of 100:1 was observed. (Reprinted from Michel et al., *Langmuir*, 18, 3281–87, 2002. With permission.)

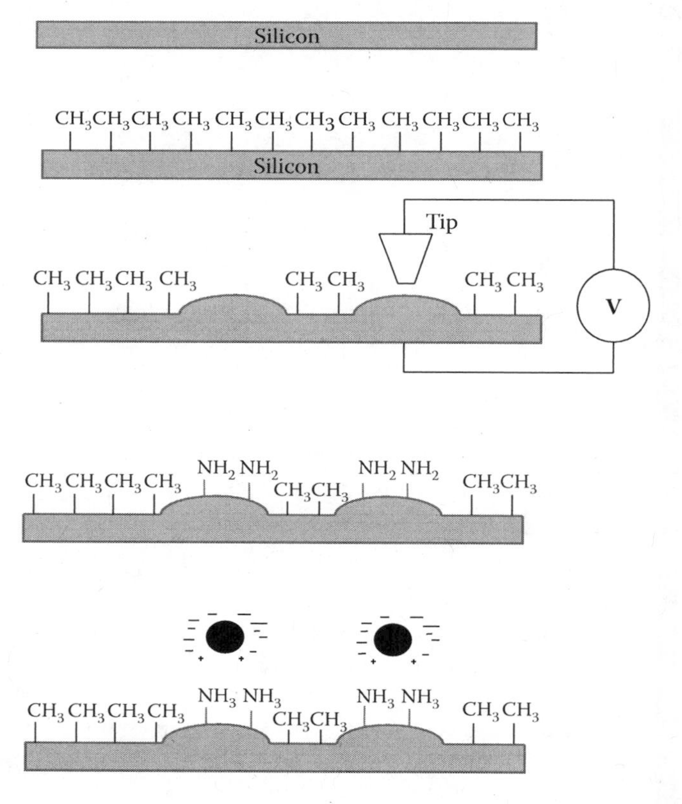

FIGURE 7.54 Experimental procedure for fabricating gold nanoparticle assemblies. (From Li et al., *Langmuir*, 19, 166–71, 2003.)

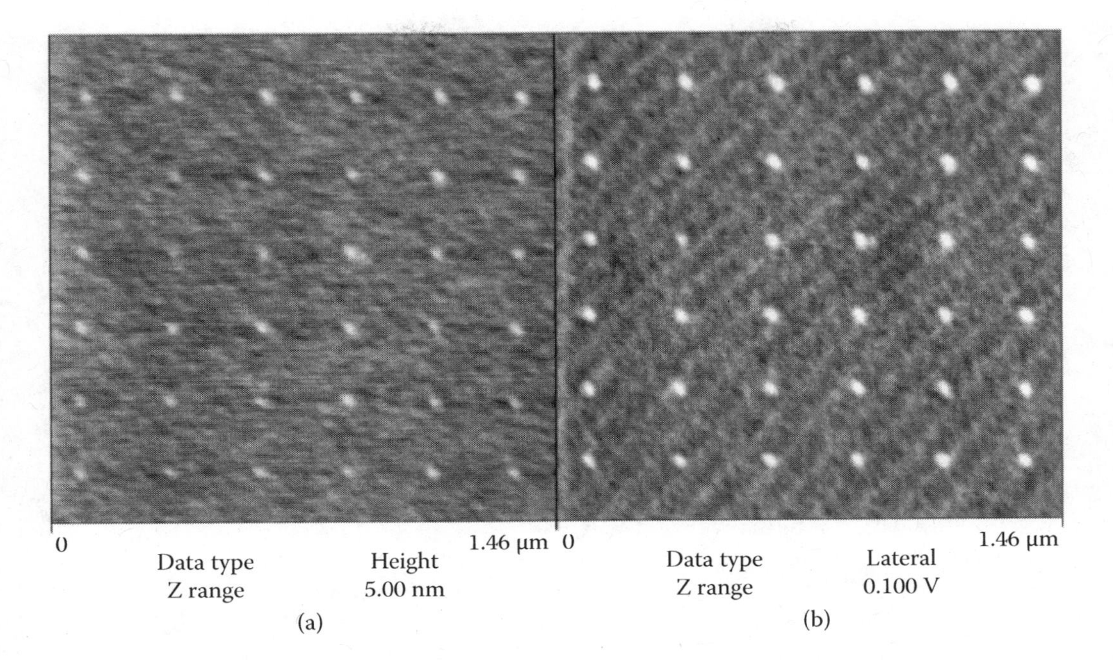

FIGURE 7.55 AFM images of the silicon oxide array fabricated by AFM nanodegradation. (a) Height image. (b) Friction image. (Reproduced from Li et al., *Langmuir*, 19, 166–71, 2003. With permission.)

7.5.14 Wetting-Driven Self-Assembly: A New Approach to Template-Guided Fabrication of Metal Nanopatterns

Wetting-driven self-assembly (WDSA) yields stable surface-immobilized nanostructures for confinement of species such as metal ions, metals, and semiconductor nanoparticles. This procedure combines the nanoelectrochemical pattern inscription on a self-assembled base monolayer with various postpatterning chemical processing operations. During these operations, only the tip-inscribed sites should be affected. One of the methods makes use of a robust methyl-terminated silane monolayer, such as n-octadecyltrichlorosilane self-assembled on silicon. The process starts with the tip-induced nanoelectrochemical oxidation of the top methyl groups of the silane molecule. Several different self-assembly and chemical modification paths are depicted in Figure 7.56.

7.5.15 Patterned Gold Films via Site-Selective Deposition of Nanoparticles onto Polymer-Templated Surfaces

In this method, polyvinyl N-methylpyridine (PVMP) is patterned by photochemical cross-linking, as shown in Figure 7.57. The positively charged surface adsorbs selectively the citrate-protected (negative) gold nanoparticles. At first, the ethanolic solution of the PVMP film is spin coated on a substrate, and then the film is cross-linked by exposure to UV light. Patterned Au films can also be prepared by prepatterning the polymer layer using lithographic or self-assembly techniques. Various postfunctionalization strategies to fine-tune surface properties can also be used. As an example, functionalization of the gold surfaces with a thiol group is shown in Figure 7.58.

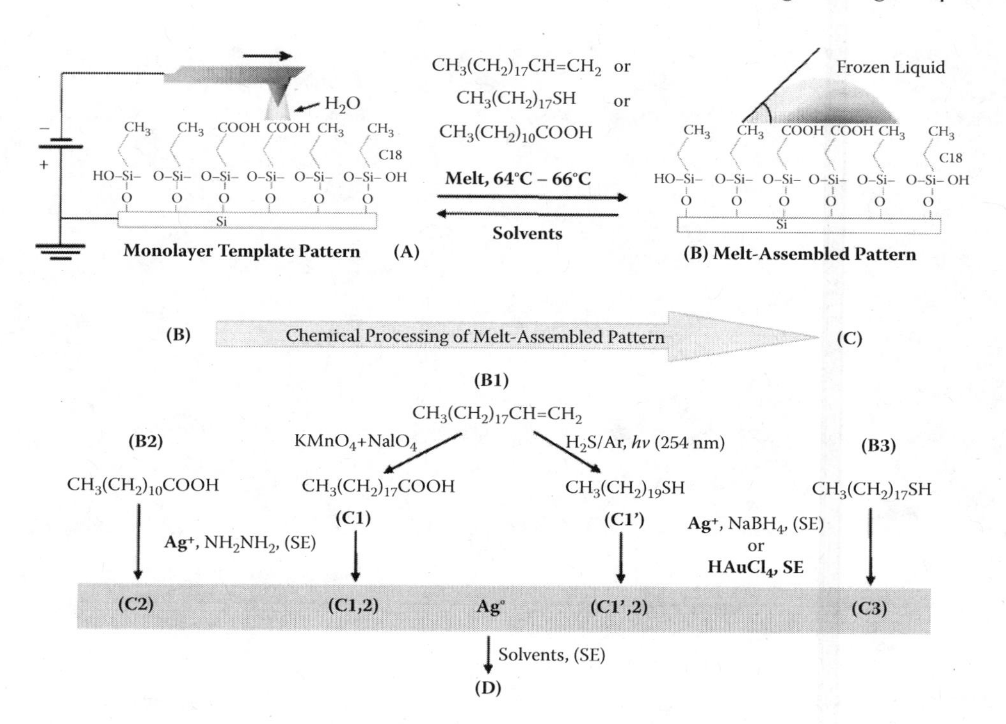

FIGURE 7.56 Schematic illustration of WDSA as a generic approach to template-guided fabrication of surface-immobilized metal nanopatterns. (Adapted from Chowdhury et al., *Nano Lett.*, 7, 1770–78, 2007.)

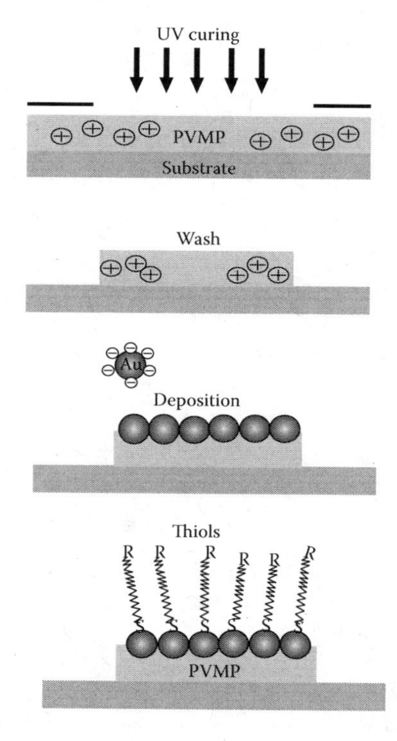

FIGURE 7.57 Fabrication of patterning gold nanoparticle surfaces via photopatterning of PVMP films. (Adapted from Xu et al., *Adv. Mater.*, 19, 1383–86, 2007.)

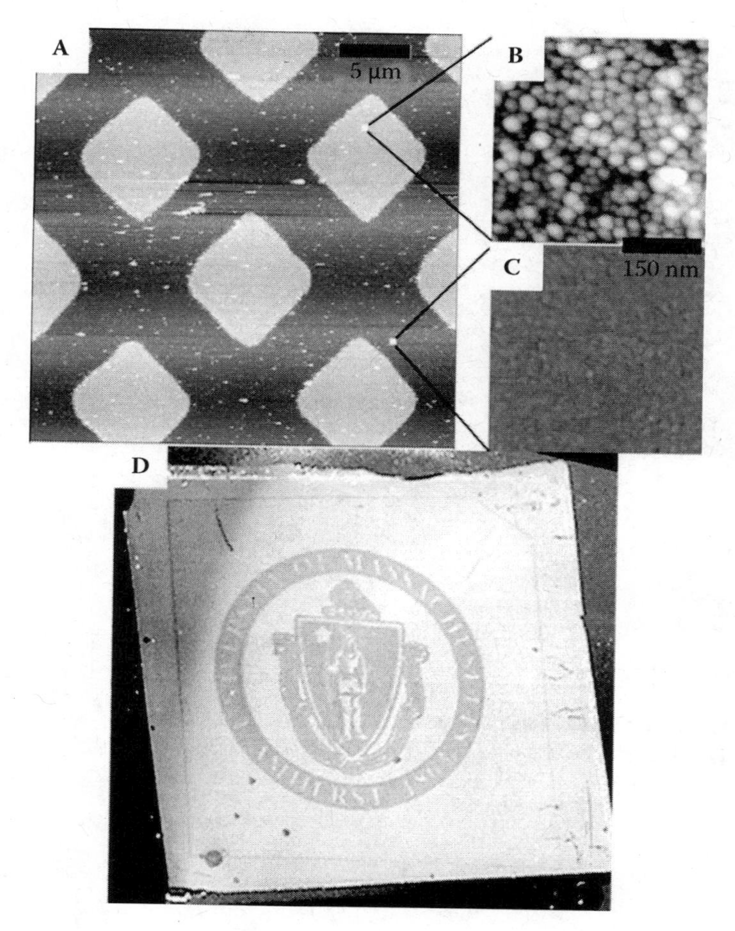

FIGURE 7.58 (A) Patterned Au nanoparticles film with zoom-in. (B) Polymer patterned. (C) Bare silicon (D) Conductive Au film with UMass logo pattern fabricated on a glass slide.. (Adapted from Xu et al. *Adv. Mater.* 19, 1383–1386, 2007.)

7.5.16 Nanopatterning by PDMS Relief Structures of Polymer Colloidal Crystals

Metals, polymers, and colloids have been patterned by using PDMS as a patterned master, and also a self-assembled monolayer stamp. The steps of the PDMS replica molding are illustrated in Figure 7.59. The colloidal crystal of polystyrene microspheres is spin coated and partially melted on a substrate. Then the chip is covered by PDMS prepolymer solution, cured, and the PDMS mold is peeled off from the colloidal crystal. The molds are subsequently coated with a silane solution, and the patterns are transferred by μCP. By dipping in a gold colloidal solution, self-assembled two-dimensional gold nanopatterns can also be obtained on a silicon substrate, as shown in Figure 7.60.

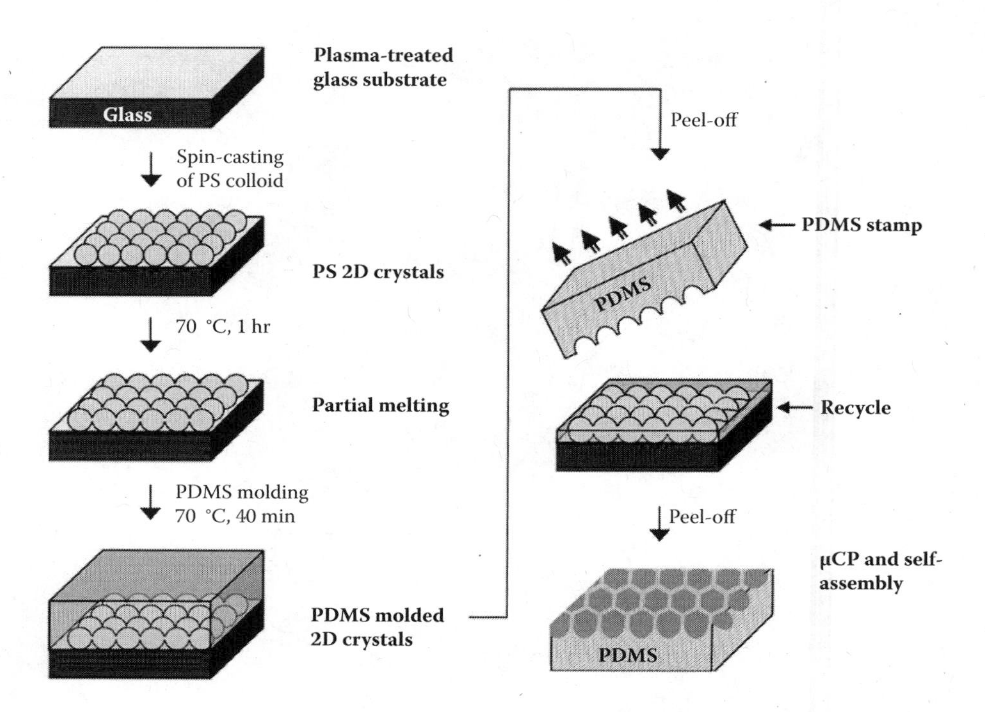

FIGURE 7.59 (See color insert.) Schematics of two-dimensional nanopatterning by PDMS relief structures of polymeric colloidal crystals. (Adapted from Nam et al., *Appl. Surf. Sci.*, 254, 5134–40, 2008.)

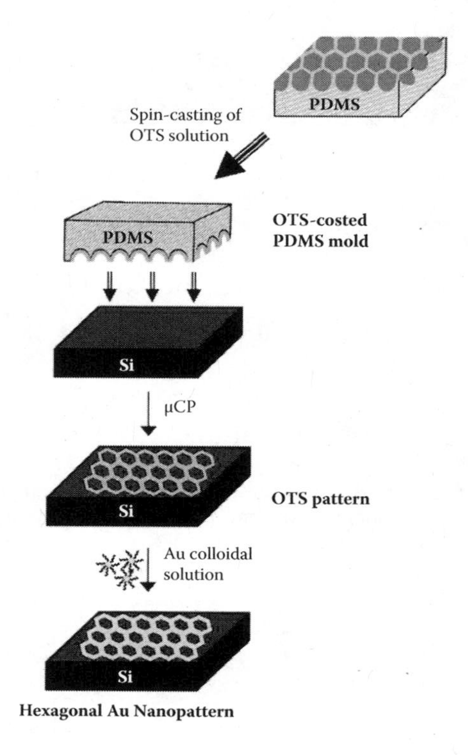

FIGURE 7.60 (See color insert.) Gold patterning by μCP. (Reprinted from Nam et al., *Appl. Surf. Sci.*, 254, 5134–40, 2008. With permission.)

REFERENCES

1. Madou, M. 2001. *Fundamentals of microfabrication: The science of miniaturization*. Boca Raton, FL: CRC Press.
2. Voldman, J., Gray, M. L., Schmidt, M. A. 1999. Microfabrication in biology and medicine. *Annu. Rev. Biomed. Eng.* 1:401–25.
3. Voevodin, A. A., Capano, M. A., Safriet, A. J., Donley, M. S., Zabinski, J. S. 1996. Combined magnetron sputtering and pulsed laser deposition of carbides and diamond-like carbon films. *Appl. Phys. Lett.* 69:188.
4. Be Bay, N. T., Pham, M. T., Ashrit, P. V., Badilescu, S., Djaoued, Y., Bader, G., Girouard, F., Le-Quang N., Truong, V.-V. 1996. Optical and electrochemical properties of vanadium pentoxide gel thin films. *J. Appl. Phys.* 80:7041–45.
5. Huy, N. D., Badilescu, S., Djaoued, Y., Girouard, F., Bader, G., Ashrit, P. V., Truong, V.-V. 1996. Preparation and characterization of lithium metaborate and lithium perchlorate doped metaborate thin films obtained by the sol-gel method. *Thin Solid Films* 286:141–45.
6. Be Bay, N. T., Tien, P. M., Badilescu, S., Djaoued, Y., Bader, G., Girouard, F., Truong, V.-V. 1995. Structure and infrared spectra of vanadium pentoxide gel films on semiconductor substrates. *Appl. Spectrosc.* 49:1279–81.
7. Badilescu, S., Ashrit, P. V., Ha, N. M., Bader, G., Girouard, F., Truong, V.-V. 1994. Study of lithium intercalation into tungsten oxide films prepared by different methods. *Thin Solid Films* 250:47–52.
8. Badilescu, S., Djaoued, Y., Ashrit, P. V., Girouard, F., Bader, G., Truong, V.-V. 1994. Metal insertion from solid substrates in sol-gel derived vanadium pentoxide thin films. *Solid State Ionics* 74:189–93.
9. Badilescu, S., Ha, N. M., Bader, G., Ashrit, P. V., Girouard, F. E., Truong, V.-V. 1993. Structure and infrared spectra of sol-gel derived tungsten oxide thin films. *J. Mol. Struct.* 297:393–400.
10. Schmidt, M. 1998. Wafer-to-wafer bonding for microstructure formation.
11. Yu, L., Tay, F. E. H., Xu, G., Chen, B., Avram, M., Iliescu, C. 2006. Adhesive bonding with SU-8 at wafer level for microfluidic devices. *J. Phys. Conference Ser.* 34:776–81. International MEMS Conference 2006, Singapore. 9–12 May.
12. Kovacs, T. A., Maluf, N. I., Petersen, K. E. 1998. Bulk micromachining of silicon. *Proc. IEEE* 86:1536–51.
13. Chandrasekaran, A., Packirisamy, M., Stiharu, I., Delage, A. 2006. Hybrid bulk micro-machining process suitable for roughness reduction in optical MEMS devices. *Int. J. Manufact. Technol. Manage.* 9:144–59.
14. He, R., Jin, C., Kim, C. J. 2007. On-wafer monolithic encapsulation by surface machining with porous polysilicon shell. *J. Micromech. Syst.* 16:463–73.
15. Whitesides, G. M. 1997. Unconventional methods and unconventional materials for microfabrication. Paper presented at TRANSDUCERS '97, 1997 International Conference on Solid-State Sensors and Actuators, Chicago, June 16–19, 1997.
16. Liu, C., Engel, J., Zou, J., Wang, X., Fan, Z., Ryu, K., Shaikh, K., Bullen, D. 2003. Polymer micromachining and applications in sensors, microfluidics, and nanotechnology. The 226th National Meeting of Am. Chem. Soc., New York, Sept. 7–11.
17. Becker, H., Gärtner, C. 2008. Polymer microfabrication technologies for microfluidic systems. *Anal. Bioanal. Chem.* 390:89–111.
18. Becker, H., Gärtner, C. 2000. Polymer microfabrication methods for microfluidic analytical applications. *Electrophoresis* 21:12–26.
19. Rötting, O., Röpke, W., Becker, H., Gärtner, C. 2002. Polymer microfabrication technologies. *Microsyst. Technol.* 8:32–36.
20. Wilbur, J. L., Kumar, Q. A., Biebuyck, H. A., Kim, E., Whitesides, G. M. 1996. Microcontact printing of self-assembled monolayers: Application in microfabrication. *Nanotechnology* 7:452–57.
21. Xia, Y., Kim, E., Mrksich, M., Whitesides, G. M. 1996. Microcontact printing of alkanethiols on copper and its applications in microfabrication. *Chem. Mater.* 8:601–3.
22. Meenakshi, U., Babayan, Y., Odom, T. W. 2007. Benchtop nanoscale patterning using soft lithography. *J. Chem. Ed.* 84:1795–98.
23. Fernández-Sánchez, C., Cadarso, V. J., Darder, M., Domínguez, C., Llobera, A. 2008. Patterning high-aspect ratio sol-gel structures by microtransfer molding. *Chem. Mater.* 20:2662–68.
24. Zhao, X.-M., Xia, Y., Whitesides, G. M. 1997. Soft lithographic methods for nanofabrication. *J. Mater. Chem.* 7:1069–74.
25. Radha, B., Kulkarni, G. U. 2009. Dewetting assisted patterning of polystyrene by soft lithography to create nanotrenches for nanomaterial deposition. *Appl. Mater. Interfaces 1*:257–60.
26. Yu, X., Wang, Z., Xing, R., Luan, S., Han, Y. 2005. Solvent-assisted capillary force lithography. *Polymer* 46:11099–103.

27. Vozzi, G., Flaim, C., Ahluwalia, A., Bhatia, S. 2003. Fabrication of PLGA scaffolds using soft lithography and microsyringe deposition. *Biomaterials* 24: 2533-2540.
28. Chen, S. C. and Lu, Y. 2005. Micro- and Nano-Fabrication of Biodegradable Polymers. In *Handbook of Biodegradable Polymeric Materials and their applications*, ed. S. Mallapragada and B. Narasimhan ,Vol. 1: 1-17.
29. Cooke, M. N., Fisher, J. P., Dean, D., Rimnac, C., Mikos, A. G. 2003. Use of stereolithography to manufacture critical-sized 3D biodegradable scaffolds for bone ingrowth. *J. Biomed. Mater. Res. B. Appl. Biomater.* 64: 65-69.
30. Balakrisnan, B., Patil, S., Smela, E.2009. Patterning PDMS using a combination of wet and dry etching. *J. Micromech. Microeng.* 19:047002.
31. Lu, Y., Chen, S. C. 2004. Micro and nano-fabrication of biodegradable polymers for drug delivery. *Adv. Drug Delivery Rev.* 56:1621–33.
32. Yi, A. Y., Lu, W., Farson, D. F., Lee, J. 2008. Overview of POLYMER micro/nanomanufacturing for biomedical applications. *Adv. Polymer Technol.* 27:188–98.
33. Salaita, K., Wang, Y. H., Mirkin, C. A. 2007. Applications of dip-pen nanolithography. *Nat. Nanotechnol.* 2:145–55.
34. Djaoued, Y., Badilescu, S., Balaji, S., Seirafianpour, N., Hajiaboli, A. R., Banan Sadeghian, R., Braedley, K., Brüning, R., Kahrizi, M., Vo-Van Truong. 2007. Micro-Raman spectroscopy study of colloidal crystal films of polystyrene-gold composites. *Appl. Spectrosc.* 61:1202–10.
35. Christman, K. L., Enriquez-Rios, V. D., Maynard, H. D. 2006. Nanopatterning proteins and peptides. *Soft Matter* 2:928–39.
36. Ormonde, A. D., Hicks, E. C. M., Castillo, J., Van Duyne, R. P. 2004. Nanosphere lithography: Fabrication of large-area Ag nanoparticle arrays by convective self-assembly and their characterization by scanning UV-visible extinction spectroscopy. *Langmuir* 20:6927–31.
37. Lu, Y., Aguilar, C. A., Chen, S. 2005. Shaping biodegradable polymers as nanostructures: Fabrication and applications. *Drug Discov. Today Technol.* 2:97–102.
38. Bäuer, D., Pedarnig, J. D., Vrejoiu, I., Peruzzi, M., Matei, D. G., Brodoceanu, D., Dinescu, M. 2005. Laser processing and chemistry: Applications in nanopatterning, material synthesis and biotechnology. *Romanian Rep. Phys.* 57:936–52.
39. Lee, M., O'Hayre, R., Prinz, F. B. 2004. Electrochemical nanopatterning of Ag on solid-state ionic conductor $RbAg_4I_5$ using atomic force microscopy. *Appl. Phys. Lett.* 85:3552–54.
40. Barton, J. E., Odom, T. W. 2004. Mass-limited growth in zeptoliter beakers: A general approach for the synthesis of nanocrystal. *Nano Lett.* 4:1525–28.
41. Michel, R., Lussi, J. W., Csucs, G., Reviakine, I., Danuser, G., Ketterer, B., Hubbell, J., Textor, M., Spencer, N. D. 2002. Selective molecular assembly patterning: A new approach to micro- and nano-chemical patterning of surfaces for biological applications. *Langmuir* 18:3281–87.
42. Li, Q., J. Zheng, Z., Liu, Z. 2003. Site-selective assemblies of gold nanoparticles on an AFM tip-defined silicon template. *Langmuir* 19:166–71.
43. Kuemmel, M., Allouche, J., Nicole, L., Boissiere, C., Laberty, C., Amenitsch, H., Sanchez, C., Grosso, D. 2007. A chemical solution deposition route to nanopatterned inorganic material surfaces. *Chem. Mater.* 19:3717–25.
44. Chowdhury, D., Maoz, R., Sagiv, J. 2007. Wetting driven self-assembly as a new approach to template-guided fabrication of metal nanopatterns. *Nano Lett.* 7:1770–78.
45. Xu, H., Hong, R., Wang, X., Arvizo, R., You, C., Samanta, B., Patra, D., Tuominen, M. T., Rotello, V. M. 2007. Controlled formation of patterned gold films via site-selective deposition of nanoparticles onto polymer-templated surfaces. *Adv. Mater.* 19:1383–86.
46. Nam, H. J., Kim, J.-H., Jung, D.-Y., Park, J. B., Lee, H. S. 2008. Two-dimensional nanopatterning by PDMS relief structures of polymeric colloidal crystals. *Appl. Surf. Sci.* 254:5134–40.

REVIEW QUESTIONS

1. What are the basic processes used to build a device?
2. Describe briefly the chemical vapor deposition method.
3. How is a substrate coated when using the spin-coating technique?
4. How is a geometrical pattern transferred onto a substrate by photolithography?
5. What is a positive and a negative photoresist?

6. Define isotropic and anisotropic etching.
7. What techniques can be used to bond a silicon and a glass substrate?
8. What is bulk micromachining and what kind of structures can be fabricated with this technique?
9. What is surface micromachining?
10. What is soft lithography?
11. What techniques are most used for biodegradable polymers?
12. Describe briefly the technique of laser ablation and show how this technique is used in nanofabrication.
13. What is dip pen nanolithography and how can it be used for patterning?
14. How is colloidal lithography used for patterning?
15. Describe the template-assisted chemical solution deposition method.

8 Introduction to Microfluidics

8.1 INTRODUCTION

Microfluidics is a multidisciplinary research field involving engineering, physics, chemistry, microtechnology, and biotechnology. It concerns the handling and manipulation of minute amounts of fluids geometrically constrained to a small, typically submillimeter scale. The typical volume of fluid handled in microfluidics is in the range of nano- to microliters, which is thousands of times smaller than a common droplet. *Lab-on-a-chip* (LOC) and *micro total analysis systems* (μTAS) are advanced technologies that integrate microfluidic components, such as microchannels, microchambers, etc., with micropumps, microvalves, etc., on a microscale chip device to miniaturize chemical and biological laboratory processes on submillimeter sizes. The earliest microfluidic application is considered to be inkjet printing.

Because there are a number of fields that share a common interest in this area, there is a considerable amount of literature on various aspects of microfluidics, such as microfabrication techniques, novel designs, integration with MEMS technologies, basic fluid mechanics, integration with microelectronics and microphotonics elements, etc.

Microfluidic systems have been shown to have great potential in diverse areas of biological applications, including diagnosis, drug delivery and therapeutics, biomolecular separations, bioassays, immunohybridization assays, and polymerase chain reactions (PCRs). Some of the applications include biosensing protein chips, pathogen detection, etc. Microchannels are also used to transport and mix biological materials. such as enzymes, proteins, DNA, cells, etc. Cell-biology-based applications range from screening of single cells to three-dimensional scaffolds for tissue engineering. A number of applications require multiple reactions to be performed in parallel, for example, sample preparation, gene expression analysis, and high-throughput assays. Microfluidic chips are also used for chemical analysis and detection, microchemical reactors, toxicity measurements, microheat exchangers, gas turbine exhaust gas analysis, breath analyzer, etc. As the reagents can be expensive and available only in limited amounts, a microfluidic-based detection of reactions would become a suitable solution. In addition, the field of microfluidics promises to integrate the functions of conventional chemistry or biochemistry laboratories into a single chip, and to change completely the way modern chemistry and biology are performed, with cheap and disposable chips, high throughput, and small handle volume.

There are many significant advantages offered by microfluidic technologies. Not only are the reagent consumption and waste products reduced, but the heat and mass transfer is rapid due to higher a surface-to-volume ratio at the microscale. As a result, heat exchange becomes very efficient at the microscale. Furthermore, microfluidic devices can be easily integrated with electronic and optical components for manipulation and sensing. Beyond realizing economies of scale and integration, the unique physics of the fluid flow and mass transport on the micrometer length scale can be exploited to enhance detector sensitivity. At these scales, the fluid dynamics is somewhat different from what one finds in everyday experience as viscous forces dominate over inertial forces. In microflows, where surface tension can be a powerful force, turbulence is nonexistent, diffusion becomes the basic method for mixing, and evaporation is rapid on exposed liquid surfaces. The effects that become dominant in microflows are laminar flow, diffusion, fluid resistance, a high surface-area-to-volume ratio, and surface tension.[1,2]

8.2 FLUID PHYSICS AT THE MICROSCALE

The most important characteristic of microfluidic devices is the dominant role of viscosity when compared to fluid velocity. The *Reynolds number* (Re) of a fluid flow, which describes the relative importance of viscous forces to inertial forces, is an important parameter that determines the flow behavior. The Reynolds number is defined as

$$\mathrm{Re} = \frac{\rho v D_h}{\mu} \tag{8.1}$$

where ρ is fluid density (kg/m^3), v is characteristic fluid velocity (m/s), μ is dynamic viscosity (kg/m.sec), and D_h is hydraulic diameter for a given channel geometry. Low Reynolds numbers indicate that the flow is laminar with dominant viscous forces. At low Reynolds numbers, mixing is difficult as turbulence is negligible. The Reynolds number could be well below 1 for the case of water, with typical velocities in the range of 1 μm/sec to 1 cm/sec and a typical channel size of 1 to 100 μm.[3] At very low Reynolds numbers, viscous forces dominate the fluid motion and inertial forces are insignificant. Microfluidic systems are typically low Reynolds number systems, and the fluid flow is governed mainly by viscous forces and pressure gradients. In the case of low Re, the flow is laminar. As a result, two fluids flowing alongside in parallel layers maintain a well-defined interface and can mix only by diffusion. One can appreciate the example of laminar flow given in Figure 8.1, created by the motion of a glacier at a very low Reynolds number. The Reynolds number for glacier motion is in the order of ~ 10^{-11} (parameters used are flow velocity of 1 m year^{-1}, length of 1 km, density to viscosity ratio of ice of ~3×10^{-7} sec m^{-1}), and the resultant laminar flow can be seen in this figure.

The mechanism of mixing at microscales is diffusion, and it is a very slow process. For example, mixing by molecular diffusion of macromolecular solutions may take considerable time, in the

FIGURE 8.1 The motion of a glacier. (Reproduced from Franke and Wixforth, *ChemPhysChem*, 9, 2140–56, 2008. With permision.)

order of tens of minutes. The molecular diffusion coefficient, D, varies between 10^{-5} cm^2 sec^{-1} for small molecules and 10^{-7} cm^2 sec^{-1} for large molecules. The typical diffusion time for diffusing half the width of the channel is around 10 sec for small molecules and 16 min for large molecules.

The mixing of fluids is poor near the wall. In order to enhance mixing, contact interfaces between two segregated fluids have to be created.

There are several devices that employ adjacent laminar flows for sensing and separating analytes. One of the commonly used devices to study reaction kinetics and measure diffusion constants of different species is a T-sensor (Figure 8.2). The different rates at which tracers of different diffusivities spread across the channel are explored in a membraneless H-filter (Figure 8.3).

The solution that is introduced in the channel contains different-sized particles with different diffusivities. The small particles diffuse to fill the channel before exiting, while the large particles remain in their half of the channel. The particles that are extracted are those having high diffusivity.

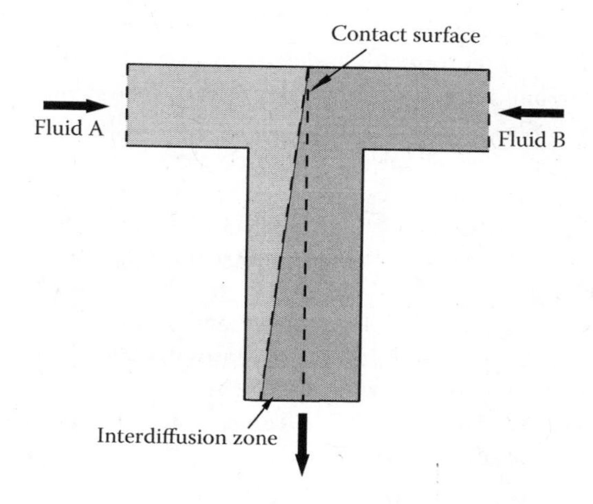

FIGURE 8.2 Diffusive mixing between two laminar streams. Different fluids are brought together at a T-junction to flow alongside each other down the channel. Confocal microscopy reveals the three-dimensional nature of the spreading of the interface in the T-sensor. (Adapted from Ismagilov et al., *Appl. Phys. Lett.*, 76, 2376, 2000.)

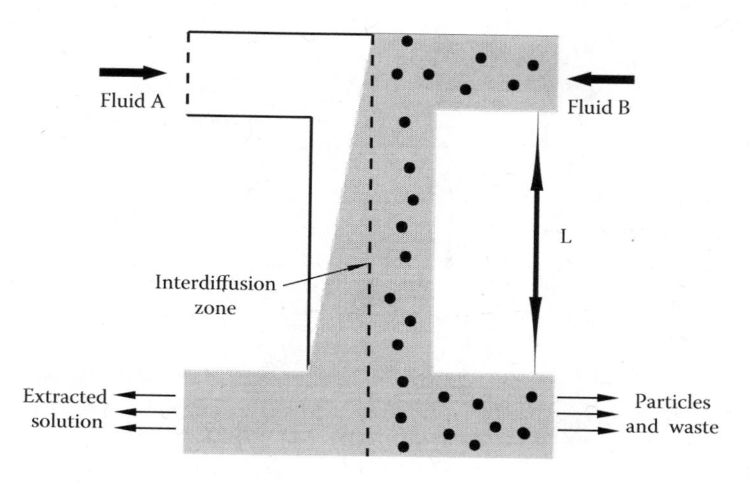

FIGURE 8.3 The membraneless H-filter. (Adapted from Brody and Yager, *Sensors Actuators A*, 58, 13–18, 1997.)

8.3 METHODS FOR ENHANCING DIFFUSIVE MIXING BETWEEN TWO LAMINAR FLOWS

It is known that turbulence in flows facilitates rapid mixing at values of Re > 2,000, which can be reached in microchannels only at very high, impractical flow velocities (~10 $msec^{-1}$). Such high flow velocities require high sample consumption and high pump pressures that are difficult to be achieved in microchips. Unfortunately, for slow microfluidic flows, diffusion alone is usually not sufficient to mix fluids, as it would require very large diffusion rates. It does not happen fast enough, especially in the case of assays that involve large particles such as cells. Various mixing principles have been applied to passive micromixers in order to achieve higher mixing efficiencies. Interdigital multi-lamination, split and recombination, reduced diffusion length design, vortex generation, and chaotic mixing are just a few of them. An efficient passive micromixer with complex three-dimensional geometries is utilized to enhance fluid lamination, stretching, and folding for optimal mixing. In a microfluidic device, there are two ways of mixing fluid streams, namely, passive mixing and chaotic advection or active mixing, as shown in Figure 8.4. *Passive mixers* use appropriate channel geometries to fold fluid streams in order to increase the area over which diffusion occurs. Examples of passive mixing include distributive mixer, static mixer, T-type mixer, and vortex mixer. The Coanda effect is used to make an in-plane micromixer that splits the fluid streams and recombines them to induce mixing. Rapid mixing with low reagent consumption can be achieved using *chaotic advection*. Chaotic advection increases the area over which diffusion occurs by continuously stretching and refolding the two fluids to achieve layers of fluids (*striations*) that become thinner and thinner until mixing becomes rapid. Some of the designs employed for passive mixing are bifurcation flow distribution structures, focusing structures for flow compression, flow obstacles within microchannels, multihole plates, tiny nozzles, and T- and Y-flow configurations.

A passive Coanda effect micromixer relies on the redirection of flow by a guiding channel that creates a new interface within the flow, as shown in Figure 8.5.

Another method to increase mixing and convective transport in microchannels is *ultrasonic agitation*, as shown in Figure 8.6.

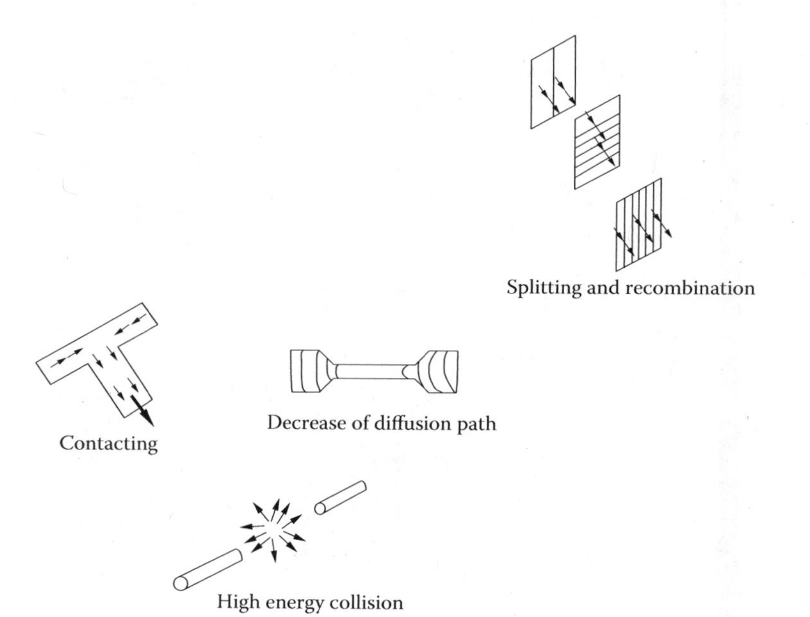

FIGURE 8.4 Schemes of selected passive and active micromixing principles. (From Hessel et al., *Chem. Eng. Sci.*, 60, 2479–501, 2005.)

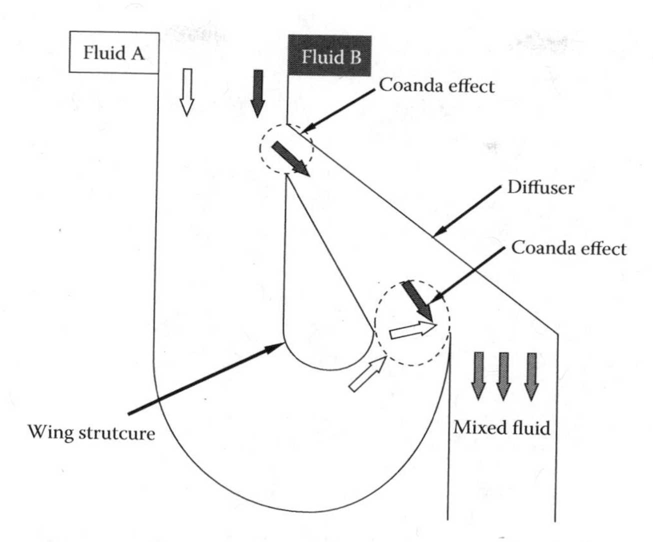

FIGURE 8.5 A schematic of a mixer using the Coanda effect, which splits the fluid streams and then recombines them. (Adapted from Hong et al., in *Micro Total Analysis Systems 2001*, Ed. Ramsey and van den Berg, Kluwer Academic, Boston, 2001, pp. 31–33.)

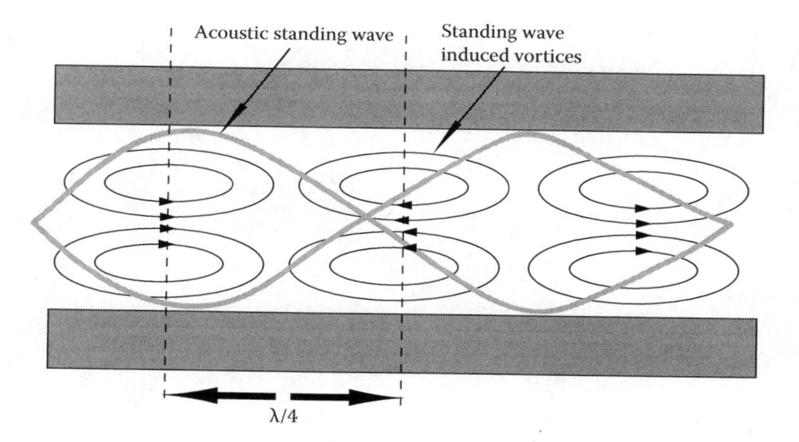

FIGURE 8.6 Schematic diagram of how an acoustic standing wave field, in a sufficiently narrow channel, induces several vortices spaced λ/4 apart (Adapted from Bengtsson and Laurell, *Anal. Bioanal. Chem.*, 378, 1716–21, 2004.)

One way to enhance mixing is using an acoustic wave field. Acoustic waves that induce rotating vortex flows in microchannels can be used (Figure 8.6). It has been shown that a lateral flow, called Rayleigh flow, which occurs in the presence of an acoustic standing wave field, may induce such vortices. The vortex flow in the microchannel will carry the reactants from the center flow to the walls, and the products of the reaction in the opposite direction.

The *streaming velocity*, u_s, at the wall, is given by Raleigh's law:

$$u_s = -\frac{3}{4\omega} V(x) \frac{dV(x)}{dx} \tag{8.2}$$

where $V(x)$ is the acoustic particle velocity outside the boundar layer and ω is the frequency of the sound field.

The feasibility of using Rayleigh flow for breaking up the laminar flow lines in microchannels was studied in the case of a parallel channel enzyme microreactor in porous silicon. Figure 8.7 shows the setup where the channels are designed to form an acoustic resonator in the vertical direction of the chip. The mixing efficiency of two fluids, A and B, was investigated in a test channel (Figure 8.8). Two fluids could be exposed efficiently to Rayleigh mixing for flow rates lower than 20 μl/min. For flow rates lower than 4 μl/min, almost full mixing was accomplished. This mixing method has been found suitable for systems based on open tubular type enzyme microreactors.

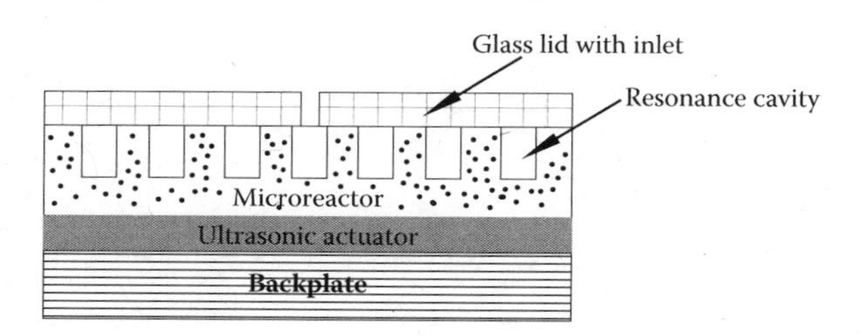

FIGURE 8.7 Schematic view of a setup with enzyme reactor, ultrasonic crystal, and polyurethane back plate. (Adapted from Bengtsson and Laurell, *Anal. Bioanal. Chem.*, 378, 1716–21, 2004.) Sufficiently narrow channel induces several vortices spaced λ/4 apart. (Adapted from Bengtsson and Laurell, *Anal. Bioanal. Chem.*, 378, 1716–21, 2004.)

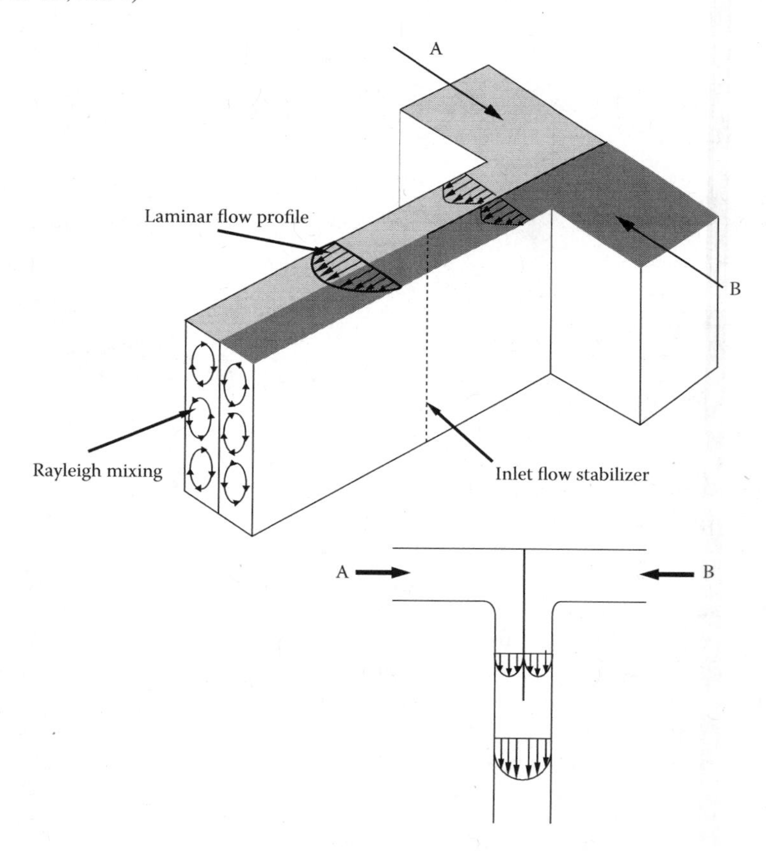

FIGURE 8.8 Drawing of the test channel. The flow divider was introduced to ensure vertical flow lamination of the two fluids (A and B in the drawing) in the high-aspect-ratio channel. The acoustic standing wave was formed between channel bottom and Pyrex lid transverse to the flow. (Adapted from Bengtsson and Laurell, *Anal. Bioanal. Chem.*, 378, 1716–21, 2004.)

An alternative method for mixing of laminar flows in an annular micromixer is *magnetohydrodynamic* (MHD) *actuation*. The behavior of two laminar flows in an annular micromixer is shown in Figure 8.9 under two different conditions. This is a low-power fluid propulsion method realized using an electric field that operates perpendicular to the magnetic field, as shown in Figure 8.10.[10]

MHD transport is realized by using a cross channel electric current in combination with a magnetic field originating from beneath the channel. The experimental setup for implementing micromixing is shown in Figure 8.10.

By this method, the two initially separated liquids are forced to pass through each other, resulting in an important increase of the interfacial area, as shown in Figure 8.9. For analyte solutions containing large molecules such as DNA and proteins, *magnetic particle actuation* can produce

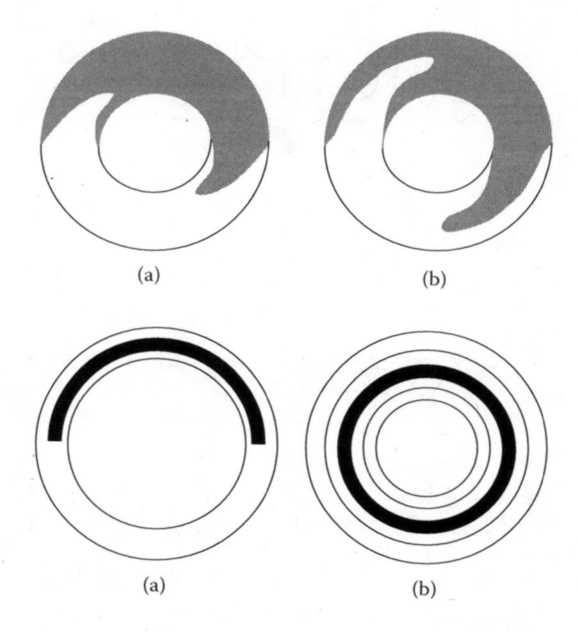

FIGURE 8.9 The two-dimensional behavior of two laminar flows in an annular micromixer at two different conditions: (a) start of mixing and (b) developed mixing. (Adapted from West et al., *Sensors Actuators B*, 96, 190–99, 2003.)

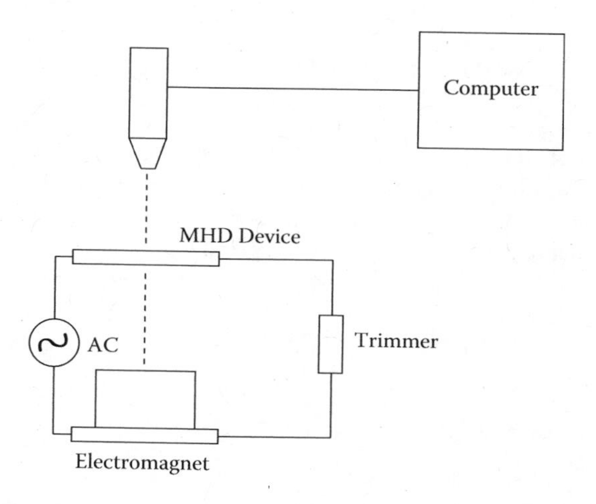

FIGURE 8.10 Apparatus and experimental setup for magnetohydrodynamic annular micromixing.

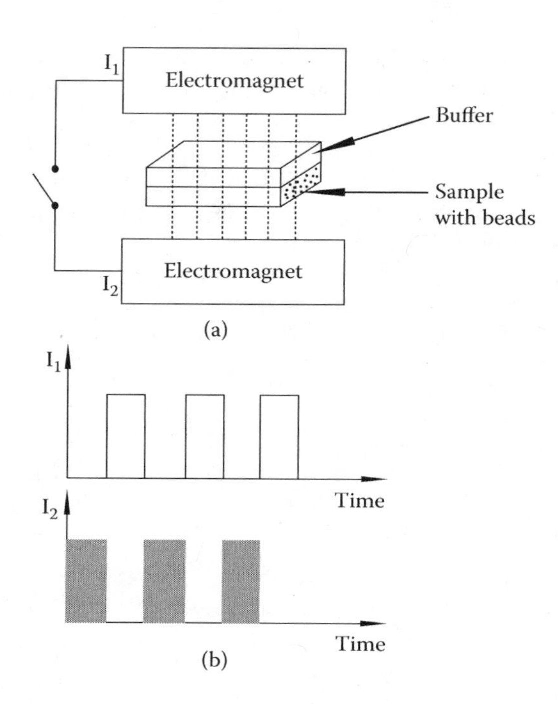

FIGURE 8.11 (a) Schematic diagram of a magnetic micromixer. (b) Time-dependent current applied to the electromagnets. (Adapted from Wang et al., *Microfluid Nanofluid*, 4, 375–89, 2008.)

excellent mixing. Actuated by magnetic forces, the suspended magnetic particles enhance significantly the mixing by adding agitation and chaotic advection in the fluid. A schematic of a magnetic particle-driven micromixer is shown in Figure 8.11.

It has been demonstrated that for a 100 μm wide channel, by applying a current of 50 mA to the electromagnet, a high mixing efficiency can be obtained within 1 sec.

The *Peclet number*, Pe, reflects the relative importance of diffusion and convection. This number is defined as

$$P_e = \frac{U_0}{D} w \tag{8.3}$$

where U_0 is the average velocity of the flow in m/sec, w is the width of the channel in m, and D is the diffusion coefficient of the molecule of interest in m^2/sec. This formula helps to calculate how far the fluids must flow before the channel is homogenized. For example, consider a T-junction in which two fluids are injected to flow alongside each other, as in Figure 8.3. The time required for the molecules to diffuse across the entire channel is given by

$$\tau \approx \frac{w^2}{D} \tag{8.4}$$

During this time, the flow would have traveled a distance of

$$Z \approx \frac{U_0}{D} w^2 \tag{8.5}$$

down the channel, so that the number of channel widths required for complete mixing would be defined as

$$\frac{Z}{w} \approx \frac{U_0}{D} w \cong Pe \tag{8.6}$$

The number of channel widths required for full mixing varies linearly with Pe. Using the numerical values of diffusivities, it can be seen that even a small protein flowing with the fluid through a 100 μm width channel at a velocity of 100 μm/sec requires about 250 channel widths (approximately 2.5 cm and 4 min) to mix completely.

It has been realized in recent years that the classical concept of microfluidics, that is, the confinement of the flow of single-phase liquids through the network of the narrow channels, has some fundamental drawbacks. The most obvious one is the fact that in small dimensions, the Reynolds number (Re) of the flow rarely exceeds unity, such that the flow is purely laminar, resulting in longer time for the completion of mixing or reaction, as shown in Figure 8.12a. Hence, there has been considerable developments toward the use of isolated water droplets suspended in an oil phase, as shown in Figure 8.12b. The main advantage considered in this context is the twisty recirculating flow pattern emerging within the droplets when they are moved through the channel system, as shown in Figure 8.12b. Mixing of two aqueous components is achieved within each droplet separately using the twisting flow pattern.

In *digital microfluidic systems*, analyte droplets of typical volume of 1 μl are transported on a planar electrode array by dielectrophoretic and electrowetting effects. The methods of drop generation and the principles of drop actuation will be discussed later.

In microchannels, surface tension forces and surface free energy are significant, and as a result, liquids can travel based on capillary forces. The large surface-area-to-volume ratios thus benefit some of the processes, such as capillary electrophoresis in microchannels.

8.4 CONTROLLING FLOW AND TRANSPORT IN MICROFLUIDIC CHANNELS

The flow of liquid in microfluidic devices is regulated by a large variety of applied forces, such as pressure differences, electrophoresis capillary forces, or Marangoni forces, by controlling spatial variations of surface tension. Thermal, electrical, or light gradients can also be used to create surface tension variations in chips. Capillary pressure gradients can also be used to move liquid in microchannels by transporting the wetting properties along the channel. However, there are two common methods, as shown in Figures 8.13 and 8.14, by which fluid actuation can be achieved: *presure-driven flow* and *electrokinetic flow*.

In pressure-driven flow, the fluid is transported by means of applied pressure differences, that is, pumped through the device by using syringe and other mechanical pumps. In the case of pressure-driven flow, the fluid velocity at the walls is zero following *no-slip boundary condition*. The no-slip boundary condition produces a parabolic velocity profile, as shown in Figure 8.13, and the average velocity is proportional to the second power of channel dimension.

The pressure drop along a channel of length L for a pressure-driven flow is

$$\Delta P = \frac{8 \mu v L}{r^2}$$

where μ is dynamic viscosity, v is velocity, and r is radius of circular flow area. The equation shows that required pressure drop to drive the flow could be too large and may be impractical for very small devices with channel widths in a range of less than 20 μm. Pressure-driven flow has been proved to be reproducible and amenable to miniaturization. Capillary driving forces due to

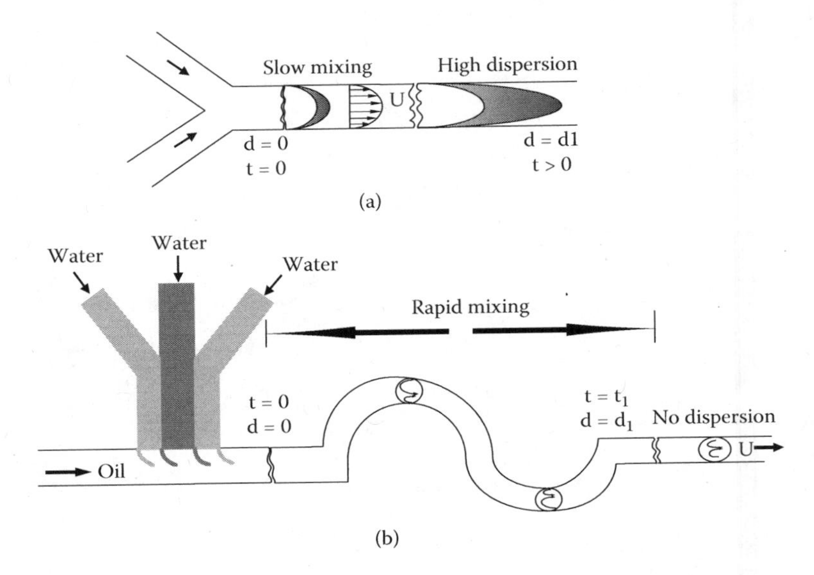

FIGURE 8.12 Schematic comparison of a reaction between two reagents conducted (a) in a standard pressure-driven microfluidic system device and (b) in the microfluidic device described here. Two aqueous reagents can form laminar streams separated by a grey "divider" aqueous stream in a microchannel. When the three streams enter the channel with an immiscible fluid, they form droplets (plugs). The reagents come into contact as the contents of the droplets are rapidly mixed. Internal recirculation within plugs flowing through channels of different geometries is shown schematically by arrows. (Adapted from Song et al., *Angew. Chem. Int. Ed.*, 42, 768–72, 2003.)

the surface tension can also lead to pressure gradients in liquids and cause fluid transport. Using pressure-driven flow is probably the most straightforward way of moving fluids in a microfluidic device. Because of the simplicity in understanding and implementing of the process, a variety of pressure-driven microfluidic devices have been developed for many applications.

There are two main approaches for pressure pumping in a microfluidic device. Either an external pump is used or the pump is incorporated into the microfluidic chip. The drawbacks of using pressure as a means of transport include the fact that the pressure-driven flow will display a nonuniform velocity profile. Also, the smoothness of the flow can be hard to maintain when integrated pumps are used to drive flow in small microfluidic channels. The high pressure needed to drive the flow in smaller-dimension channels can be dangerously high, which might even damage the bond strengths of chips, resulting in leakage failure of them.

The second common technique is based on *electrokinetic pumping*. If the walls of the channel are charged, two ionic layers of fluid are created adjacent to the wall. The ions in the two layers, called double layer, will move toward the electrode of opposite polarity, producing a motion of the fluid near the walls. In the case of electrokinetic flow, the velocity profile is uniform across the width of the channel.

The schematic of velocity profiles for the pressure-driven flow and electrokinetic flow is shown in Figure 8.13.

A significant disadvantage of the electrokinetic flow is the requirement of high voltages. However, electrokinetic pumping has been explored as one of the leading methods for driving and controlling the flow of fluids in microchannels. An important advantage compared to pressure-driven flow is that electrokinetic flow does not need moving parts as a part of a microfluidic system. Electrokinetic flows can be easily controlled in interconnected channels by switching voltages without the need for valves.

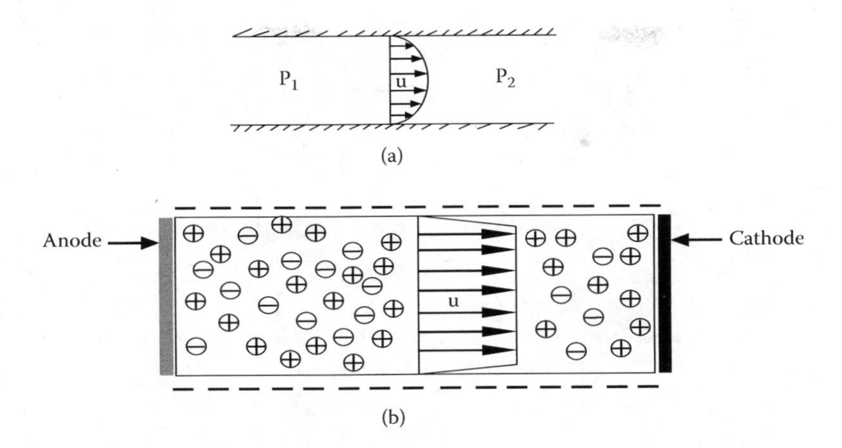

FIGURE 8.13 Schematic of velocity profiles for (a) pressure-driven flow and (b) electrokinetic flow. (Adapted from Bayraktar and Pidugu, *Int. J. Heat Mass Transfer*, 49, 815–24, 2006.)

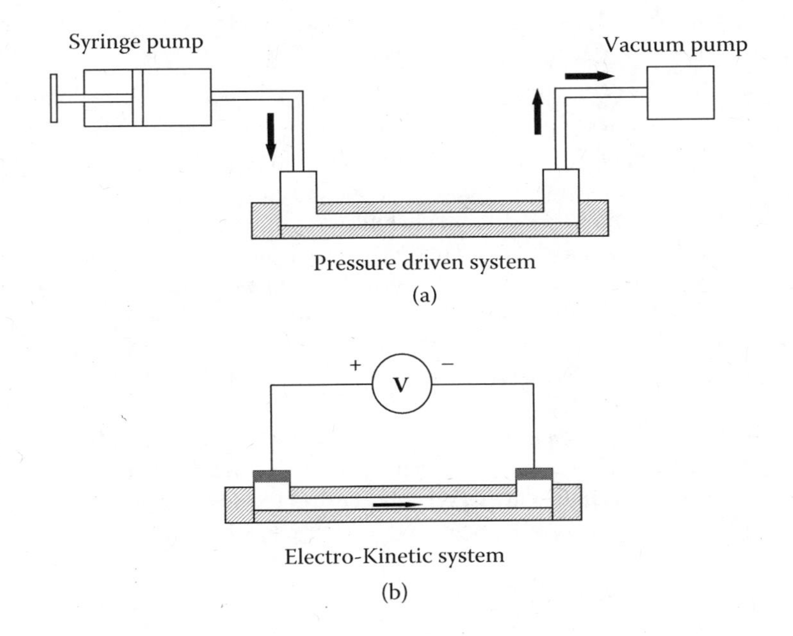

FIGURE 8.14 Methods of driving fluid. (a) Pressure-driven system using a vacuum pump at the outlet or a pump at the inlet. (b) Electrokinetic system with high voltage applied across the channel. (Adapted from Sia and Whitesides, *Electrophoresis*, 29, 3563–76, 2003.)

Most fluidic devices that produce internal flows use electrophoresis, electroosmosis, electrohydrodynamics, magnetohydrodynamics, centrifugation, or pressure gradient.

Electrokinetics is the study of the bulk motion of conductive fluids, containing electrolytes or selected particles distributed in fluids, under the influence of electric fields (Figure 8.15). With the recent developments in microfabrication, electrokinetics provides effective manipulation of various biological objects in the micro- and nanodomains. The ability to manipulate objects down to molecular levels opens new avenues to explore biological and chemical technologies. Understanding of the fundamental characteristics and limitations of electrokinetic forces becomes important for successful applications of this mechanism.

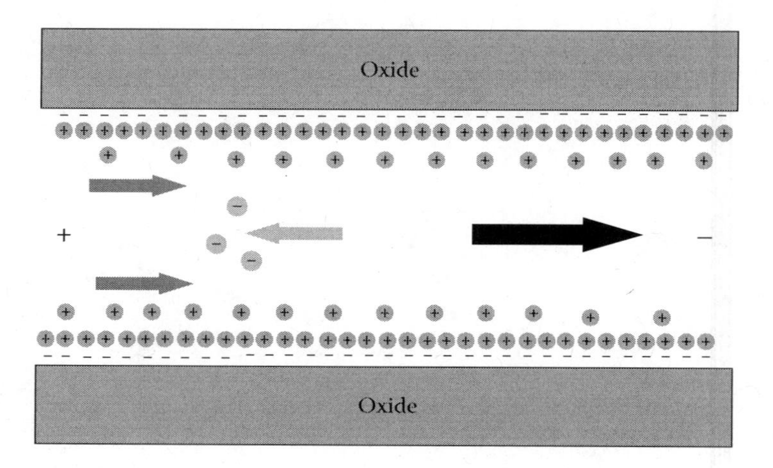

FIGURE 8.15 Motion of a bulk fluid containing electrolytes.

8.4.1 Physical Processes Underlying Electrokinetics in Electroosmosis Systems

The spontaneous separation of charges at solid-liquid interfaces, determined by the ionization of surface groups, is central to electrokinetic actuation of flows in microfluidic devices. Many microfluidic substrates behave as weak acids due to the reactivities of surface groups (amines, carboxylic acids, or oxides) in aqueous solutions, For example, surface silanol groups of glass substrates can be deprotonated in aqueous solutions, leaving a negative surface charge. Some of the charging mechanisms include ionization, ion adsorption, and ion dissolution. The electrokinetic properties of such systems depend largely on the pH of the solution. The surface charge generates an electric field, which attracts counterions toward the surface, and repels coions away from it. As a result, counterions preferentially concentrate near the surface, shielding the bulk of the solution from the surface charge. The combination of attraction, repulsion, and mixing due to random thermal motions of ions leads to the formation of an *electrical double layer* (EDL). A schematic diagram of the EDL is shown in Figure 8.16. The electric double layer is a region close to the charged surface in which there is an excess of counterions over coions in order to neutralize the surface charge, and these ions are spatially distributed in a diffuse manner. There is no charge neutrality within the double layer because the number of counterions will be large compared with the number of coions. The bulk of the double layer is made of counterions. The thickness of the EDL layer is approximated by the Debye length, λ:

$$\lambda = \left(\frac{\varepsilon RT}{e^2 \sum z^2 c_i^b} \right)^{1/2} \tag{8.7}$$

where ε is the dielectric permittivity of the medium, c^b_i is the bulk concentration of the ith ion, R is the gas constant, T is the absolute temperature, and z is the charge. For the range of concentrations used in practice, λ is around 10 nm. At distances beyond λ, the fluid is electrically neutral. The Debye length is defined at the $1/e$ value of the wall potential, φ_w.

The EDL is comprised of a Stern layer and a diffuse layer. The Stern layer that is adjacent to the surface consists of counterions that are immobilized on the surface, and its thickness is dictated by the size of the ions. The Stern layer is fixed to the surface due to ionic affinity, resulting in a no-slip flow boundary condition. The diffuse layer is above the Stern layer, and it is responsible for the electrokinetic phenomena relevant to microfluidic devices. The diffuse layer consists of both counterions and coions. As the number of counterions is large compared to coions, the movement

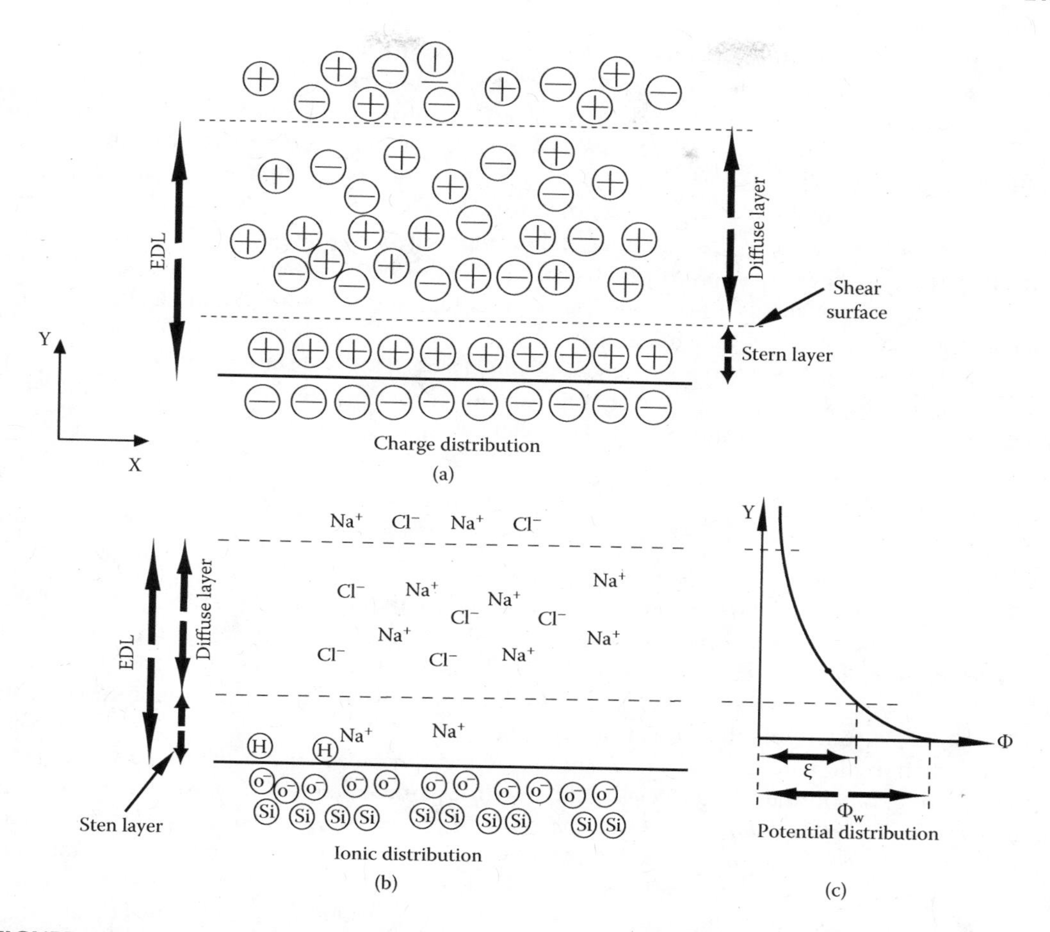

FIGURE 8.16 Scheme of the electrical double layer. (a) Charge distribution. (b) Ionic distribution. (c) Potential variation. (Adapted from Tandon et al., *Electrophoresis*, 29, 1092–101, 2008.)

of this layer is decided by the movement of counterions that form the bulk of this layer. Counterions are attracted by a cathode electrode, resulting in the movement of the bulk of the fluid. The interface between the diffuse layer and the Stern layer is called the *shear surface*. The electrical potential at the shear surface is called the zeta potential (ζ), and at the wall, the wall potential (φ_w). The strength of the attraction between ions and the charged channel walls is determined by the *zeta potential* for the given material and solution. When an electric field is applied along the length of the microfluidic channel, the ions that compose the shear layer of the EDL are attracted toward the electrode of opposite polarity.

Since the ions are solvated, they will effectively drag the solution with them, resulting in *electroosmotic flow* (EOF). The electroosmotic fluid flow rate can be expressed as

$$Q_{eof} = \frac{\pi \varepsilon \zeta E d^2}{4\mu} \tag{8.8}$$

where ε is the electrical permitivity of the solution, ζ is the zeta potential of the microchannel inner wall, E is the electric field applied across the microchannel, d is the equivalent diameter of the microchannel, and μ is the dynamic viscosity of the solution.

Electroosmotic flow in the microchannels is based on the existence of an excess of ions in the fluid near solid walls. When an electric field is applied in the axial direction of the channel, the electrical forces acting on the excess of ions drag the surrounding liquid, and in this way the electroosmotic flow develops. The techniques exploiting the physicochemical properties of the solid-electrolyte interface, as described above, are referred to as *electroosmosis*. Because pressure-driven flow is not possible for smaller dimensions (less than around 10 μm), electroosmosis is commonly used to actuate flows in microfluidic channels. Both the electrical field and charge density determine the driving force of EOF. The net charge density is dependent on the EDL field or on zeta potential, which is a function of ionic valence, ionic concentration of electrolyte solution, and surface properties of the microchannel wall. Zeta potential is constant for a system with a simple electrolyte solution and homogenous channel walls. However, the charge on the wall can be varied along the channel by surface modification to produce a different zeta potential distribution. Pressurized flow in a cylindrical capillary can be expressed as:

$$Q = \frac{\pi \cdot r^4}{8 \cdot \eta \cdot l} \cdot \Delta P \tag{8.9}$$

Where Q is the volume flow rate, r, the radius of the capillary, η, the dynamic viscosity of the fluid, and ΔP is the pressure difference along a cylindrical capillary.

When immiscible fluids having different electrical properties are present simultaneously, surface tension forces become dominant. In this case, the electric field can inject electrostatic energy into the interface and, as a result, the contact angle can be modified. The consecutive changes in the contact angle will result in fluid movement. *Electrowetting* is a technique that explores the above-mentioned wetting force through the electric field. *Electrowetting on dielectric* (EWOD), also known as *electrowetting on insulator-coated electrodes* (EICE), is a technique that uses a thin insulation layer between the electrode and the fluid in order to avoid electrolysis at high voltages. In summary, electrophoresis exploits the electrically generated body force, while electroosmosis utilizes surface forces at the solid-electrolyte interface, and EWOD modulates wetting forces at the contact line. Electrophoresis has been largely used to manipulate biochemical species, while EWOD has been used to manipulate the carrier of the biochemical species.

In addition to being driven through electroosmotic force, motion could also be supplemented with pressure force, as shown in Figure 8.17. When there is a charge separation at the fluid-solid interface, pressure-driven flow results in a bulk motion of unbalanced charges in the fluid, leading to the generation of a net electrical current called streaming current (I_{stream}), as shown in Figure 8.18. *Streaming current* under constant pressure gradient for a circular channel depends on several parameters, like

$$I_{stream} = \frac{\varepsilon \zeta \Delta P A}{\mu L} \tag{8.10}$$

where A is the cross-sectional area of the channel, $\Delta P = P_{downstream} - P_{upstream}$, and μ is dynamic viscosity.

In small channels, the nonuniform ion distribution and surface conduction may have significant effects.

One important consequence of creating EDL in pressure-driven flows through microchannels is the development of *streaming potential*, due to the pressure-driven ionic transport in the moving part of the EDL toward the downstream. This induces the streaming current that flows in the direction of the fluid motion. The resultant accumulation of ions in the downstream section of the channel creates an induced electrical field, known as the *streaming potential*, and the resulting current flows back against the direction of the pressure-driven flow and balances the streaming current at

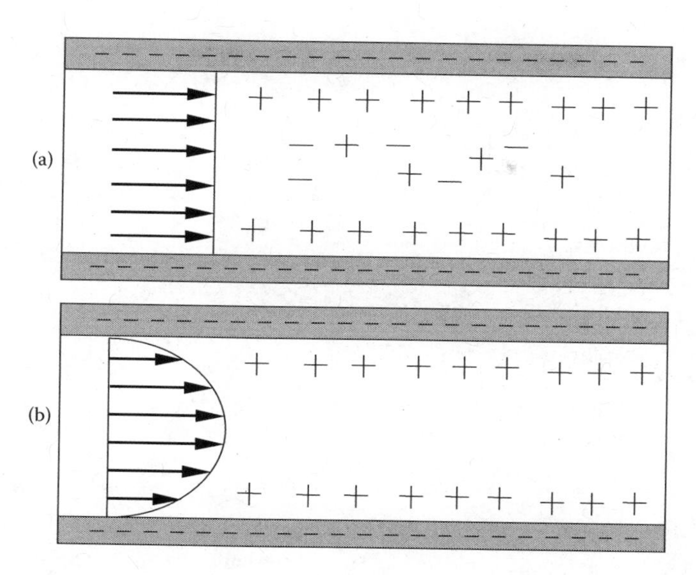

FIGURE 8.17 (a) Uniform electroosmotic flow (EOF) resulting from an applied electric field. (b) Redistribution of ions in pressure-driven flow in a microchannel. (Adapted from Tandon et al., *Electrophoresis*, 29, 1092–101, 2008.)

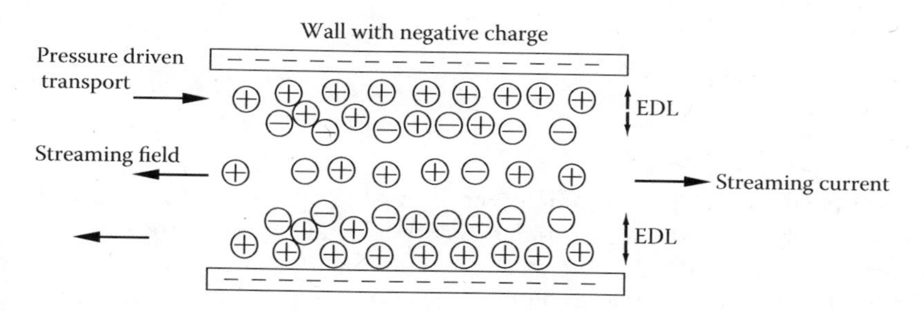

FIGURE 8.18 A schematic diagram for streaming field and streaming current. (From Chakraborty and Das, *Phys. Rev. E*, 77, 037303-1-4, 2008.)

steady state. The net electrical current becomes zero, resulting in a pure pressure-driven flow condition. The experimental setup that was used to demonstrate the pressure-assisted electroosmotic flow is given in Figure 8.19. One important consequence of the EDL-induced counteracting ionic migration mechanism in pressure-driven flows through narrow fluidic confinements is an enhanced viscous resistance that opposes the forward. The increase in resistance due to backward movement in EOL can be considered equivalent to enhanced effective viscosity. This is usually referred to as the *electroviscous effect*.

By monitoring the changes in electrokinetic parameters, the streaming potential technique can be used as a method for characterizing heterogeneous surfaces, for example, due to protein adsorption.

8.4.2 Droplet Actuation Based on Marangoni Flows

Microfluidic drop formation is important for a variety of applications. Microfluidic technologies have enabled the use of droplets as reaction containers, switches for microelectrical actuation, etc. The first droplet generator used to generate aqueous droplets in an organic continuous phase is T-junction geometry. The main channel carries the continuous fluid, while the to-be-dispersed fluid is sent through the

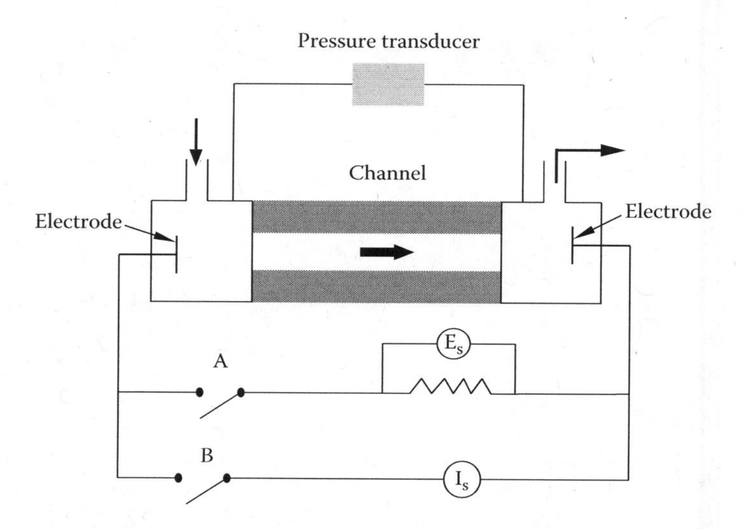

FIGURE 8.19 Typical streaming potential and streaming current apparatus (A, open; B, closed for streaming current measurement; A, closed; B, open for streaming potential measurement). (From Erickson and Dongqing, *J. Coll. Interface Sci.*, 237, 283–89, 2001.)

main channel before forming droplets. The capillary number (C_a) that describes the ratio of viscous to interfacial stresses is normally in the order of $C_a \sim 10^{-2}$ and larger. The capillary number is defined as

$$C_a = \frac{\mu v}{\nu} \tag{8.11}$$

where ν is interfacial tension, μ is dynamic viscosity, and v is velocity. The mechanism of droplet formation is driven by normal stresses, rather than tangential stresses, exerted on the emerging droplet by the continuous fluid. In spite of the fact that interfacial stresses dominate both shear stresses and gravitational effects, the size of the droplets generated in the T-junction is independent of C_a and is a function only of the ratio of flow rates of two immiscible fluids.[18]

There are two types of droplet makers, depending on inlet channel geometry: T-junction (TJ) and flow-focus (FF) drop makers. The TJ drop makers have two inlets, while FF drop makers use three inlets and a symmetrical junction. The properties of the droplets produced in the two types of drop makers are different. The inlet geometry of drop makers influences the range of capillary numbers and flow rate ratios over which monodisperse emulsions can be formed. TJ drop makers form monodisperse emulsions at low and moderate capillary numbers, while FF drop makers need moderate and high capillary numbers.

Fluid motion can be generated by controlling the spatial variations of surface tension, called Marangoni stresses, when a liquid-liquid interface is present. Thermal, chemical, electrical, or light gradients can be used to create variations of surface tension. Some TJ-based droplet maker and images of droplets produced are shown in Figure 8.20.

In addition to the design of drop makers shown in Figure 8.12, a new channel geometry was developed that focuses the droplet break-off location to a single point. Both the geometry of channel junctions and flow rates are used to control the droplet sizes. Monodispersed picoliter- to femtoliter-sized droplets can be generated by using the channel geometry shown in Figure 8.21. The droplet breakup occurs at a fixed point located at the orifice where the shear force is the highest.

The channel is made by bonding the oxidized PDMS to glass.

Figure 8.22 shows 52.9 μm diameter primary droplets and corresponding 2.8 μm satellite droplets. Satellite droplets, if not desirable, can be removed by using a sorting technique.

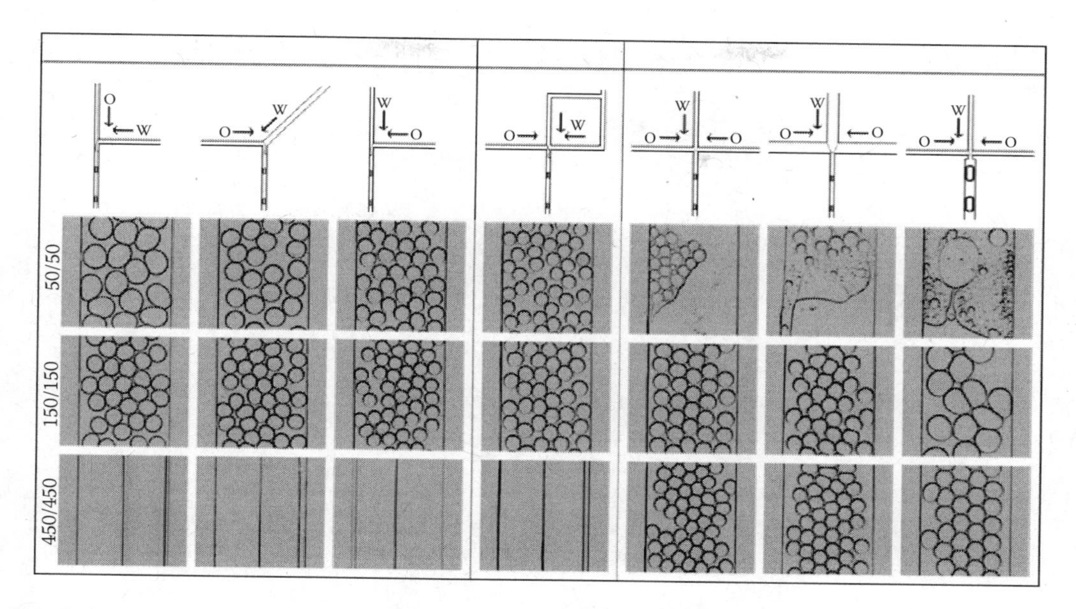

FIGURE 8.20 Schematics of drop makers with different inlet channel geometries are shown in the top row (w, water; O, oil). Example images of drops formed by each device are shown in the lower rows. (Reproduced from Abate et al., *Phys. Rev. E*, 80, 026310-5, 2009. With permission.)

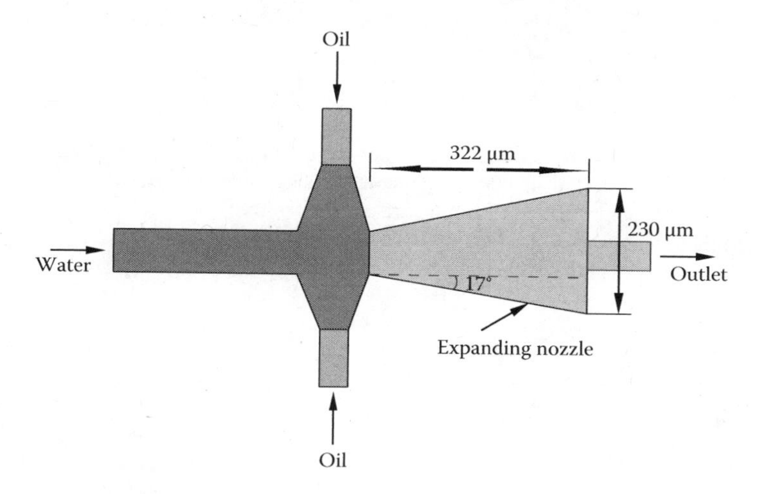

FIGURE 8.21 Schematic of the droplet generation channel geometry. The water phase enters from the middle channel and is sheared by two oil streams at the orifice of the expanding nozzle. The depth of channel is 40 μm. (Reproduced from Tan et al., *Sensors Actuators B*, 114, 350–56, 2006. With permission.)

Marangoni flows occur under the gradient of surface tension created by temperature gradient or chemical concentration. Droplet actuation can be accomplished using Marangoni flows that are generated by an array of heaters kept above the oil layer, as shown in Figure 8.23. When a small heat source is placed above a layer of oil, temperature increases locally and surface tension decreases. This results in a Marangoni flow oriented away from the heat source on the fluid surface, and inward below the surface. The recirculating flows can trap, filter, and pump aqueous droplets in oil. The droplets can be moved if the heat source is moved laterally with a scanning stage.

The wall of the microchannel contains alternating ribs and cavities perpendicular to the flow direction. If the spacing between the cavities is small, the liquid will not enter the cavities for

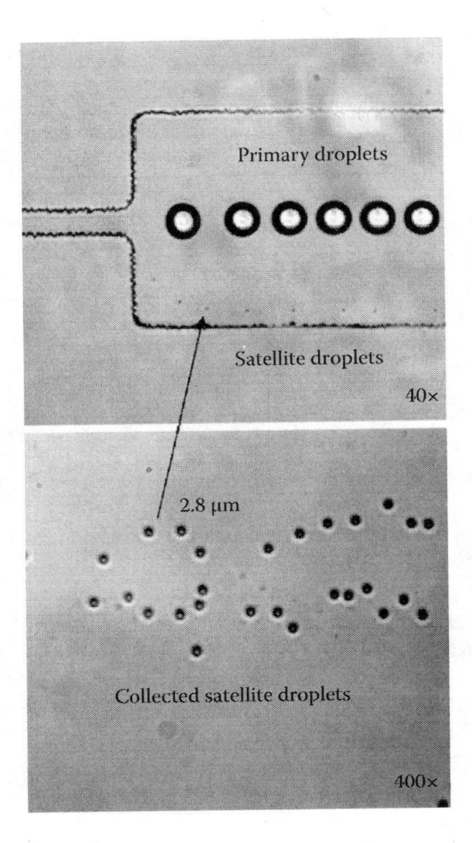

FIGURE 8.22 Size of satellite droplet under different magnifications. Top: The satellite droplets travel to the side of monodispersed primary droplets after generation. The diameters of the primary droplets are 52.9 μm. Bottom: Under 400× magnification, the monodispersed satellite droplets collected at the output have diameters of 2.8 μm. (Reproduced from Basu and Gianchandani, *Appl. Phys. Lett.*, 90, 034102.1–2.3, 2007. With permission.)

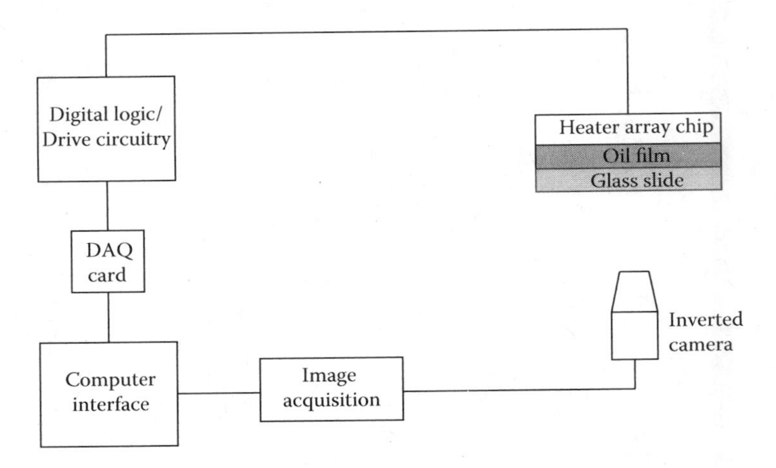

FIGURE 8.23 A complete system for programmable, noncontact droplet actuation based on Marangoni flows, including a 128-pixel array of resistive heaters suspended above the oil layer, control circuitry, and a graphical user interface. (Adapted from Anna et al., *Appl. Phys. Lett.*, 82, 364–66, 2003.)

hydrophobic surfaces (Figure 8.24a). However, when the spacing is larger (Figure 8.24b), the liquid will fully wet the grooves. When oil is introduced into the channel replacing water, as in Figure 8.24c, stationary water droplets are formed in the oil phase (Figure 8.24d).

Figure 8.25 shows the flow-focusing geometry integrated into a planar microchannel design that has been used to study the drop formation in liquid-liquid systems. Oil was used as a continuous phase and water as the dispersed phase. When the two liquid phases are forced to flow through the small orifice, due to the pressure and viscous stresses, the inner fluid (water) breaks inside or downstream of the orifice. The flow rate of the oil is always chosen to be greater than the flow rate of water. The breakup sequences with the formation of nearly monodisperse suspension of water droplets are shown in Figure 8.26.

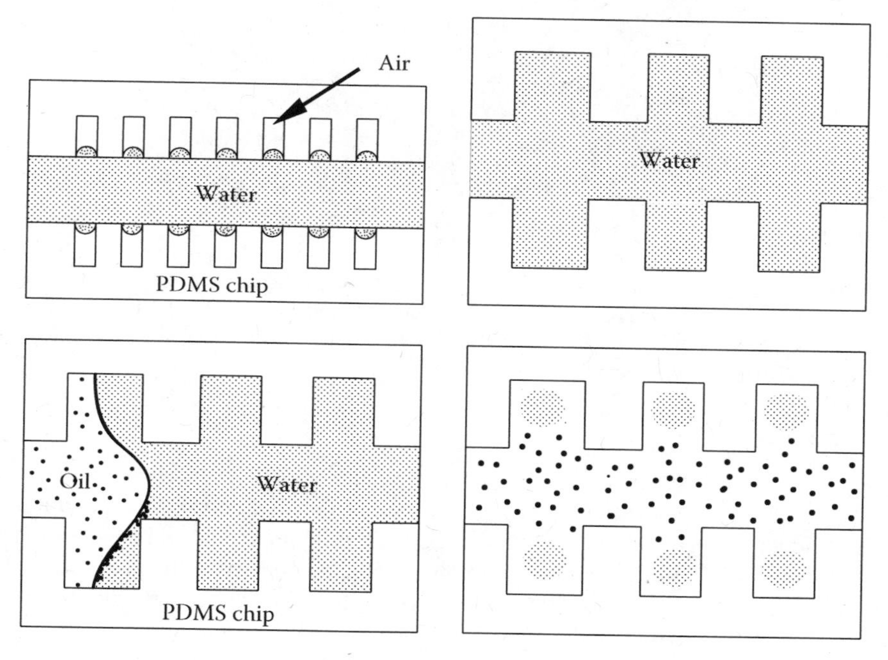

FIGURE 8.24 Schematics of droplet generation. (Adapted from Anna et al., *Appl. Phys. Lett.*, 82, 364–66, 2003.)

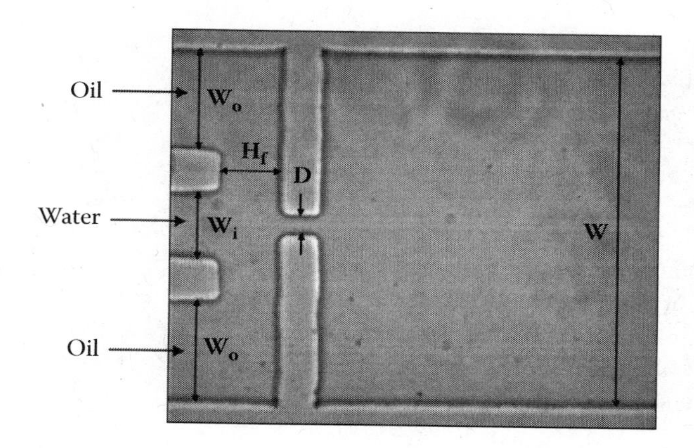

FIGURE 8.25 Flow-focusing geometry. (Adapted from Anna et al., *Appl. Phys. Lett.*, 82, 364–66, 2003.)

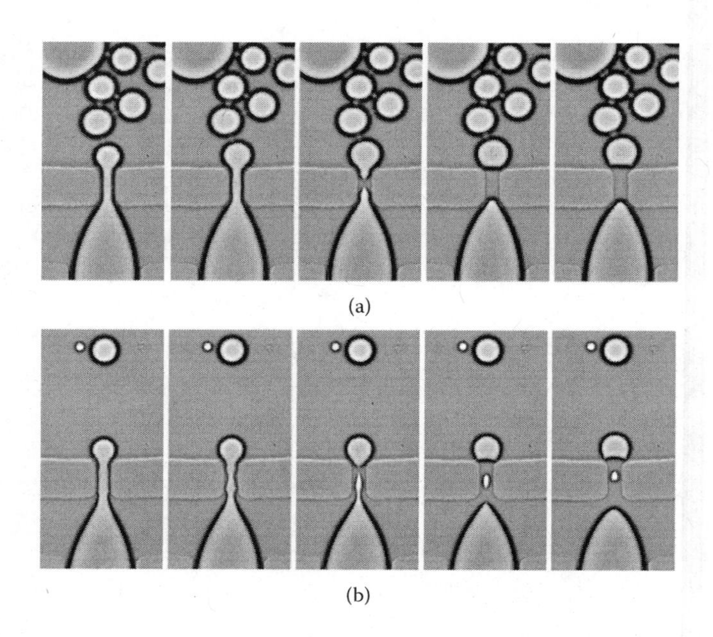

FIGURE 8.26 Images of drop breakup sequences occuring inside the flow-focusing orifice. (a) Uniform-sized drops are formed without visible satellites; breakup occurs inside the orifice. (b) A small satellite accompanies each large drop. Breakup occurs at two corresponding locations inside the orifice. (Adapted from Anna et al., *Appl. Phys. Lett.*, 82, 364–66, 2003.)

8.4.3 Electrowetting

The wetting behavior of a conductive liquid droplet, in contact with a hydrophobic insulated electrode, is modified by an electric field. The phenomenon is named *electrowetting*, and it consists of electric-field-induced interfacial tension changes between a liquid and a solid conductor.

A unit cell of an EWOD-based digital microfluidic biochip consists of two parallel glass plates, as shown in Figure 8.27. The bottom plate contains a patterned array of individually controllable electrodes, and the top plate is coated with a continuous ground electrode. A dielectric, for example, parylene C, coated with a hydrophobic film of Teflon AF, is added to the plates to decrease the wettability of the surface and to add capacitance between fluids and the control electrode. The droplet containing biochemical samples and the filler medium, such as silicone oil, are sandwiched between the plates. The droplets travel inside the filler medium. In order to move a droplet, a control voltage is applied to an electrode adjacent to the droplet, and at the same time, the electrode just under the droplet is deactivated. By sequencing the electrical potential along a linear array of electrodes, electrowetting can be used to move liquid droplets along this line of electrodes. The velocity of the droplet can be controlled by adjusting the control voltage between 0 and 90 V, and droplets can be moved at a speed above 20 cm/sec. Droplets can also be transported, in user-defined patterns and under clocked-voltage control, over a two-dimensional array of electrodes.

Microliter- and nanoliter-size droplets have been manipulated in electrowetting-based systems. Among the advantages offered by a digital-microfluidic platform, it can be mentioned that the need for moving parts such as pumps, valves, etc., is eliminated, and because the electrowetting force is localized at the surface, droplets can be controlled independently. In addition, because of using indium tin oxide (ITO) transparent electrodes, digital microfluidic systems based on electrowetting are compatible with microscopy. Droplet speeds up to about 25 cm/sec can be obtained by using this method.

Figures 8.28 and 8.29 show the EWOD-driven droplet manipulation and the on-chip integration of optical detection.

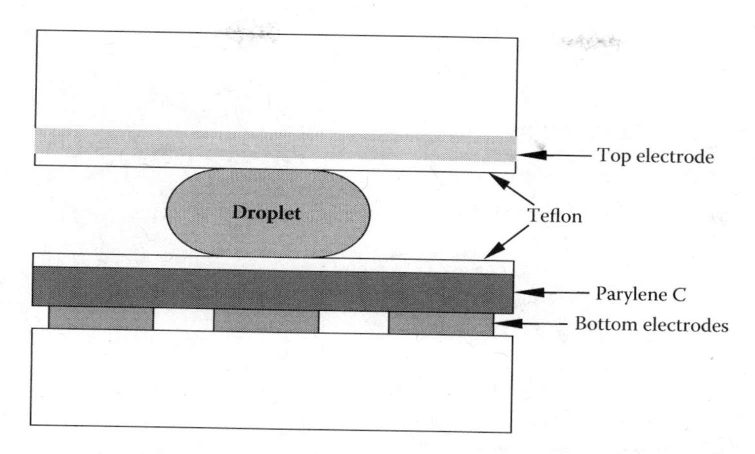

FIGURE 8.27 Electrowetting effect on a digital-microfluidic platform. Because of the conductive top plate and the individually addressed buried electrodes in the bottom plate, applying a voltage can actuate the droplet from one electrode position to the next. (Adapted from Charabarty and Zeng, *ACM J. Emerg. Technol. Comput. Syst.*, 1, 186–223, 2006.)

As shown in Figure 8.30, the electrical potential applied between the electrodes will modify the charge distribution at the solid-liquid interface and the surface tension. The relationship between the applied potential (V) and the surface tension (γ) is known as the Lippmann-Young equation:

$$\cos\theta = \cos\theta_0 + \frac{\varepsilon_0 \varepsilon_r V^2}{2d\gamma_{LG}} \tag{8.12}$$

where γ_{LG} is the liquid/gas surface tension, θ is the contact angle, θ_0 is the initial contact angle before voltage application, ε_0 is the dielectric permittivity of vacuum, ε_r is the dielectric constant of the layer, and V is the applied potential. The applied potential causes the lowering of the surface tension at the liquid-gas interface and modifies the contact angle.

8.4.4 Thermocapillary Pumping

Thermocapillary pumping is another method for moving nanoliter- and picoliter-sized drops of liquids to move discrete drops within microchannels. The method is based on heating one end of a drop to create a surface tension difference between the ends of a drop, as shown in Figure 8.31.

Drop motion results from heating-induced capillary pressure differences between the ends of the drop. Heating occurs (a) at the receding end in hydrophilic systems (b) and at the advancing end in hydrophobic systems.

It can be shown that the thermocapillary pumping velocity is proportional to the difference in temperature drop between the ends. Velocities up to 20 mm/min were measured for several liquids at temperature differences between 10 and 70°C. Thermocapillary pumping offers a simple mechanism for moving drops of liquid within microchannels.

8.4.5 Surface Electrodeposition

Surface electrodeposition on solid electrolytes can also be used to control the flow in microfluidic devices. An electrodeposited silver layer changes the solid-liquid interaction and the movement of the fluid. Figure 8.32 shows the structure of the electrodeposit, and Figure 8.33 shows how the contact angle of a water and methanol drop changes in the presence of the silver deposit. The contact

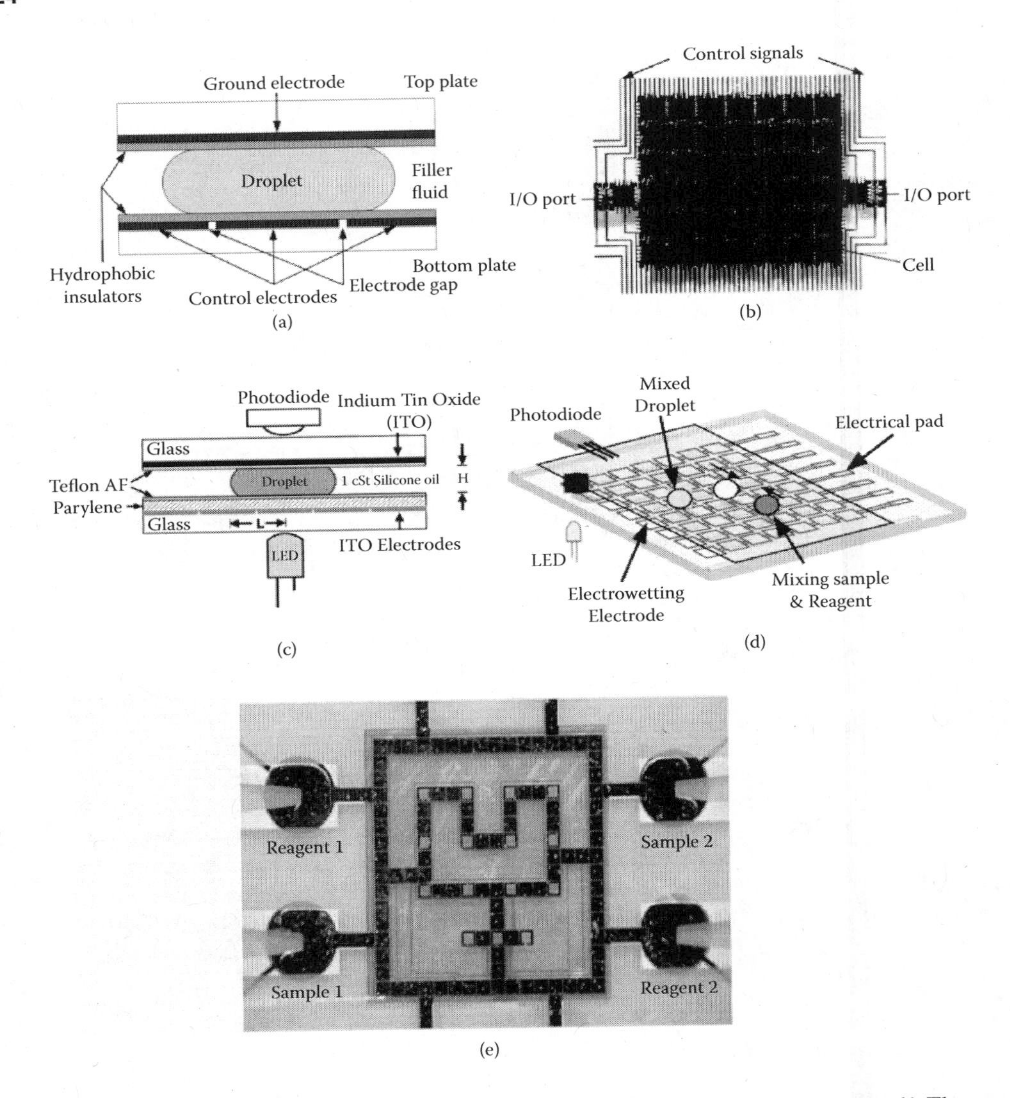

FIGURE 8.28 EWOD-driven digital microfluidic biochip used for colorimetric assays. (a, b) The on-chip droplet manipulation: the basic unit cell (a) and the two-dimensional array architecture (b). (c, d) The on-chip integration of optical detection: the basic unit cell (c) and the two-dimensional array architecture (d). (e) The micrograph of the fabricated microfluidic array used for multiplexed bioassays. (Reproduced from Moon and Kim, *Sensors Actuators A*, 130–31, 537–44, 2006. With permission.)

angle has increased from 65° to 85° for DI water and 7° to 34° for methanol through electroplating the surface.

The changes observed in the contact angles are sufficient to modify, retard, or stop the flow.

8.5 MODELING MICROCHANNEL FLOW

8.5.1 INTRODUCTION

Numerical simulation of microfluidic systems is a valuable tool for research, design, and optimization. As the fabrication processes are slow, numerical simulations can serve to reduce the time from concept to device. In parametric analysis, prior to expensive experiments, numerical models

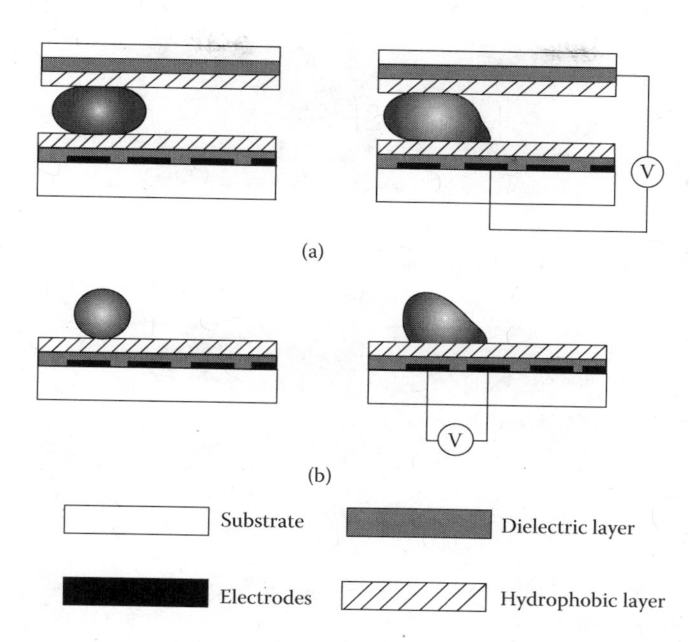

FIGURE 8.29 Schematics of an EWOD system configuration: (a) with isolated additional ground electrode and (b) without additional electrode.

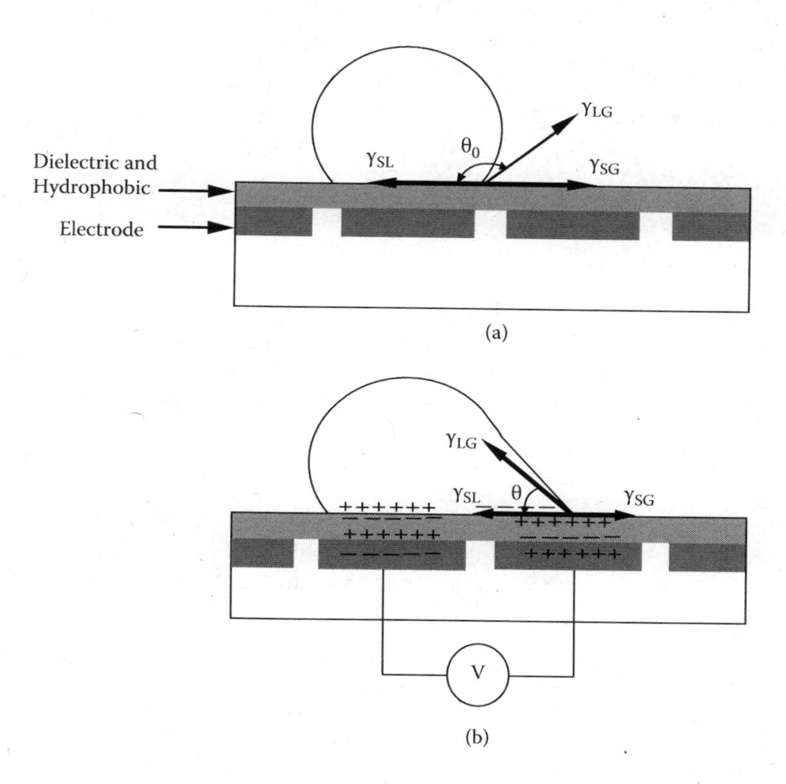

FIGURE 8.30 Charge distribution of our device: (a) no voltage applied and (b) after voltage application. (From Sammarco and Burns, *AIChE J.*, 45, 350–66, 1999.)

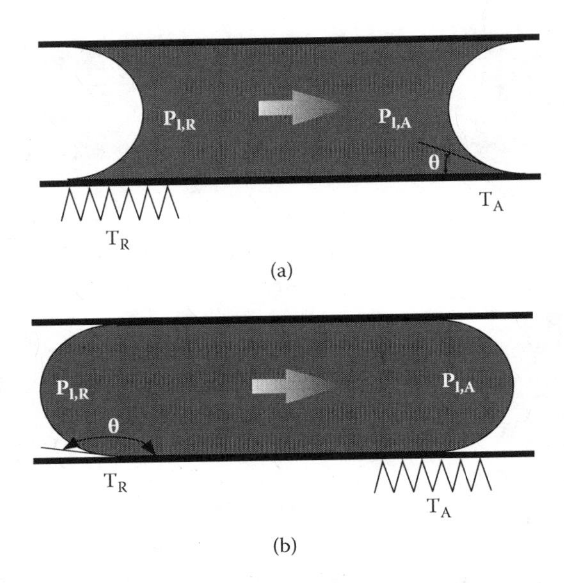

FIGURE 8.31 Thermocapillary pumping (TCP) of drops of liquid in (a) hydrophilic ($\theta < 90°$) and (b) hydrophobic ($\theta > 90°$) channels. (Adapted from Kozicki et al., *Superlattices Microstruct.*, 34, 467–73, 2003.)

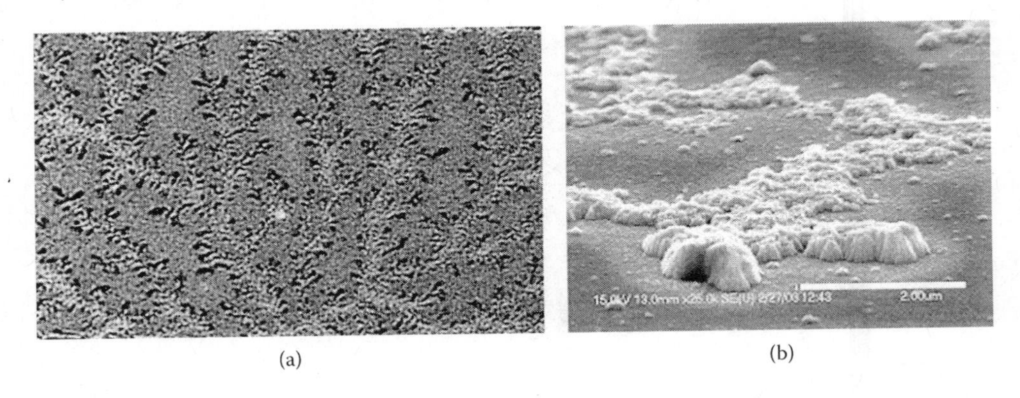

FIGURE 8.32 (See color insert.) (a) An optical micrograph of a silver electrodeposit on Ag-Ge-Se solid electrolyte. The field of view is approximately 200 μm width. (b) A scanning electron micrograph of one branch of the electrodeposit. The size bar is 2 μm. (From Przekwas and Makhijani, "Mixed-Dimensionality, Multi-Physics Simulation Tools for Design Analysis of Microfluidic Devices and Integration Systems," in *Technical Proceedings of the 2001 International Conference on Modeling and Simulation of Microsystems*, 2001, www.cr.org.)

provide an effective method in prediction and initial designs. Numerical studies offer a less expensive and fast way to investigate the physics behind phenomena. With the advancement of numerical simulation techniques, one recent trend is to perform a numerical study for the microchannel flow problem, such that any difficulties associated with micromanufacturing and instrumentation could be avoided.

When modeling the flow behavior, the selection of physical and mathematical models, boundary conditions, and solution procedure is quite critical. The design of microfluidic systems requires expertise in materials, microfabrication, chemistry, biology, and engineering, and an understanding of the complex interplay between variables that influence and limit the system performance. Modeling tools are needed to simulate the complex underlying phenomena, such as electroosmosis, electrophoresis, mixing, etc., and these tools must be easily usable by the microfluidic community,

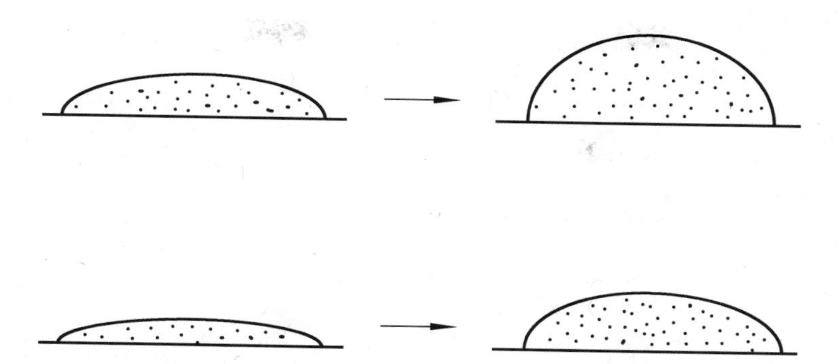

FIGURE 8.33 (a) A DI water drop on a solid electrolyte surface with a contact angle of 65°. (b) A DI water drop on a silver-electroplated surface with a contact angle of 85°. (c) Methanol on a solid electrolyte surface with a contact angle of 7°. (d) Methanol on a silver-electroplated surface with a contact angle of 34°. (From Kozicki et al., *Superlattices Microstruct.*, 34, 467–73, 2003.)

which is comprised of scientists and engineers from a variety of disciplines. The most challenging aspect of computational simulation of microfluidic devices is the multiphysics nature that combines fluidics, transport, and thermal mechanics, electronics, and optics with chemical and biological thermodynamics and reaction kinetics.

Simulation and design of microfluidic systems require at least two levels of modeling:[29]

- High-fidelity models for multiphysics design and optimization of a particular component or device
- System level modeling of integrated microsystems with many components for which reduced or compact models of the components are used

Modeling of fluid behavior depends on the nature of fluid, whether it is a continuum or molecular. The continuum of the fluid breaks down when the dimension or the size becomes very small. For a fluid to be considered continuum, its state and fluid properties have to be continuum in the size considered. The approximate hypothetical length scale for continuum is 1 μm for gases and 10 nm for liquids. These limits can be obtained when the properties become invariant with increase in size, and also become continuous and independent of size. It is better to have larger dimensions to ensure continuum.

The nondimensional number, called the Knudsen number, will be useful in defining the continuum:

$$K_n = \frac{\lambda}{L} \tag{8.13}$$

where λ is the mean free path and L is the length scale of the channel.

Fluid can be considered safely as a continuum for a Knudsen number less than 0.1. For $K_n > 0.1$, the fluid is considered noncontinuum, and molecular approaches have to be applied for modeling its flow behavior. The popular molecular statistical approaches for modeling are molecular dynamics (MD) and direct simulation Monte Carlo (DSMC).

For dimensions satisfying the continuum requirement, the continuum approach can be applied for modeling fluid flow in microdevices.

The *continuum approach* can be applied especially if the fluid is a liquid. In the framework of continuum fluid mechanics, fluid velocity u, pressure p, and electric fields are governed by the following set of coupled equations, known as Navier-Stokes equations. They are derived from the principles of conservation of mass, momentum, and energy and can be expressed in partial differential form as

$$\frac{\delta\rho}{\delta t}+\frac{\delta\rho}{\delta x_i}(\rho u_i)=0 \tag{8.14}$$

$$\frac{\partial}{\partial t}(\rho u_i)+\frac{\partial}{\partial x_i}(\rho u_j\, u_i)=\rho F_i-\frac{\partial p}{\partial x_i}+\frac{\partial}{\partial x_j}\tau_{ji} \tag{8.15}$$

$$\frac{\partial}{\partial t}(\rho e)+\frac{\partial}{\partial x_i}(\rho u_i e)=-p\frac{\partial u_i}{\partial x_i}+\tau ji\frac{\partial u_j}{\partial x_j}+\frac{\partial q_i}{\partial x_i} \tag{8.16}$$

where u_i represents the flow velocity in the ith coordinate, ρ is the density, p is the pressure, τ is the stress tensor, e is the specific internal energy, F is the body force, and q is the heat flux. The first equation (8.14) is often called the mass continuity equation, while Equation 8.15 is called the conservation of momentum equation. The last equation (8.16) expresses the first law of thermodynamics and conservation of energy.

In the case of microfluidics, the fluid is mostly assumed incompressible and the fluid density is constant. The above equations can be simplified further for Newtonian and incompressible fluid. The energy equation can be removed from these equations when a constant temperature assumption is valid. The effects of the applied electric field forces have to be included when the flow transport is initiated by application of a high electric field.

The conservation of mass, momentum, and energy equations for the microflow of a Newtonian, incompressible, isotropic fluid in the presence of an external electric field can be written as follows:

$$\frac{\partial u_i}{\partial x_i}=0 \tag{8.17}$$

$$\rho\left(\frac{\partial u_i}{\partial t}+u_j\frac{\partial u_i}{\partial x_j}\right)=-\frac{\partial P}{\partial x_i}+\frac{\partial}{\partial x_i}\left[\mu\left(\frac{\partial u_i}{\partial x_j}\right)+\frac{\partial u_j}{\partial x_i}\right]+\rho_E E_i \tag{8.18}$$

$$\rho C_v\left(\frac{\partial T}{\partial t}+u_i\frac{\partial T}{\partial x_i}\right)=\frac{\partial}{\partial x_i}\left(k\frac{\partial T}{\partial x_i}\right)+\rho_E E_i u_i \tag{8.19}$$

where μ is the dynamic viscosity, ρ_E is the electric charge density, E is the external electric field, C_v is the specific heat at constant volume, and k is the thermal conductivity. The term $\rho_E E_i$ in the momentum equation is the Lorentz body force, and $\rho_E E_i u_i$ in the energy equation is the corresponding work term.

For one-dimensional electroosmotic flow, the middle equation (8.18) can be written as

$$-\frac{dP}{dx}+\mu\frac{\partial^2 u}{\partial y^2}-\varepsilon\frac{\partial^2\Phi}{\partial y^2}E_x=0 \tag{8.20}$$

where ε is the permittivity and φ is the electric potential.[13]

8.5.2 The Finite Element Method

The complexity of solving the above equations will vary depending upon the geometries and multidisciplinary nature. There are many numerical techniques, such as finite difference method, finite volume method, and finite element method (FEM), to numerically solve the above equations. Out of these methods, FEM is the most popular due to its versatility to model multidisciplinary problems involving fluid mechanics, thermal micromechanics, electrofluidics, chemical kinetics, and optics.

The *finite element method* (FEM) can be summarized as follows. The solution domain is first discretized by constructing a series of interlocking nodes and elements, resulting in a finite number of elements that approximate the geometry. The elements can be of different shapes, such as linear, triangular, and rectangular, depending upon complexities. By assuming suitable functional variation within its elements, equations are rewritten in terms of unknown nodal values using a variational form:

$$KU = F \tag{8.21}$$

where K is the global system matrix, or the stiffness matrix, U is the global matrix of unknowns (velocities, pressures, etc.), and F is the force matrix that contains effects of all relevant body forces and boundary conditions. The solution to the above system of equations (8.19) yields the unknown nodal values or functional values at the nodes. Functional values in between the nodes will be obtained by using the assumed interpolation functions between nodes. In its application, the object or system is represented by a geometrically similar model consisting of multiple, linked, simplified representations of discrete regions, i.e., finite elements on an unstructured grid. While being an approximate method, the accuracy of the FEM method can be improved by refining the mesh or element size in the model using more elements and nodes.

There are many commercial softwares available based on FEM. Some of them include Ansys, Adina Comsol, Coventer, and Intellisuite. Some examples of microfluidic modeling with FEM using Comsol[29] software are given below.

8.5.3 Simulation of Flow in Microfluidic Channels: Case Studies

8.5.3.1 Case 1: Silicon Microfluidic Platform for Fluorescence-Based Biosensing

A simple silicon microfluidic device used for fluorescence-based biosensing is provided in Figures 8.34 and 8.35. The flow behavior prediction using finite element analysis of a microfluidic device consists of two channels, namely, the detection channel and the rinsing channel. The effect of channel width has been studied by solving the continuity and Navier-Stokes equations for several channel geometries, assuming two-dimensional no-slip pressure-driven flow conditions. The flow is considered to be invariant with respect to the depth of the channel. The flow is assumed to have constant velocity at the input and ambient pressure at the outlet. It was possible to predict stagnation flow zones within the microfluidic channels. The predicted flow variation is shown in Figure 8.35.

8.5.3.2 Case 2: Numerical Simulation of Electroosmotic Flow in Hydrophobic Microchannels: Influence of Electrode's Position

In this study, modeling of the influence of the electrode's position in a low voltage micropump is presented. Low-voltage electroosmotic micropumps are attractive since they can be easily integrated for the realization of portable micro total analysis systems. Absence of any moving parts in the electroosmotic micropumps makes it attractive for application. The fabrication of electroosmotic

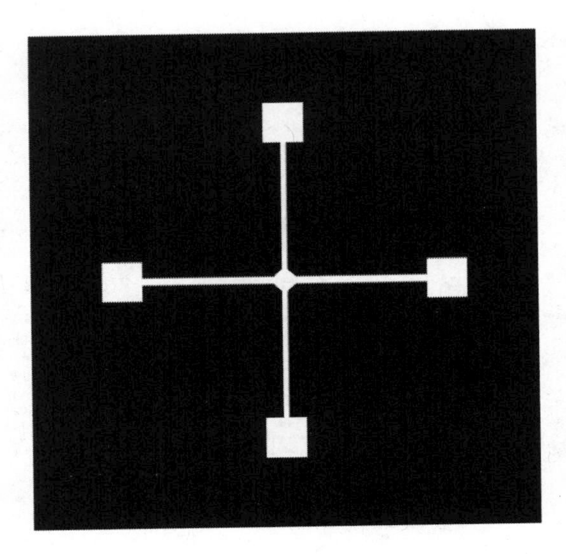

FIGURE 8.34 Schematic of microfluidic channels. (From "Use of Comsol Multiphysics in Undergraduate Research Projects to Solve Real-Life Problems," http://faculty.washington.edu/finlayso/Finlayson_paper_611d.pdf.)

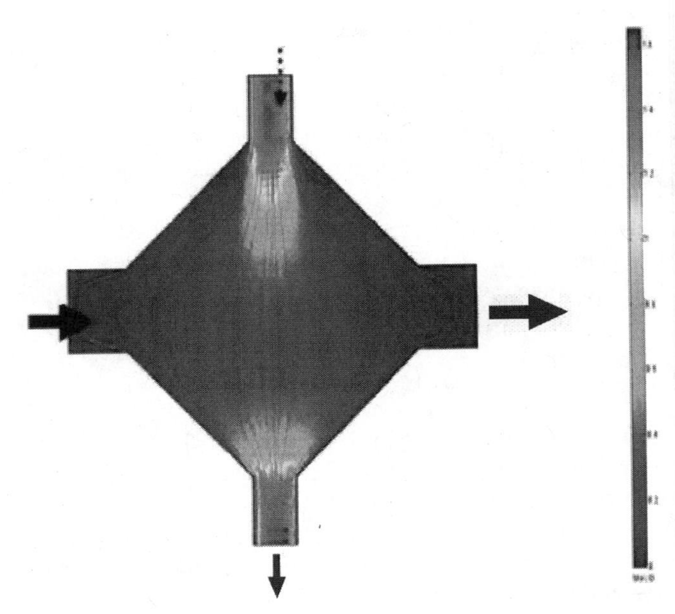

FIGURE 8.35 (See color insert.) Predicted flow behavior in the chamber using FEM simulation.

micropumps could become difficult, depending upon complexities involved with the positions of electrodes. It is possible to model the influence of the position of electrodes with FEM. Different possible positions of electrodes in low-voltage electroosmotic micropumps are investigated in this case study.[31] Figure 8.36 shows the three possible positions of electrodes that are considered the design of the electroosmotic micropump, in which the TOP electrode configuration is easily fabricable by patterning the electrode on a dielectric surface and bonding to the microchannel of an electroosmotic micropump. The influence of position of the electrode configuration and geometrical parameters were analyzed by using COMSOL Multiphysics FEM software.[31]

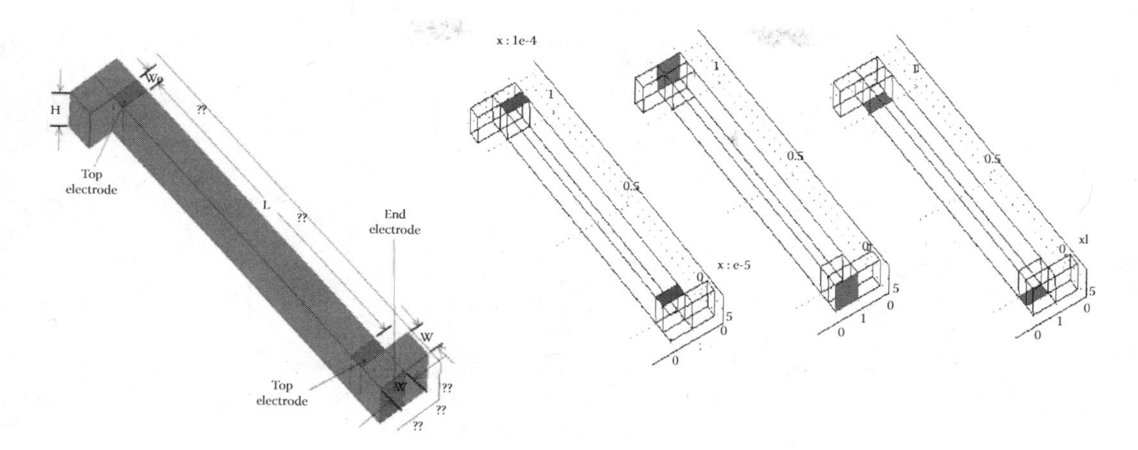

FIGURE 8.36 (a) Electroosmotic pump geometry used for the COMSOL modeling. *H*, height; *W*, width; *L*, length of the channel; *We*, width of the TOP electrode. (b) Three different possible positions of the electroosmotic micropump. (From Ozhikandathil et al., "Modeling and Analysis of Low Voltage Electro-Osmotic Micropump." paper presented at Proceedings of ASME 2010 3rd Joint US-European Fluids Engineering Summer Meeting and 8th International Conference on Nanochannels, Microchannel and Minichannels, FEDSM ICNMM2010-31213, Montreal, Canada, 2010.)

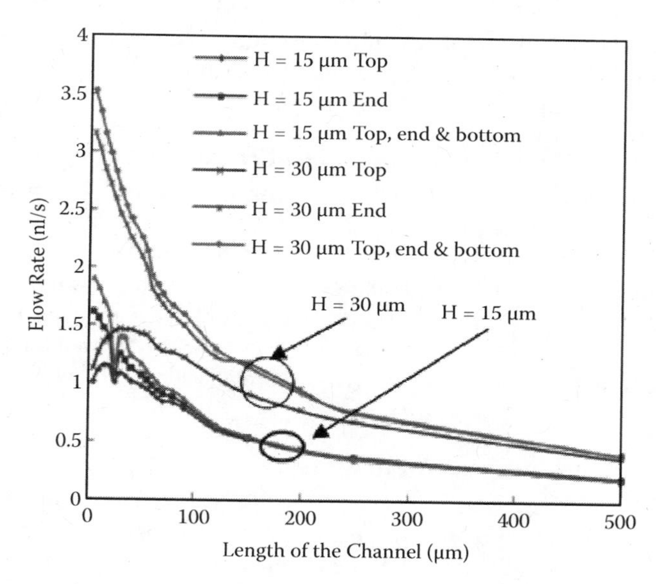

FIGURE 8.37 (See color insert.) Influence of electrode position on flow rate with varying the length of the pump. (From Ozhikandathil et al., "Modeling and Analysis of Low Voltage Electro-Osmotic Micropump." paper presented at Proceedings of ASME 2010 3rd Joint US-European Fluids Engineering Summer Meeting and 8th International Conference on Nanochannels, Microchannel and Minichannels, FEDSM ICNMM2010-31213, Montreal, Canada, 2010.)

The influence of the electrode's position on flow rate was investigated by varying the length (L) of the micropump geometry. As can be seen in Figure 8.37, the flow rate became independent of the position of the electrode when the length of the channel was increased. A significant increment in flow rate was observed when height was increased and length was decreased. This concludes that the flow rates of longer pumps have lesser influence on the electrode configuration, and hence they can be realized by using the easily fabricable TOP electrode configuration. Figure 8.38 shows

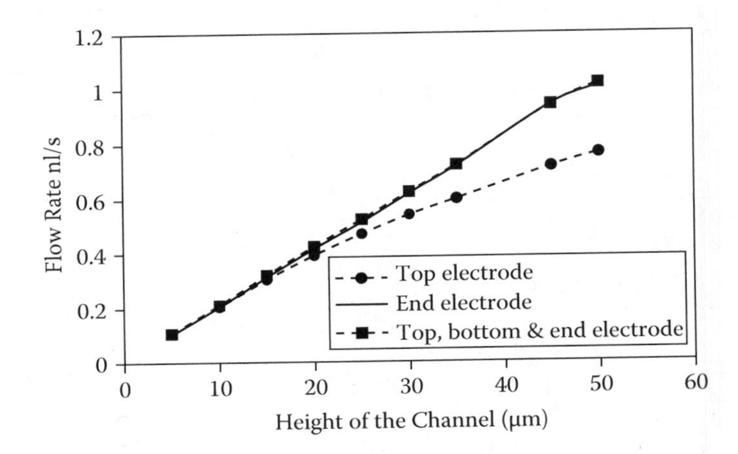

FIGURE 8.38 Influence of electrode position on flow rate with varying the height of the pump. (From Ozhikandathil et al., "Modeling and Analysis of Low Voltage Electro-Osmotic Micropump." paper presented at Proceedings of ASME 2010 3rd Joint US-European Fluids Engineering Summer Meeting and 8th International Conference on Nanochannels, Microchannel and Minichannels, FEDSM ICNMM2010-31213, Montreal, Canada, 2010.)

the simulation result, showing the influence of the electrode position on flow rate with the varying height of the channel. This figure indicates that the TOP electrode configuration is suitable for the lower-aspect-ratio (W/H) pumps.

8.5.3.3 Case 3: Prediction of Intermittent Flow Microreactor System[32,33]

A microreactor device fabricated on a microfluidic platform, as shown in Figures 8.39 and 8.40, was used to study an enzymatic reaction.

The prototype of the microreactor system is shown in Figure 8.40.

The two-dimensional finite element modeling was carried out with COMSOL 3.1 software solving an incompressible Navier-Stokes equation. The finite element results are shown in Figure 8.41. The finite element model is able to capture the recirculating flows that were used for enhancing mixing.

The predicted flow patterns for different Reynolds numbers showed that the size of the recirculation zone increased with the Reynolds number.

8.5.3.4 Case 4: Modeling of Electrowetting Flow

In electrowetting on dielectric (EWOD), the interfacial energy of a liquid-solid interface changes by applying electrical potential, resulting in altering of the droplet contact lines. This alteration leads to movement of the droplet toward the electrode supplied with electrical potential. The geometry of electrodes, applied voltage, and switching sequence of electrical potential of the electrodes[34,35] have important effects on the efficiency of electrowetting (EW) systems. Hence, it would be very useful to simulate the effect of these parameters before fabricating the devices. For example, to increase the flow rate of such a digital microfluidic (MF) system, one way is to increase the number of transported droplets in unit time by increasing the droplet velocity. Simulations using finite element analysis will be of great help.

For simulation of electrowetting, Surface Evolver was used for the study of surfaces shaped by surface tension and other energies. Here a surface is implemented using triangular elements. For a facet with edges $\vec{S}_0$ and $\vec{S}_1$, the facet energy due to surface tension γ can be calculated by

$$E = \frac{\gamma}{2}\left|\vec{S}_0 \times \vec{S}_1\right| \tag{8.22}$$

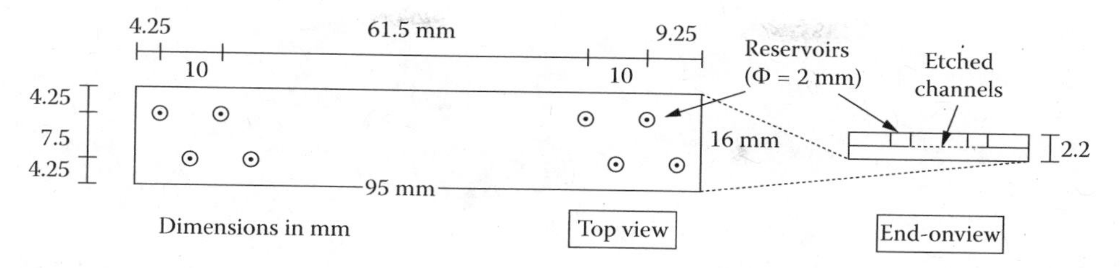

FIGURE 8.39 Layout of the microfluidic device used for the study. (From Chandrasekaran and Packirisamy, *IEE Proc. Nanotechnol.*, 153, 137–43, 2006; Chandrasekaran and Packirisamy, *IET Nanobiotechnol.*, 2, 39–46, 2008.)

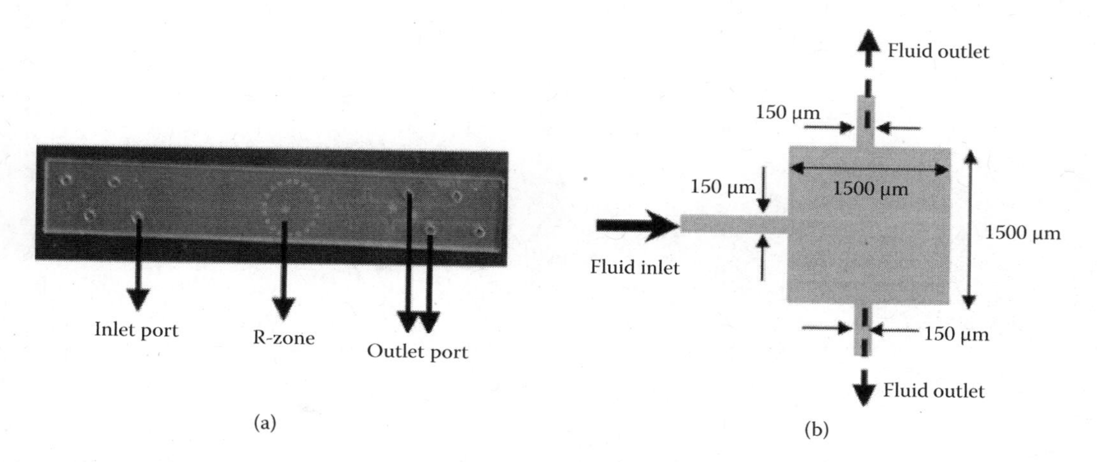

FIGURE 8.40 Prototype of the microreactor system. (From Chandrasekaran and Packirisamy, *IEE Proc. Nanotechnol.*, 153, 137–43, 2006; Chandrasekaran and Packirisamy, *IET Nanobiotechnol.*, 2, 39–46, 2008.)

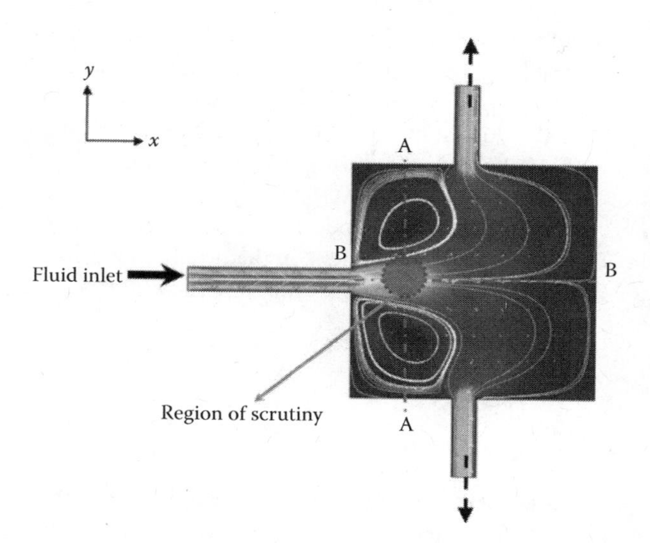

FIGURE 8.41 FEM prediction of recirculation flows under high Reynolds numbers. (From Chandrasekaran and Packirisamy, *IEE Proc. Nanotechnol.*, 153, 137–43, 2006; Chandrasekaran and Packirisamy, *IET Nanobiotechnol.*, 2, 39–46, 2008.)

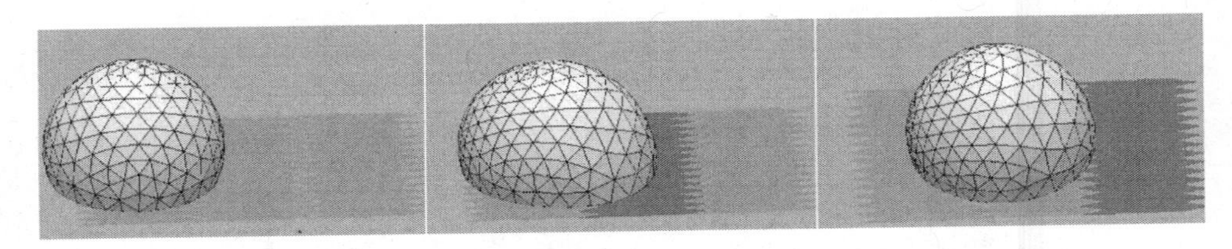

FIGURE 8.42 (See color insert.) A typical simulation: droplet at rest (left) and during movement toward activated electrode (right). (From SadAbadi et al., in *Microsystems and Nanoelectronics Research Conference, 2009*, MNRC 2nd, Ottawa, Canada, 2009, pp. 76–79.)

The surface energy of the free droplet surface can be calculated by integration over the surface of the droplet, and the surface energy on the substrate is calculated by line integral along the contact line using the divergence theorem:

$$E_{SL} = \int_S \gamma_{SL} dA = \int_S (\nabla.\vec{v}) dA = \int_{\partial S} \vec{v}.dl \tag{8.23}$$

with $\vec{v}$ such that $\nabla.\vec{v} = \gamma_{SL}$.

Figure 8.42 shows a typical simulation of EWOD on a series of three linear electrodes.

The resulting force in each position can be estimated by $F = -\partial E / \partial x \approx -\partial E / \partial x_c$, where E is the total energy or total surface energy and x_c is the droplet centroid, which can be calculated by calculating the integral

$$x_c = \int_V x dv \Big/ V$$

So by plotting the droplet energy vs. its centroid x position, also called energy curve, one can easily estimate the resulting force acting on the droplet.

For example, the FEA results of the effect of applied voltage are shown in Figure 8.43. The results show that for low voltages, the droplet cannot move in response to the third electrode. The reason for this is that the resulting electrical field is not enough to change the contact angle high enough to cause motion of the droplet. However, for higher voltage (more than 60 V), the same result is obtained. This is because the droplet in higher voltage tries to expand its area over the electrode instead of moving to the third electrode. Thus, for a specific geometry there is a range of voltage over which the electrowetting devices can work effectively.

The effect of an electrode-switching sequence is also very important. Improper switching of electrical potential between two electrodes could stop the droplet. Thus, the simulation results can show the best position of the droplet for switching the face-forward electrode, which is where the droplet just passes over that electrode.

8.6 EXPERIMENTAL METHODS

8.6.1 Flow Visualization at Microscale

As the area of microfluidics develops and finds new applications, the ability to perform accurate measurements inside microfluidic systems becomes very essential. Depending upon the requirements, many methods are being used and developed. Methods that were earlier used for bioanalysis, such as fluorescence microscopy, optical coherence tomography (OCT), and confocal microscopy, are presently extended for microfluidic applications. Similarly, particle image velocimetry (PIV),

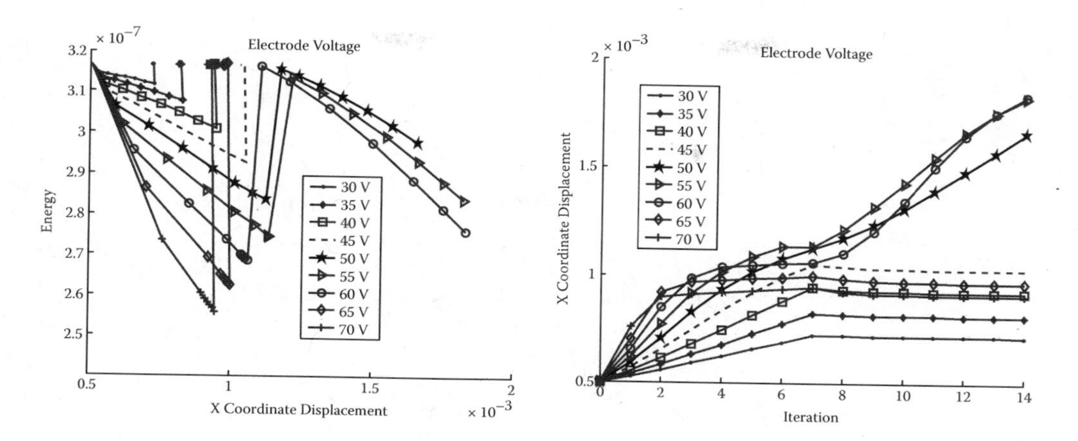

FIGURE 8.43 Effects of applied voltage on the energy curve (left) and droplet position diagram (right). (From SadAbadi et al., in *Microsystems and Nanoelectronics Research Conference, 2009*, MNRC 2nd, Ottawa, Canada, 2009, pp. 76–79.)

which was used for analyzing flows in macrostructures, is now popularly used for microflow measurements. In line with this development, micro particle image velocimetry (μPIV), an optical nonintrusive measurement technique, was developed by Santiago in 1998.[36] Among all experimental techniques available to study microscale flow phenomena, the micro PIV technique is considered the most appropriate to instantaneously characterize the local fluid motion at several locations in a field of microscopic dimensions. The flow velocities that can be measured with this technique range from nanometers per second to meters per second. Some of the flow visualization methods are briefly discussed in the following sections.

8.6.2 Fluorescent Imaging Method

The simple method of determining qualitatively the flow streamlines is established through *visual imaging* of flow with fluorescent particles. In this case, the fluid is mixed with fluorophore particles of a few microns in size, and the flow is recorded using an epifluorescence microscope. When the fluorophores are excited with light corresponding to the excitation wavelength, they emit fluorescing light that can be imaged using a high-speed camera attached to the microscope. Suitable optical filters are used in order to pass only the fluorescent wavelength of fluorophores onto the camera, as shown in Figure 8.44.

Epifluorescence microscopy is an ideal tool that aids in the flow visualization of microfluidic systems. The flow visualization setup could incorporate an epifluorescence microscope, imaging camera, and data acquisition software. The epifluorescence microscope consists of an OSRAM continuous-arc-discharge high-pressure mercury lamp power supply used as an excitation source. The broadband illumination is spectrally filtered with a bandpass emission filter. Green or other fluorescent polymer microspheres of ~5 μm diameter can be used as tracer particles. The green fluorescent microspheres emit a bright and characteristic color at 508 nm when illuminated by a powerful light of 468 nm wavelength. The highly contrast-colored microspheres are clearly visible relative to background materials, offer improved sensitivity of observation, and could be analyzed using the epifluorescence microscope. Streak images of the flowing particles are obtained by altering the shutter speed of the imaging video camera accordingly.

Figure 8.45 shows the comparison of the fluid flow contours and streamlines between the FEM simulation results and the experimental results. Figure 8.45a represents the predicted streamlines, while the experimental images in Figure 8.45b are the streak lines of flowing green microspheres within the microchamber. This method will be useful for identifying qualitatively the flow behavior, recirculation zone, stagnant zones, etc.

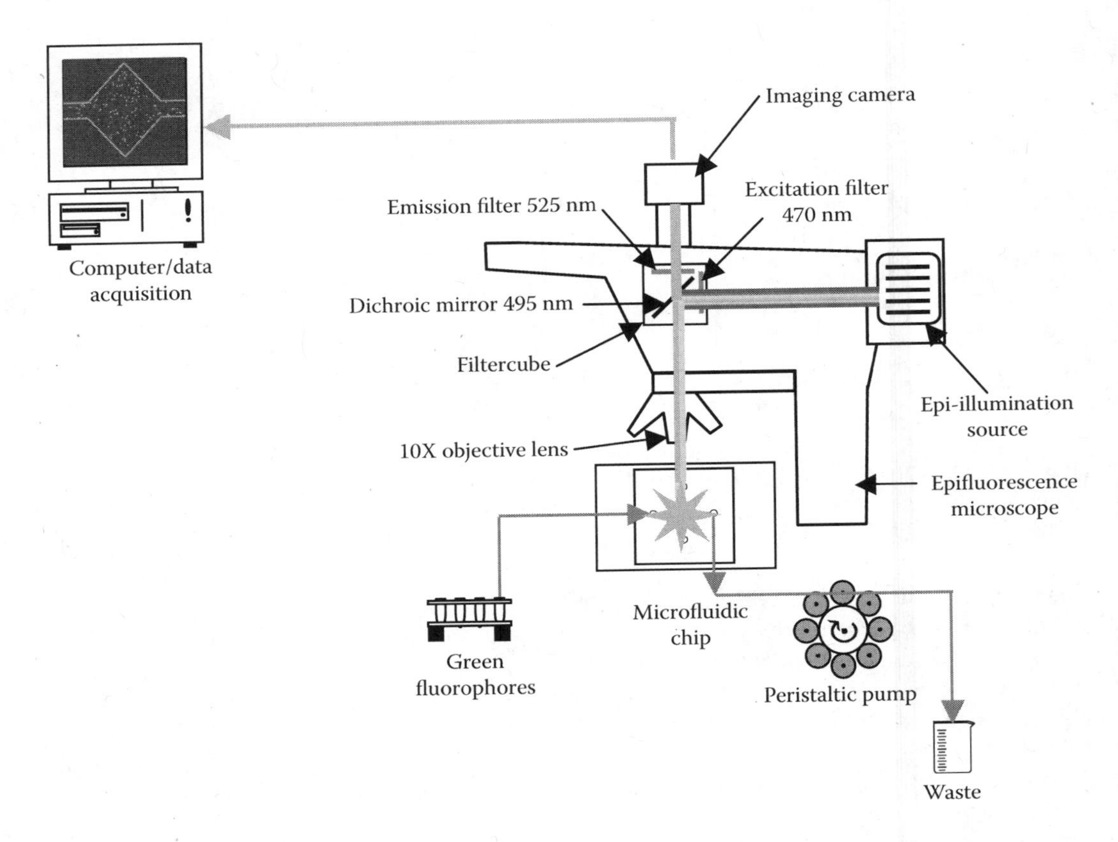

FIGURE 8.44 (See color insert.) Schematic of epifluorescence microflow visualization setup. (From Acharya, "Modeling, Fabrication and Testing of OLED and LED Fluorescence Integrated Polymer Microfluidic Chips for Biosensing Applications," MASc thesis, Concordia University, 2007.)

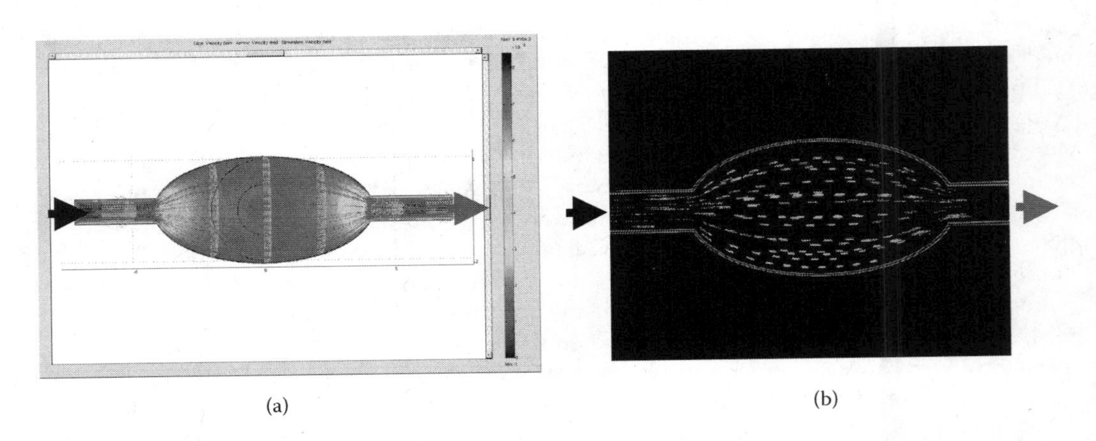

FIGURE 8.45 Flow lines in an elliptical microchamber with (a) predicted and (b) experimental results. The width of the channel is 100 μm. (From Acharya, "Modeling, Fabrication and Testing of OLED and LED Fluorescence Integrated Polymer Microfluidic Chips for Biosensing Applications," MASc thesis, Concordia University, 2007.)

8.6.3 Particle Streak Velocimetry

The *particle streak velocimetry* (PSV) technique can be used to find the experimental flow lines within the chamber. In PSV, the displacement of a flowing particle is recorded in a single image over a period of time by using a slow-shutter-speed camera that integrates the signal over time as the particle flows and captures streak images that clearly depict the nature of the flow. Quantitative particle velocity parameters could be obtained with relatively little infrastructure by using the PSV technique. Brody et al.[38] used an intensified CCD camera to image particle streaks in a cross section of a $11 \times 72\ \mu m^2$ silicon channel. Taylor and Yeung[39] have utilized particle streak imaging of fluorescent tracer particles for electrokinetic and pressure-driven flows.

8.6.4 Particle Tracking Velocimetry

One can calculate the fluid velocity if the position of the particles is known at two known instants of time, as in the case of particle tracking velocimetry (PTV). With the known time interval between the successive positions of the particle, the velocity can be estimated using the distance between the positions as measured from the images. The velocity measurements determined by particle streaks are less reliable, less accurate, and also have less spatial resolution than PIV measurements. The particle imaging setup shown in Figure 8.44 can also be used for PTV measurements. In this case, the pulsed excitation or high-speed video camera is used to capture the images of fluorophore particles at known instants of time. These sequential images taken at different times can be used to estimate the flow velocities. However, the image acquisition and velocimetry calculation process are extremely time-consuming, and the images acquired can be disturbed, with high amounts of noise. For example, the positions of a fluorophore particle at different times are shown in Figure 8.46. One can estimate the velocities of the flow by measuring the distance between these positions. But, this method is not suitable for high velocities and sharp directional changes.

8.6.5 Micro Particle Imaging Velocimetry (μPIV)

Particle imaging velocimetry (PIV) technology,[40,41] a noncontact method, was first applied for estimating transient macroflows. In order to improve the accuracy of measurements of microflows, it is important to produce consecutive laser pulses with very short intervals in the order of nanoseconds

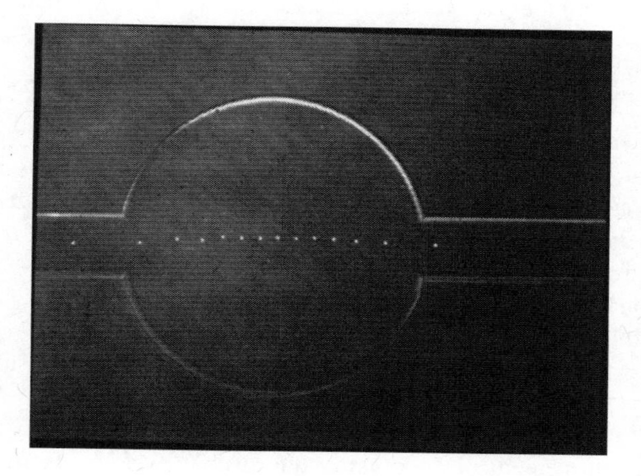

FIGURE 8.46 Particle tracked along the mid-axis at $\Delta t = 66.7$ msec within circular microchamber. (From Acharya, "Modeling, Fabrication and Testing of OLED and LED Fluorescence Integrated Polymer Microfluidic Chips for Biosensing Applications," MASc thesis, Concordia University, 2007.)

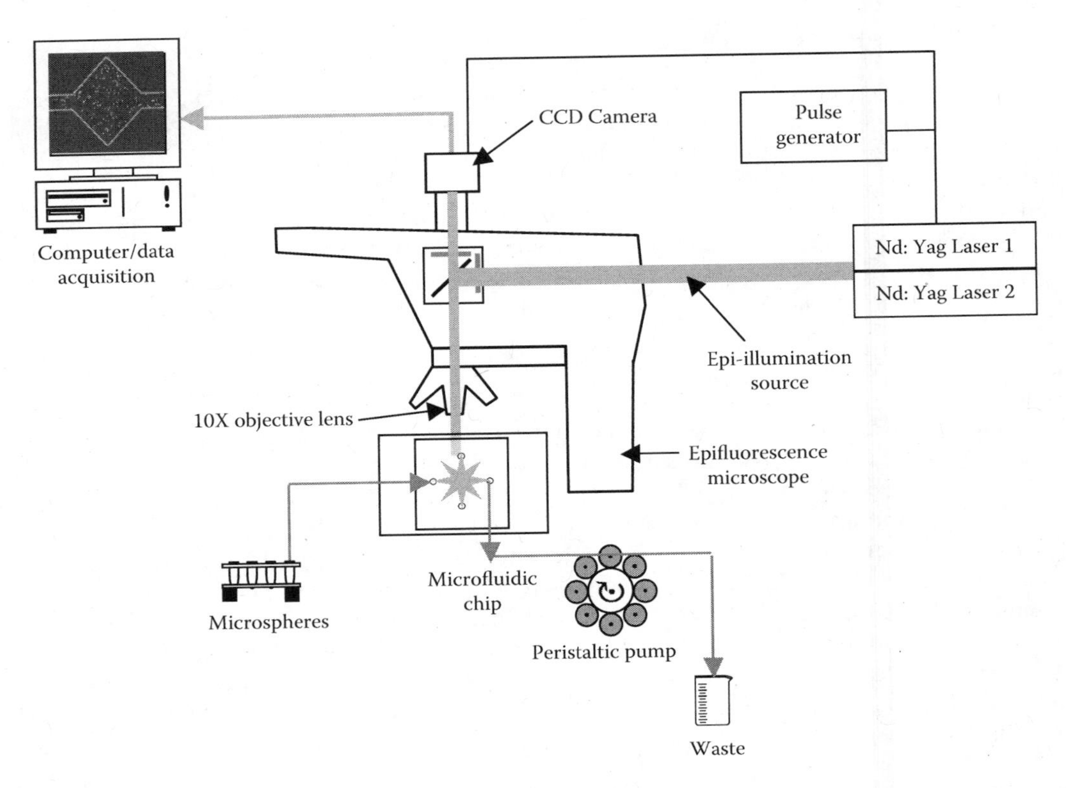

FIGURE 8.47 Setup for μPIV experiment platform. (From Acharya, "Modeling, Fabrication and Testing of OLED and LED Fluorescence Integrated Polymer Microfluidic Chips for Biosensing Applications," MASc thesis, Concordia University, 2007.)

to picoseconds, and also to synchronize the laser source with the video camera. As shown in Figure 8.47, the main components of the system are microscope equipped with fluorescence filters, a CCD camera, a light source (a pulsed light source), and beam expanders. Fluorescent tracer particles are introduced into fluid inside the microchannel and are illuminated and imaged through the microscope objective onto the CCD camera. Fluorescent polystyrene beads that have a density close to that of the fluid are used for seeding.

Within the system, there are fluorescent particles in flow that absorb the light with shorter wavelength arousing from the light source, and emit fluorescent light with a certain longer wavelength. As shown in Figure 8.48, the microscope filters and enlarges the image, and the CCD camera shooting device captures images at known time periods. Correlation algorithms are used to estimate the velocity parameters from the positions of the particles in the images.[42] In order to evaluate the data, a spatial correlation analysis is performed.

The two-dimensional flow velocities and distribution can be calculated from two frames of particle images captured one after another. As an example, Figure 8.49 shows the comparison of flow velocity along the mid streamline, between FEM results and μPIV results for a microchamber configuration.

Applications of the method include velocity measurements in near-wall, particle migration in electrokinetic flow, movement of cells using fluorescent dye, etc. Both *in vivo* and *in vitro* measurements have been performed, and further biomedical applications are expected in the future. The μPIV technique has also been used to study the effects of surface roughness on flow behavior.[43]

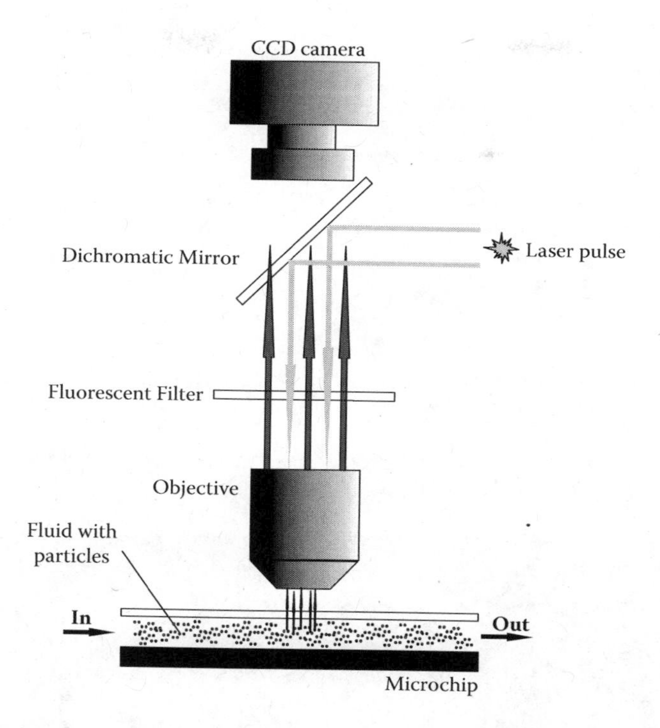

FIGURE 8.48 (See color insert.) Optical detection setup used in other studies by μPIV. (From Fang, Q., "Optical Fiber Coupled Low Power Micro-PIV Measurement of Flow in Microchambers: Modeling, Fabrication, Testing and Validation," MASc thesis, Concordia University, 2009.)

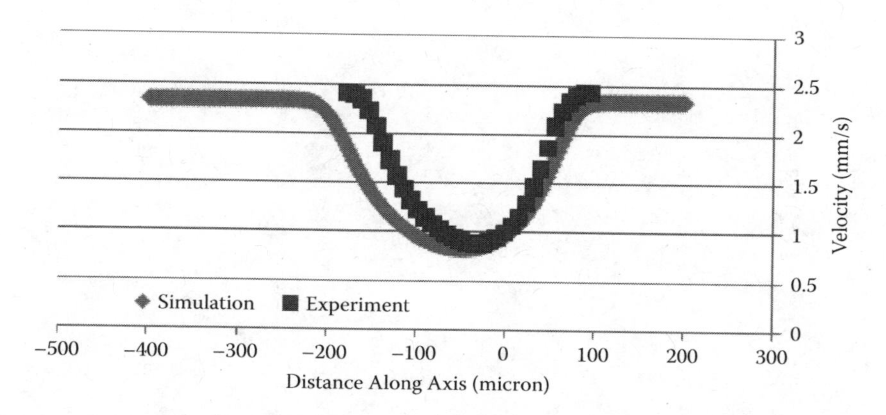

FIGURE 8.49 Velocity distribution along the mid streamline at Re = 6.87e-2. (From Fang, Q., "Optical Fiber Coupled Low Power Micro-PIV Measurement of Flow in Microchambers: Modeling, Fabrication, Testing and Validation," MASc thesis, Concordia University, 2009.)

8.6.6 Micro-Laser-Induced Fluorescence (μLIF) Method for Shape Measurements

The schematics of the three-dimensional shape measurement for microdroplets are shown in Figure 8.50. It consists of an inverted epifluorescent microscope with a mercury lamp, a CCD camera, and an optical filter set.[43,44]

The method involves measuring the dependency of the emission of the fluorescent dye solution on the thickness of a microdroplet. This method can also be used to investigate a gas-liquid two-phase flow.

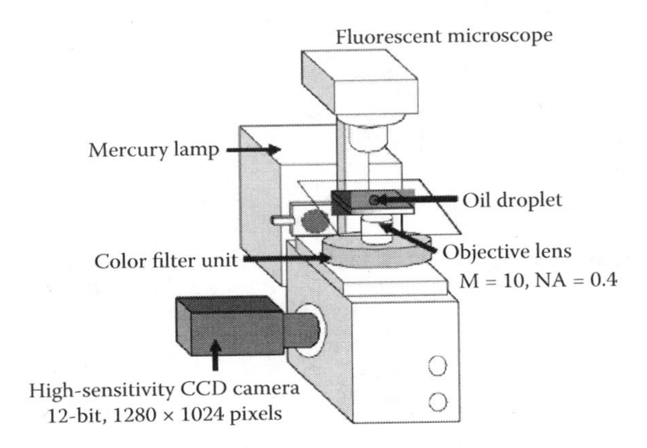

FIGURE 8.50 Schematic of the setup used for droplet-shape measurement. (Reproduced/adapted from Sugii and Horita, *J. Fluid Sci. Technol.*, 3, 956–64, 2008.)

FIGURE 8.51 Numerical simulation (top) and bleached fluorescence (bottom) analysis of dispersion caused by a constant radius turn in an electrophoresis channel. (From Lundy, *Microscopy Today*, 14, 8–14, 2006.)

8.6.7 Caged and Bleached Fluorescence

The *caged fluorescence* method uses a fluorescence dye that initially, when introduced into the flow, is in a nonfluorescent ("caged") state. An activating pulse of ultraviolet light breaks the bonds that cage the fluorescence group, but only in the illuminated region. After activation, a pulse of laser light will excite the fluorescent dye and an imaging device will capture its distribution. Figure 8.51 shows the results obtained for electroosmotically driven flows. As discussed above, in electroosmosis, the dispersion is minimal, while in pressure-driven flow, the velocity is maximum in the center of the channel and zero at the wall.

The figures show that by modifying the inner radius of the turn in folded channels, the unwanted dispersion in the analyte can be reduced.

REFERENCES

1. Squires, T. M., Quake, S. R. 2005. Microfluidics: Fluid physics at the nanoliter scale. *Rev. Mod. Phys.* 77:977–1026.
2. Beebe, D. J., Mensing, G. A., Walker, G. M. 2002. Physics and applications of microfluidics in biology. *Annu. Rev. Biomed. Eng.* 4:261–86.
3. Hansen, C., Quake, S. R.2003. Microfluidics in structural biology: Smaller, faster ... better. *Curr. Opin. Struct. Biol.* 13:538–44.
4. Franke, T. A., Wixforth, A. 2008. Microfluidics for miniaturized laboratories on a chip. *ChemPhysChem* 9:2140–56.
5. Ismagilov, R. F., Stroock, A. D., Kenis, P. J. A., Whitesides, G., Stone, H. A. 2000. Experimental and theoretical scaling laws for transverse diffusive broadening of two-phase laminar flows in microchannels. *Appl. Phys. Lett.* 76:2376.
6. Brody, J. P., Yager, P. 1997. Diffusion-based extraction in a microfabricated device. *Sensors Actuators A* 58:13–18.
7. Hessel, V., Löve, H., Schönfeld, F. 2005. Micromixers—A review on passive and active mixing principles. *Chem. Eng. Sci.* 60:2479–501.
8. Hong C, Choi J, Ahn C. 2001. A novel in-plane passive micromixer using Coanda effect. In *Micro total analysis systems 2001*, ed. Ramsey J, van den Berg A, 31–33. Boston: Kluwer Academic.
9. Bengtsson, M., Laurell, T. 2004. Ultrasonic agitation in microchannels. *Anal. Bioanal. Chem.* 378:1716–21.
10. West, J., Gleeson, J. P., Alderman, J., Collins, J. K., Berney, H. 2003. Structuring laminar flows using annular magnetohydrodynamic actuation. *Sensors Actuators B* 96:190–99.
11. Wang, Y., Zhe, J., Chung, T. F., Dutta, P. 2008. A rapid magnetic particle driven micromixer. *Microfluid Nanofluid* 4:375–89.
12. Song, H., Tice, J. D., Ismagilov, R. F. 2003. A microfluidic system for controlling reaction networks in time. *Angew. Chem. Int. Ed.* 42:768–72.
13. Bayraktar, T., Pidugu, S. B. 2006. Characterization of liquid flows in microfluidic systems. *Int. J. Heat Mass Transfer* 49:815–24.
14. Sia, S. K., Whitesides, G. M. 2003. Microfluidic devices fabricated in poly(dimethylsiloxane) for biological studies. *Electrophoresis* 24:3563–76.
15. Tandon, V., Bhagavatula, S. K., Nelson, W. C., Kirby, B. J. 2008. Zeta potential and electroosmotic mobility in microfluidic devices fabricated from hydrophobic polymers. 1. The origin of charge. *Electrophoresis* 29:1092–101.
16. Chakraborty, S., Das, S. 2008. Streaming field-induced convective transport and its influence on the electroviscous effects in narrow fluidic confinement beyond the Debye-Huckel limit. *Phys. Rev. E* 77:037303-1-4.
17. Erickson, D., Dongqing, L. 2001. Streaming potential and streaming current methods for characterizing heterogeneous solid suraces. *J. Coll. Interface Sci.* 237:283–89.
18. Garstecki, P., Gañán-Calvo, A. M., Whitesides, G. M. 2005. Formation of bubbles and droplets in microfluidic systems. *Bull. Polish Acad. Sci. Technical Sci.* 53:361–72.
19. Abate, A. R., Poitzsch, A., Hwang, Y., Lee, J., Czerwinska, J., Weitz, D. A. 2009. Impact of inlet channel geometry on microfluidic drop formation. *Phys. Rev. E* 80:026310-5.
20. Tan, Y.-C., Cristini, V., Lee, A. P. 2006. Monodispersed microfluidic droplet generation by shear focusing microfluidic device. *Sensors Actuators B* 114:350–56.

21. Basu, A. S., Gianchandani, Y. B. 2007. Shaping high-speed Marangoni flow in liquid films by microscale perturbations in surface temperature. *Appl. Phys. Lett.* 90:034102.1–2.3.
22. Anna, S. L., Bontoux, N., Stone, H. A. 2003. Formation of dispersions using "flow focusing" in microchannels. *Appl. Phys. Lett.* 82:364–66.
23. Fair, R. B., Khlystov, A., Tailor, T., Ivanov, V., Evans, R. D., Griffin, P. B., Srinivasan, V., Pamula, V. K., Zhu, J., Pollack, M. G. 2007. Chemical and biological applications of digital-microfluidic devices. *IEEE Design & Test of Computers*, January–February, 10–24.
24. Charabarty, K., Zeng, J. 2006. Design automation for microfluidics-based biochips. *ACM J. Emerg. Technol. Comput. Syst.* 1:186–223.
25. Moon, I., Kim, J. 2006. Using EWOD (electrowetting-on-dielectric) actuation in a microconveyor system. *Sensors Actuators A* 130–31:537–44.
26. Sammarco, T. S., Burns, M. 1999. Thermocapillary pumping of discrete drops in microfabricated analysis devices. *AIChE J.* 45:350–66.
27. Kozicki, M. N., Maroufkhani, P., Mitkova, M. 2003. Flow regulation in microchannels via electrical alteration of surface properties. *Superlattices Microstruct.* 34:467–73.
28. Przekwas, A., Makhijani, V. B. 2001. Mixed-dimensionality, multi-physics simulation tools for design analysis of microfluidic devices and integration systems. In *Technical Proceedings of the 2001 International Conference on Modeling and Simulation of Microsystems*, www.cr.org.
29. http://www.comsol/showroom/gallery/279/.
30. Use of Comsol multiphysics in undergraduate research projects to solve real-life problems. http://faculty.washington.edu/finlayso/Finlayson_paper_611d.pdf.
31. Ozhikandathil, J., Packirisamy, M., Stiharu, I. 2010. Modeling and analysis of low voltage electro-osmotic micropump. Paper presented at Proceedings of ASME 2010 3rd Joint US-European Fluids Engineering Summer Meeting and 8th International Conference on Nanochannels, Microchannel and Minichannels, FEDSM ICNMM2010-31213, Montreal, Canada.
32. Chandrasekaran, A., Packirisamy, M. 2006. Absorption detection of enzymatic reaction using optical microfluidics based intermittent flow microreactor system. *IEE Proc. Nanotechnol.* 153:137–43.
33. Chandrasekaran, A., Packirisamy, M. 2008. Enhanced bio-molecular interactions through recirculating microflows. *IET Nanobiotechnol.* 2:39–46.
34. SadAbadi, H., Stiharu, I., Packirisamy, M., Wuthrich, R. 2009. A parametric study of interdigital electrodes for achieving high droplet speed in electrowetting. In *Microsystems and Nanoelectronics Research Conference, 2009*, MNRC 2nd, Ottawa, Canada, pp. 76–79.
35. SadAbadi, H., Packirisamy, M., Dolatabadi, A., Wuthrich, R. 2010. Effects of electrode switching sequence on EWOD droplet manipulation: A simulation study. Paper presented at ASME International Conference on Nanochannels, Microchannels, and Minichannels, 2010, ICNMM 8th, Montreal, Canada.
36. Santiago, J. G., Wereley, S. T., Meinhart, C. D., Beebe, D. J., Adrian, R. J. 1998. A particle image velocimetry system for microfluidics. *Experiments in Fluids* 25: 316-319.
37. Acharya, A. 2007. Modeling, fabrication and testing of OLED and LED fluorescence integrated polymer microfluidic chips for biosensing applications. MASc thesis, Concordia University.
38. Brody, J. P., Yager, P., Goldstein, R., Austin, R. H. 1996. Biotechnology at low Reynolds numbers. *Biophys. J.* 71:3430–41.
39. Taylor, J. A., Yeung, E. S. 1993. Imaging of hydrodynamic and electrokinetic flow profiles in capillaries. *Anal. Chem.* 65:2928–32.
40. Mielnik, M. M., Saetran. 2004. Micro particle image velocimetry—An overview. *Turbulence* 10:83–90.
41. Lindken, R., Rossi, M., Grosse, S., Westerweel, J. 2009. Micro-particle velocimetry (μ-PIV): Recent developments, applications, and guidelines. *Lab Chip* 9:2551–67.
42. Fang, Q. 2009. Optical fiber coupled low power micro-PIV measurement of flow in microchambers: Modeling, fabrication, testing and validation. MASc thesis, Concordia University.
43. Leal, N., Semião, V. 2008. Effect of the wall roughness on fluid flow inside a microchannel. Paper presented at 14th International Symposium on Applications of Laser Techniques to Fluid Mechanics, Lisbon, Portugal, July 7–10, 2008.
44. Sugii, Y., Horita, R. 2008. Three-dimensional shape measurement method for micro droplet utilizing micro LIF technique. *J. Fluid Sci. Technol.* 3:956–64.
45. Lundy, T., 2006. Advanced confocal microscopy—An essential technique for microfluidic development. *Microscopy Today* 14:8–14.

REVIEW QUESTIONS

1. Which are the disciplines that share an interest in microfluidics?
2. What are lab-on-a-chip and micro total analysis systems (μTAS) and how are they related to microfluidics?
3. What are the most important advantages of microfluidics?
4. What is the Reynolds number (Re) and what do small Reynolds number indicate?
5. How can the diffusive mixing between two laminar flows be enhanced?
6. Define the Peclet number (Pe).
7. What are the two common methods by which fluid actuation can be achieved?
8. Explain the formation of the electrical double layer (EDL).
9. What is the difference between electrowetting and electrowetting on dielectric?
10. Give an example of a droplet generator.
11. Explain how droplet actuation can be accomplished using Marangoni flows.
12. How can droplets be actuated in electrowetting-based systems?
13. Explain how the drop motion results by using thermocapillary pumping.
14. Describe briefly the finite element method generally used for solving microfluidics problems.
15. Give two examples of problems related to biological applications that can be solved through simulation.
16. Explain how micro particle image velocimetry (μPIV) works.

9 BioMEMS
Life Science Applications

BioMEMS devices have become an enabling technology for a variety of life science applications due to their size relative to biospecies, increased sensitivity, small volume requirement, high throughput, reduced test time, performance tenability, batch fabrication feasibility, low cost, portability, disposability, microelectronics and microphotonics integration possibility, low power consumption, automation capability for remote and hazardous environments, reduced manual labor, close proximity to biointeractions, etc. In view of their immense potential, the applications have expanded from biosensing to drug delivery and single-cell manipulation.

The portability and disposability aspects of BioMEMS devices have contributed tremendously to the development of diagnostic biochips for *point-of-care testing* (POCT) and *point-of-need* (PON) applications. The *lab-on-a-chip* (LOC), *micro total analysis systems* (μTAS), and *biochips* that are used for POCT and PON applications have great and immediate potential for developing countries that have inadequate laboratory and hospital facilities, and high medical demand. The application areas of BioMEMS expand into bioterrorism, defense, diagnosis, drug discovery and delivery, pharmaceuticals, nutrition, public health, agriculture, and other life science areas.

BioMEMS devices have recently their established relevance to biosensing, diagnosis, drug delivery, and therapeutic applications. The demand from these applications has expanded from protein and cellular to tissue levels, harnessing the potential of BioMEMS in terms of size and sensitivity. The BioMEMS devices could be developed based on microfluidics, micromechanical, microelectronics, and microphotonics elements, or combinations of these, depending on the application methods and requirements. The recent developments in the area of transportation at the microlevel, such as electrophoresis, electroosmosis, electrowetting, droplet manipulation, etc., have strengthened the application and maturity of BioMEMS devices for life science applications.

Applications at the molecular level include microarrays for diagnosis, microreactors, micromixers, enzymatic assays, etc. The recent exciting application of BioMEMS at the cellular level incorporates single-cell manipulation, research on stem cells, cell sorting, cell diagnosis and characterization, cell culture, cellular interactions with the environment, etc. As the behavior of cells could represent the behavior of macrobiospecies, BioMEMS devices for cellular studies have vast potential in terms of diagnosis, therapeutics, preventive health, and individualized medicine.

Other important applications involve the gene chips for DNA identification and sequencing, DNA coding and storage, gene isolation, and gene transformation for many clinical applications. Similarly, BioMEMS devices are also applied for the development of proteomic chips, immunoassays, separation of proteins and amino acids, etc. The BioMEMS devices are also explored for *in vitro* and *in vivo* applications. The recent success in chemical synthesis of nanoparticles under the BioMEMS environment opens up a vast potential for diagnosis, drug delivery, and pathogens detection.

In the market, there are books available, dedicated to the life science applications of BioMEMS. In addition, the literature available for the life science applications is vast, and the reader is advised to read those books if more information is needed. Here, only a few typical examples are shown to demonstrate the capability of BioMEMS for life science applications.

9.1 INTRODUCTION TO MICROARRAYS

A microarray is a collection of spatially addressable probes, immobilized as spots on a surface or substrate. The increased throughput in microarray is achieved due to the small spot size, typically in the range of 150 to 200 μm diameter, which facilitates a large number of spots per microarray.

Thousands of probes printed on the microarray surface can be used in solution to detect a specific target in one microarray experiment. In addition to the high throughput achieved, microarray assays are highly sensitive and require extremely small amounts of samples. The increase in sensitivity in microarray-based methods is due to the miniature format, which leads to an increase in the signal density (signal intensity/area). Compared to the microtiter plate format employed in enzyme-linked immunosorbent assay (ELISA), a typical microarray spot is more than twenty-five times smaller. This improves the signal density and enhances the intensity. The amount of sample required to saturate a microarray spot also decreases in proportion to its surface area, and hence, typically a few nanograms is sufficient for several microarray experiments. When compared to ELISA, an equal or greater sensitivity can be achieved with microarrays using only a fraction (down to 1/1,000th) of the sample size required for ELISA. The advantages of high sensitivity and high throughput make microarray a potentially powerful technology for biodetection and diagnosis.

As shown in Figure 9.1, biomolecule arrays are divided into three categories: antibody arrays, protein/peptide arrays, and DNA arrays. DNA arrays are used to characterize cDNA libraries by DNA hybridization with single DNA probes, and to determine gene expression patterns by hybridization with complex hybridization probes. Multiplexed genotyping is achieved by primer extension of arrayed oligos using reverse transcription with complex RNA mixtures as the transcription templates. The main applications of arrays of antibodies, RNA aptamers, or plastibodies with known binding specificities will be used for the detection and quantification of biomolecules, such as proteins, peptides, or chemical compounds from complex mixtures such as clinical samples. In contrast, arrays consisting of recombinant proteins or synthetic peptides are mainly used to identify and characterize interactions or biological activities of proteins with various kinds of biomolecules. An example could be the screening of an arrayed expression library with an antibody of unknown binding specificity.

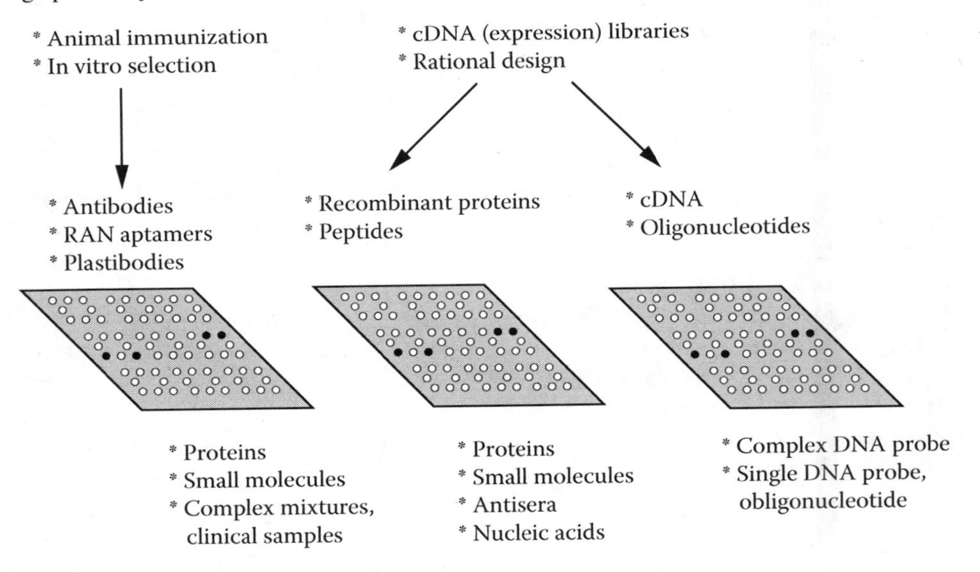

FIGURE 9.1 Types of high-density biomolecule arrays. (Adapted from Büssow et al., *Am. J. Pharmacogenomics*, 1, 1–6, 2003.)

9.2 MICROARRAYS BASED ON DNA

9.2.1 INTRODUCTION TO DNA CHIPS

DNA chips are the miniaturized high-density arrays of molecular probes on which the targets (unknown and to be detected DNA single strands) could be hybridized. The structure and composition of nucleic acids were discussed in Chapter 3. The DNA array is an ordered collection of microspots, each spot containing a single defined species of a nucleic acid or a fragment of nucleic acid.

Nucleic acids, DNA, and RNA are molecules holding the genetic information in all life cells, and also are the fundamental components of genes, monitoring the functioning of living organisms. DNA chips are considered new and powerful tools that could combine the integration ability of microelectronic devices for simultaneous analysis of thousands of nucleic acids. In the coming years, DNA microarray technology will substantially support molecular biological research. The main challenge of the postgenomic era is to use genomic structural information to display and analyze biological processes on a genome-wide scale, and to assign the gene function. The availability of the complete human genome and of several other organisms allows the application of microarray technologies to several model organisms. Microarray technology has evolved from the old method of Southern blotting, where fragmented DNA is attached to a substrate and then probed with a known gene or fragment. Early gene arrays were made by spotting cDNA onto filter paper with a pin-spotting device. The use of microarrays for gene expression profiling was first reported in 1995.[2] The nucleic acid microarray may use short oligonucleotides (15 to 25 nt), long oligonucleotides (50 to 120 nt), or polymerase chain reaction (PCR)–amplified cDNAs as array elements (the unit nt indicates nucleotides). The array elements are derived from individual genes located at defined positions on a solid support, enabling the analysis of thousand of genes in parallel, by specific hybridization.

9.2.2 PRINCIPLES OF DNA MICROARRAY: THE DESIGN, MANUFACTURING, AND DATA HANDLING

The core principle behind microarrays is the hybridization between two DNA strands. *Hybridization* is the property of complementary nucleic acid sequences to specifically pair with each other by forming hydrogen bonds between the complementary nucleotide bases. A high number of complementary base pairs in a nucleotide sequence gives rise to stronger noncovalent bonding between the two strands. A glass slide is spotted or “arrayed” with DNA fragments, oligonucleotides that represent specific gene coding regions or RNA. Purified RNA is then fluorescently or radioactively labeled and hybridized to the slide. In some cases, hybridization is done simultaneously with a reference RNA to facilitate the comparison of data across multiple experiments.

There are three kinds of chips. In the first two methods, both cDNA and oligonucleotide arrays can be used to analyze the patterns of gene expression, even though the methods are different. The long oligonucleotides give strong hybridization signals and good specificity. For the cDNA arrays, the nucleic acid fragments are spotted using robots. The cDNAs used for spotting are usually derived by PCR amplification of cDNA libraries. For oligonucleotide arrays, *in situ* synthesis produces short up to 25 nucleotides by photolithography, or lengths up to 60 nt (nucleotides) by ink-jet technology. A third type of array is made by spotting presynthesized oligonucleotides on glass slides. The elements of one array are each around 100 μm in diameter. Data analysis bioinformatics software and databases are essential for data acquisition and tracking of data points. Known probes are located on the solid surface at a predefined position in the array. The device is kept in contact with a solution containing the target to be detected. When probes and targets are complementary, hybridization occurs on the solid support, indicating biochemical recognition. The detection on the matrix is indicated at areas where hybridization is complete. The kinetics of DNA hybridization is strongly influenced by the fact that DNA probes are immobilized on a solid support. The main steps of a microarray experiment are probe generation and microarray design, target preparation and hybridization, data generation, and analysis, as schematically shown in Figures 9.2 and 9.3.

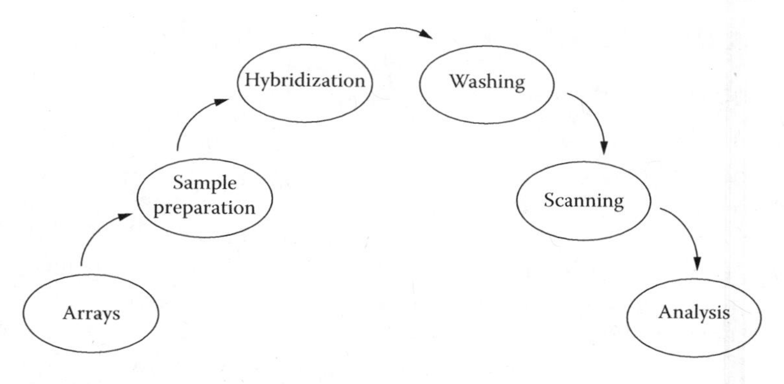

FIGURE 9.2 Steps of a microarray experiment.

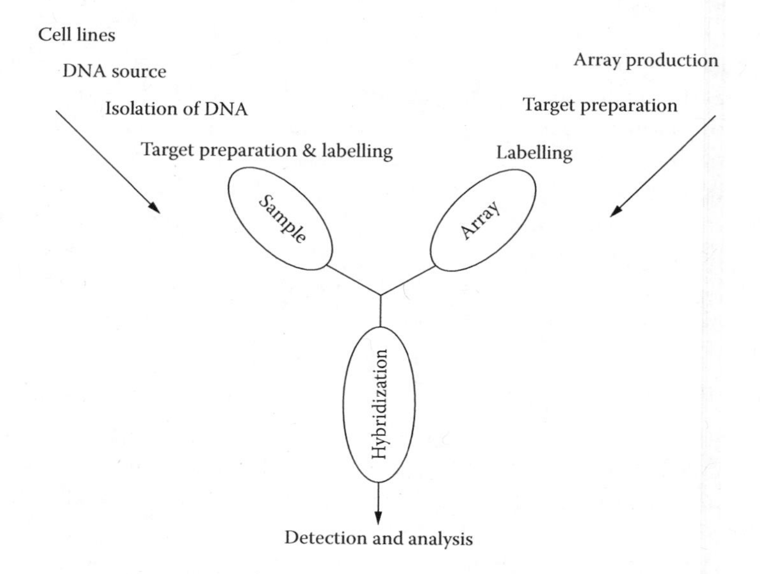

FIGURE 9.3 Process sequence used for RNA expression analysis using gene chip microarrays.

High-throughput DNA preparation, done by robotic systems, consists of tens of thousands of PCRs. Purified PCR products representing specific genes are spotted on a matrix by a robot that deposits nanoliters of PCR product onto the matrix in a sequence. Microscope slides have the advantage of two-color fluorescence labeling with low background fluorescence.

To perform a microarray experiment, as shown in Figure 9.4, RNA from the experimental sample(s) is first isolated and purified. The purified RNA is then reverse transcribed in the presence of labeled nucleotides. Reverse transcription (RT reaction) is a process in which single-stranded RNA is reverse transcribed into complementary DNA.

In the case of custom-made arrays, fluorophores like Cy3 and Cy5 are typically used. The two-color hybridization strategy permits simultaneous analysis of two samples on a single array, as shown in Figure 9.5. For high-density commercial arrays, nonfluorescent biotin, labeled by staining with a fluorescent streptavidin conjugate, is used. The labeled probe is fragmented and hybridized to the array, and then the array is washed and stained. The signal intensity, proportional to the amount of bound probe, is measured by scanning with a confocal laser. For each spot on the array, the background signal is subtracted from the average signal intensity to generate a quantitative image. The relative abundance of each transcript can be determined, as the sequence of each cDNA or oligonucleotide on the grid is known. Data are normalized across experiments by calculating the variance of all genes in the sample.

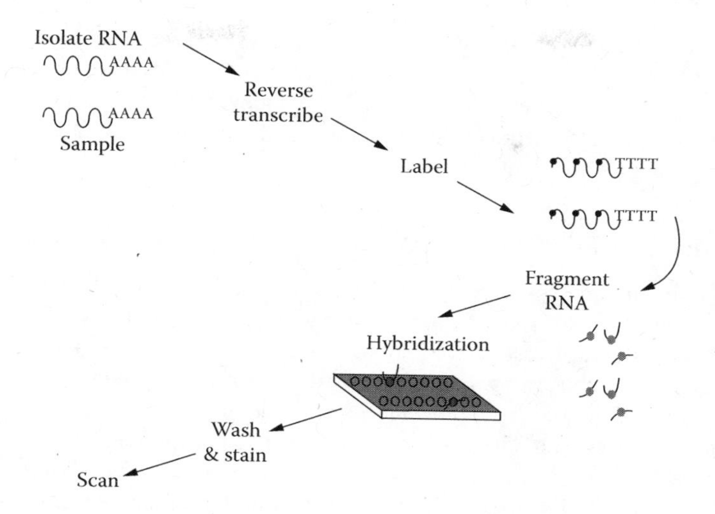

FIGURE 9.4 Microarray experimentation.

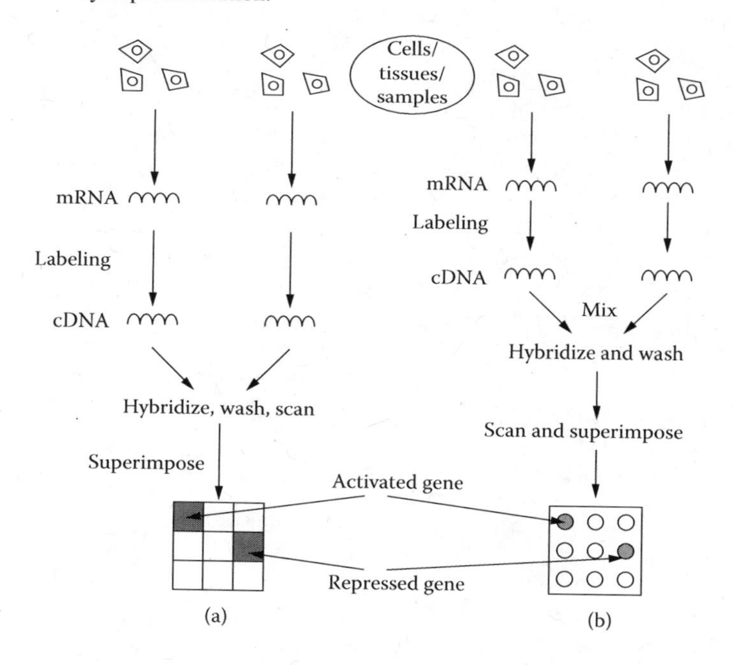

FIGURE 9.5 Gene expression analyses by microarray. (a) One-color expression analysis uses a single fluorescent label and two arrays to generate expression profiles for two cell or tissue samples (test and reference samples). Activated and repressed genes are obtained by superimposing images obtained by the two arrays. (b) Two-color expression analysis uses two different fluorescent labels and a single array to generate expression profiles for the test and reference samples. (Adapted from Lettieri, *Environ. Health Perspect.*, 114, 4–9, 2006.)

Activated and repressed genes are obtained by superimposing images generated in different channels on a single array. Since an array can contain tens of thousands of probes, a microarray experiment can accomplish many genetic tests in parallel. In standard microarrays, the probes are attached to a gold surface through covalent bonding to a chemical matrix formed with epoxy silane, aminosilane, lysine, or polyacrylamide. The solid surface can be a glass or a silicon chip, in which case the microarray is commonly known as a *gene chip*. After the nonspecific bonding sequences are washed off, only strongly paired strands will remain hybridized. Fluorescently labeled target sequences that bind to a probe sequence generate a signal that depends on the strength of the hybridization, determined by the number of paired bases. The identity of the features is known by their position. Microarrays use relative quantification in which the intensity of a feature is compared with a reference feature under a different condition.

9.2.3 Applications of DNA Microarrays

The field of *genomics* refers to the comprehensive study of genes and their function. Arrays are used to evaluate the extent of a *gene expression* under various conditions. This technology has been successfully applied to investigate simultaneous expression of many thousands of genes in a highly parallel and comprehensive manner, and to the detection of mutations or polymorphisms. The expression levels of thousands of genes are simultaneously monitored to study the effects of certain treatments, diseases, and developmental stages on gene expression.

The pattern of gene expression produced, known as *expression profile*, shows the subset of gene transcripts expressed in a cell or tissue. Combining PCR and microarray technology, gene expression profiling can be used to identify genes whose expression is changed in response to pathogens, by comparing gene expression between infected and uninfected cells or tissues. Among the applications of microarray technology, the study of drug-induced hepatotoxicity deals with changes in gene expression associated with toxic exposure.

Microarray studies and other genomic techniques are also stimulating for the discovery of new targets for the treatment of disease, immunotherapeutics, and gene therapy. For mutation analysis, the genes may differ from each other even by a single nucleotide base. A single base difference between two sequences is known as *single nucleotide polymorphism* (SNP), and identification of it is known as SNP detection.

The study of the relationship between the therapeutic response to drugs and the genetic profiles of patients is called pharmacogenomics. Comparing genes from a diseased and a normal cell will help not only the identification of the biochemical constitution of the proteins synthesized by the diseased genes, but also drug discovery. DNA microarrays are also used for the study of the impact of toxins on the cells and changes in the genetic profiles of cells exposed to toxicants. Figure 9.6 summarizes the applications of microarray technologies.

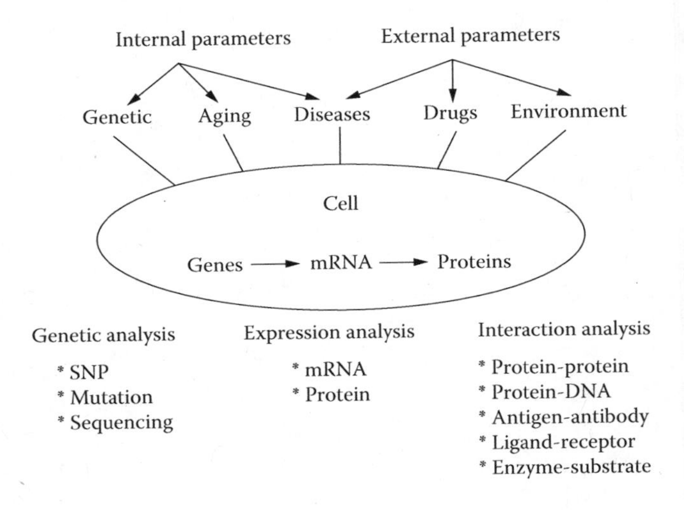

FIGURE 9.6 Microarrays for genomics and proteomics. The physiological state of a cell is influenced by external and internal parameters. Microarray technology can be applied to monitor intracellular gene and protein expression mechanisms. (Adapted from Templin et al., *Trends Biotechnol.*, 20, 160–66, 2002.)

9.3 POLYMERASE CHAIN REACTION (PCR)

9.3.1 Introduction

DNA amplification is the production of multiple copies of a sequence of DNA by using a process called *polymerase chain reaction* (PCR). PCR is the *in vitro* multiplication of defined strands of DNA with the use of enzymatic polymerase molecules. The technique allows a small amount of DNA to be amplified exponentially in order to produce enough DNA to be adequately tested. Since its first introduction in 1985, PCR-based techniques have become the most important part of DNA diagnostic laboratories because nearly all molecular biological operations need PCR to enrich the nucleic acid samples. At present, PCR is a widely utilized scientific tool in DNA sequencing, genotyping, and new drug discovery. Due to its high specificity, PCR will play a major role in the future in the success of individualized medicine and *point-of-care* (POC) diagnostics. The discovery of this technology earned its inventor, K. B. Mullis, a Nobel Prize in Chemistry in 1993. While qualitative PCR has become a well-established technology, the quantification of specific target DNA sequences has been a difficult task. The concentration of a DNA species is nearly doubled in every PCR process, involving stepping through three different temperatures. In this way, the DNA concentration can be multiplied more than a million-fold by twenty to thirty cycles of temperature, and the DNA/RNA inside the PCR mixture will be rich enough for further analysis. Conventional PCR devices are suitable for amplifying DNA, but they are sizable devices and require long cycling times. Recently, *miniaturized PCR* devices have attracted great interest because of their advantages over conventional PCR devices, such as portability, low cost, fast thermal cycling speed, reduced reagents/sample consumption, low power consumption, and low cost of fabrication. Portability is of great advantage for environmental screening, as well as in the medical sector. In addition, smaller sample volumes have less heat capacity, allowing rapid temperature change. PCR microfluidic technologies have facilitated DNA amplification with improved performances.

9.3.2 PCR Process

PCR is an *in vitro* enzymatic amplification process of nucleic acid sequences based on the choice of specific primers. The basic principle in PCR is the repetition of three temperature-specific steps: *denaturation* (94 to 96°C), *annealing* (50 to 65°C), and *extension* (68 to 72°C). In the first step, at elevated temperatures of about 90 to 96°C, the original double-stranded DNA "melts" into two single-stranded DNA molecules. In a second step, after denaturation, the temperature is lowered to a degree specific for the primer (oligonucleotide), which generally lies between 55 and 70°C. The primer then takes its place at the specific DNA recognition sequence (according to their complementarity) of the single-stranded DNA template sequence. Then, the temperature is increased to 72°C, the optimal temperature for the polymerase enzyme to copy the DNA strands. The polymerase molecules attach themselves to the single-strand DNA and form the complementary DNA strand (*primer elongation*), leading to an identical copy of the original double-stranded DNA. These two molecules then undergo the same cycle again and again, which leads to an exponential growth of the number of identical DNA strands. Following the cycles of temperature zones, theoretically, a DNA segment can get amplified 2^n times after n cycles of PCR. The three steps of the amplification process are schematically shown in Figures 9.7 and 9.8.

There are several methods to perform PCR. In the stationary chamber-based PCR amplification format, the temperature of the reaction chamber is cycled between different temperatures while the PCR solution is kept stationary. They can be classified as single-chamber stationary PCR chip, multichamber stationary PCR chip, and virtual chamber PCR chip (Figure 9.9). The single-chamber chip can perform very well in terms of fluidic and thermal controls, and it offers beneficial properties such as reduced thermal and fluidic crosstalks between chambers. However, it cannot realize high throughput and cannot readily be used for special purposes, such as single-cell gene expression

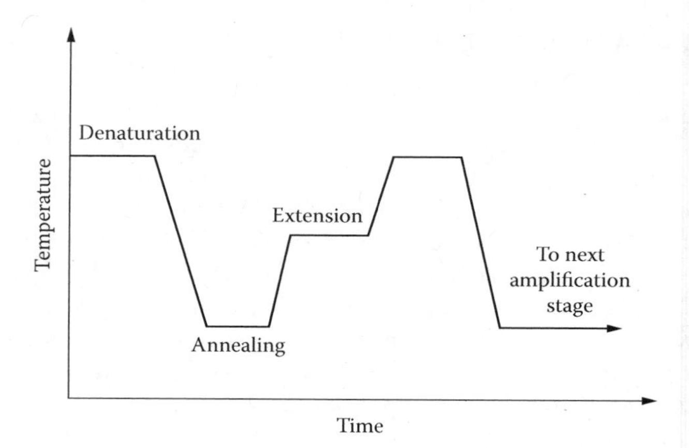

FIGURE 9.7 The DNA amplification process of PCR.

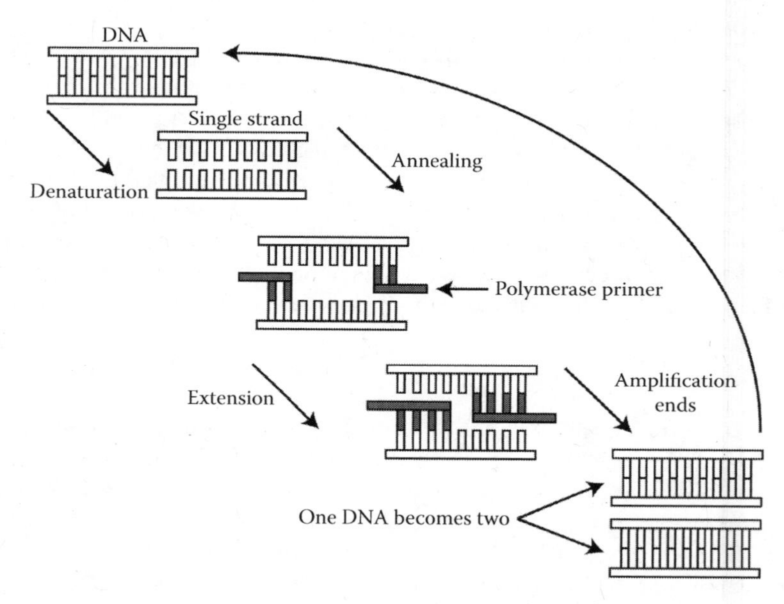

FIGURE 9.8 Schematic of the PCR process. (Adapted from Gärtner et al., *Proc. SPIE*, 6465, 646502, 2007.)

analysis. In the case of virtual chamber PCR, samples are entrapped in oil drops that act as virtual chambers. All these PCR chips are integrated with thin-film heaters through microfabrication.

Another approach, called continuous flow PCR, is realized by transporting the sample in a microchannel through spatially fixed temperature zones. In this approach, only the moving liquid has to be heated and cooled along the flow path in a PCR chip, as shown in Figures 9.10 and 9.11. This format avoids temperature cycling of the entire device and leads to more rapid heat transfer and faster throughput than stationary PCR microfluidic chambers. This technique allows for a continuous flow of the process instead of the batch processing of conventional thermocyclers.

A PCR chip can be integrated with a microheater, microthermal sensor, and cooling channels for PCR amplification. The PCR microchip shown in Figure 9.11 has three layers: the top microfluidic polydimethylsiloxane (PDMS) layer with a chamber of 20 μl volume, the middle glass support layerl and the bottom PDMS layer integrated with heater, temperature sensor, and cooling microchannels. One can implement many PCR cycles through appropriate microfluidic design. This microchip can

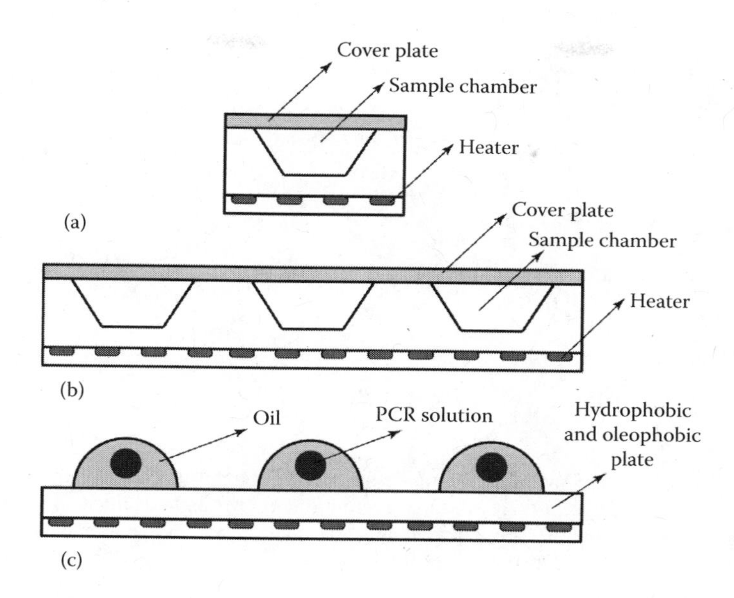

FIGURE 9.9 Stationary chamber-based PCR chip. (a) Single-chamber PCR. (b) Multichamber PCR. (c) Virtual reaction chamber (VRC)–PCR. The PCR sample is introduced into the single, multiple, or virtual chamber. The chip is then heated and cooled to undergo thermocycling. (Adapted from Zhang and Xing, *Nucleic Acids Res.*, 2007, 1–15, 2007.)

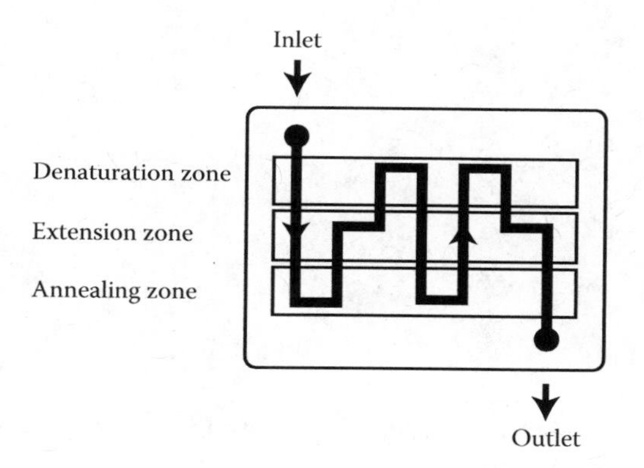

FIGURE 9.10 Schematic of a continuous flow PCR chip. Many cycles of PCR can be incorporated through microfluidic design.

also be integrated with other functions to realize an inexpensive and disposable LOC capable of carrying out many analytical steps.

The classical lab equipment shown in Figure 9.12 in which PCR is carried out is called a *thermocycler*. The DNA sample, along with primer and polymerase molecules in buffer solution, is temperature cycled in a small plastic tube through denaturation, annealing, and extension phases, typically between fifteen and twenty-five times.

Different arrangements of continuous flow PCR chip using serpentine and spiral channels passing through three thermal copper blocks are shown in Figure 9.13.

Various heating methods are used for performing rapid PCR in volumes ranging from 1 pl to 50 μl. They include resistive heating, microwave heating, infrared-mediated thermal cycling, and Joule heating. The high surface-area-to-volume ratio in PCR chips facilitates rapid heat transfer and temperature change due to smaller sample volumes, less heat capacity, and high heat transfer area.

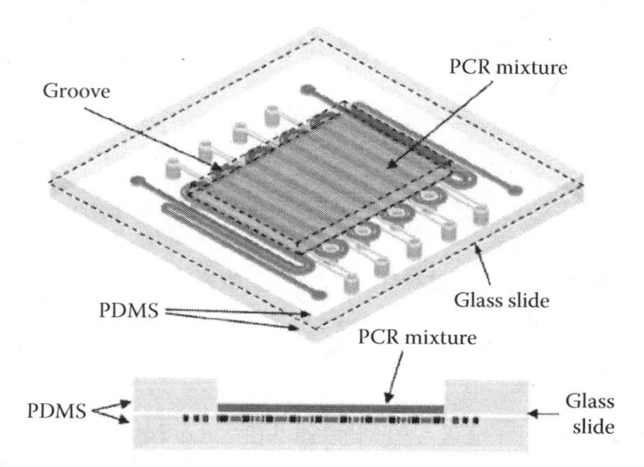

FIGURE 9.11 (a) A schematic view of PCR chip composed of three layers. The upper layer is PDMS with a 13 × 9 mm cavity filled with 20 μl of PCR mixture. The middle layer is a glass slide of 140 μm thickness. The lower PDMS layer is integrated with a microheater, microthermal sensor, and cooling channel. (Adapted from Wu et al., *Biomicrofluidics*, 3(1), 012005-1–7, 2009.)

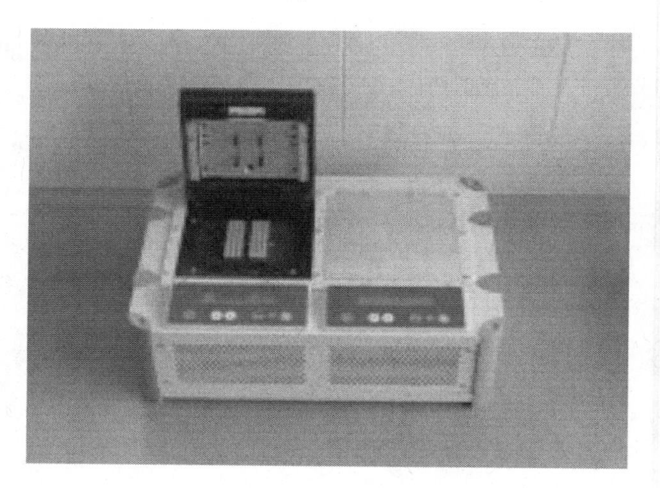

FIGURE 9.12 Classical thermocycler unit equipped with two PCR chips. (Reproduced from Gärtner et al., *Proc. SPIE*, 6465, 646502, 2007. With permission.)

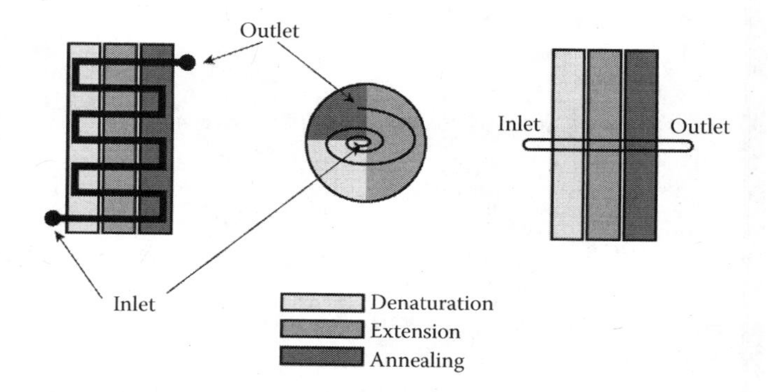

FIGURE 9.13 Different arrangements of continuous flow PCR. (A) Serpentine channel continuous flow PCR. (B) Spiral channel continuous flow PCR. (The sample is pumped from the inlet to the outlet.) (C) The straight channel oscillatory flow PCR. (Adapted from Zhang and Xing, *Nucleic Acids Res.*, 2007, 1–15, 2007.)

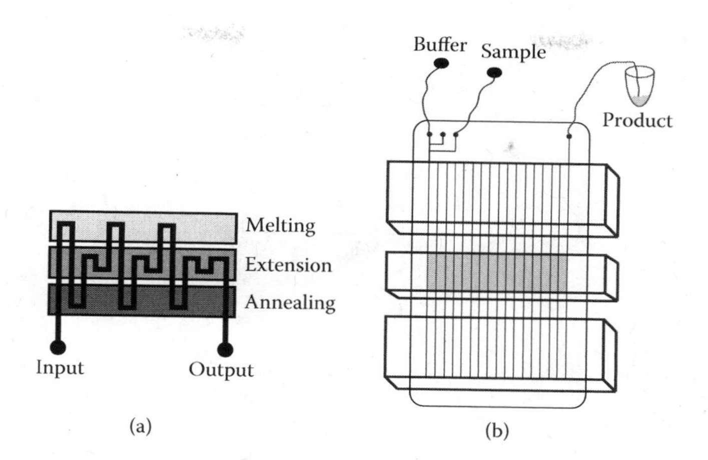

FIGURE 9.14 Schematic diagram of continuous flow PCR chip. (a) layout of three temperature zones for PCR thermal cycles that are maintained at 95, 77, and 60°C. (b) Device layout: Three inlets on the left and one outlet on the right. Two inlets are used to inject sample and buffer. (Adapted from Zhang and Ozdemir, *Anal. Chim. Acta*, 638, 115–25, 2009.)

A continuous flow PCR chip with a pressure-driven flow of sample through a single channel etched into a glass chip is shown in Figure 9.14.

Fabrication of PCR microchambers or microchannels is usually done by photolithography and chemical etching on silicon and glass substrates. The microchambers and microchannels in PCR chips are mostly fabricated from silicon or glass as substrate materials. Silicon has superior thermal conductivity, thereby allowing fast temperature response. In addition, heaters and sensors also can be patterned on silicon through microfabrication. However, silicon is not transparent in the visible range of wavelengths, limiting the real-time optical detection capability. Glass, on the other hand, possesses some beneficial characteristics, such as well-defined surface chemistries and superior optical transparency. However, due to the high cost of fabrication, the PCR microfluidics made from silicon or glass material are not disposable. There is no single substrate material that can satisfy all requirements, such as ease of fabrication, optical transparency, disposability, and biocompatibility. A new class of polymer materials has proved to be superior to silicon or glass as substrates for PCR microfluidic substrates. Many polymer-based PCR microfluidic devices have been synthesized using a variety of materials, such as polydimethylsiloxane, polymethylmethacrylate (PMMA), polycarbonate, polyethyleneterephathalate (PET), photoresists, SU8, and others. In order to take full advantage of the properties of silicon, glass, and polymers, researchers have developed hybrid platforms for PCR chips using combinations such as silicon-glass, polymer-silicon, and polymer-glass.

Droplet-based PCR chips that combine the advantages of both microchamber PCR chip and continuous flow PCR chip have been recently developed. The schematic structure of a micro-oscillating-flow PCR chip, together with a stationary chamber-based PCR chip, is shown in Figure 9.15. A completely different approach to handle small-scale sample using a hydrophobic or oleophobic surface to provide virtual fluid confinement is shown in Figure 9.16. The concept of virtual reaction chamber (VRC) shown in Figure 9.9c, where each PCR sample droplet is covered with a drop of mineral oil, was applied for the first time in 2007.

Droplets as water-in-oil emulsions are indeed one of the most attractive candidates for microfluidic reactors in the application of lab-on-a-chip methods for high-throughput assays because they are stable and can be easily generated and spatially arrayed on a solid substrate, without crosstalk between neighboring droplets. Low-power (~30 mW) infrared laser radiation is used as an optical heating source for high-speed real-time polymerase chain reaction (PCR) amplification of DNA in nanoliter droplets dispersed in an oil phase. The infrared laser heats only the droplet, not the oil or

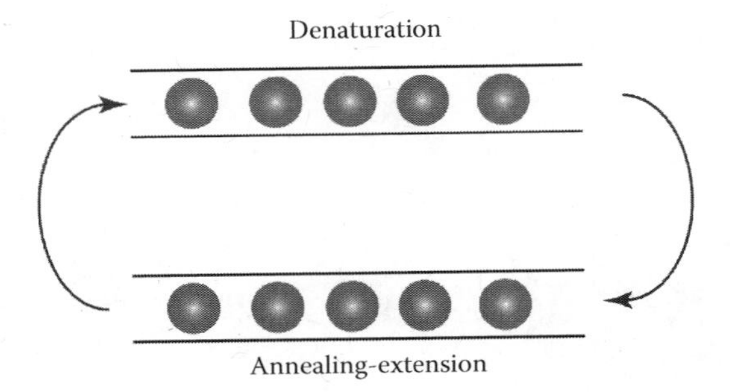

FIGURE 9.15 Illustration of microfluidic PCR in droplets: The PCR mixture is contained in discrete droplets moving through different temperature zones in the microchannels for DNA denaturation, annealing, and extension.

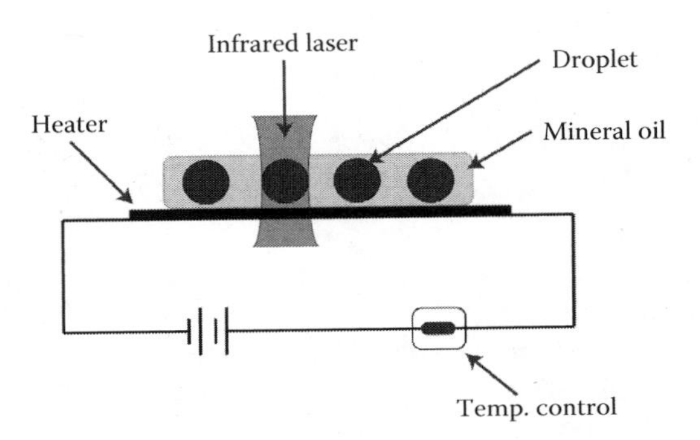

FIGURE 9.16 Illustration of laser-irradiated droplets dispersed in mineral oil for PCR. (Adapted from Kim et al., *Opt. Express*, 17, 218–27, 2009.)

plastic substrate, providing fast heating with a capacity to complete forty cycles of PCR in 370 s. PCR can be performed in one droplet without affecting the neighboring droplets.

Typical amplification curves obtained using the droplet PCR technique for different concentrations are shown in Figure 9.17.

The microfluidic PCR in droplets has many advantages compared to conventional PCR with single-phase flows: elimination of carryover contamination between successive samples, absence of adsorption at the surface, dilution of samples due to diffusion in flows, prevention of the synthesis of artifacts, rapid thermal response for fast PCR process, low consumption of reagents, and easy integration as a function into a μTAS. In addition, individual droplets can contain samples, so that microfluidic PCR in droplets is particularly suitable for single-cell and single-molecule amplification.

The throughput can be increased by reducing further the reaction volume and increasing the process capability. In limiting dilution PCR, a high-throughput microfluidic chip that encapsulates PCR reagents in millions of picoliter droplets in a continuous oil flow has been designed.[9] The oil stream conducts the droplets through alternating denaturation and annealing zones, resulting in rapid and efficient PCR amplification. Inclusion of fluorescent probes in the PCR mix enables the amplification process to be monitored within individual droplets, at specific locations within the microfluidic chip. In addition to providing high-throughput PCR, this platform generates droplets that are uniform in diameter, allowing for constant reaction rates across the droplets and highly reproducible amplification. *Limiting dilution PCR* has become a useful technique for determining the total number of initial DNA targets present in a complex mixture, such as cells isolated from a tumor. By partitioning microliter scale samples into discrete, picoliter droplets, the LOC increases the number of reactions that can be performed from thousands

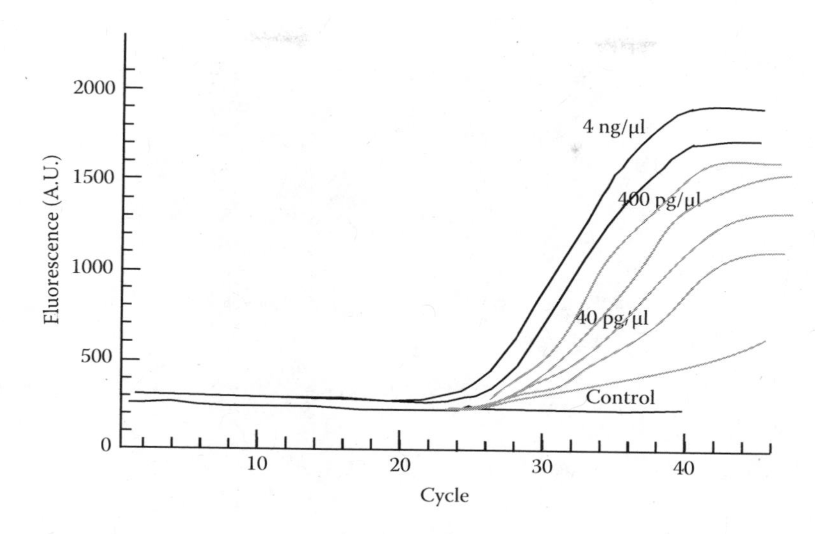

FIGURE 9.17 Amplification curves of droplet-based PCR with laser heating at different concentrations. The 4 ng/μl, 400 pg/μl, and 40 pg/μl concentrations correspond to roughly 14,000, 1,400, and 140 target copies per droplet, respectively. (Reproduced from Kim et al., *Opt. Express*, 17, 218–27, 2009. With permission.)

per day to millions per hour, offers great advantages in terms of reagent savings, and reduces process time and costly manual preparation steps. The configuration of a limiting dilution PCR is shown in Figure 9.18.

Reagents are encapsulated in picoliter-sized droplets that travel along long serpentine channels spatially arranged such that the droplets pass through alternating denaturation and annealing zones for efficient amplification. An experimental setup used for implementing limiting dilution PCR is shown in Figure 9.19.

By using a method of sample partitioning into monodisperse picoliter droplets emulsified in oil-on-chip, one can achieve a six-order magnitude reduction in size from commercial PCR systems. The described method requires only eighteen cycles for single copy. Similarly, microflow channels for continuous flow of water-in-oil nanoliter droplets can also be designed in a radial pattern, as shown in Figure 9.20. In this design, the droplets flow through alternating temperature zones in a radial pattern for denaturation, annealing, and extension steps. Highly efficient amplification of thirty-four cycles of PCR in only 17 min was achieved even at low concentrations. Analysis of the product by gel electrophoresis, sequencing, and real-time PCR showed that the amplification is specific, and amplification up to 5×10^6 times could be obtained, compared to the amplification factors obtained in a benchtop PCR machine.

9.3.3 On-Chip Single-Copy Real-Time Reverse Transcription PCR in Isolated Picoliter Droplets: A Case Study

In this lab-on-a-chip system, the isolation of RNA is followed by reverse transcription (RT) and PCR amplification. A complete analysis of samples containing pathogens requires the ability to detect both DNA- and RNA-based organisms. While extensive work has been done on microfluidic PCR-based assays, less has been done on RT-PCR systems. RT-PCR has also been performed in batch-generated emulsions. Batch-generated emulsion generation of droplets in a microfluidic channel offers advantages such as control over size and quantity of reactants in droplets, compared to bulk emulsion methods. Figure 9.21 shows an on-chip digital microfluidic real-time RT-PCR instrument for generating monodisperse isolated picoliter reactors. It is used for reverse transcription PCR and real-time detection of single RNA genome copies. A crucial challenge in reaching this level

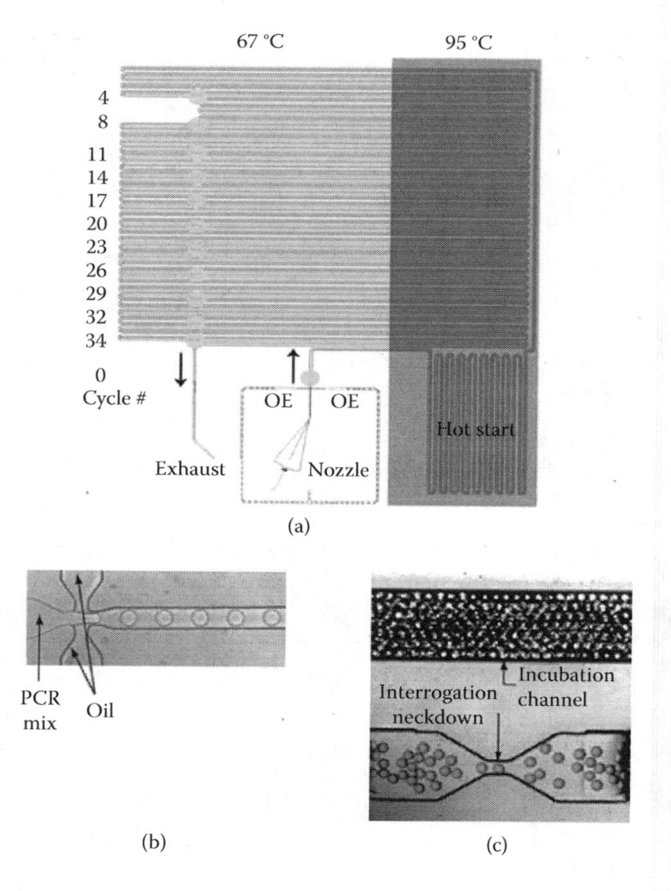

FIGURE 9.18 (See color insert.) The layout of PCR chip. (a) With different temperature zones for thermal cycling. (b) The PCR mixture is injected into the main nozzle, while oil is injected through the side nozzles for droplet generation at the nozzle. (c) Generated uniform picoliter droplets are seen in the downstream channel and neck areas. (Reprinted from Kiss et al., Anal. Chem., 80, 8975–81, 2008. With permission.)

of performance is the physical isolation of the reaction volumes to prevent evaporation and fluidic crosstalk between adjacent containers during thermal cycling.

In the system shown in Figure 9.21, RNA was isolated in picoliter droplets, and the reverse transcription takes place within the individual droplets. Droplets are emulsified in oil to generate isolated chemical reactors for nucleic acid detection. Subsequently, amplification is detected in each interrogated droplet. The method required around twenty-three cycles for single-copy reverse transcription from RNA, amplification, and detection.

9.4 PROTEIN MICROARRAYS

9.4.1 Introduction

As shown in the previous section, DNA microarrays are now an established technology in biological and pharmaceutical research, providing a wealth of information essential for understanding the biological processes and helping drug development. *Protein microarrays* are emerging as a follow-up technology that offers high-throughput profiling of cellular proteins. They can produce large amounts of data that must be analyzed in order to yield information that can lead to novel drug targets and biomarkers. Protein microarrays, as shown in Figure 9.1, are grids that contain small amounts of

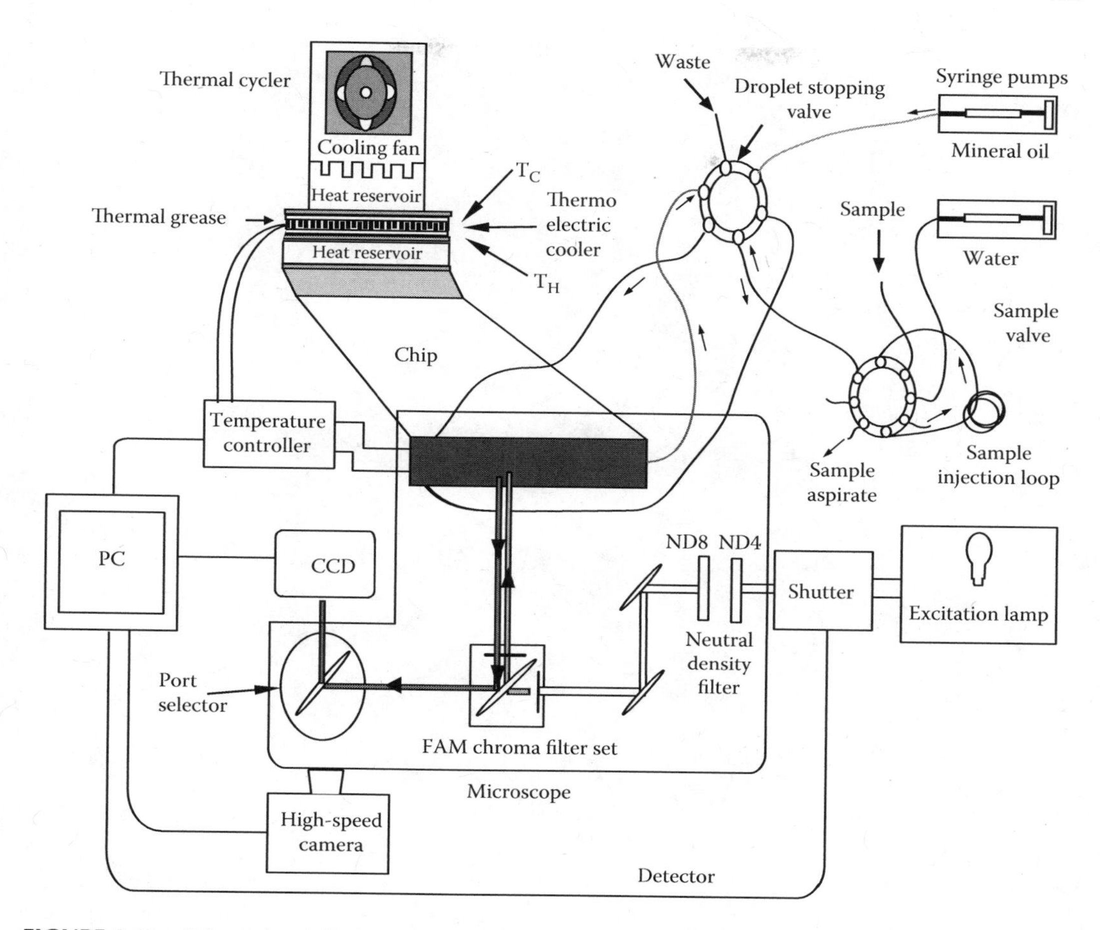

FIGURE 9.19 Schematic of the instrument used for real-time PCR in picoliter droplets showing the integrated droplet generator, thermal cycler, and fluorescence detector. (Adapted from Beer et al., *Anal. Chem.*, 79, 8471–75, 2007.)

purified proteins or antibodies, affixed in an ordered manner at separate locations on a solid support, forming a microscopic array. A typical array may contain hundreds of spatially distinct elements within a total area of 1 cm^2. A slide or "chip" could be spotted with thousands of known antibodies or peptides. When a biological sample is spread over the chip, any resultant binding at those spots can be detected. Generally, all protein arrays are composed of a substrate, which constitutes the underlying core material of the array, and a protein attachment layer providing a chemical interface.

Analogous to DNA microarrays, protein arrays can screen thousands of immobilized biomolecules at a time using small amounts of sample. However, technical difficulties have caused protein microarray development to lag behind that of DNA microarrays. While immobilization of DNA to a solid support is implemented through a simple charge interaction, the immobilization of a protein is more difficult, as it has to maintain the folding of the molecules or functional conformation. Development of protein arrays is technically difficult, mostly due to the structural diversity and complexity in proteins compared to nucleic acids. In addition, as proteins cannot be amplified using PCR, technologies for high-throughput protein production and purification are necessary to generate protein microarrays. Many different types of proteins can be simultaneously detected on the same chip. In spite of some success with protein microarray in profiling the proteome of the cell, it has many limitations that prevent protein microarray technology from reaching its full potential.

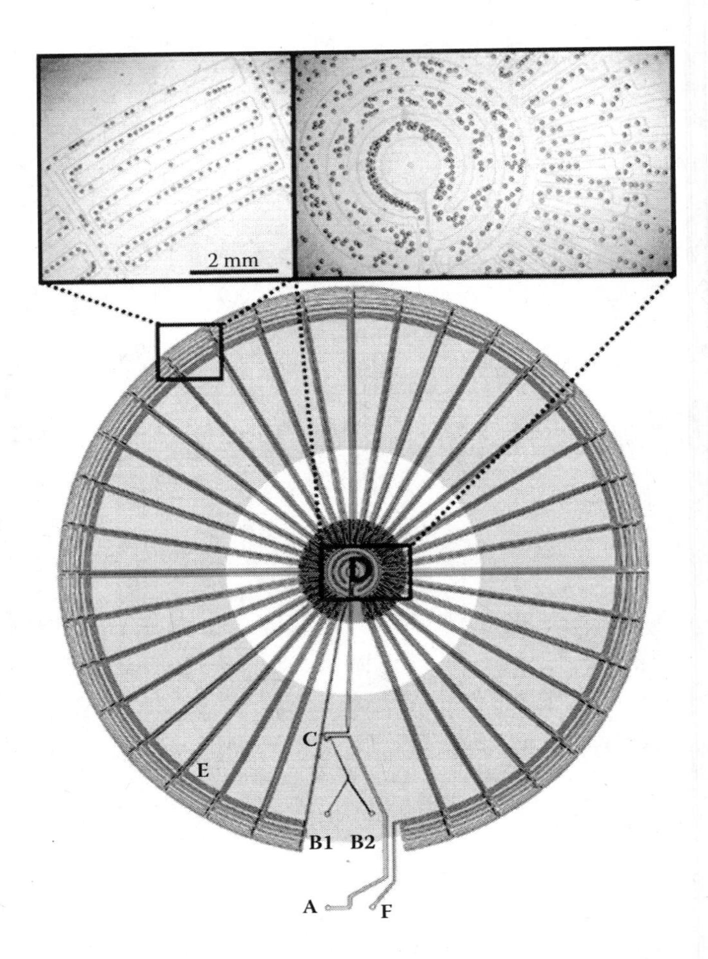

FIGURE 9.20 (See color insert.) Schematic diagram of a radial PCR device. Droplets are generated by sending carrier fluid oil at inlet A, and the aqueous phase at two inlet channels, B1 and B2. Droplets are generated at the T-junction denoted by C. Initial denaturation is implemented at location D, while annealing and extension are implemented in the peripheral zone, E, where primer annealing and template extension occur. The droplets are collected at exit F after thirty-four PCR cycles. The heating is implemented using the Peltier heat module. (Reprinted from Zhang and Ozdemir, *Anal. Chim. Acta*, 638, 115–25, 2009. With permission.)

Protein and nucleic acid microarrays are similar in some aspects, but important differences are found in the physical properties and stability of proteins and DNA at interfaces. Unlike DNA, proteins have a three-dimensional structure that is critical to their function, and they are known to adsorb nonspecifically to commonly used substrates, resulting in false signals. These factors can limit the specificity and sensitivity of microarrays, particularly when complex biological fluids are used. In addition, surface modifications provide homogenous and efficient activation of the entire surface of the substrate used. Detection of a bound target is considerably more complex than with nucleic acid arrays (Figure 9.22). Even if they can be labeled with different fluorophores, reproducibility of these reactions is poor.

In the postgenomic era, proteomics has enormous potential in biology and medicine. In spite of the technical problems, protein microarray technology has emerged as a promising approach for profiling the protein expression in diagnostic, prognostic, and disease progression monitoring. Some of the applications of protein microarray include identification of protein-protein interactions, protein-phospholipid interactions, small-molecule targets, and substrates of protein kinases. In order to provide insights into the mechanisms of biological processes and to get detailed information about

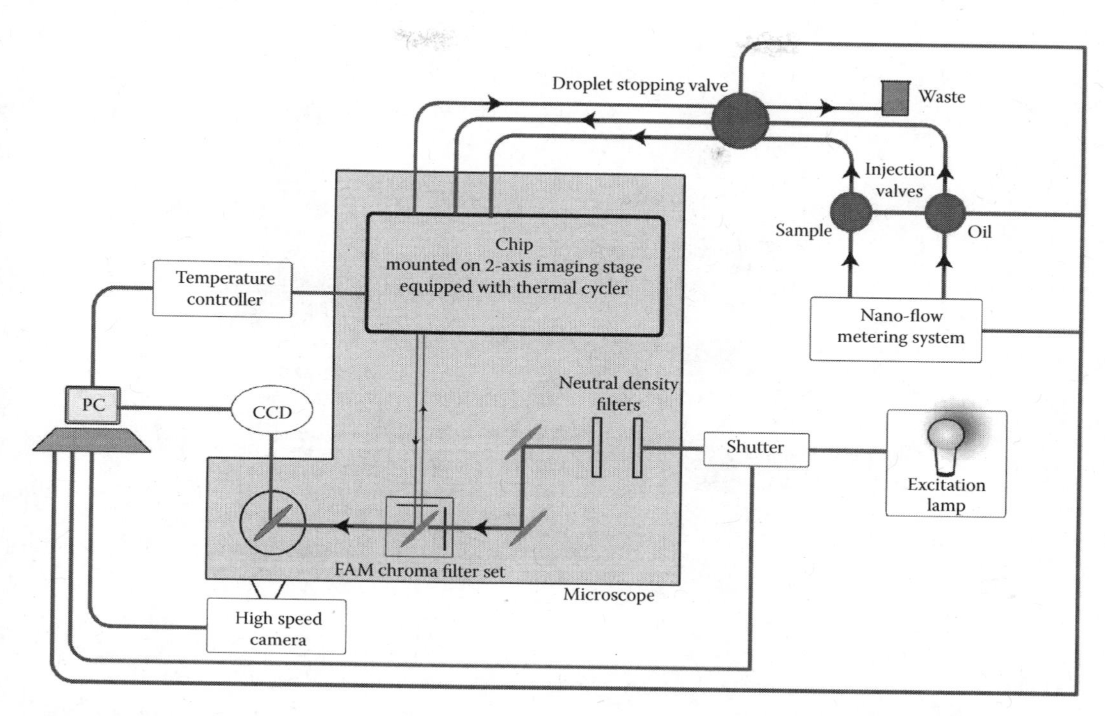

FIGURE 9.21 Schematic of a RT-PCR LOC device. (Adapted from Zhang and Ozdemir, *Anal. Chim. Acta*, 638, 115–25, 2009.)

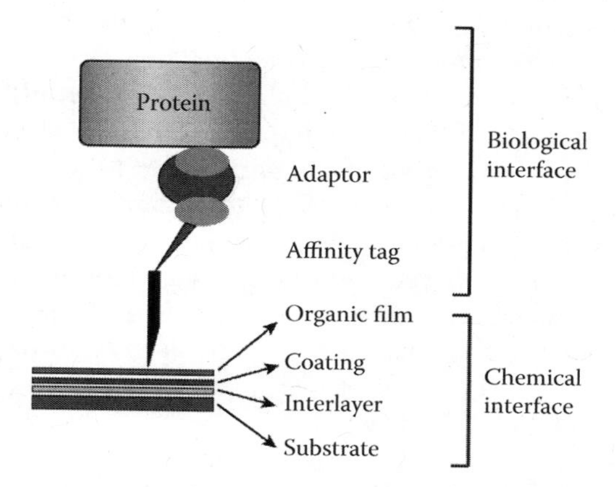

FIGURE 9.22 Microarray architecture. The protein is immobilized directly or through an optional affinity tag onto a thin organic layer formed on the array substrate. Each microarray has a two-dimensional addressable grid of microscale spots of biological molecules immobilized on an organic thin film coated on a substrate. (Adapted from Talapatral et al., *Pharmacogenomics*, 3, 1–10, 2002.)

a complex cellular system, the identification and analysis of each of its components and the determination of their biochemical activities are required. The most common protein microarray is the *antibody microarray*, as shown in Figure 9.23, where antibodies are spotted onto the chip and used as *capture molecules* to detect proteins from cells.

Protein microarrays, currently used to study the biochemical activities of proteins, can be classified into three categories: analytical microarrays, functional microarrays, and reverse phase microarrays. *Analytical microarrays*, mostly antibody microarrays, are used to measure binding

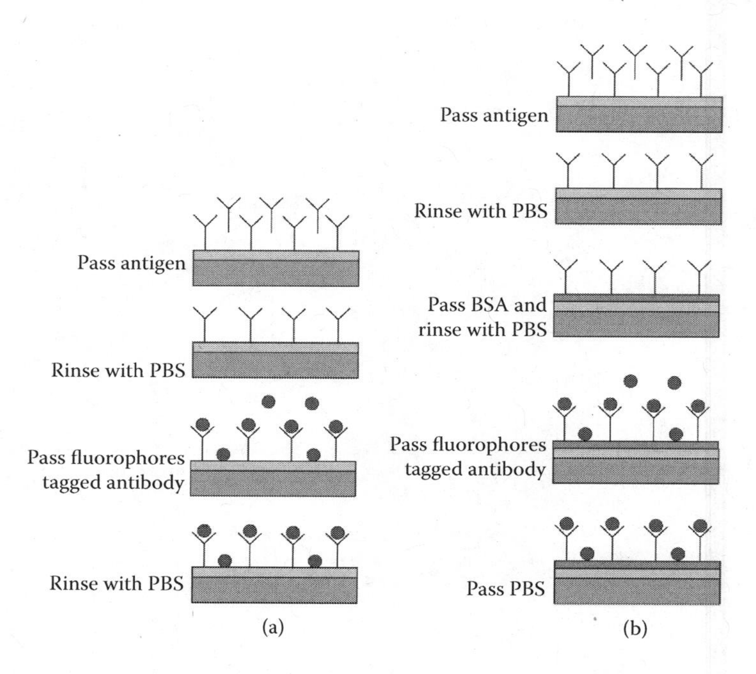

FIGURE 9.23 Types of antibody microarray-based detection. (a) Fluorescence-linked immunosorbent assay. (b) Enzyme-linked immunosorbent assay (ELISA). (Reproduced from http://wapedia.mobi/en/Antibody_microarray.)

affinities, specificities, and protein expression levels of proteins in a mixture of proteins. A library of antibodies or aptamers is arrayed on a glass microscope slide, and the array is probed with a protein solution. *Functional protein microarrays* are composed of arrays containing full-length functional proteins or protein domains. They are used to study the biochemical activities of an entire proteome in a single experiment. A third type of protein microarray, which is related to analytical microarrays, is known as *reverse phase protein microarray* (RPMA). In RPMA, cells isolated from various tissues of interest are lysed, and the lysate is arrayed onto a nitrocellulose slide, together with reference peptides. The slides are then probed with antibodies against the target protein of interest, and the bonded antibodies are detected with chemiluminescent, fluorescent, or colorimetric methods. These types of microarrays can be used to monitor differential expression profiles, for example, profiling responses to environmental stress. The three types of protein microarrays are shown in Figure 9.24.

9.4.2 Fabrication of Protein Microarrays

There are different strategies for the fabrication of protein microarrays. Microarray formats are based on different immobilization chemistries, detection methodologies, and methods of capturing of the analyte. Protein arrays can be either *forward phase* (antibody immobilized onto surface) or *reverse phase* (antigen immobilized onto the surface). Many different types of proteins can be simultaneously detected on the same chip.

As shown in Figure 9.25, protein microarrays come in a variety of formats. These include standard protein microarrays that use purified recombinant proteins, antibody microarrays, and reverse protein microarrays that use cell lysates. Although the applications of protein microarrays can differ widely, the general concept to detect interaction partners is the same in all. Binding partners are incubated with the arrayed proteins, and binding is detected by using a label, bound to either the interaction partner or a secondary antibody.

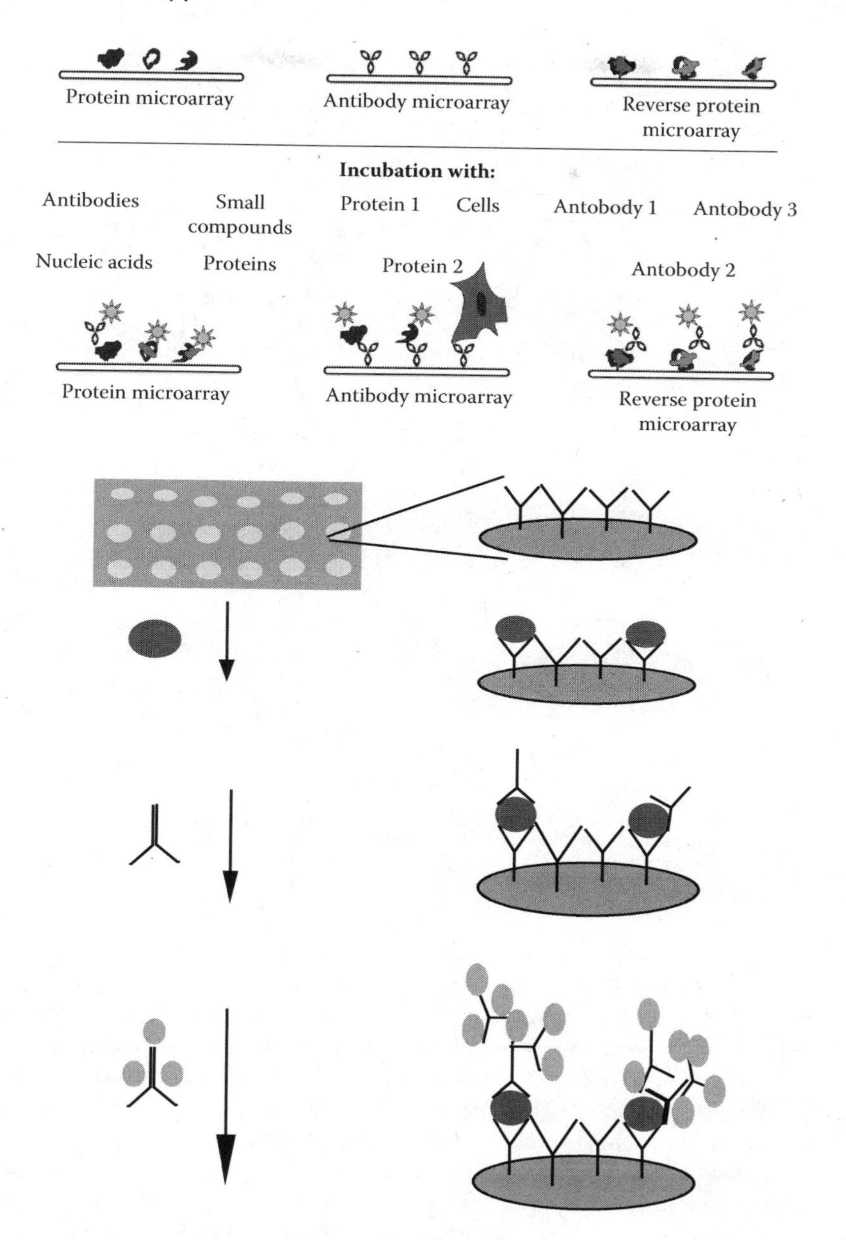

FIGURE 9.24 Types of protein microarrays and their possible applications. (a) Common types of arrays: protein microarrays (PMAs; consisting of individual recombinant proteins), antibody microarrays (AMAs; consisting of antibodies or fragments of antibodies), and reverse protein microarrays (RPMAs; consisting of whole or fractionated protein lysates/extracts). (b) Screening applications of the three array types with known or unknown labeled interaction partners. (Adapted from Hultschig et al., *Curr. Opin. Chem. Biol.*, 10, 4–10, 2006.)

Forward phase arrays immobilize antibodies designed to bind with specific analytes in a mixture of sample proteins. The bound analytes are detected by the second sandwich antibody, or by labeling the analyte directly. Reverse phase arrays immobilize the analytes on the solid phase that are designed to bond with a specific antibody. Bound antibodies can be detected with fluorophore tagging.

Different attachment methods are used, depending on the orientation of proteins required on the surface desired. Random attachment can be achieved when the slide surface is functionalized with

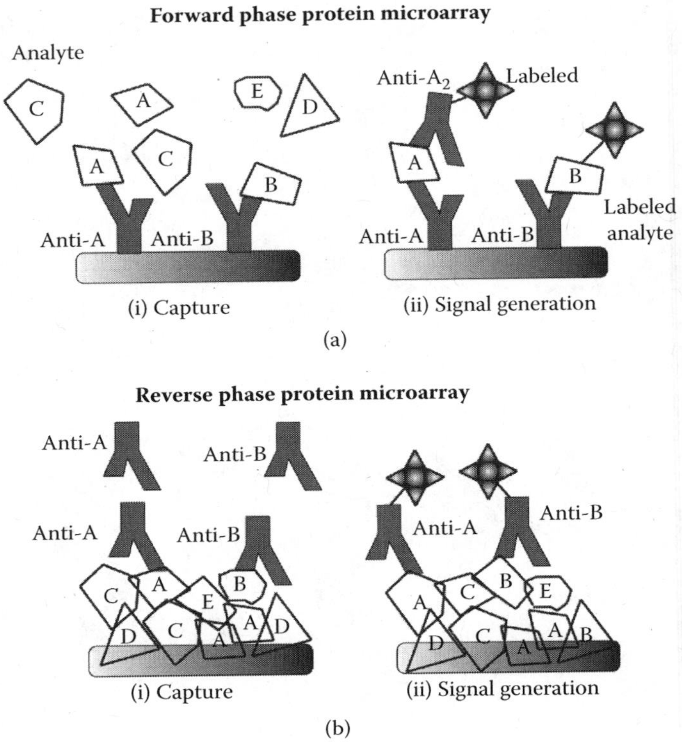

FIGURE 9.25 Types of protein microarray platforms based on immobilizing elements. (a) Forward phase protein microarray. (b) Reverse phase protein microarray. (Adapted from Liotta et al., *Cancer Cell.*, 3, 317–25, 2003.)

aldehydes, amines, or epoxy derivatives as shown in Figure 9.26 (top). Uniform attachment can be achieved only through a ligand, as shown in Figure 9.26 (bottom).

The format and preparation of protein microarrays depend on the nature of the immobilized biomolecule and the intended application. The technique of immobilization depends on both concentration and orientation of immobilized proteins or antibodies on the surface. Proteins that are soluble in their native environments may precipitate on chip surfaces. Consequently, it can be difficult to select the surface chemistry that permits diverse proteins to retain their original folded conformation and biological activity. For the study of protein-protein, protein-DNA, protein-RNA, and protein-ligand interactions, proteins can be screened in a high-throughput fashion, using either *pin-based spotting* or *liquid microdispensing*. Several types of chips have been designed, such as glass slides, porous gel pad slides, and microwells. In addition, standard microarrayers and scanners used for DNA chips can be used for proteins as well. A variety of methods have been reported, including the adsorption onto charged or hydrophobic surfaces, covalent cross-linking, or specific binding through tags. The glass surface can be activated with a cross-linking agent that reacts with primary amines. In order to prevent dehydration, proteins are often spotted in a 40% glycerol solution. When proteins or antibodies are immobilized in gel pads, the gel is activated with glutaraldehyde. The capacity of protein immobilization in the gel pads is much higher than on a two-dimensional glass or plastic surface. The microwell protein chip is particularly suitable for biochemical assays because proteins in microwells are well separated, and they are kept in liquid that minimizes the sample drying. With fluorescent markers or other methods of detection, protein microarrays are used as powerful tools in high-throughput proteomics and drug discovery. Different types of protein microarrays are shown in Figure 9.27.

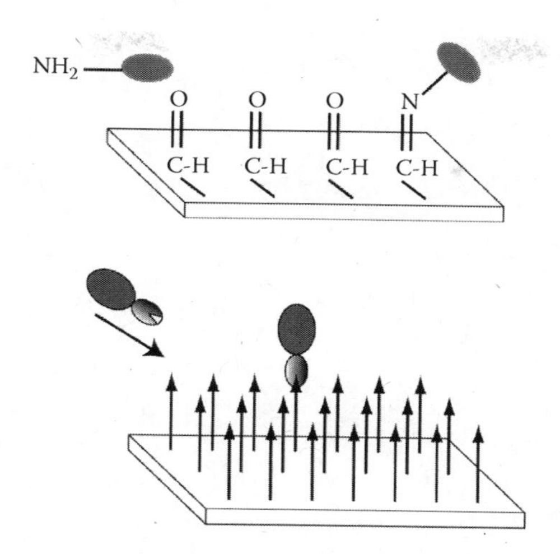

FIGURE 9.26 Different protein attachment schemes. Proteins can be attached randomly using different chemistries, including aldehyde- and epoxy-treated slides that covalently attach proteins by their primary amines (upper figure). Proteins can be uniformly orientated onto slides coated with a ligand. Biotinylated proteins can be attached to streptavidin-coated slides. This leads to attachment through the tag and presumably orientates the protein away from the slide surface. (Adapted from Hall et al., *Mech. Ageing Dev.*, 128, 161–67, 2007.)

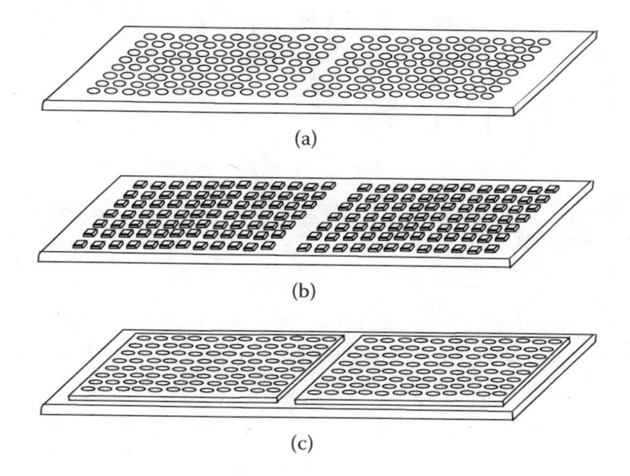

FIGURE 9.27 Three types of protein microarrays: (a) plain glass slide, (b) three-dimensional gel pad chip, (c) and microwell chip. (Adapted from Zhu and Snyder, *Curr. Opin. Chem. Biol.*, 5, 40–45, 2001.)

The density of immobilized protein molecules is determined by the surface structure. A flat surface offers less binding capacity than the three-dimensional structure of a polyacrylamide gel. The gel pads can provide an aqueous environment and accommodate proteins up to 400 kilodaltons (kDa) (1 Da = 1 g/mol). Proteins and other macromolecule molecular weights are usually measured in kilodaltons (that is, 1,000 Da) in size. Microarray manufacture is a highly automated process that involves imprinting of "capture" molecules on a bioreactive film on an array or slide surface in a two-dimensional array format. The sample to be immobilized can also be introduced using microfluidic channels in a continuous flow stream. The spacing between spots depends on the size of the capture agents and antibody arrays, and they are typically of 375 μm spacing. Alternatively, peptides have been synthesized on planar surfaces in an array format using photolithography or SPOT technology. SPOT synthesis is an easy technique for positionally addressable, parallel chemical

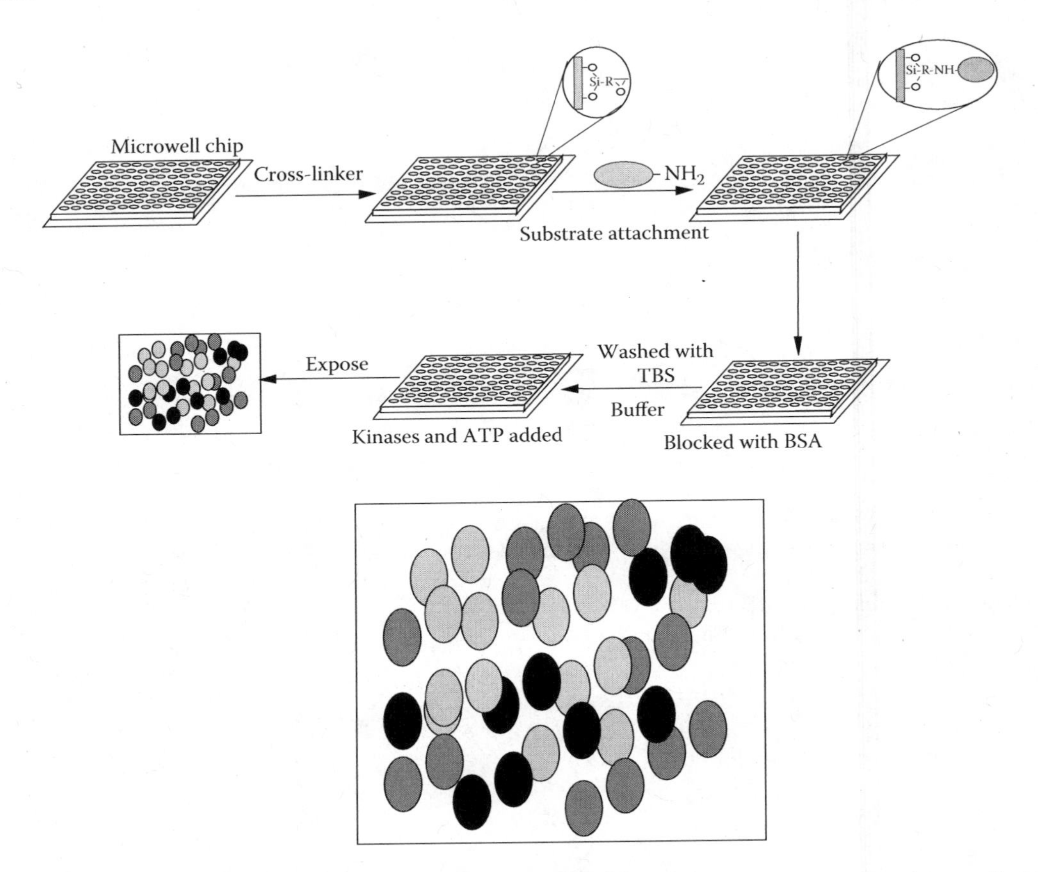

FIGURE 9.28 Application of protein microwell chips in kinase assays. The surface of the microwell chip is activated with a cross-linker 3-glycidoxypropyltrimethoxysilane (GPTS) and cured at 130°C for an hour. Protein substrates are added to the wells and incubated in a humidity chamber for 3 h on ice. Wells are blocked with 1% bovine serum albumin (BSA) and washed with tris-buffered saline (TBS) buffer, before kinase, 33P-γ-ATP, and buffer are added. After incubation for 30 min at 30°C, the chips are washed and exposed to a phosphoimager. Bottom: An enlarged image of the resulting phosphorylation using poly-Glu-Tyr as a substrate. (Adapted from Zhu and Snyder, *Curr. Opin. Chem. Biol.*, 5, 40–45, 2001.)

synthesis of peptides on a membrane support. Using photolithography, miniature wells are etched on the surface of silicon arrays, capable of detecting up to ten thousand proteins in parallel (Figure 9.28). Protein array technology has to face several technical problems, such as the acquisition, arraying, and stable attachment of proteins to array surfaces, and the detection of interacting proteins.

Antibodies are the best protein binding reagents, with high affinity, high selectivity, and manufacturability in both high quantity and purity for arraying purposes. The biggest challenge for antibody arrays is to obtain antibodies against about one hundred thousand proteins that form the human proteome. Currently, there are antibodies available for a mere fraction of the proteome, and their specificity is poorly documented. Protein binding molecules other than antibodies, namely, aptamers and fibronectin-based peptide scaffolds, are useful alternatives for antibodies. *Aptamers* are protein binding RNA molecules, which are relatively easy to synthesize and produce in array. Analyte-selective binding and specific retention on the array surface are enabled by thermodynamically driven binding mechanisms. Detection of a bound target is considerably more complex than the detection of nucleic acid arrays. Although the proteomes under comparison can be labeled in a comparable fashion with different fluorophores, the reproducibility of these chemical reactions is poor, and interferences with the protein-antibody interactions present an additional complexity.

To locate reactive proteins on a proteome chip, small-molecule probes are labeled with either fluorescent, affinity, photochemical, or radioisotope tags. Fluorescent labels are generally preferred, as they are compatible with readily available microarray laser scanners. However, probes can also be labeled with affinity tags or photochemical tags. Regardless of the type of label used, the label itself may interfere with the probe's ability to interact with the target protein. A number of label-free detection methods have recently been developed that not only overcome the problem of a label interference, but also allow for the collection of kinetic binding data.

9.4.3 Applications of Protein Arrays

Protein microarray applications include expression profiling, serum-based diagnostics, protein-protein binding assays, drug-target binding, receptor epitope binding, etc. Protein array technology provides a powerful and versatile tool for the genome scale analysis of gene function, such as enzyme activity, protein-protein and protein–nucleic acid interaction, and small molecule–drug interactions, directly at the protein level. The ultimate goal of proteomics is to study the biochemical activities of every protein encoded by an organism or proteome.

A *kinase* is a protein enzyme that adds a phosphate onto a molecule, though typically only proteins. The phosphorylated molecule can be another protein, the kinase itself (autophosphorylation), or any other molecule. The source for the phosphate is terminal phosphate, which is called the gamma (g) phosphate from ATP.

Arrays can be engineered to address protein identification, quantification, and affinity studies. A *profiling array* quantifies the levels of specific proteins on a global scale. An *affinity array* probes the interaction of peptides, proteins, oligonucleotides, oligosaccharides, or small molecules with immobilized proteins, which are typically receptors, enzymes, or antibodies. Arrays can be further classified as finite or chemical and self-replicating or biological. *Finite (chemical) arrays* are involved in specific binding analyses, enzymatic assays, or high-throughput screening for discrete biochemical activities. *Biological (self-replicating) affinity arrays* utilize living cells and facilitate functional studies of complex biological processes in a cellular environment. Profiling disease-related tissues using protein microarrays will facilitate discovery of novel biomarkers and potential drug targets. Although protein arrays hold many promises, the principal limitation to their utility is that there is not always a direct correlation between protein abundance and activity. Coupled with mass spectrometric identification, protein chips might also have a wide application in drug discovery and protein-protein interactions. Proteins and small-molecule ligands can be bound to proteins immobilized on a chip, and the bound molecules can be identified using matrix-assisted laser desorption/ionization time-of-flight (MALDI-TOF) mass spectroscopy. By analyzing a large number of samples, the molecules and proteins that bind specifically to many different proteins can be identified. Small molecules can then be used to probe protein function; protein-protein interactions can be used to deduce molecular networks and pathways. Different ways of sample introduction and affinity binding of protein in affinity array chips are schematically shown in Figure 9.29.

A representative sample of the different assays that have been performed on functional protein microarrays is shown in Figure 9.30. Proteins are immobilized at high spatial density onto a microscope slide, and the slide can then be probed for various interactions. While Cy5 is the fluorophore shown, many other fluorophores can be used for detection.

9.5 CELL AND TISSUE-BASED ASSAYS ON A CHIP

The quantification of biomarkers on tissue microarrays (TMAs) provides numerous technical challenges for the development of automated systems capable of reading microarrays and translating the image information into useful data. Unlike DNA/RNA expression arrays, each spot, called a histospot on a TMA, represents a miniature histologic section of tissue that contains complex spatial information that can dramatically affect the quantitative analysis of biomarkers (Figure 9.31). The

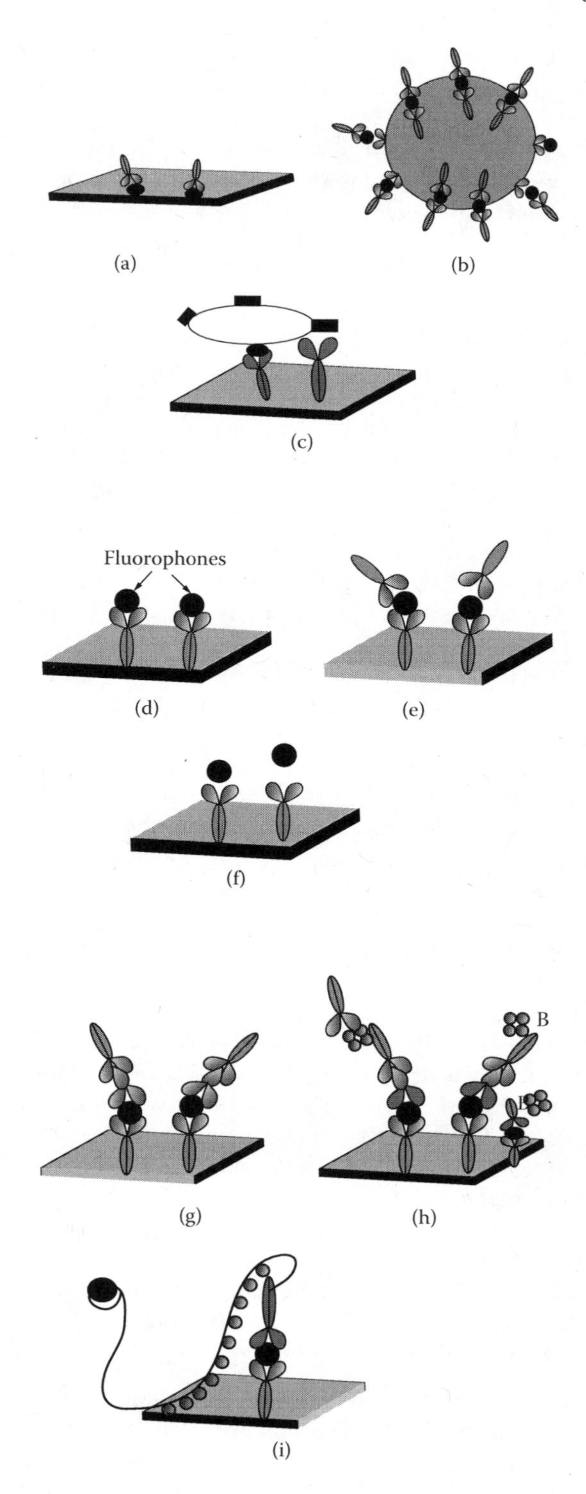

FIGURE 9.29 Schematics of the use of affinity protein microchips. Upper panel: Sample introduction and affinity binding of proteins of interest. (a) Reverse phase protein array. (b) Bead-based array. (c) Whole cell array. (d) Direct labeling. (e) Sandwich labeling. (f) Label-free. (g) Tertiary Ab. (h) Biotinylated Ab. (i) Rolling circle amplification. (Adapted from Spisak et al., *Electrophoresis,* 28, 4261-73, 2007.)

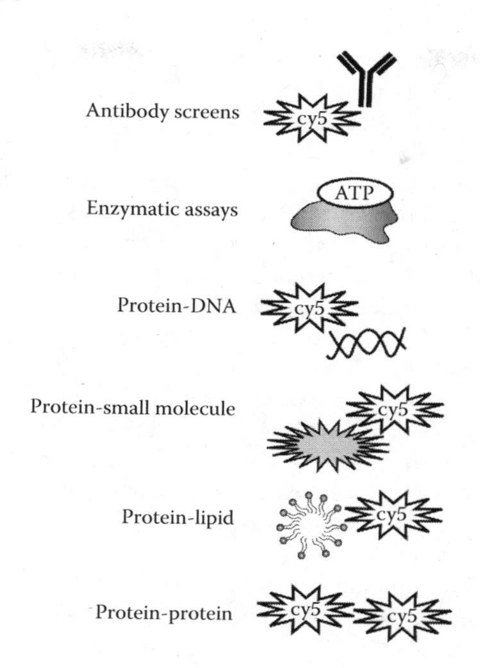

FIGURE 9.30 Applications of functional protein microarrays. (Adapted from Hall et al., *Mech. Ageing Dev.*, 128, 161–67, 2007.)

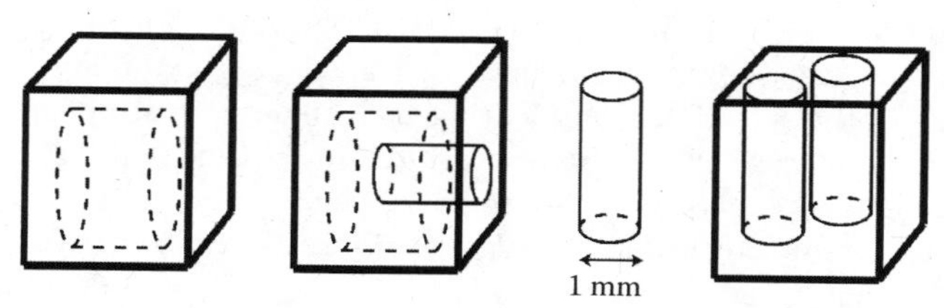

FIGURE 9.31 Tissue microarray element. A representative tissue cylinder is extracted from a donor block and placed into an acceptor block.

location and expression level of biomarkers on TMAs is generally determined using target-specific antibodies tagged with fluorescent markers. In some cases, a given biomarker may only be represented in a subset of cells within a TMA spot (e.g., tumor or stromal cells) or may be expressed differently, in different subcellular compartments within cells (e.g., membrane or cytoplasm or nucleus of a cell). In some cases, the element of interest may only represent a small percentage of the analyzed area. As a result, an analysis of biomarker intensity within the entire area is not representative of the results. Accurate quantification, therefore, requires accurate measurement of both biomarker expression and location. Preserving the spatial information has proven to be an essential component of automated TMA analysis and is incorporated, to some degree, into most of the systems that are currently available for automated TMA analysis.

The goal underlying the use of cells and tissues for a sensor or diagnostic system is to obtain the information based on biological activity or mechanisms. These systems could have a number of important applications, including screening large libraries of drugs for activity or toxicity or diagnostic information from clinical samples. The technical challenges associated with the successful fabrication of these microarrays include selection of source material for cells and tissues, sample preparation, integration of cellular or tissue materials with two- and three-dimensional materials, stability of cells and tissues, and bioinformatics tools for the extraction of information from the response libraries.

Potential sources of cells and tissues include primary cell cultures from animal donors. To date, there has been considerable development of cells and tissue-based technologies and devices from a sample of cell and tissue types, from both primary and transformed sources. Cell types that have received attention are principally those of neural origin, because neural tissue can be extracted from primary sources and maintained in culture, and their electrical activity directly related to cellular function can be achieved on microelectrode-integrated substrates. For medical diagnostic samples, the presence of serum components or other molecular or cellular interferents from blood, urine, or mucosal samples must be considered. Selective surface adhesion of cells has been demonstrated with surface patterning techniques such as photolithography with self-assembled monolayers, microcontact printing, and other microfabrication techniques. This has been demonstrated on a number of materials, including silicon, glass, and plastic. Tissue-based arrays have also been created in silicon, using focused ion beam technology, reactive ion etching, or other microfabrication techniques. Three-dimensional matrices of materials have also been used to create detection or diagnostic cell and tissue systems (Figure 9.33). Biocompatible polymers such as alginates and collagen have been used to embed cells and tissues for addressing functional responses. The interaction of various materials on cell function must be extensively evaluated prior to incorporation into sensor platforms, as the extracellular environment can influence cell signaling. Stabilizing cells and tissues for working devices presents two significant challenges: (1) preservation of cells or tissues for long-term shelf life and (2) stabilization of sensor functional performance. In order to overcome these difficulties, *cryopreservation techniques* exist, particularly for transformed cells. However, cryopreservation of primary cell and tissue still presents a significant challenge in providing sensor elements with long shelf life for these devices. Cellular signaling can be captured using a variety of detection and reporting schemes. Two common methods of transducing cellular responses are optical and electrical. The advent of cellular engineering to include optical or luminescent reporting elements such as green fluorescent protein has resulted in the ability to use optical detection to transduce cellular signaling events. Optical interfaces can be used to detect visible, fluorescent, or luminescent signals from cells or tissues. Electrically active cells or tissues can be interfaced with microelectrodes embedded in silicon or plastics, which allow the capture of extracellular spikes or impedance changes associated with cellular or tissue response. Bioinformatic tools can be used to compare the measured responses to a library of known responses. User interfaces that effectively communicate information from these complex systems will be required to enable insertion of useful sensor or diagnostic technologies.

A key constraint for the above technology is the requirement of considerable manual effort to prepare a source plate for printing. In cell chips, cells are grown under normal tissue culture conditions. As a result, pin-based printing requires very high cell density in the source plate. The use of other printing technologies, such as inkjet or other microspray methods, enables microplate-based cell culture growth and treatments more compatible with cell chip printing. Such developments of cell chip technology could make it readily applicable for functional genomics and chemogenomics.

In a cell chip process, cells are grown and treated under normal cell culture conditions. This allows the storage of cells for several weeks, before resuspending in phosphate-buffered saline (PBS) and transferring to a source plate for printing. Then, the cells are printed onto streptavidin-coated slides using a robotic microarray spotting device. Each slide is probed for immunofluorescence for the selected target and imaged as shown in Figure 9.32.

The three-dimensional cell microarray is a high-throughput platform that enables quantification of on-chip cellular protein levels using the interactions of the cells upon addition of small molecules (Figure 9.34). An immunofluorescence-based assay for high-throughput analysis of target proteins on a three-dimensional cellular microarray platform has been developed. This process integrates the use of three-dimensional cellular microarrays that mimic the cellular microenvironment, with sensitive immunofluorescence detection, and provides quantitative information on cell function.

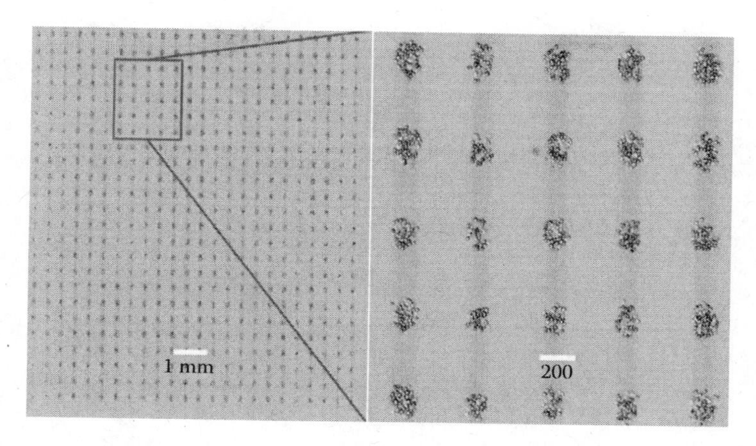

FIGURE 9.32 An outline of the spotted cell chip process. (Adapted from Hart et al., *PLoS One*, 4(10), e7088, 2009.)

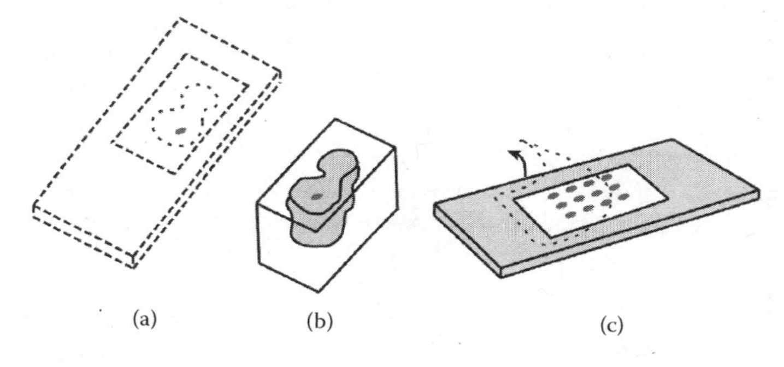

FIGURE 9.33 TMA process: (a) A cylindrical tissue, usually 0.6 mm in diameter, is removed from a donor block using a tissue microarrayer. (b) These are released into premade holes of an empty recipient block. (Adapted from Sauter et al., *Nature Rev. Drug Discov.*, 2, 962–72, 2003.)

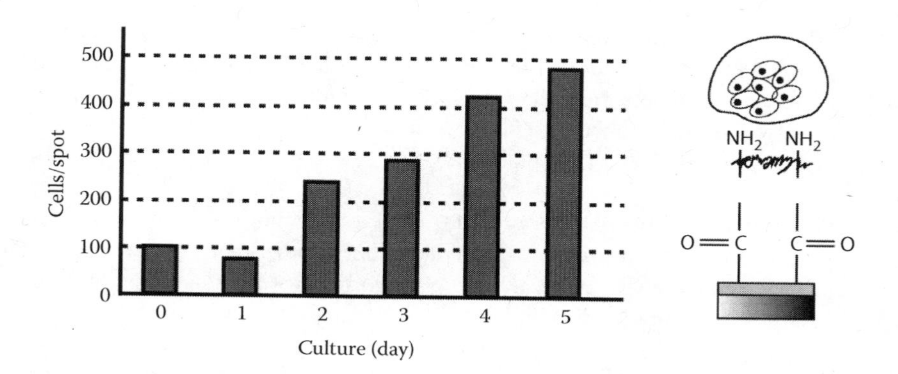

FIGURE 9.34 Three-dimensional monitoring of a cellular microarray for the analysis of target proteins. Variation of a number of cells in a spot against culture duration. In this study, cells are encapsulated on a functionalized glass slide and spotted using a microarray spotter. (Adapted from Fernandes et al., *Anal. Chem.*, 80, 6633–39, 2008.)

9.6 MICROREACTORS

9.6.1 Introduction

Microchips have the ability to miniaturize current benchtop experiments for implementing reaction systems. As seen earlier, microfluidic systems can manipulate microliter to nanoliter amounts of fluid using channels with dimensions of tens to hundreds of micrometers. Microfluidic technologies have provided chemists with powerful platforms for performing simple and complex reactions under flowing conditions. They open up possibilities for new concepts and provide new opportunities for advanced chemical synthesis and biological analysis.

Miniaturized reactor platforms offer important advantages over the conventional system used in chemistry and biology. Devices are designed to manipulate fluids in microchannels with reduced requirements for reagents and solvents. Due to high surface-area-to-volume ratios, microfluidic reactors demonstrate high efficiency in heat and mass transfer. The synthesis parameters are better controlled because of efficient mixing and rapid chemical reactions under microdimensions. Fluid flow as well as mass and heat transport are more easily controlled on the microscale, often resulting in enhanced yields compared to conventional reactors. Microreactors are environmentally friendly, as the process consumes less material and reduced hazardous chemicals. Their small size makes them suitable for field analysis, providing a definite advantage for environmental applications. The possibility of producing microreactors using microfabrication reduces the time to market and cost.

In order to perform reactions within microfluidic devices, the device should be able to perform the procedures normally conducted for reactions on the macroscale. The procedures include the controlled addition of reagents to a reaction mixture, the mixing of reagents, monitoring the reaction, transduction, etc. In addition, multiple interactions between different process variables can also be implemented.

The large surface area provided by microchannel systems is advantageous for many chemical processes, such as extractions, catalytic reactions, highly exothermic reactions, and reactions where hazardous compounds are generated. Microreactors have also found applications in a number of fields, such as environmental and toxicity analysis, catalyst screening, and synthetic chemistry.

Microsystems are typically designed in microwell or microfluidic configurations. The microwell systems are widely applied in analytical and screening applications, and have become popular tools in biochemical engineering.

For performing reactions on a small scale, droplets can also be used as chemical microreactors (Figure 9.35).

Segmented flows are composed of two immiscible phases, a continuous and a dispersed phase. Discrete liquid droplets are encapsulated by a carrier fluid that wets the microchannel. The reactions that occur in these droplets are called plugs. Systems that use plugs differ from segmented flow injection analysis in that reagents in plugs do not come into contact with the microchannel. Enzymatic reactions on a chip can take place either in discrete reservoirs/spots or within microfluidic channels.

Enzyme-based microchips offer an array of attractive opportunities, in areas as diverse as new reaction discovery, optimization of enzyme activity under synthetically relevant conditions, new compound discovery, biodiagnosis, and biosensing. Microreactors offer a number of advantages for enzyme analysis, including smaller dead volumes, shorter analysis time, controlled residence time, lower cost, and greater sensitivity and flexibility in use. Enzymes can be studied in solution using a microfluidic reactor; however, the immobilization of enzymes is often preferred, as it allows the reuse of enzyme. Immobilized enzymes show low rates of denaturation or inactivation, and as a result, secondary reactions can be minimized.

In a similar way, microfluidic devices can also be used for nanoparticle synthesis. Enhanced processing accuracy and efficiency can be obtained.

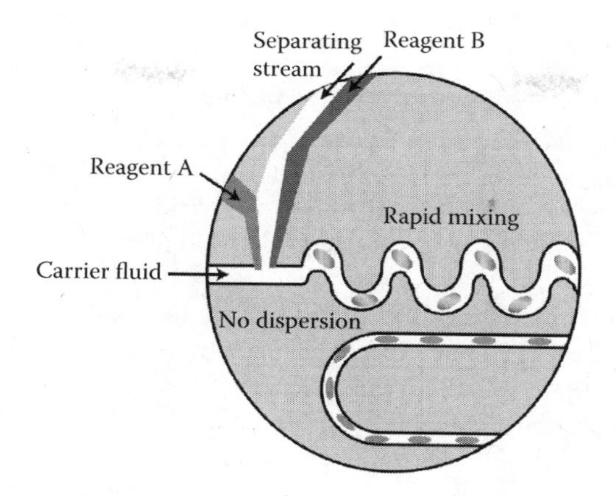

FIGURE 9.35 Droplets formed within microfluidic channels can serve as microreactors. In this example, the reactions are performed within aqueous droplets, which contain reagent A, reagent B, and a separating stream containing buffer. The droplets are encapsulated by a layer of a fluorinated carrier fluid and transported through the microchannels. (Adapted from Song et al., *Angew. Chem.*, 115, 792–96, 2003.)

9.6.2 Microchannel Enzyme Reactors

Many enzymatic conversion processes have been developed, but they are not still available as standard technologies at the commercial level.

There are two types of techniques used for enzymatic conversions: *solution phase reactions* and *solid phase immobilization***.** Both *continuous* and *pulsed flow reactions* can be performed on a chip type microreactor using solution phase methods with highly improved yields and accelerated reactions. Enzymes can be immobilized within microchannels on various solid supports. The solid support can be magnetic, glass, polystyrene, and agarose beads that can be carried in the flow or the surface of microchannels. The microchannel surfaces are immobilized by introducing various functional groups. For example, biotinylated polylysine was immobilized on glass to capture the streptavidin-conjugated enzyme. Enzymes can also be immobilized on a membrane embedded within the microchannel. Polyethyleneglycol-based hydrogels can be used to immobilize different enzymes on the substrate.

Reactions in microfluidic channels can be made to occur in two types of discrete flows. In the first case, reaction occurs in discrete liquid plugs encapsulated by an immiscible continuous-phase-based carrier fluid, as shown in Figure 9.36. In the second case (not shown here), the reaction occurs in the discrete pockets of the continuous liquid phase separated by bubbles. In the first case, reagents in droplets are not in contact with the microchannel.

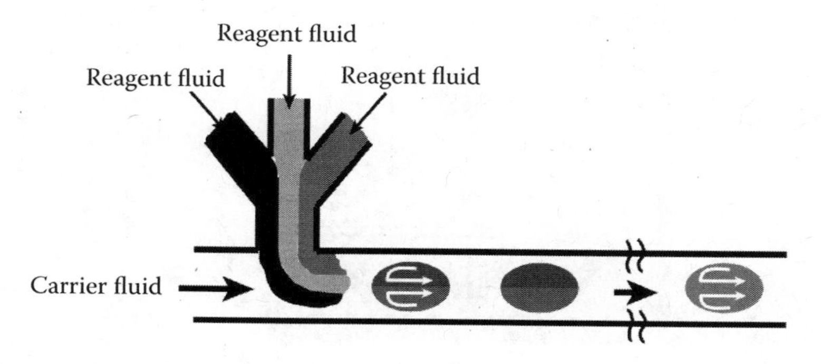

FIGURE 9.36 Discret droplet microreactor. (Adapted from Song et al., *Angew. Chem.*, 45, 7336–56, 2006.)

9.6.3 Enzymatic Conversions: Case Studies

9.6.3.1 Glycosidase-Promoted Hydrolysis in Microchannels

Glycosyltransferase and glycosidase are carbohydrate-related enzymes and have been used for oligosaccharide synthesis. Bioactive glycoconjugates are essential for biological reactions such as cell adhesion and migration. Enzymatic oligosaccharide synthesis, including the forward reaction, called hydrolysis, and the reverse reaction, called transglycosylation, can be performed in microfluidics.

The microfluidic chip is made from PMMA with very long (40 cm) and deep (200 μm) channels used for enzymatic study, as shown in Figure 9.37. In a study the enzymatic hydrolysis reaction between p-nitrophenyl-β-D-galactopyranoside and β-galactosidase was found to be about five times faster in the microreaction channel than in the standard lab chip. A comparative study on the rate of hydrolysis reaction between microfluidic chips and other standard techniques is shown in Figure 9.38. This result shows that the hydrolysis reaction is complete in a short duration of 10 min in microchips. It also shows that enzymatic oligosaccharide synthesis can be performed rapidly and efficiently using a microchip.

9.6.3.1.1 Influence of Glass Beads on Hydrolysis in Microchannels

One can influence the chemical reactions in microfluidics, either through appropriate design of microchannels such that the chemical reactions are enhanced through proper mixing, higher residence time, etc., or by adding catalysts to promote the chemical reaction. This can be implemented by adding flow restrictions to induce turbulence and mixing. In this example, glass beads are packed at the intersection region of the channel to enhance mixing and the rate of the reaction in the hydrolysis of *o*-nitrophenyl-β-D-galactopyranoside by β-galactosidase, as shown in Figure 9.41.[30] The rate of hydrolysis reaction is enhanced if the residence time is made longer by designing a channel with sharp loops, as shown in Figure 9.39.

NaOH is mixed with a colored indicator solution and the degree of mixing can be evaluated by *in situ* spectrometry. The experiments indicate that the degree of mixing increases with the use of microchannels and many sharp bends, as shown in Figure 9.39.

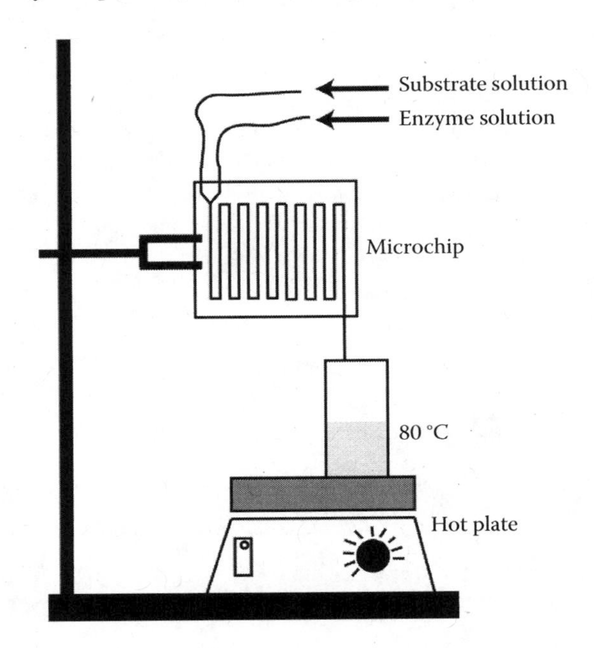

FIGURE 9.37 Microchannel reactor system. (Adapted from Kanno et al., *Lab Chip*, 2, 15–18, 2002.)

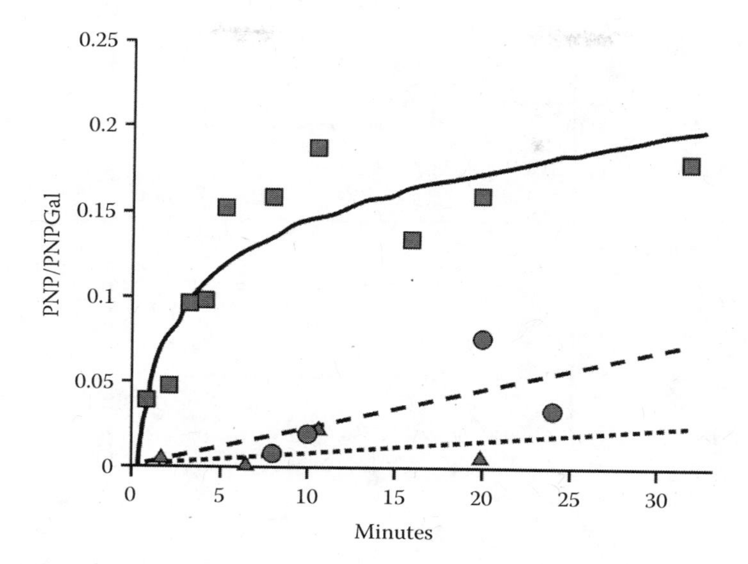

FIGURE 9.38 Initial hydrolysis reaction: hydrolysis in a microchip (▪), in a commercial micro-test tube (•), and with a pretreated chip (Δ). (Reproduced from Kanno et al., *Lab Chip*, 2, 15–18, 2002. With permission.)

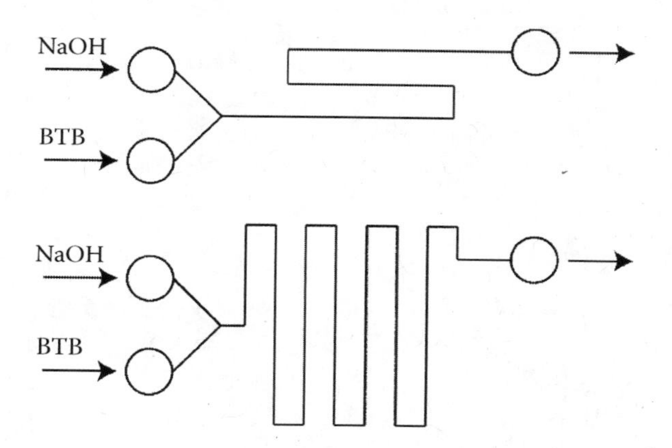

FIGURE 9.39 Microchannels (400 μm width, 200 μm depth) with and without sharp loops. (Adapted from Sotowa et al., *Kor. J. Chem. Eng.*, 22, 552–55, 2005.)

The microchannels are incorporated with projections called dams, and the glass beads are trapped between them to enhance mixing and reaction. Gaps above the dams are made shorter than the size of the beads to entrap them (Figure 9.40).

Mixing experiments were performed with NaOH and colored indicator solutions, and the degree of mixing was evaluated through the change of absorbance measured by *in situ* spectrometry.

Figure 9.41 illustrates the enhanced mixing with microchips, while Figure 9.39 shows two arrangements of microchannels packed with glass beads in which the enzymatic reaction was established. The enzymatic reaction was studied in microfluidic chips at 52°C under a low flow rate of 6 ml/h. For comparison, reactions were carried out in batches or flasks of 100 ml as well. There is clear evidence of improvement with microfluidic chips compared to macrobatches, as shown in Figure 9.41. The results shown in Figure 9.41 confirm that enzyme reactions are much faster in the bead-packed microchannels than in the batch reactor and unpacked microchannels.

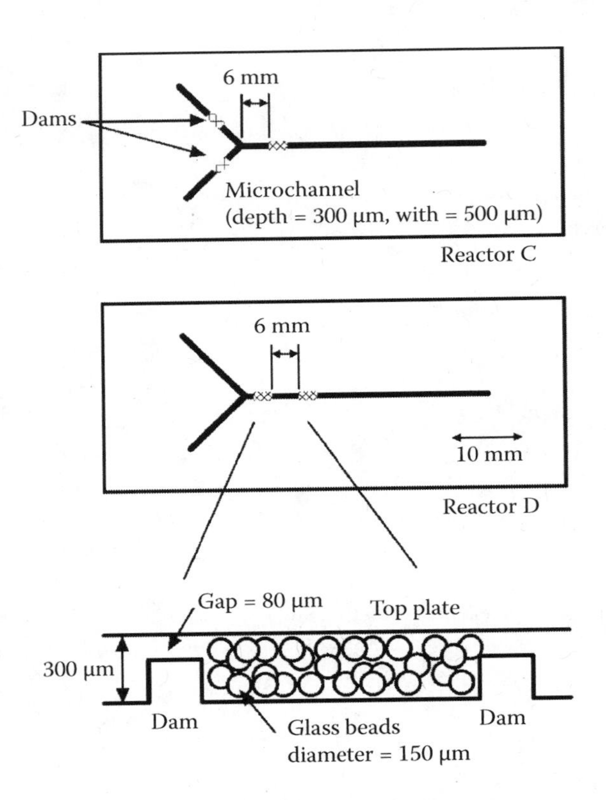

FIGURE 9.40 Schematic of microchannels packed with glass beads between dams. Top: Reactor C. Bottom: Reactor D. (Adapted from Sotowa et al., *Kor. J. Chem. Eng.*, 22, 552–55, 2005.)

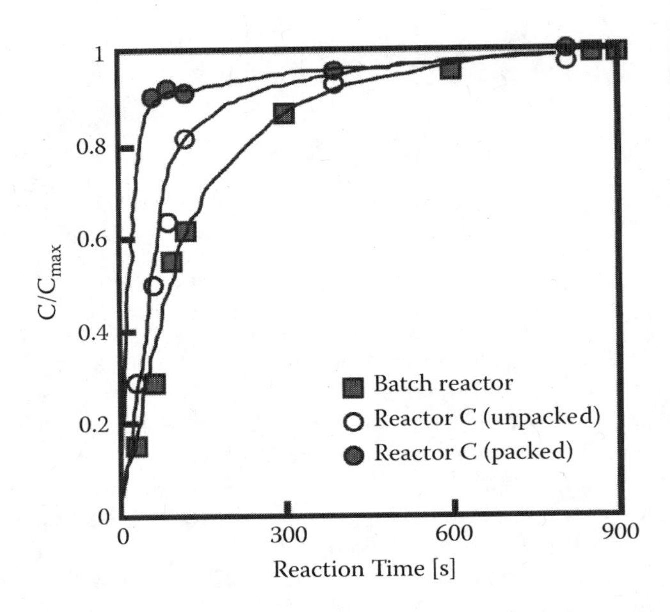

FIGURE 9.41 Comparison of reaction rates in different microreactors. (Adapted from Sotowa et al., *Kor. J. Chem. Eng.*, 22, 552–55, 2005.)

9.6.3.2 Lactose Hydrolysis by Hyperthermophilic β-Glycoside Hydrolase with Immobilized Enzyme

Continuous flow kinetic studies can be easily carried out using microfluidic chips fabricated in PDMS with many channels and chambers, and with enzymes covalently immobilized on the walls that are aminosilanized. The method is useful for the production of glucose and galactose from lactose. For example, the microchip and the temperature-controlled housing for testing the microchip are shown in Figures 9.42 and 9.43, respectively. The total reaction volume is 167 μl. The enzyme is immobilized onto the surface under flow conditions by passing the enzymes attached to the free aldehyde groups of glutaraldehyde over the silanized surface. The results show that in the surface-bound enzyme, the retention of activity is not more than 3 to 4%, and the free energy of activation for the lactose hydrolysis appears to be decreased as well.

9.6.3.3 Photopatterning Enzymes inside Microfluidic Channels

Photopatterning of enzymes at desired locations can be carried out in a microfluidic network. The three possible enzymes for photopatterning are glucose oxidase (GOD), horseradish peroxidase (HRP), and alkaline phosphatase (ALP). The photoactive molecules are adsorbed from aqueous solutions and attached to protein-coated interfaces, as shown in Figure 9.44. The photoimmobilization of enzymes is schematically shown in Figure 9.44 using a photobleaching process. The process involves passivation of the microfluidic surface with fibrinogen, photobleaching of selected patterns using B4F, immobilization of enzymes through linking with streptavidin, and monitoring the enzymatic reaction at the patterned areas.

As can be seen, enzymes will bind to the surface only at the locations exposed to light and will form patches. The surface of the microchannel is passivated with a fibrinogen monolayer before the fluorophore's attachment.

The fluorophore molecules are bleached by irradiation with an Ar^+/Kr^+ laser, and this photobleaching creates highly reactive species that will attach themselves to the adsorbed proteins. The enzymes linked to streptavidin or avidin are then immobilized and the reaction can be monitored. The results show that the fibrinogen coating results in a high density of specifically bound streptavidin molecules and low nonspecific absorption.

This method permits precise control over enzyme patterning, in addition to being fast.

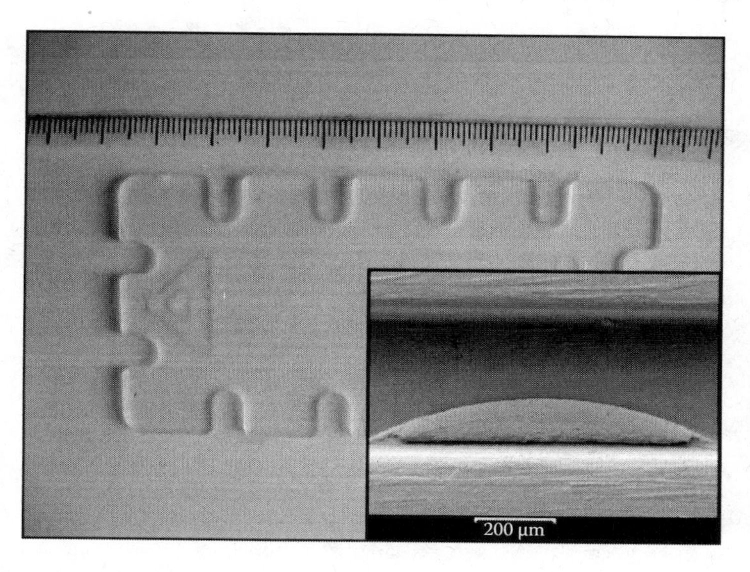

FIGURE 9.42 Photograph of PDMS microchip with multichannel. Inset: A SEM picture of a passive mixing element. (Adapted from Thomsen and Nidetzky, *Eng. Life Sci.*, 8, 40-48, 2008.)

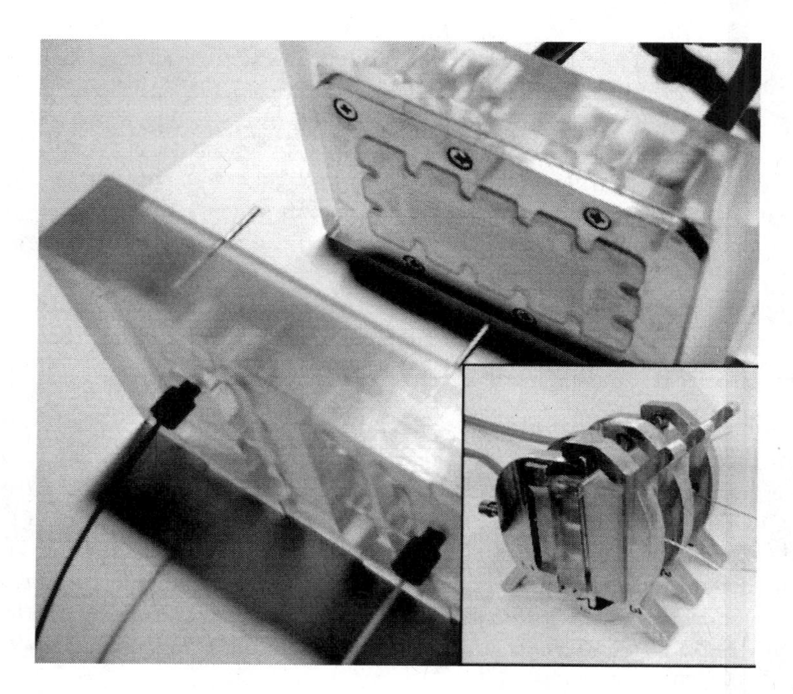

FIGURE 9.43 Housing and assembly of a microreactor. The PDMS microchip is placed between the PMMA parts of the housing and clamped. Inset: The assembled microreactor. (Adapted from Thomsen and Nidetzky, *Eng. Life Sci.*, 8, 40–48, 2008.)

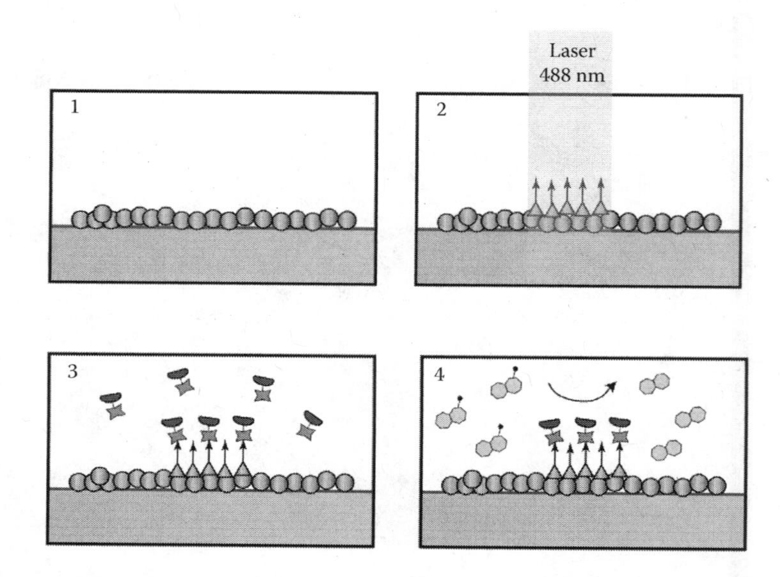

FIGURE 9.44 Photoimmobilization of enzymes. (1) Passivation of microfluidic surface with the flow of fibrinogen. (2) Surface is attached with biotin-4-fluorescein (B4F). Selective exposure to laser irradiation at 488 nm through mask or x,y scanning. This step is called photobleaching. (3) Exposure to streptavidin-linked enzyme. Binding happens at the patterned areas of B4F. (4) Patterned enzymes are used for monitoring the reaction process. (Adapted from Holden et al., *Anal. Chem.*, 76, 1838-43, 2004.)

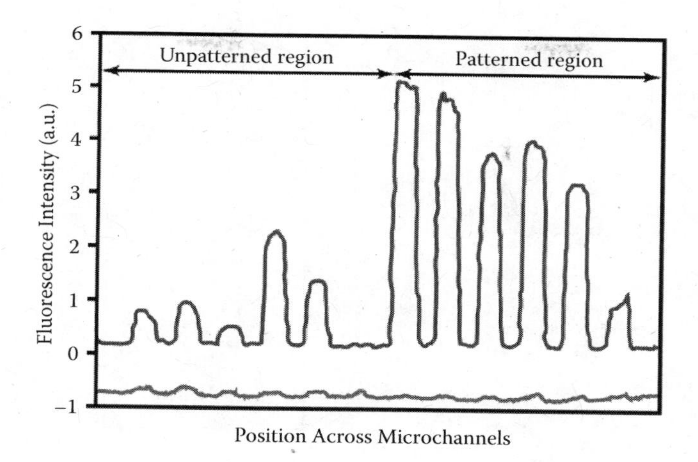

FIGURE 9.45 Fluorescence emanating from a protein-patterned region of microchannels. Fluorescence intensities from unpatterned and patterned regions are shown. Fluorescence indicates specific adsorption. (Adapted from Holden et al., *Anal. Chem.* 76, 1838-43, 2004.)

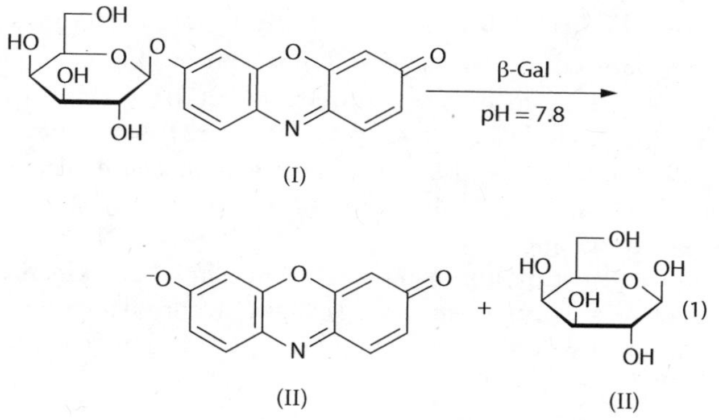

β-Gal
pH = 7.8
(I)
+
(1)
(II)
(II)

FIGURE 9.46 The general on-chip enzymatic reaction between β-galactosidase (β-Gal) and RBG (I-RBG, III-resorufin).

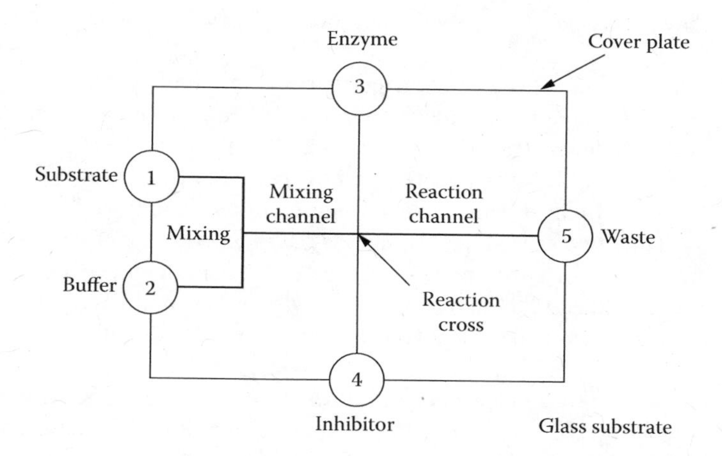

FIGURE 9.47 Schematic of an enzyme analysis chip: 1 = substrate, 2 = buffer, 3 = enzyme, 4 = inhibitor, 5 = waste. (Adapted from Hadd et al., *Anal. Chem.* 69, 3407-12, 1997.)

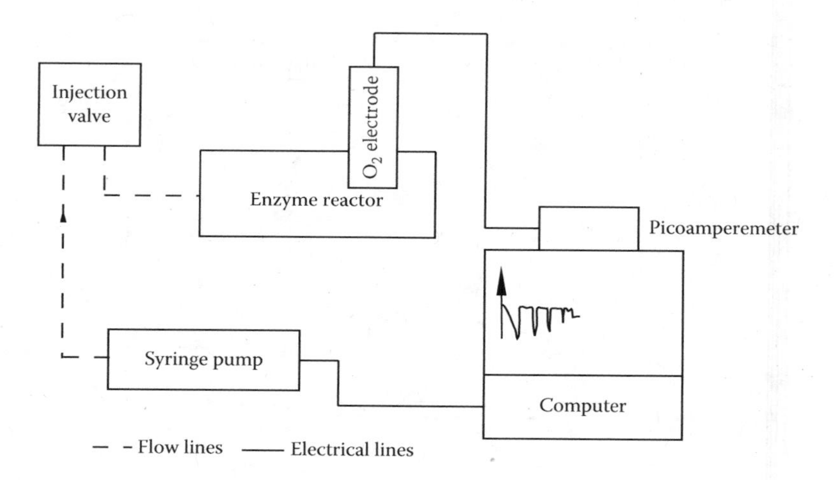

FIGURE 9.48 The flow injection set-up. (Adapted from Drott et al., *J. Micromech. Microeng.*, 7, 14-23, 1997.)

9.6.3.4 Integrated Microfabricated Device for an Automated Enzymatic Assay

Electrokinetic flow can be used to control the dilution and mixing of the reagents used in an enzyme assay. The reaction of hydrolysis between a fluorogenic substrate, resorufin β-D-galactopyranoside (RBG), and β-galactosidase (β-Gal) (Figure 9.46) is studied by monitoring the fluorescence of resorufin, which is the product of the reaction. The reagents are mixed by using electrokinetic pumping, and the microfluidic control is realized by adjusting the potentials applied at each reservoir, as shown in Figure 9.48a.

The experiment revealed that the diffusion associated with electrokinetically induced flow is rapid, and the diffusion across 17 μm of the mixing channel occurred approximately in 0.5 sec. The amount of reagents consumed was reduced by four orders of magnitude over a conventional assay using this method.

9.6.3.5 Silicon Microstructured Enzyme Reactor with Porous Silicon as the Carrier Matrix

Porous materials have a high surface-to-volume ratio, and they can enhance surface-area-dependent activities.

In this case, porous silicon is used as a carrier matrix to immobilize glucose oxidase due to the increased surface area available for enhanced coupling and enzyme activity. The enzyme activity is monitored by using a colorimetric assay and compared with a microreactor without the porous layer. A hydrogen peroxide (Trinder) reagent containing peroxidase and glucose is pumped through the reactor (Figure 9.49). An absorbance detector is used to measure the color shift that measures the enzyme activity. The same procedure was carried out in the absence of the porous silicon layer. Measurements were performed at different glucose concentrations at increasing flow rates. Glucose measurements were performed in the flow injection mode, as shown in Figure 9.48.

The oxygen consumption, which is the measure of the conversion of glucose, was monitored with a Clark type electrode.

The results of this study have demonstrated the effect of the pore size on the activity of the enzyme.

The injection volume was 0.5 μl. Oxygen consumption is monitored with a Clark type electrode when glucose in PBS is injected into the flow.

The turnover rate is defined as the number of catalyzed glucose molecules per unit time and is estimated from the recorded absorbance shift. An increase in enzyme activity by a factor of one

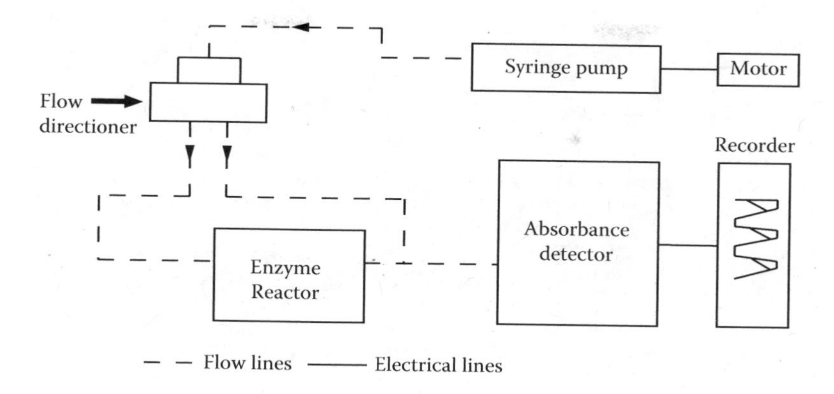

FIGURE 9.49 Schematic of the system setup for the enzyme activity determination (-----, flow lines; ______, electrical lines). (Adapted from Drott et al., *J. Micromech. Microeng.* 7, 14-23, 1997.)

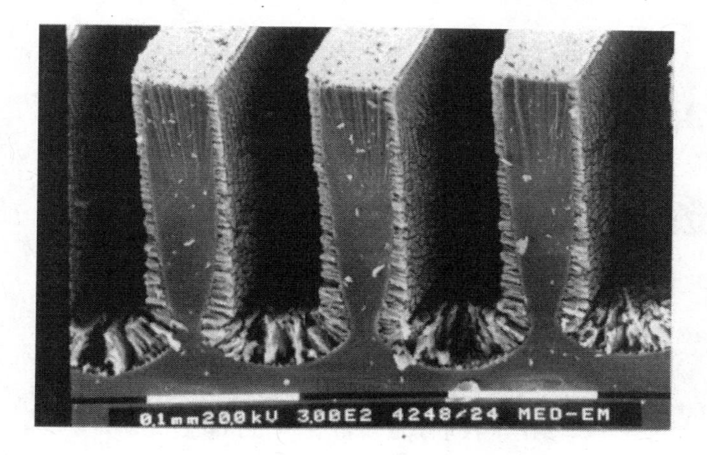

FIGURE 9.50 Scanning electron micrograph of the porous channel wall. (Adapted from Drott et al., *J. Micromech. Microeng.* 7, 14-23, 1997.)

hundred was achieved for the reactor anodized at 50 mA/cm^2, when compared to the nonporous reference microreactor. The use of porous silicon in the enzyme reactors proved to give better results than with microbeads serving as the carrier matrix. Figure 9.50 shows the scanning electron microscopy (SEM) view of the microreactor.

9.6.3.6 Enzymatic Reactions Using Droplet-Based Microfluidics

On the microfluidic scale, glucose analyzers are usually based on continuous flow, implemented by using syringe pumps or electroosmotic pumping. Microdialysis sampling has also been coupled with continuous flow microfluidic devices for real-time monitoring of glucose. The most commonly used biorecognition element is the immobilized glucose oxidase, and the enzymatic assays are performed in solution phase coupled with electrophoretic separation. Detection is typically implemented using electrochemical methods such as amperometry or optical methods such as absorbance or chemiluminescence. An alternative method toward microfluidic systems is to manipulate the liquid as discretized microdroplets. This approach has several advantages over continuous flow systems, the most important being the ease of fabrication, reconfigurability, and the scalability of architecture.

Electrowetting-based digital microfluidics platforms to assay glucose in physiological samples have been developed. A colorimetric enzyme-kinetic method based on Trinder's reaction is used for the determination of glucose concentration. Glucose is enzymatically oxidized to gluconic acid and

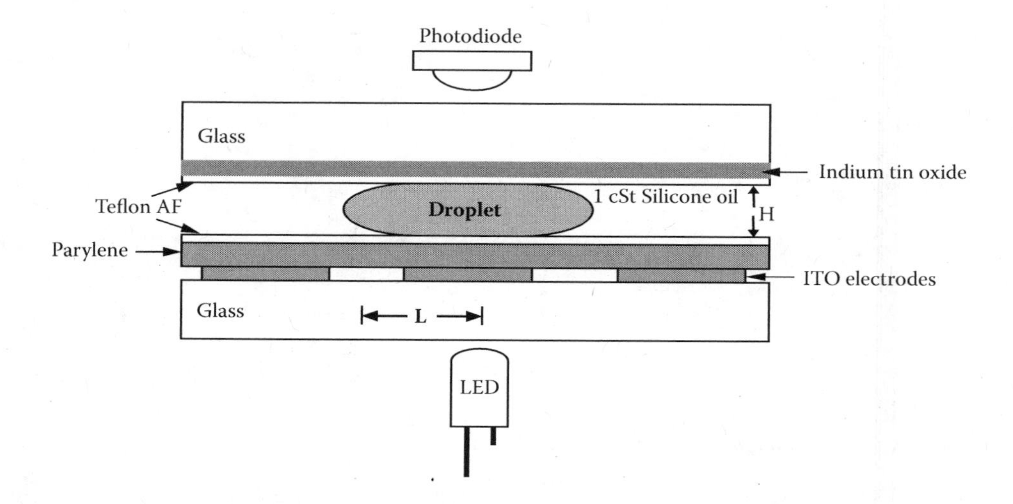

FIGURE 9.51 Schematic of electrowetting chip with optical detection. (Adapted from Srinivasan et al., *Anal. Chim. Acta* 507, 145-50, 2004.)

hydrogen peroxide in the presence of glucose oxidase. The hydrogen peroxide reacts with 4-amino antipyrine (4-AAP) and *N*-ethyl-*N*-sulfopropyl-*m*-toluidine (TOPS) in the presence of peroxidase (EC 1.11.1.7) to form violet-colored quinoneimine, which has an absorbance peak at 545 nm, as shown in Figure 9.52.

The Trinder's reactions are given as follows:

$$\text{Glucose} + H_2O + O_2 + \text{glucose oxidase} \rightarrow \text{gluconic acid} + H_2O_2$$

$$2H_2O_2 + \text{4-AAP} + \text{TOPS peroxidise} \rightarrow \text{quinoneimine} + \text{4H2O}$$

The rapid mixing and biocompatible interfacial chemistry make droplet-based microfluidics a suitable platform to measure single-turnover kinetics of enzyme ribonuclease A (RNase A) within millisecond resolution.

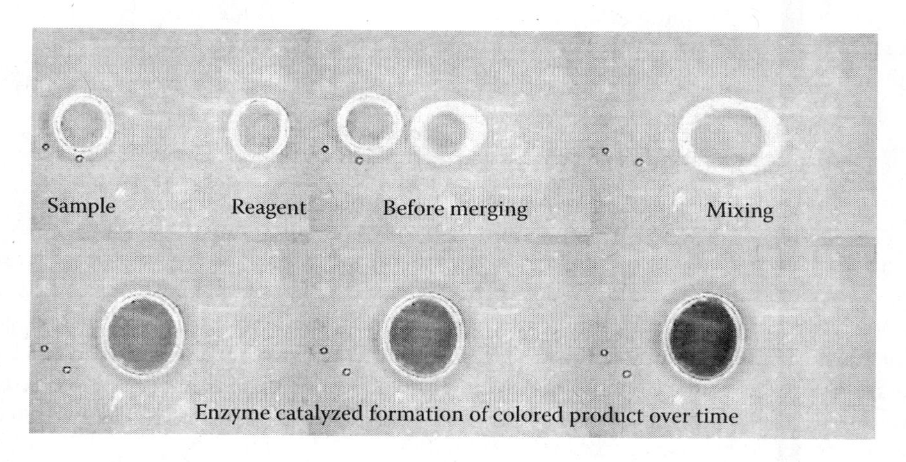

FIGURE 9.52 Snapshots of the sample and reagent droplet from the top during a glucose assay. The concentration of the sample droplet is 800 mg/dl. (Reproduced from Srinivasan et al., *Anal. Chim. Acta* 507, 145-50, 2004. With permission.)

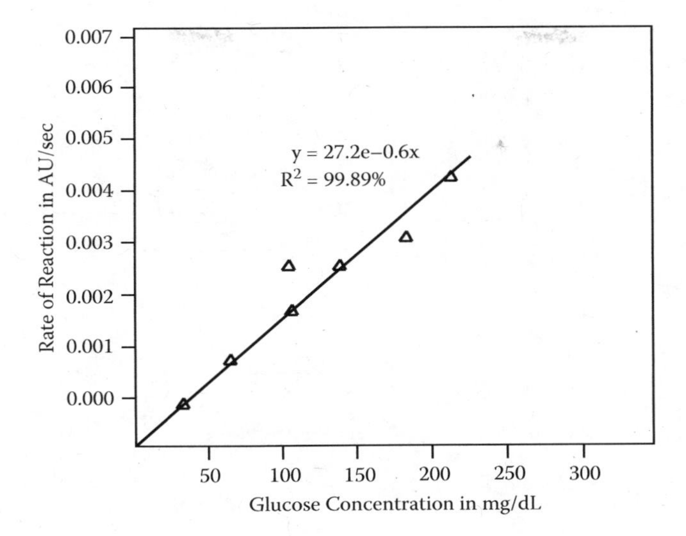

FIGURE 9.53 Variation of rate of reaction against glucose concentration. (Reproduced from Srinivasan et al., *Anal. Chim. Acta*, 507, 145–50, 2004. With permission.)

As shown in Figure 9.54, measurements of the kinetics of RNase A turnover were performed for three substrate concentrations and with only 66 nl. A kinetic trace is obtained from time-averaged images of the fluorescence intensity. Enzymatic reactions can be performed on timescales from seconds to hours by controlling the dispersion. For example, the conversion of alkaline phosphatise was also studied within plugs on a millisecond timescale, while the activity of luciferase within the droplets of viscous solutions was studied within a millisecond resolution.

9.6.4 Synthesis of Nanoparticles and Biomaterials in Microfluidic Devices

Microfluidic devices have been developed for synthesizing nanoparticles and biomaterials with diameters of nanometers to micrometers. Homogeneous metal nanoparticles are generated in continuous flow microfluidic reactors by taking advantage of the effective mixing and good heat and mass transfer. The effective mixing in short periods of time will ensure the one-time nucleation necessary for obtaining a narrow size distribution of nanoparticles.

Microfluidic chemical reactors were used for the continuous sol-gel synthesis of colloidal silica particles as well. The change in particle sizes and their distribution was studied when the linear flow velocities were varied. Figure 9.56 shows one of the PDMS reactor configurations examined in this work. In this design, micromixing is accomplished by fluid "layering." Spherical silica particles are synthesized by mixing a diluted solution of TEOS in ethyl alcohol with a solution of ammonia, water, and ethyl alcohol.

9.6.5 Microfluidic Devices for Separation

Cell or particle sorting, depending upon their sizes, has a vast potential of applications. Particle or cell sorting can be implemented in many ways, by either using a mechanical sieve of different sizes, or hydrodynamically by using flow geometries or electrostatic or magnetic effects on cells.

The most widely used separation principles in the macroworld are chromatography and electrophoresis. Chromatography is based on different distributions of compounds in a mixture of two phases: one called stationary and the other mobile. Electrophoresis is based on the difference in the movement of charged components under an electrical field. The first scaled-down

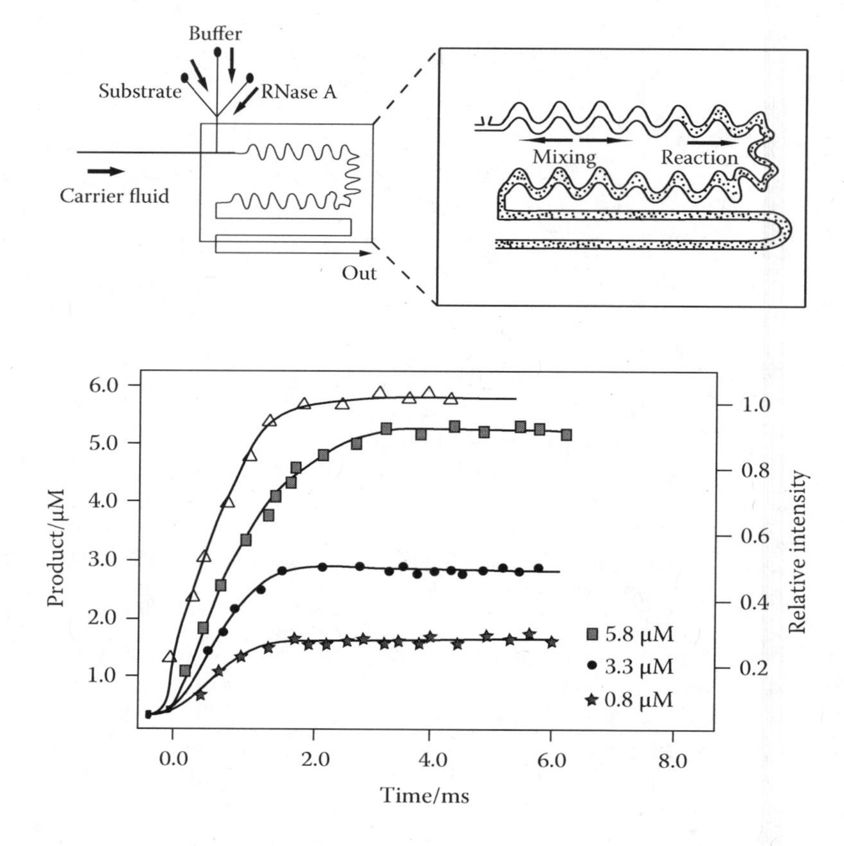

FIGURE 9.54 Kinetic analysis of RNase A in plugs in millisecond timescale. (a) Left: Experimental setup. Right: Scheme that shows the time-averaged intensity of aqueous plugs and carrier fluid moving through the microchannel. (b) Graph of the experimental kinetic data for three substrates. (Reprinted with permission from H. Song and Ismagilov, *J. Am. Chem. Soc.*, 125, 14613–19, 2003.)

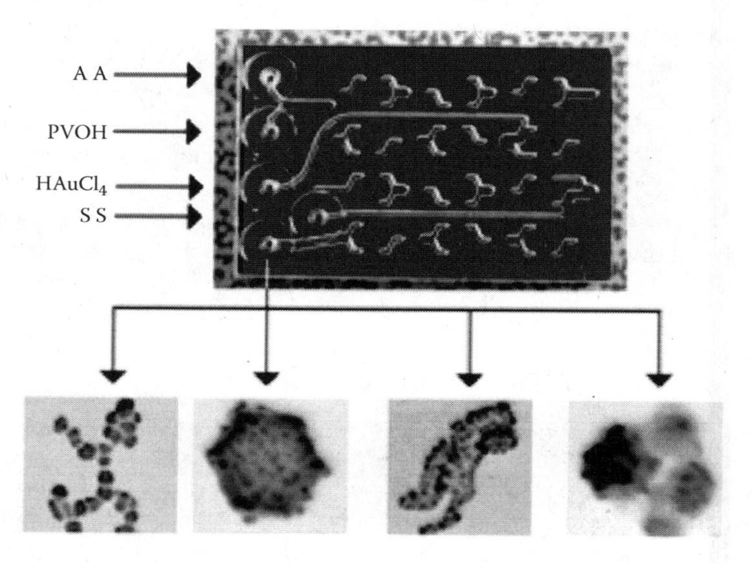

FIGURE 9.55 Microfluidic mixer design with a three-step static micromixer. Isolated and clustered Au nanoparticles were synthesized. (Adapted from Hung et al., *J. Med. Biol. Eng.* 27, 1-6, 2007.)

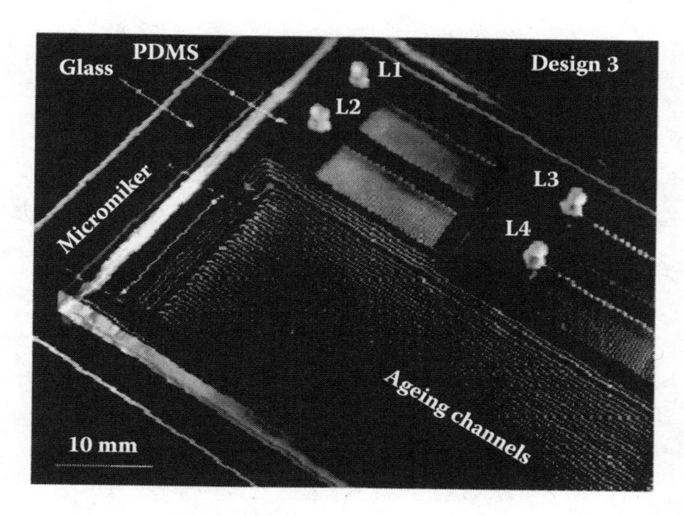

FIGURE 9.56 Photograph of a segmented flow Lab-on-a-chip with micromixer and ageing channels for the colloidal synthesis of silica nanoparticles. Reagents are passed through four liquid inlets into a micromixer. Gas is used to create gas-liquid plugs. (Reprinted from Jacobson and Ramsey, *Anal. Chem.*, 68, 720–23, 1996. With permission.)

electrophoresis instrument was launched in 1999 by Agilent (2100 Electrophoresis Bioanalyzer), and this was the first commercial microfluidics-based platform for the analysis of nucleic acids and cells.

9.6.5.1 Separation of Blood Cells

Magnetophoresis using continuous diamagnetic capture (DMC) can be utilized for whole cell purification, and electrical impedance spectroscopy (EIS) for electrophysiological analysis of purified cells. This method is based on the inherent magnetic properties of cells as shown in Figure 9.57.

When an external magnetic field is applied normally to the microchannel, the red blood cells act as paramagnetic particles and are forced away from the ferromagnetic wire, while the white blood cells act as diamagnetic particles and are drawn closer, as shown in Figure 9.60. Magnetic cell separation is generally considered to be nonlethal to cells.

Whole blood was purified using magnetophoresis in the continuous diamagnetic capture (DMC) mode to increase the concentration of white blood cells. The results show that 89.7% of the red

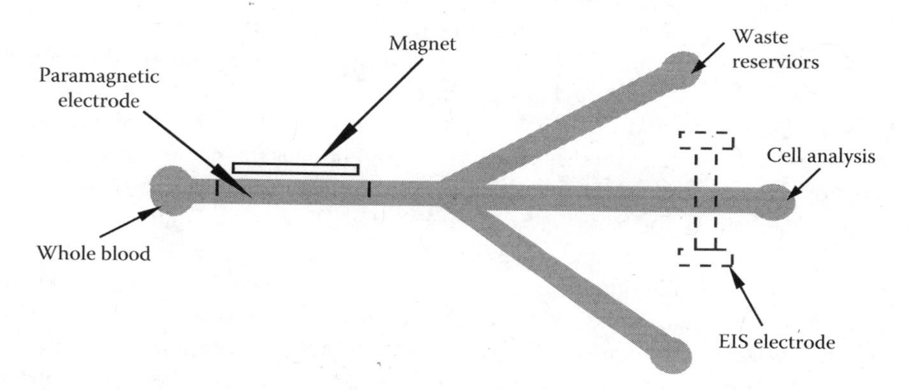

FIGURE 9.57 Schematic of a LOC for blood sample analysis. (Adapted from Han et al., *J. Semicond. Technol. Sci.* 5, 1-9, 2005.)

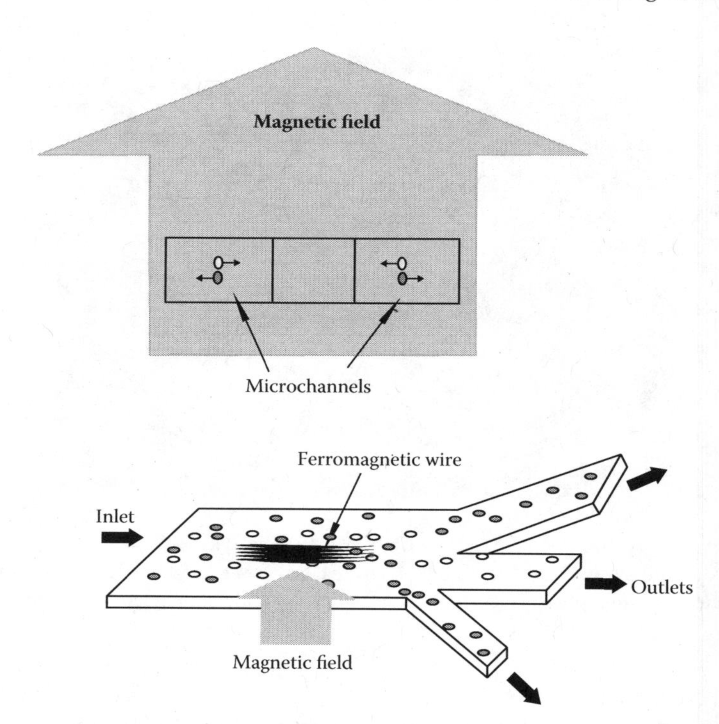

FIGURE 9.58 Schematic view of the LOC used for blood sample analysis (a) •, red blood cell (paramagnetic); °, white blood cell (diamagnetic). (Adapted from Han et al., *J. Semicond. Technol. Sci.* 5, 1-9, 2005.)

blood cells (RBCs) could be continuously separated out from a whole blood sample within 5 min using an external magnetic flux.

9.6.5.2 Cell or Particle Sorting

In the cross-flow filtration shown in Figure 9.59, particles smaller than lateral sieve or micropore size will enter the side channel, while the main flow at the center channel will eventually have particles of sizes larger than the pores.

The laminar flow in the bottom half of the channel aligns the particles to the top half of the channel, as shown in Figure 9.59b. The particles travel along different trajectories, depending upon the size or mass, and are collected at different channels.

When a laminar flow occurs through arrays of posts or obstacles, larger particles tend to deviate laterally from the flow, while the smaller particles stay close to the initial lateral position. The amount of lateral shift depends on the relative size of the particle in relation to the geometry and pitch of the flow obstacles, as shown in Figure 9.59c.

Both cross-flow filtration and laminar flow methods are considered to be highly portable, as they only require fluidic pumps (or other pressure sources) besides the microfluidic separator chip. Cell and microparticle separation in microfluidic systems has recently gained significant attention in sample preparations for biological and chemical studies. Microfluidic separation is typically achieved by applying differential forces on the target particles to guide them into different paths. Various strategies of microfluidics have been used for cell and particle separation by making use of dielectrophoretic, optical, magnetic, and acoustic forces. Each approach proved effective in separating binary mixtures of microscale particles and, in some cases, even the separation of polydisperse mixtures was possible. The method to be used is decided based on the needs of the application, including separation criteria, resolution, efficiency, and throughput, as well as physiological effects. Although physiological effects due to noninertial forces in microfluidic cell separation are

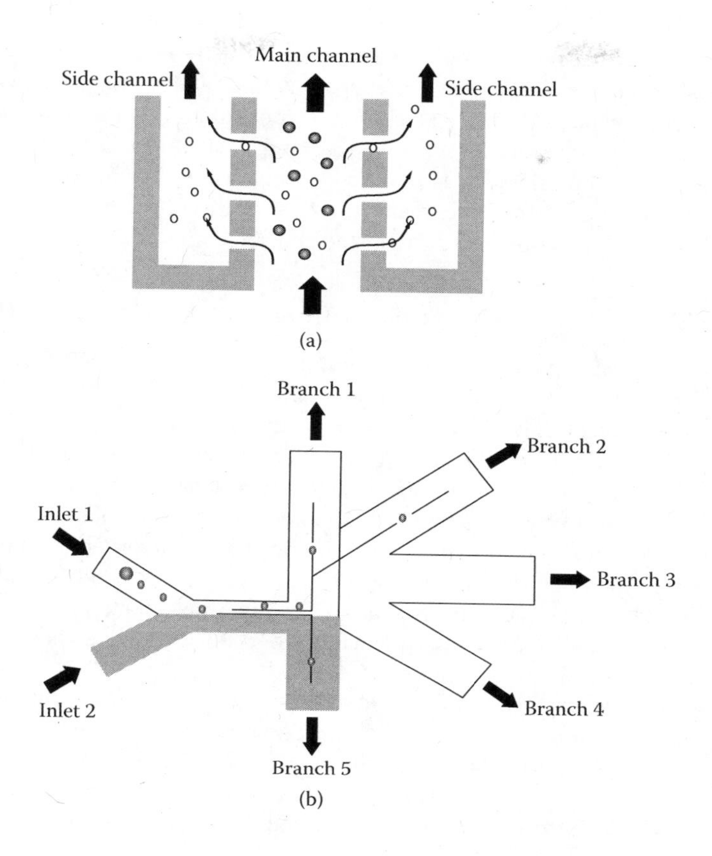

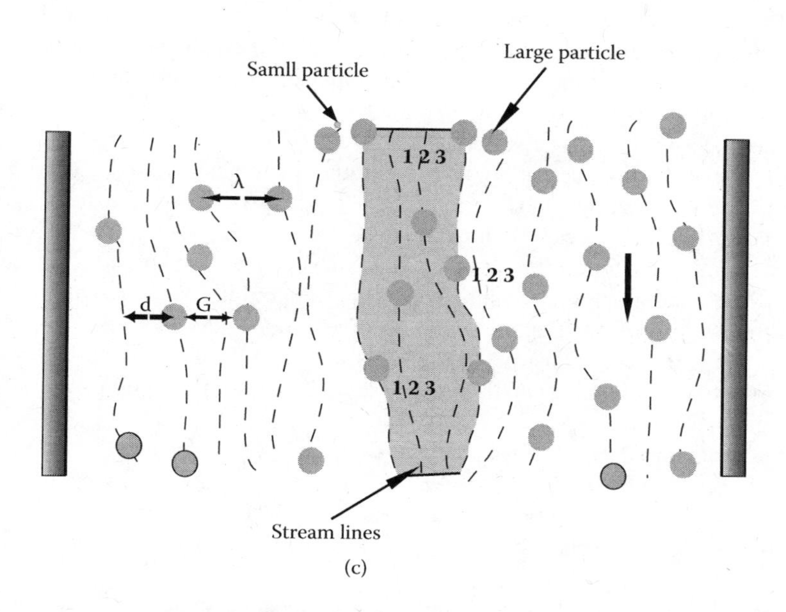

FIGURE 9.59 Particle separation with cross-flow filtration and laminar flow. (a) Cross-flow filtration. (b) Particle sorting using asymmetric pinched flow. (c) Deterministic lateral displacement. (Adapted from Tsutsui et al., *Mech. Research Comm.* 36, 92-103, 2009.)

not considered harmful, further investigation should be performed to elucidate the significance of such forces, specifically under microfluidic separation conditions. Technological advances allowed the realization of precise channel geometries, leading to separation of cells and particles through a combination of microstructures and laminar flow around them. In addition, force transducers were used to manipulate microparticles by noninertial forces. Noninertial forces include dielectrophoretic (DEP) force, optical gradient force, magnetic force, and acoustic primary radiation force, leading to numerous microchip designs for particle and cell separation.

9.7 MICRO TOTAL ANALYSIS SYSTEMS (μTAS) AND LAB-ON-A-CHIP (LOC)

Miniaturized analytical devices fabricated by integrating multiple BioMEMS functionalities are called micro total analysis systems (μTAS) or lab-on-a-chip (LOC). μTAS are capable of performing the sampling, sample transport, filtration, dilution, chemical reactions, separation, and detection requirements on a micron scale, as multiple functionalities are integrated on a single module. μTAS have many advantages, as they are able to perform simultaneous analysis of a large number of biomolecules of interest in a short time by using a small amount of sample volumes. Since the μTAS are miniaturized systems, they can be packaged and interfaced with integrated circuits to make portable systems facilitating the point-of-care testing (POCT) and point-of-need (PON) applications. If the reader is interested, a large literature is available on the implementation and application of μTAS and LOC for further study. Only a few applications of LOC are covered here for demonstration purposes.

The concept of μTAS was proposed by Manz et al.[35] in 1990. In this work, the theoretical performance of miniaturized systems facilitating flow injection analysis, chromatography, and electrophoresis is compared with that of the existing macroscopic scale systems. Miniaturization of total analysis systems was the aim due to the enhanced performance, rather than the reduction of size. A generalized block diagram showing different process flows involved in μTAS is shown in Figure 9.60.

The main processes that are generally involved in μTAS are illustrated in Figure 9.62. Samples from different chemical or biological reagents will be taken for the pretreatment and passed through different components; for example, it can be a mixer. After the reagents are mixed, they can be introduced to a chamber where the reaction will take place. The reaction is followed by molecule separation and detection. Signal processing and analysis can also be carried out with the help of an integrated circuit incorporated in the system.

A summary of transduction, actuation, and integration methods that are available for the development of μTAS and LOC is given in Figure 9.61. One could use these methods depending upon the application, sensitivity, and material platform.

μTAS based on an optical microfluidic system has also been developed for cell detection by integrating a liquid-chromatography-based molecules separation unit.[36] The system consists of a flow cell fabricated by anisotropic etching and bonding of a silicon wafer. A three-dimensional microfluidics flow system with an ion-sensitive field effect transistor (ISFET)[37] was used for the detection of pH and phosphate measurements. This system had two silicon micropumps connected with the ISFET using silicon rubber.

In the case of LOC for DNA fragment analysis, samples of DNA and enzymes were injected into the reaction chamber by electrophoresis, and a CCD camera was used to image the particles with laser-induced florescence.[38] The fluorescence measurements were carried out using the photomultiplier tube. This device was a microfluidic chip with optical detection and flow visualization tools integrated externally.

A μTAS was fabricated by integrating multiple components, such as a nanoliter liquid injector, sample mixer and positioning system, temperature-controlled reaction chamber, electrophoretic particle separation system, and fluorescence detection system on a silicon wafer. This device was fabricated by monolithic integration of multiple modules and a photodetector onto a silicon

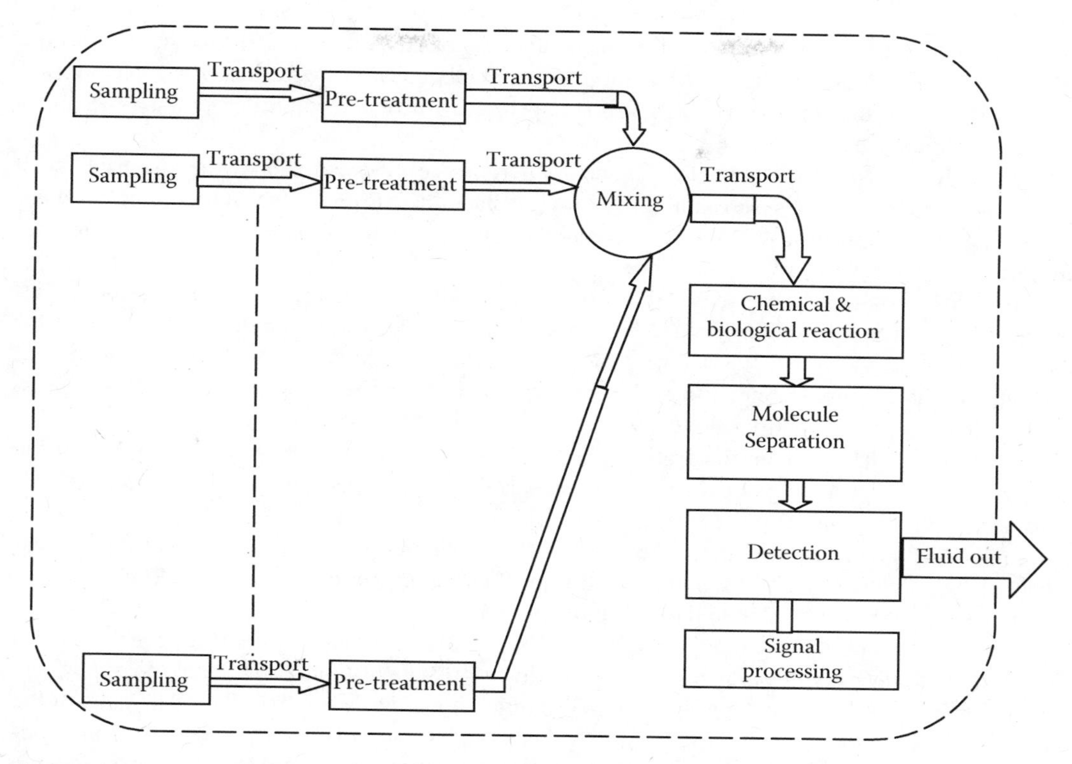

FIGURE 9.60 A generalized block diagram of process flow in μTAS.

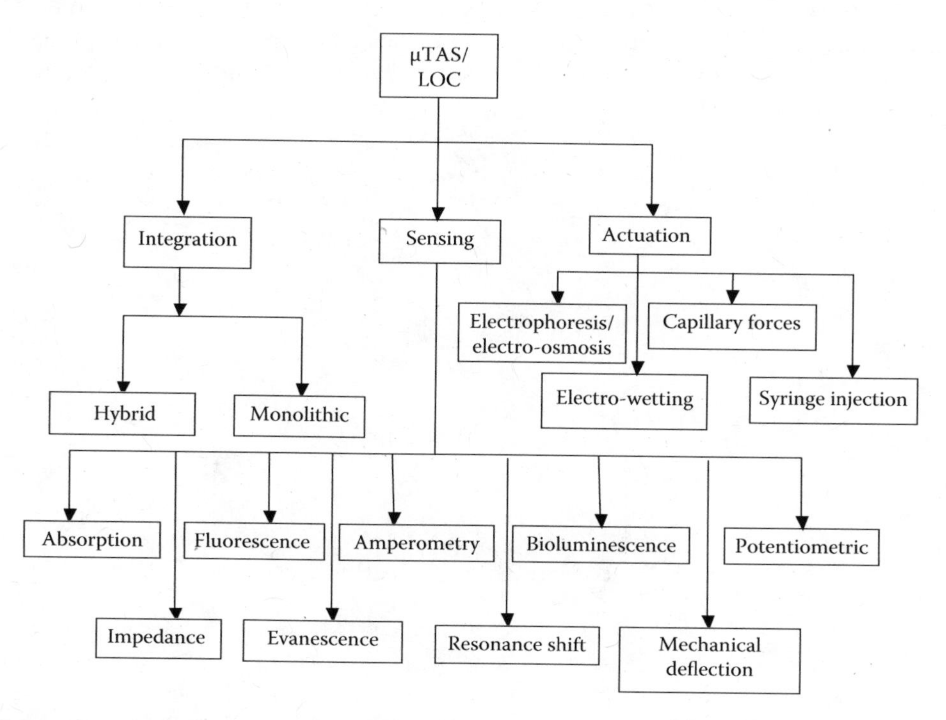

FIGURE 9.61 Classification of μTAS/LOC based on the integration process, transduction, and actuation methods.

platform, and demonstrated to be useful for the analysis of DNA at nanoliter volumes. Moreover, this device was able to operate at lower voltages, thereby being suitable for portable applications.[39] Most importantly, the novel idea of utilizing a temperature sensor and controller for the μTAS applications was proposed in the work.

As an alternate to silicon for the fabrication of microdevices, some of the polymer materials, such as polydimethylsilaxane (PDMS) and polymethylmethacrylate (PMMA), are found to be excellent for μTAS fabrication because of the cost-effective fabrication methodologies and excellent material properties supporting μTAS functionalities.

Miniaturization science had sensed the possibilities of potential application of electroosmotic phenomena for microfluidic transportation. Molecule separation capabilities of electroosmotic flow for applications in μTAS were first tested in 2001. Guenat et al.[40] realized a μTAS by integrating an electroosmotic micropump. Herein, a three-layer microfluidic structure was fabricated on Pyrex glass, and two reagents were electroosmotically driven through two microchannels into a micromixer.

Integration of μTAS on an integrated optical platform has several advantages, mainly that the semiconductor fabrication processing technologies could be extended for the fabrication of microphotonics and microfluidic parts. A microfluidic channel on a silica-on-silicon waveguide, sealed with polysilicon by anodic bonding, was used in 2001 for flow cytometry. Absorption and fluorescence detection were carried by pumping 100 μm flourophores to the microchannel and exciting with a laser of 488 nm coupled to the waveguide.

Several microphotonics components, such as waveguides, a beam splitter, an S-bend, and a Y-branch on a silica-on-silicon platform by flame hydrolysis deposition (FHD), pholithography, and silica micromachining were integrated into μTAS,[41] as shown in Figure 9.62. The schematic of the device is shown in Figure 9.63. Two thick PDMS layers were bonded on the waveguide for incorporating the fluidic reservoir. The liquid containing the 24 μM Cy5 dye was injected by using a syringe pump in the microchannel, and a fluorescence measurement was taken.

A polymer μTAS for the detection of oxygen, lactate, and glucose in blood was also reported.[42] Herein, microfluidics channels were fabricated on a cyclic olefin copolymer (COC) by using injection molding. For fabricating the device, two types of fluidic systems have been proposed. One

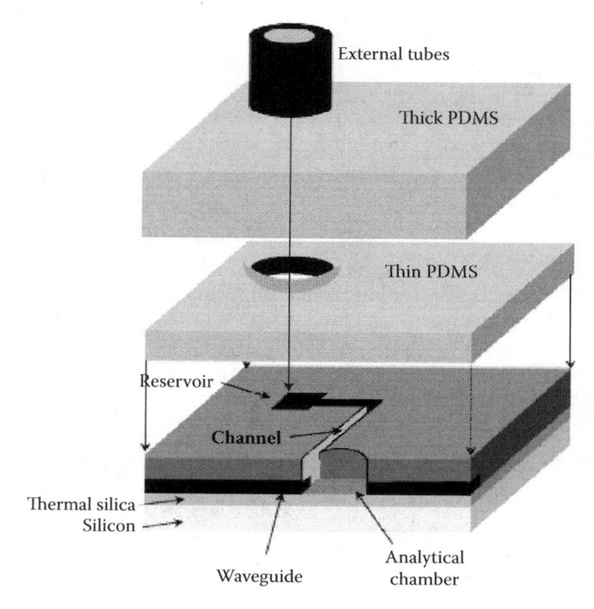

FIGURE 9.62 Schematic of the device fabricated by Ruano et al. (Adapted from Ruano et al., *Biosensors Bioelect.*, 18, 175–84, 2003.)

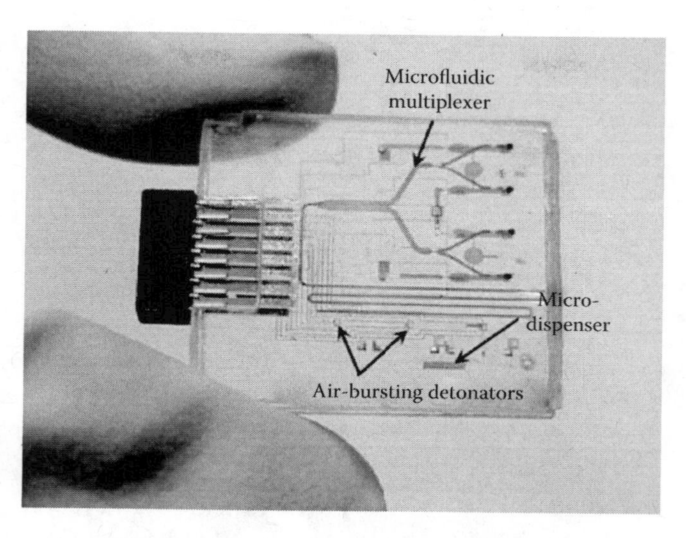

FIGURE 9.63 μTAS with sPROM microfluidic control systems and air-bursting detonators. (From Ahn et al., *Proc. IEEE*, 92, 154–73, 2004.)

of them is the smart programmable microfluidic system (sPROM), comprised of microvalves and microdispensers. The second one had air-bursting detonators, wherein the fluid was pumped into the channel by pressurized gas stored in the chamber, while a microheater was embedded in the channel to pressurize the gas. The device is shown in Figure 9.35.

Integration of a dye laser, waveguides, microfluidic channel, diffusion mixer, and photodetectors onto μTAS has also been demonstrated, as shown in Figure 9.64. The waveguides and microfluidics are fabricated in SU8 on a silicon platform. Fluorescence detection was achieved by the integrated photodetector and dye laser.

Several hybrid and monolithic integrated μTAS[53,54] were also fabricated. A μTAS platform supporting genetic analysis, heaters, temperature sensors, and valves for controlling two nanoliter reactors was also fabricated. An electrophoretic separation unit was also integrated on glass and silicon. It was demostrated for the detection of the sequence-specific A/LA/1/87 strain of the influenza virus.[44]

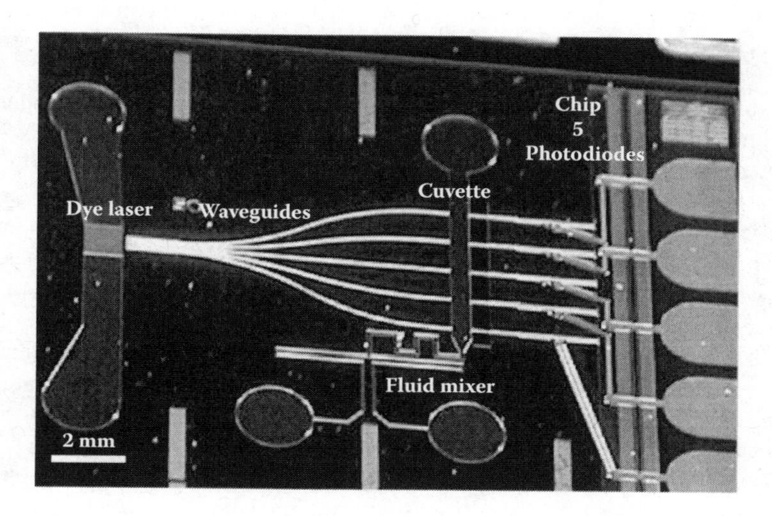

FIGURE 9.64 Fully integrated μTAS. (From Balslev et al., *Lab Chip*, 6, 213–17, 2006.)

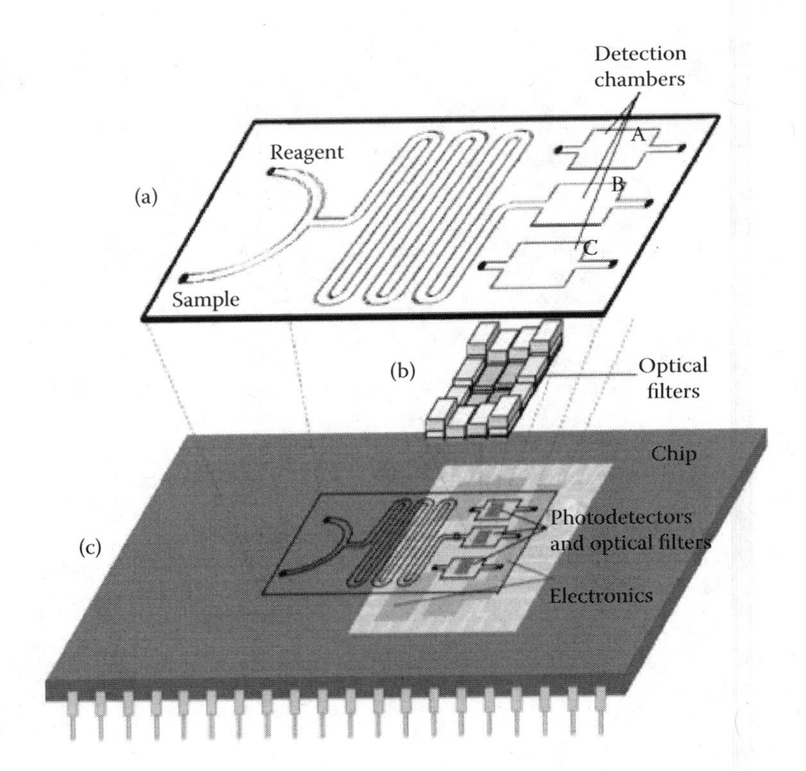

FIGURE 9.65 Layout of the device fabricated by Minas et al. (From Minas et al., *Lab Chip*, 5, 1303–9, 2005.)

A μTAS for photospectrometric analysis comprised of a microfluidic channel of SU8, thin-film Fabry-Perot filters, detection, and readout units was also fabricated by the CMOS microelectronics process.[45] The layout of the device is shown in Figure 9.65. Herein, detection and concentration estimations were performed through the absorption of light. Use of a Fabry-Perot filter facilitates the use of a white light source. A microchannel containing the biomolecule and a specific reagent was illuminated from the top of the device. The transmitted light intensity was modulated by the reagent-biomolecule reaction, and hence the quantification of the concentration of the biomolecule was carried out. The CMOS readout system was designed to give the signal in digital format.

Chandrasekaran and Packirisamy[46] integrated micropumps and a Mach-Zender interferometer (MZI) with a silicon-on-insulator (SOI) waveguide. A SEM image of the μTAS is shown in Figure 9.66. A fluorescence detection system by the hybrid attachment of the silicon microfluidic system with a spectrometer-on-chip was also demonstrated.[47] Silicon's surface properties, in particular, the surface affinity to biomolecules, were investigated through fluorescence measurements from fluorescence-tagged sheep antibody-antigen reaction in the silicon microfluidic channel. As a result, silicon was found to have good bioaffinity dependent on pH values.

The optical properties of gold film were used for the fluorescence enhancement in LOC. Herein, the microchannel in the form of a V-groove was fabricated by anisotropic machining of silicon, as shown in Figure 9.67. A thin gold layer was deposited on the V-groove, fluorescence-tagged biomolecules were injected into channel, and the fiber for the excitation and collection path was equipped on top of the channel.

Recently, a simple and cost-effective microfluidic channel integration technique on a silicon-on-silicon (SOS) waveguide was established by using a wafer dicing saw, in which the dicing depth was adjusted in order to get a 100 μm channel on an SOS chip. This chip was integrated on a PDMS

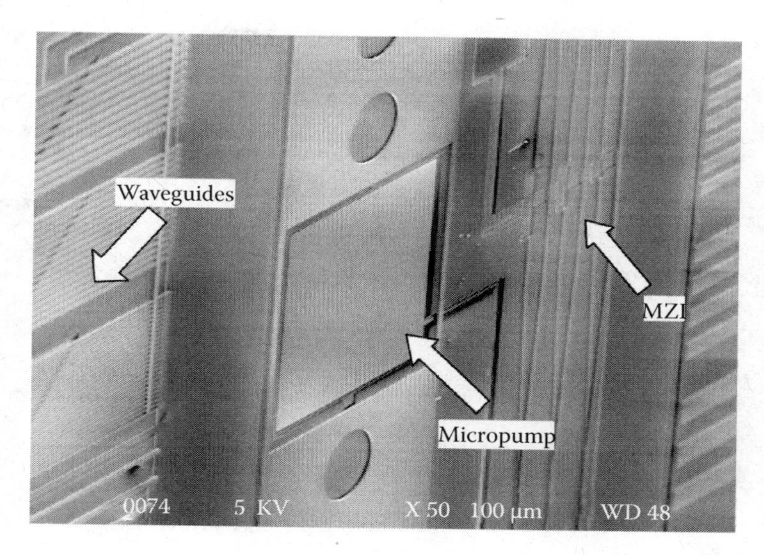

FIGURE 9.66 SEM image of μTAS having a micropump, waveguide, and Mach-Zender interferometer on the SOI platform. (From Chandrasekaran and Packirisamy, *Sens. Rev.*, 28, 33–38, 2008.)

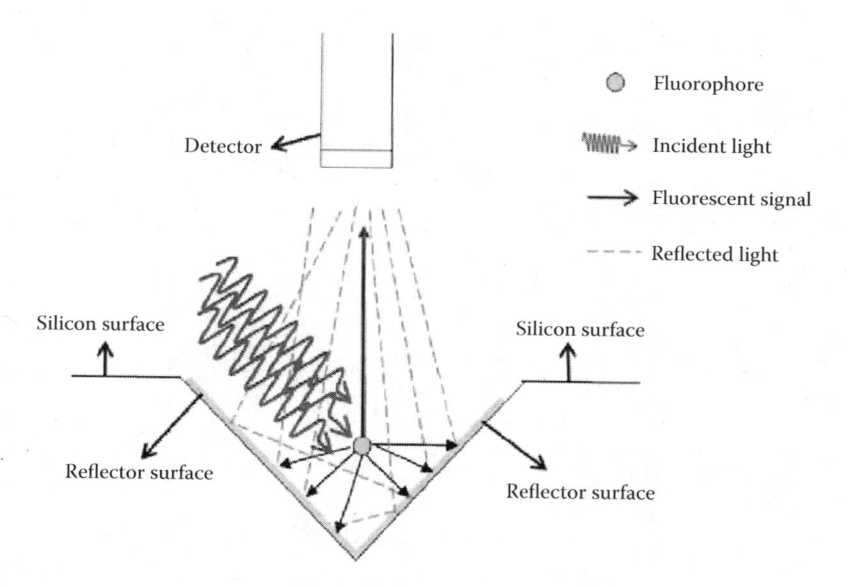

FIGURE 9.67 (See color insert.) Fluorescence enhancement from gold-sputtered V-groove.

platform containing microfluidic parts and sealed with a PDMS layer. This SOS-PDMS device was demonstrated for the fluorescence detection of quantum dots. A photograph of the SOS-PDMS and the optical characterization setup of the device are shown in Figure 9.68.

The fully integrated biochip developed, shown in Figure 9.69, is very small and incorporates all the functionality required for testing clinically relevant parameters from human blood. Furthermore, the analyzer developed in this work is of immediate relevance to portable biochemical detection applications. In addition, the disposable plastic biochip has successfully been tested for the measurements of partial oxygen concentration, glucose, and lactate level in human blood using an integrated biosensor array.

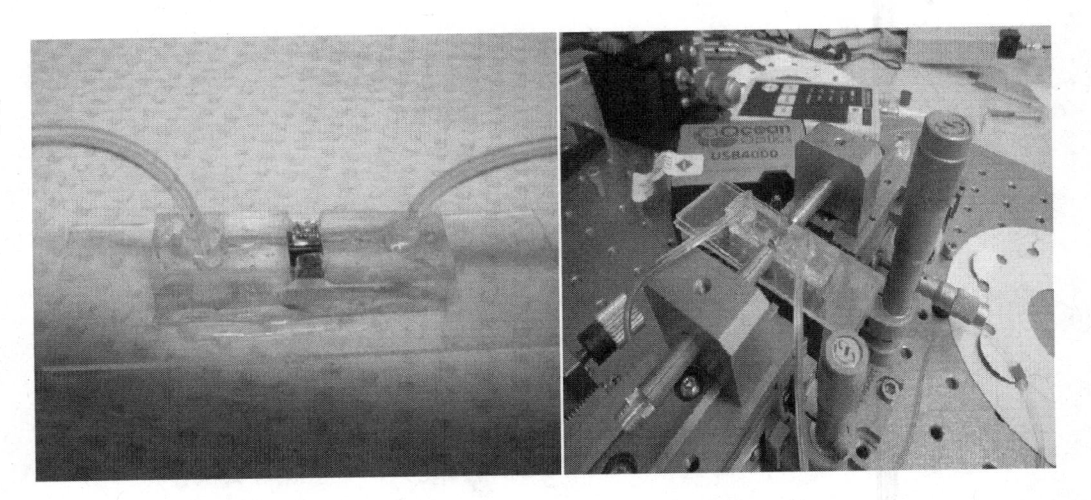

FIGURE 9.68 (See color insert.) SOS-PDMS platform for the fluorescence detection of QD[48] and optical characterization setup.

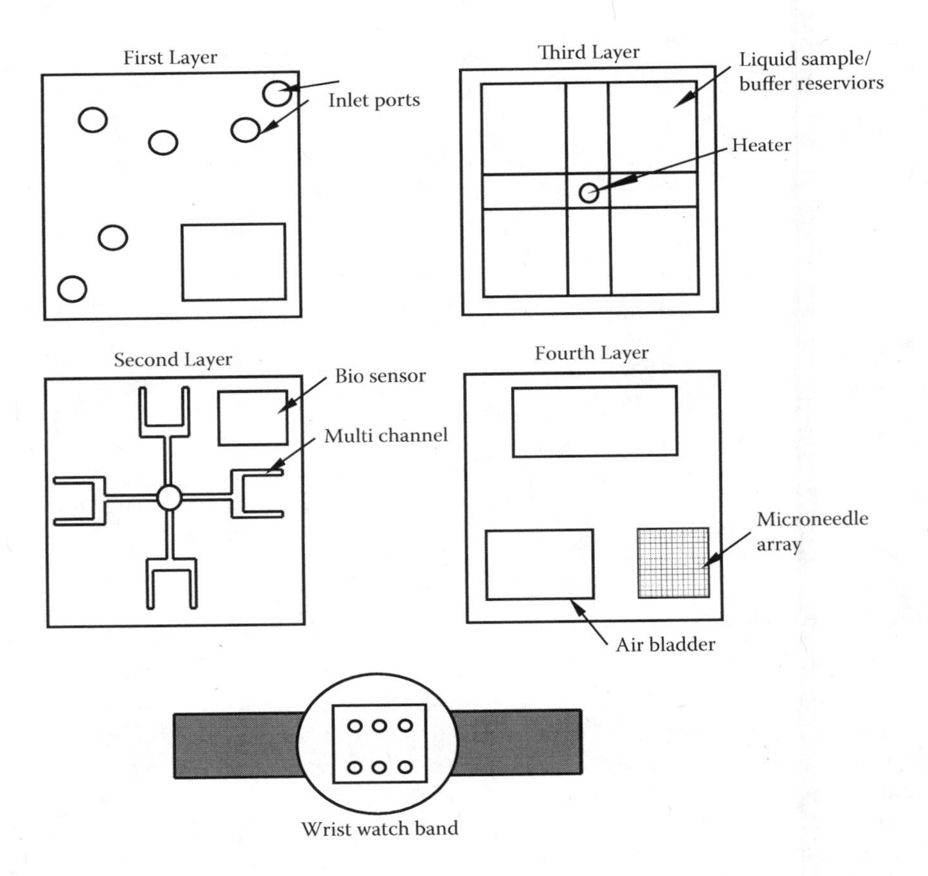

FIGURE 9.69 Schematic sketch showing (a) details of a multilayer plastic disposable biochip and (b) a wristwatch analyzer for detecting point-of-care testing with a biochip. (Adapted from Ahn et al., *Proc. IEEE*, 92, 154–73, 2004.)

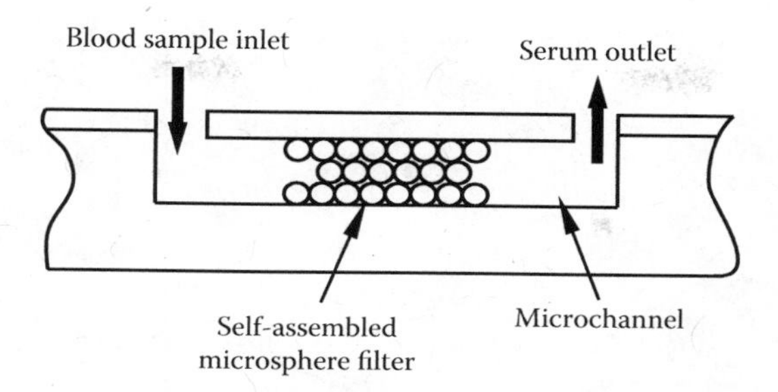

FIGURE 9.70 Design of a blood serum separator using *in situ* self-assembled silica microspheres. (Adapted from Han et al., "An On-Chip Blood Serum Separator Using Self-Assembled Silica Microspheres," paper presented at Filter Transducers: The 13th International Conference on Solid-State Sensors, Seoul, Korea, June 5–9, 2005.)

9.8 LAB-ON-A-CHIP: CONCLUSION AND OUTLOOK

The lab-on-a-chip and miniature chemistry concept depends to a great deal on the development of excellent precision fabrication capabilities.

The fully integrated biochip developed in this work is very small and incorporates all the functionality required for testing clinically relevant parameters from human blood. Furthermore, the analyzer developed in this work is of immediate relevance to portable biochemical detection applications. In addition, the disposable plastic biochip has successfully been tested for the measurements of partial oxygen concentration, glucose, and lactate level in human blood using an integrated biosensor array.

Current application areas include analysis (μTAS), high-throughput screening, chemical synthesis, microreactors, the development of assays, and analysis of cell functions and genetics. Figure 9.70 shows a blood serum separator using self-assembled silica microspheres.

The silica microsphere filter reported in this work is designed to replace membranes that are not easily integrated with a lab-on-a-chip. The microchannel used for the self-assembly has a width of 250 μm, a depth of 100 μm, and a length of 4 cm, and the substrate material is a cyclic olefin copolymer. Silica microspheres were assembled in a hexagonal closed packing arrangement by evaporation of the solvent. By using the self-assembled microsphere filter, one could successfully separate serum from diluted human blood. The filters are suitable for integrating with disposable polymer lab-on-a-chips for point-of-care testing.

9.9 MICROCANTILEVER BIOMEMS

9.9.1 INTRODUCTION

It has been seen in the previous chapters and sections that BioMEMS devices are used for diagnosis, sensing, and delivery, at various levels, including tissues, cells, molecules, and protein levels, with many application focuses on POCT, LOC, and μTAS areas. In this section, some of the recent developments in micro- and nanomechanical sensors, their fabrication, implementation, and main applications, will be discussed. The basic principles of cantilever-based biosensors in static and dynamic modes are presented, and the cantilever-based detection of proteins, DNA, and other biomolecules is also discussed.

The BioMEMS device will be used not only as a stand-alone device at the microlevel, but also as an integration platform between the nano- and macrolevel, as schematically shown in Figure 9.71. The advent of nanomedicine and nanobiotechnology areas will greatly benefit from the BioMEMS device as a mediating device between the nano- and macrolevels, as shown in Figure 9.71.

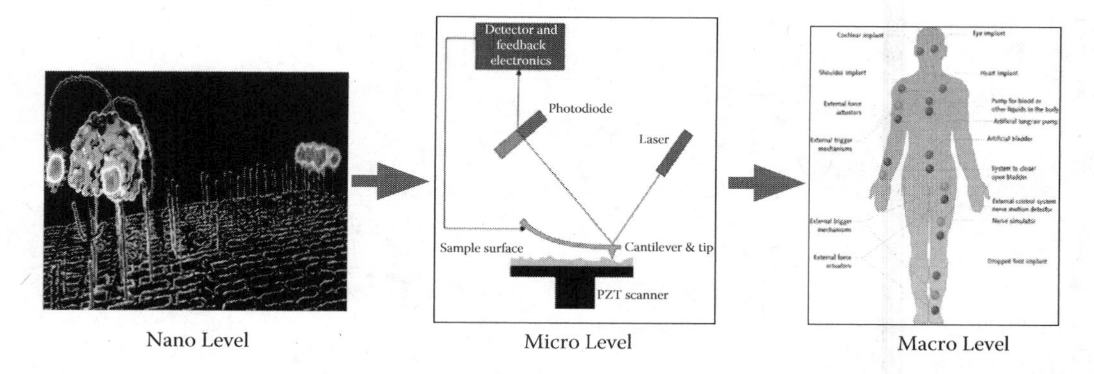

FIGURE 9.71 Schematic diagram of the BioMEMS mediating platform between nano- and macrolevels.

There exist many diagnosis methods for biosensing, as seen in Chapter 6. The biorecognition elements could include tissues, cells, nucleic acids, enzymes, antibodies, proteins, and other biospecies, while the transduction methods could be electrical, optical, mechanical, and any other method. There are two broad categories of biosensing methods: labeled approach and label-free approach. As seen earlier, the labeled approach[50] includes branch peptide amplification (BPA), enzyme-linked immunosorbent assay (ELISA), and tagging of electromagnetic beads, quantum dots (QDs), and fluorophores to biospecies.[51] Some of the label-free sensing methods include a quartz crystal mass (QCM) resonator, field effect transistor (FET), bioimpedance, surface plasmon resonance (SPR), micromechanical sensors, optical methods based on absorption, evanescence, transmission, etc. While QCM works on the principle of measuring the change in natural frequency of the structure due to mass variation of biospecies, micromechanical sensors can be designed to measure the change in any of the following: deflection, shape, resonance frequency, mode shapes, etc. As the micromechanical device has the advantages of tuning capability in terms of performance, sensitivity, and ruggedness, and ease of fabricating using microfabrication methods, this chapter will focus on microcantilever-based BioMEMS devices, as microcantilevers are being widely used for biosensing. The microcantilever beams become capable of detecting small mechanical stresses and mass additions. Microcantilever-based BioMEMS devices offer promising prospects for chemical, physical, and biological sensing with high sensitivity.

Whenever the mechanical structures are subjected to variations in stress, loading, and mass, their mechanical properties change, resulting in deflection, bending,[52] and variation in equivalent stiffness and mass properties. The loading of microcantilevers could be due to thermal stress, internal stress, surface stress, fluid viscous or pressure forces, electrostatic loading, electromagnetic loading, biomechanical interactions, etc., as shown in Figure 9.72. As microcantilevers are sensitive to finer variations of its structural properties, they can be used to translate a number of different phenomena,[53] such as change of mass, temperature, heat, or stress, into bending or deflection, as in the case of static measurement (static mode), or into change in resonant frequency and mode shapes, as in the case of dynamic measurement (dynamic mode), as shown in Figure 9.72.

9.9.2 Basic Principles of Sensing Biomechanical Interactions

Any micromechanical structure, such as a cantilever, beam, or diaphragm, or any other freestanding microstructure can be used for the detection of biomechanical interactions. Microcantilevers are widely used for biomechanical interaction studies due to their simplicity, versatility, tunability of elastic stiffness, sensitivity, easy fabrication, and suitability for many transduction methods. Due to these properties, microcantilevers are used in a variety of applications, ranging from BioMEMS to integrated optical microsystems.[54–60] Due to their high sensitivity, a microcan-

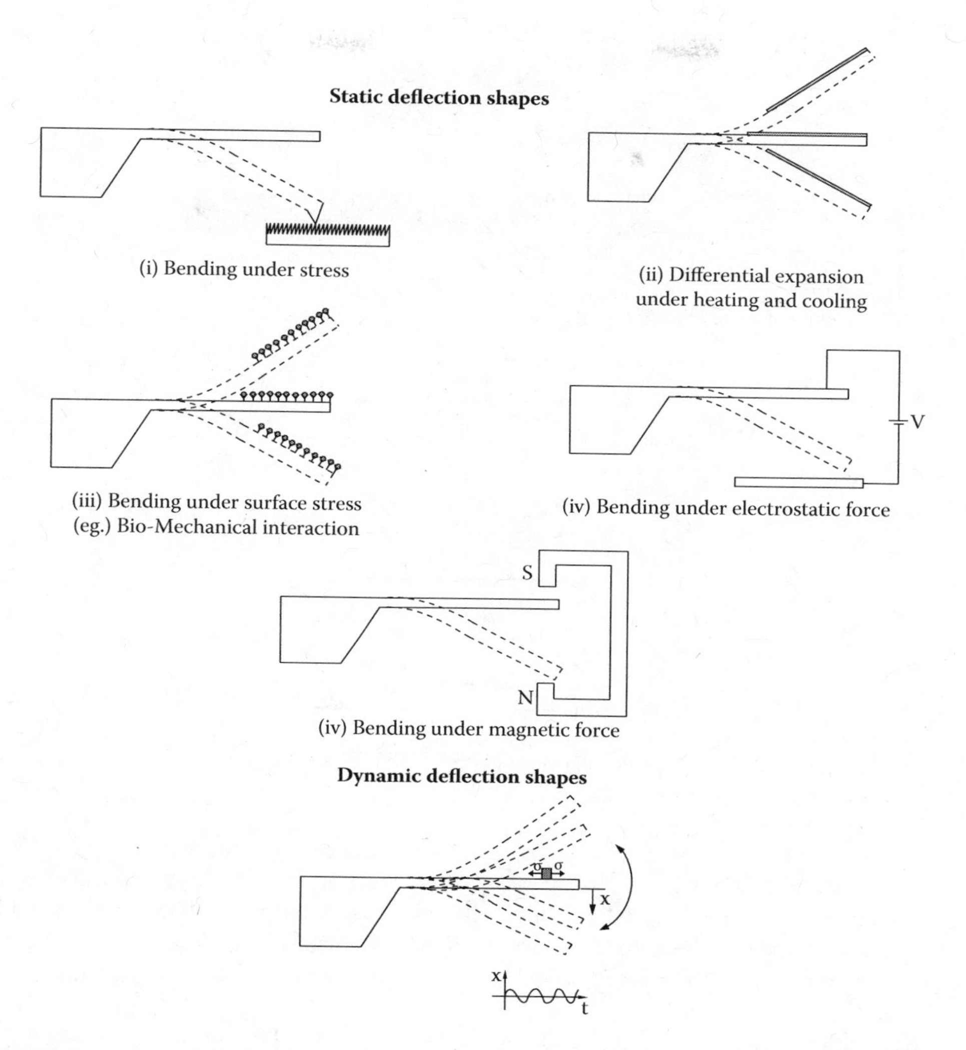

FIGURE 9.72 Microcantilevers for measuring (i) force, (ii) temperature and heat, (iii) biomechanical or microfluid-structure interactions, (iv) electrostatic force, (v) magnetic force, or (vi) change in elastic and mass properties as a vibration sensor.

tilever can sense the change in environment in terms of static deflection or change in natural frequency.[61–63]

Consider a microcantilever subjected to biointeractions on its surface, as shown in Figure 9.73. During the biointeraction between biospecies, they undergo conformational changes or structural changes. When these biointeractions occur on a mechanical surface, they produce significant stress on it. The examples of biointeractions will include bioaffinity-based interactions between antigen and antibody, enzyme and antienzyme, DNA pairs, etc. Different living species, like cells, also interact differently with mechanical surfaces when they are in close contact with the surfaces. The change in mechanical properties of the structure due to biointeractions that occur in the vicinity of the structure is called *biomechanical interaction.*

The change in static and dynamic performance of a structure occurs due to the change in mechanical properties and geometries of the structure, and also due to the biomechanical interactions, as shown in Figures 9.73 and 9.74.

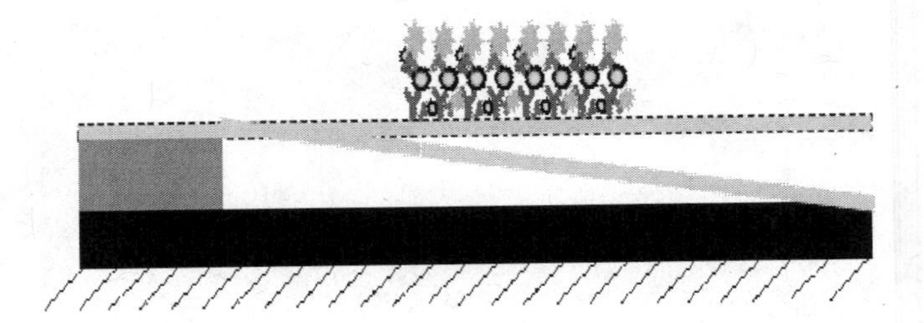

FIGURE 9.73 Schematic diagram of biointeractions occurring on the cantilever.

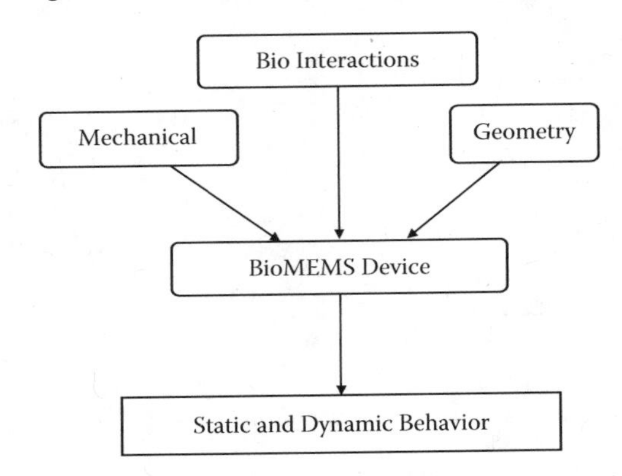

FIGURE 9.74 Parameters that influence the static and dynamic behavior of BioMEMS devices.

When the biospecies are added on the mechanical surface, they create two effects: (1) local increase in mass at the place of biointeractions and (2) creation of surface stress on the structure due to biomechanical interactions, which will modify the elastic stiffness of the structure and loading distribution on the structure. In summary, the biomechanical interaction effect can be classified into *mass-dominant* effects and *stiffness or stress-dominant* effects for the structure shown in Figure 9.75.

One can compare the biospecies on a microcantilever with a person on a diving board, as shown in Figure 9.76.[64] In this way, a cantilever configuration with biospecies is similar to a macroscale diving board with a person on it. As vibration frequency and deflection of the board depend on the mass and position of the person, the deflection and vibrational behavior of the microcantilever also depend on the mass and position of the biomechanical interactions on the cantilever.

The mass effects and stiffness or stress effects can be measured through change in static or dynamic behavior of the structure. Depending upon the detection method, microcantilever-based BioMEMS devices operate in two modes: the static mode and the dynamic mode. In the static mode, the bending deflection of the cantilever due to the stresses induced by biomechanical interactions is measured. The change in the surface stress due to the specific biomolecular interactions between the probe and target on the surface of a microcantilever results in physical bending of the microcantilever beam. More precisely, microcantilevers translate the molecular recognition of biomolecules into nanomechanical motion. In the dynamic mode, the change in vibrational behavior is mainly studied. In the dynamic mode, the change in resonance frequencies and mode shapes of the cantilever, due the biomechanical interaction-dependent mass and stiffness variations, is measured. The two measurement modes[65] are depicted schematically in Figure 9.77.

In the case of diagnosis based on bioaffinity between targets and analytes, it is important to functionalize one surface of the cantilever in such a way that a given target molecule will be preferentially bound to the functionalized surface upon its exposure. As the measurement is quantitative,

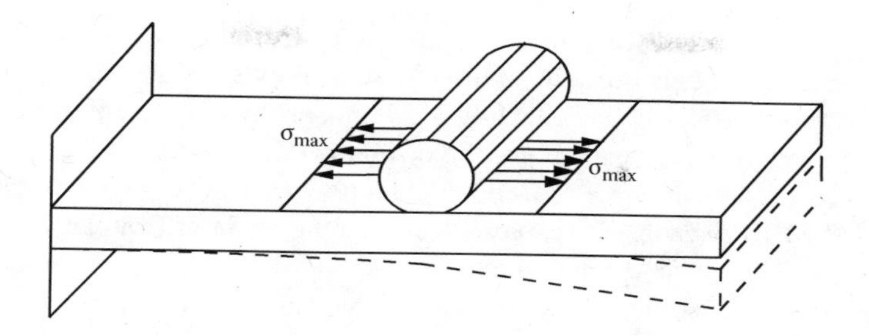

FIGURE 9.75 Generation of surface stress near the biospecies that also results in a local increase of mass.

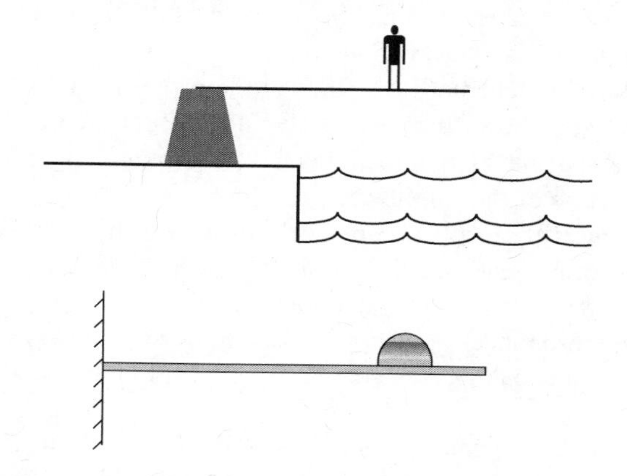

FIGURE 9.76 Comparison between macroscale and microscale cantilevers.

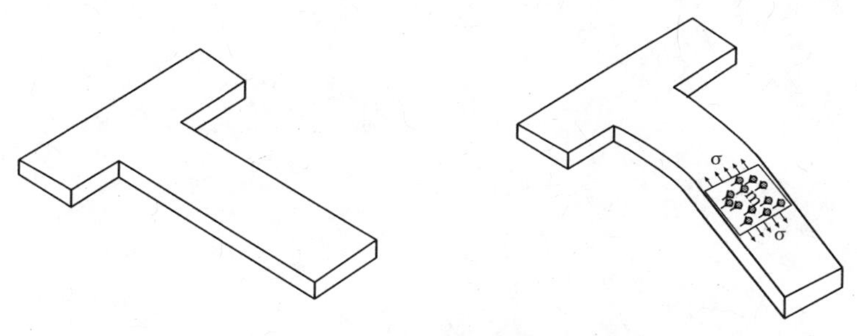

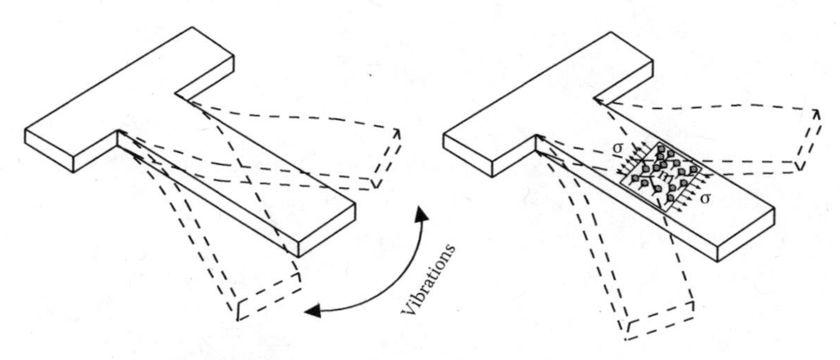

FIGURE 9.77 Cantilevers in static and dynamic modes of diagnosis.

this method can be used to quantify the concentration of the analyte, degree of bioaffinity, and strength of biomechanical interactions, in addition to selectivity among analytes.

The adsorption process onto two chemically different surfaces of the cantilever induces differential surface stress that tends to bend up or bend down the microcantilever due to the nature of the induced surface stresses, which may be tensile or compressive. The deflection of the cantilever can be measured by MOSFET, piezoresistive, piezoelectric, imaging, laser Doppler velocimetry (LDV), and other optical methods, such as the optical lever method. Piezoresistive readout is based on the changes observed in the resistivity of the material of the cantilever as a consequence of a surface stress change. A piezoresistive sensor measures the variation in film resistance with respect to surface stress caused by the specific binding of molecules. Similarly, in the piezoelectric method, the voltage generated by the piezoelectric material in consequence to the change in material stress due to the biomechanical interaction is measured.

The optical lever method[66] is schematically shown in Figure 9.78. In this method, the position of the light reflected from the cantilever moves to different positions, depending upon the deflection and slope of the cantilever. The position of the reflected laser light on the position-sensitive detector (PSD) quantifies the deflection of the cantilever.

Microcantilevers are physical sensors that respond to surface stress variation due to the specific biomechanical interaction of molecules. Adsorption of molecules on one of the surfaces of the cantilever results in a differential surface stress due to adsorption-induced forces, which manifests as deflection. In addition, the resonance frequency of the cantilever can vary due to mass loading and surface stress. The two signals, cantilever bending and adsorption-induced frequency change, can be monitored simultaneously in some applications.

In mechanical terms, the biomechanical interactions of bio or chemical species on a cantilever can be represented by mass and stress effects, as shown in Figure 9.79. When these biointeractions do not happen in a closed environment, the evaporation of liquids also plays an important role in

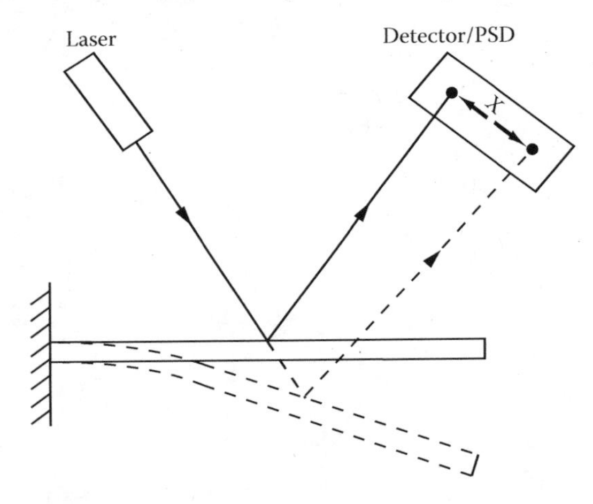

FIGURE 9.78 Optical lever method with a laser and position-sensitive detector (PSD).

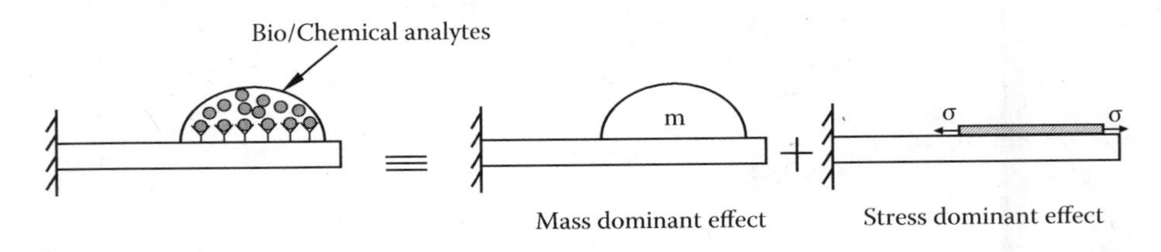

FIGURE 9.79 Scheme for the inherent mass and stress effects of biomechanical interaction.

deciding the time-varying static and dynamic behavior. The surface stress that results in bending of the structures actually modifies the stiffness of the structure. The relative contribution of mass and stiffness toward the behavioral change of the structure is dependent on the location and strength of biointeractions in relation to the mechanical stiffness of the structure. The contribution of biomechanical interactions in terms of mass and stress effects is schematically shown in Figure 9.79.

In a typical situation of microsystems, the intermolecular forces can change the stiffness in the order of 1 mN/m, electrostatic forces can modify the stiffness in the order of 0.5 N/m, surface forces like surface tension can modify the structural stiffness in the order of 20 to 100 mN/m, while the stiffness change due to biomechanical interactions can vary in the order of 10 N/m, compared to the mechanical stiffness of the structure in the order of 10 to 100 N/m. Hence, it can be seen that intermolecular and electrostatic forces will be important at the nanolevel, while biomechanical forces play a significant role at the microlevel.

9.9.3 Detection Modes of Biomechanical Interactions

9.9.3.1 Static Mode

In static mode, the deflection is a result of two mechanisms: the mass of the adsorbed molecule and the differential surface stress. The deflection of the free end of a cantilever beam under applied concentrated force F is shown in Figure 9.80a, and due to uniform loading w_0, as shown Figure 9.80b, is

For point loading:
$$\delta = \frac{F}{6EI}(2L^3 - 3L^2a + a^3) \tag{9.1}$$

For distributed loading:
$$\delta = \frac{w_0 L^4}{8EI} \tag{9.2}$$

The moment of inertia for a cantilever of rectangular cross section with constant thickness t is

$$I = \frac{wt^3}{12} \tag{9.3}$$

where E is the Young's modulus of elasticity, L is the cantilever length, and I is the area moment of inertia. In this case, the force F could be the gravity force or other external local forces.

In the case of bending due to biomechanical stress, the bending of the cantilever can be related to the differential surface stress by Stoney's equation:[67,68]

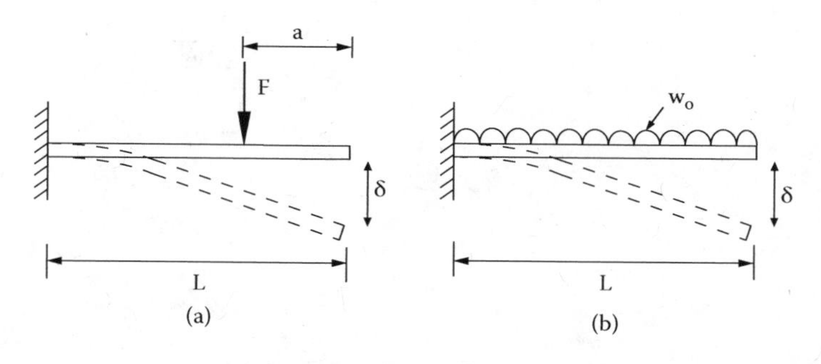

FIGURE 9.80 Scheme for point loading and uniform loading on cantilever.

$$\frac{1}{r} = \frac{6}{Et^2}\Delta\sigma \tag{9.4}$$

where r is the radius of curvature, $\Delta\sigma$ is the difference between surface stresses (N/m) at the top and bottom surface of the cantilever, and t is the thickness of the cantilever. When the plane strain condition is applied, this equation may be rewritten[68,69] as

$$\frac{1}{r} = \frac{6(1-\nu)}{Et^2}(\Delta\sigma_1 - \Delta\sigma_2) \tag{9.5}$$

where $\Delta\sigma_1$ and $\Delta\sigma_2$ are surface stresses (N/m) at the top and bottom surfaces of the cantilever, ν is Poisson's ratio, E is Young's modulus, and t is the beam thickness. By applying geometrical relations, the vertical deflection of the beam can be obtained from the above equations as

$$r^{-1} = \frac{2\Delta z}{L^2} \tag{9.6}$$

where Δz is the vertical deflection.

This equation shows that only the differential surface stress across the thickness of the cantilever will result in bending of the cantilever. As a result, only one surface of the cantilever is usually functionalized (activated) for adsorption of the molecules, while the other surface is passivated for the biointeractions. One can also apply finite element analysis for more accurate prediction. Hence, by measuring the deflection of the cantilever, it is possible to estimate the surface stress induced by biomechanical interactions.

The cantilever can be modeled as a simple mass-spring system with equivalent stiffness of the cantilever and equivalent mass of the cantilever. During biointeractions, the mass, $m(t)$, of the biospecies changes with time due to evaporation, and the force generated due to biomechanical interactions, $F(t)$, also changes with time. As shown in the Figure 9.81, deflection is obtained with the force balance on the structure.

If deflection due to the weight of the cantilever is neglected, then the time-varying deflection of the cantilever only, due to biomechanical interaction, is obtained after the static force equilibrium as

$$\delta(t) = \frac{F_{bc}(t) + m(t)g}{K_M} \tag{9.7}$$

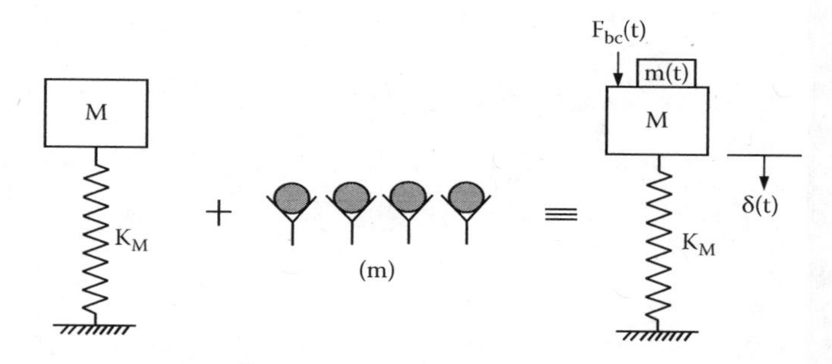

FIGURE 9.81 The lumped representation of a microcantilever under static equilibrium with biomechanical interaction.

where $\delta(t)$ is the time-varying deflection of the cantilever, $m(t)$ is the time-varying mass of the added biospecies, $F_{bc}(t)$ is the equivalent force generated due to biomechanical interactions, and K_M is the equivalent mechanical stiffness of the cantilever.

The pattern of time-varying deflection becomes indicative of the specific nature of the biomechanical interaction, and also the measure of specific bioaffinity interaction.

9.9.3.2 Dynamic Mode

Similar to the static mode, the biomechanical interactions could be identified in dynamic mode also through vibrational behavior under dynamic equilibrium. The natural frequencies and mode shapes of vibration become a good measure to characterize the dynamic behavior of the cantilever under biomechanical interactions. The vibration of a microcantilever can be studied by the lumped modeling of it as a mass-spring-damper system.

Because our interest is to find the natural frequencies, the damping is neglected for further studies for simplicity. Hence, the cantilever subjected to biointeractions can be modeled as a damped mass-spring system, as shown in Figure 9.82. When a simple mass-spring system of a cantilever with equivalent mass M and equivalent mechanical stiffness K_M is subjected to biointeractions, the resultant biomechanical system is schematically shown as in Figure 9.82. In this figure, $m(t)$ represents the mass of the added biospecies. The added mass of biospecies undergoes evaporation and changes with time as biointeraction progresses. The surface stress produced by the biomechanical interaction also changes in time as the biointeraction progresses and results in a time-varying moment that can be represented by an equivalent stiffness change. Biointeraction-dependent equivalent stiffness[70,71] is denoted as $K_{bc}(t)$ in this figure.

The equation of motion for the vibration of the biomechanical system under biointeraction is simplified as

$$[M + m(t)]\,\frac{d^2x}{dt^2} + [K_M + K_{bc}(t)]\,x = 0 \tag{9.8}$$

where x is the dynamic displacement. The natural frequency for such a system in rad/sec is

$$\omega_n(t) = \sqrt{\frac{K_M + K_{bc}(t)}{M + m(t)}} \tag{9.9}$$

The above equation is generally valid for any micromechanical structure subjected to biointeractions. M is the effective mass of the structure, K_M is the effective mechanical stiffness of the structure, m is the mass of the bio or chemical species added, K_{bc} is the time-varying equivalent stiffness

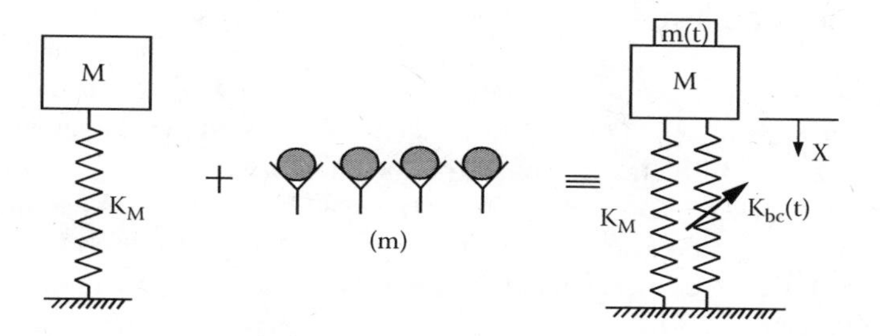

FIGURE 9.82 The lumped representation of a microcantilever under dynamic equilibrium with a biomechanical interaction.

due to bio and chemical mechanical interactions, x is the time-varying equivalent motion of the structure, and ω is the time-varying cyclic natural frequency in rad/sec.

As the net stiffness and net mass change with time during biomechanical interactions, time-dependent variation of the natural frequency $\omega(t)$ becomes a measure of biointeractions and also the specificity of biointeractions.[72–75] Equation 9.9 can be applied to any mechanical structure, including cantilevers, beams, plates, diaphragms, etc.

In the case of a cantilever where a layer of biomolecules is absorbed (or coated) uniformly on the entire surface of the cantilever, the added mass would be $m(t)$. The new resonant frequency, $\omega_n(t)$, due to increasing of the mass is[68,69,72]

$$\omega = \frac{\alpha^2}{\sqrt{3}}\sqrt{\frac{K_M}{M+m}} \quad (9.10)$$

Such a layer can also increase the stiffness of the beam by increasing the thickness of the cantilever. Therefore, by considering both changes, the frequency will be[68,74]

$$\omega = \frac{\alpha^2}{\sqrt{3}}\sqrt{\frac{K_M}{M+m} + \frac{3E_{ads}I_{ads}}{L^3(M+m)}} \quad (9.11)$$

$$I_{ads} = \frac{wt_{ads}^3}{12} + wt_{ads}\left[\left(\frac{t_{ads}}{2}\right) - t_b - y_{cm}\right]^2 \quad (9.12)$$

where y_{cm} is the centroid of the cross section in the presence of an adsorbed layer given by

$$y_{cm} = \frac{Et^2 + E_{ads}(2t_{ads}t + t_{ads}^2)}{2E_{ads}t_{ads} + 2Et} \quad (9.13)$$

where the subscript *ads* refers to the adsorbed layer on the surface, $\alpha = 1.875$, t is the thickness of the beam, t_{ads} is the thickness of the adsorbed layer, E is Young's modulus of the cantilever, E_{ads} is Young's modulus of the adsorbed layer, L is length of the beam, and w is the width of the beam.

When the biospecies do not cover the entire cantilever surface and cover only part, as shown in Figure 9.75, then the cantilever has to be modeled as a continuous system using other methods, such as FEM or energy methods like Rayleigh-Ritz.[70,76]

The resonance frequency during biointeractions can be measured using many testing methods, such as the optical lever method, LDV, stroboscopic interferometry, and piezoresistive and piezoelectric methods.

We have seen that the changes in stiffness and mass due to biointeractions influence the natural frequency. The magnitude of shift in natural frequency depends on the change in stiffness relative to the change in mass due to biointeractions. The location of the biointeraction on the cantilever influences significantly the relative change in stiffness with respect to the change in mass. One can explore the effect of location on changes in mass and stiffness to improve the sensitivity and increase the output of the device. Hence, in the next sections, the effect of location on the vibration behavior will be discussed.

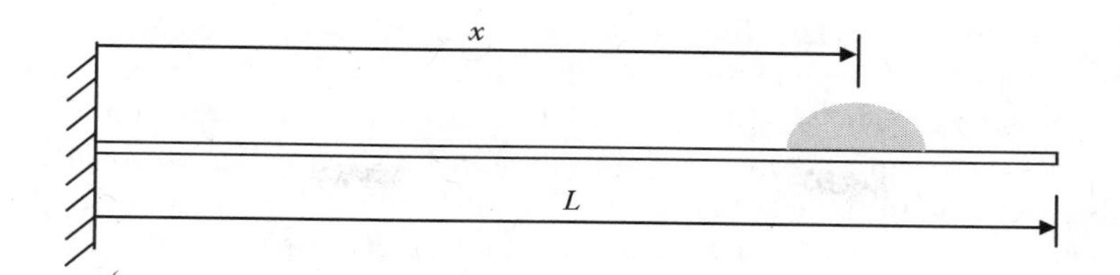

FIGURE 9.83 Scheme showing the effect of location of added mass on cantilever.

9.9.4 Location of Interaction in the Case of Mass-Dominant BioMEMS Devices

In applications where the effect of mass is considered to be significant in determining the frequency shift, as in the case of the QCM resonator, it is important to find the optimum range of location of biointeraction on the cantilever for a maximum shift in natural frequency. One can understand this effect[70] only if the effect of the location on mass is studied. Consider a cantilever loaded with a drop of liquid at a given location, as shown in Figure 9.83.

The natural frequency has been predicted for different positions of the added mass equivalent to biospecies, and it is normalized with respect to the natural frequency of the original cantilever, without the addition of mass. The normalized natural frequency is presented in Figure 9.84 against the normalized position ($\xi = x/L$) of the added mass. As per the equation for natural frequency, one would expect the natural frequency to decrease with the addition of mass, as depicted in Figure 9.84. But, the shift in natural frequency is high when the mass is added near the tip or free end of the cantilever. The change in frequency is less than 20% when the location of added mass is within 40% of the length from the fixed end of the cantilever, i.e., for $\xi \leq 0.4$. Hence, in order to have a higher frequency shift and sensitivity, it is important to design the biointeraction zone near the tip portion of the cantilever for mass-dominant BioMEMS devices that rely more on the mass effect for detection.

9.9.5 Location of Interaction for Stress-Dominant BioMEMS Devices

It is also important to understand the effect of location of interaction on the frequency shift for stress- or stiffness-dominant BioMEMS devices. One could anticipate from the equation of natural frequency that a reduction in stiffness will reduce the natural frequency, while a reduction in mass

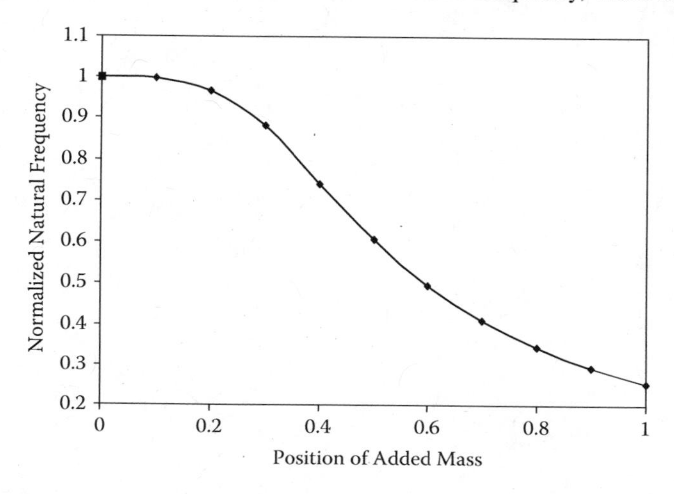

FIGURE 9.84 Effect of the location of added mass on natural frequency. (Adapted from Rinaldi et al., *Sensors Rev.*, 29, 44–53, 2009.)

will increase the natural frequency. In order to understand the simultaneous influence of stiffness and mass, the effect of slot location and size are studied.[76] The mass of the cantilever is reduced through the introduction of slots, as shown in Figure 9.85. In the configuration called *left to right* (LR), the slot is extended from the fixed end to the free end, while the slot is extended from the free end to the fixed end in the *right-to-left* (RL) configuration. The natural frequency has been predicted for these configurations, and the corresponding eigenvalue is presented against the length of the slot in Figure 9.86. The eigenvalue (λ) is defined as

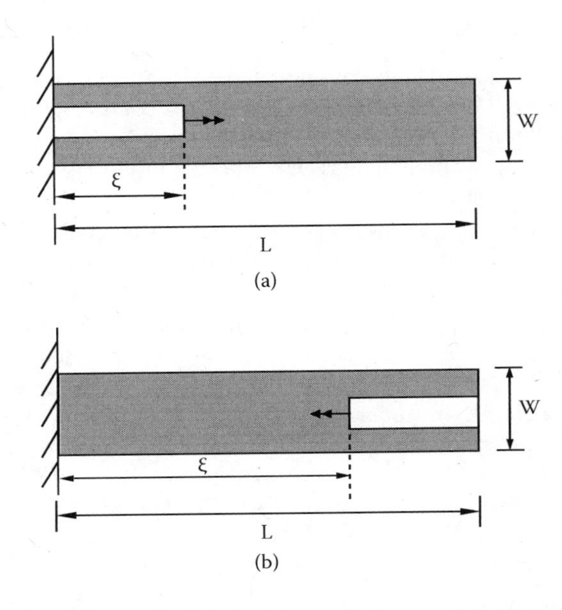

FIGURE 9.85 Configurations of cantilevers with slots: (a) left to right (LR) and (b) right to left (RL). (Adapted from Rinaldi et al., *Int. J. COMADEM*, 10, 13–19, 2007.)

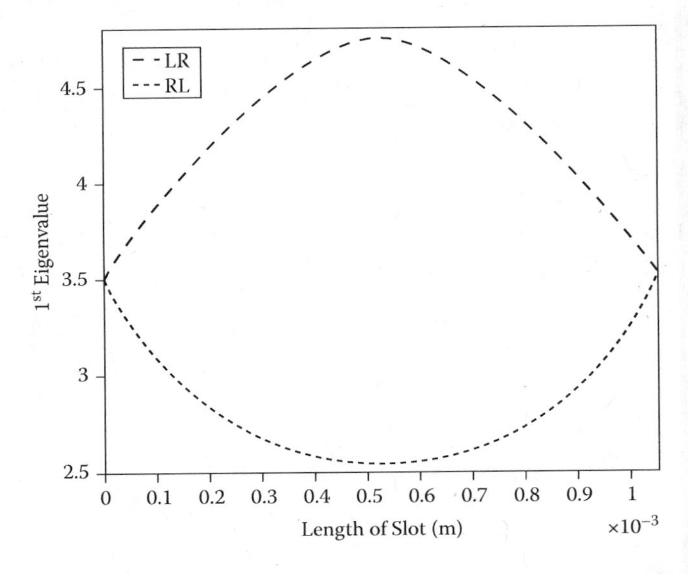

FIGURE 9.86 Predicted variation of eigen value of cantilever. (Adapted from Rinaldi et al., *Int. J. COMADEM*, 10, 13–19, 2007.)

$$\lambda = \frac{\omega^2 \rho A_0 L^4}{EI} \tag{9.14}$$

where $A_0 = w\,t$,

$$I = \frac{w\,t^3}{12}$$

ω is the natural frequency, w is the width of the cantilever, t is the thickness of the cantilever, I is the area moment of inertia, ρ is the density of the cantilever, and E is Young's modulus of elasticity.

In the case of the LR configuration, the frequency reduces rapidly for the slot length, up to 40% of the length, even when the mass is reduced. This shows that the stiffness effect is dominant for $\xi \le 0.4$, that is, in the region closer to the fixed end. For the slot extending beyond 60% of the length, the increase in frequency is rapid, with the decrease in mass indicating that the mass effect is dominant in the region closer to the free end for $\xi \ge 0.6$. These results are confirmed by the LR configuration, in which the reduction of mass closer to the free end with a slot length for $1.0 \ge \xi \ge 0.6$ increases the natural frequency, while the extension of slot length closer to the fixed end reduces the natural frequency. In summary, one can conclude that it would be advantageous for mass-dominant BioMEMS devices if the biointeraction happened near the free end ($0.6 \le \xi \le 1.0$). Similarly, it would increase the output if the biointeraction happened near the fixed end ($0 \le \xi \le 0.4$) for stress-dominant BioMEMS devices.

9.9.6 Fabrication and Functionalization of Microcantilevers

Depending on the application, the cantilevers can be made from different materials. In some cases, atomic force microscope (AFM) tips are used as silicon cantilevers for diagnosis. Silicon, silicon nitride, and silicon oxide cantilevers with different shapes and sizes, batch fabricated using thin-film processing technologies, are available commercially. As the sensitivity of a microcantilever depends on the spring constant, one could obtain these cantilevers with different spring constants, depending upon the requirement. Cantilevers are also made out of polymers like PDMS and SU8, using microfabrication techniques, and integrated into BioMEMS devices. Polymers, with a Young's modulus about forty times lower than that of silicon, have been shown to be very sensitive. But, it is difficult to achieve a stable immobilization of the receptor on polymers.

In order to have a high reproducibility and selectivity, the immobilization should be uniform, compact, and allow accessibility by the target molecules. The cantilever sensor usually has one surface coated with a thin layer of evaporated or sputtered gold, and biomolecules are adsorbed either on silanes on silicon substrates or on self-assembled monolayers of thiols on gold, through a sulfur linkage.

One of the strategies employed for proteins is covalent immobilization of carboxylate-terminated alkanethiols, followed by esterification of the carboxylic group with 1-ethyl-3-(3-dimethylamminopropyl)-carbodiimide (EDC) and N-hydroxysuccimide (NHS). EDC catalyzes the formation of amide bonds between carboxylic acids and amines by activating carboxyl to form an O-urea derivative. Langmuir-Blodgett deposition can also be used to create organic layers on the microcantilever surface.

The microcantilever-based detection can be carried out in a closed environment, like inside the microchannels, or in an open environment. In the case of cantilevers inside the microchannels, the testing has to be carried out in an aqueous environment, and as a consequence, the cantilever deflection is influenced by flows and microfluidic structure interactions.

In the field of genomics, arrays of microcantilevers are used for DNA detection. Single-strand DNA hybridization can also be monitored using microcantilevers. It is possible to identify two oligonucleotides with a single-base mismatch using cantilevers.

In proteomics, antibodies are used as receptors for detecting proteins. An example includes a prostate-specific antigen, which is a marker for the early detection of prostate cancer. Detection of different pathogens such as *Salmonella enterica* and pesticides (DDT, etc.) has also been performed using microcantilevers.

Cantilever-based BioMEMS can be used for the diagnosis of biointeractions at various levels, from the protein level to the molecular level to the cellular level. Similarly, this technique can be extended for the detection of chemical interactions as well. Even though there are many papers available in the area of cantilever MEMS for different applications, only a few sample applications are given here as case studies for demonstration purposes.

9.9.6.1 Case 1: Detection of Interaction between ssDNA and the Thiol Group Using Cantilevers in the Static Mode[77]

In this study, the interaction between ssDNA and thiol groups functionalized on the surface of cantilevers is identified using microcantilvers in the static mode. The experiments carried out using functionalized and unfunctionalized cantilvers with ssDNA indicate different stress patterns created due to the interactions. In the case of functionalized cantilevers, the sulfur atom interacts with 2lM ssDNA (27 mer) and generate a time-varying pattern of surface stress during DNA immobilization, which is expressed by the time-varying deflection pattern shown in Figure 9.87. In the absence of thiol groups on the surface, no deflection was obtained, indicating the absence of a thiol-DNA interaction.

9.9.6.2 Case 2: Specific Detection of Enzymatic Interactions in the Static Mode

In order to demonstrate the selectivity of the cantilever method, studies on two enzymes are presented here.

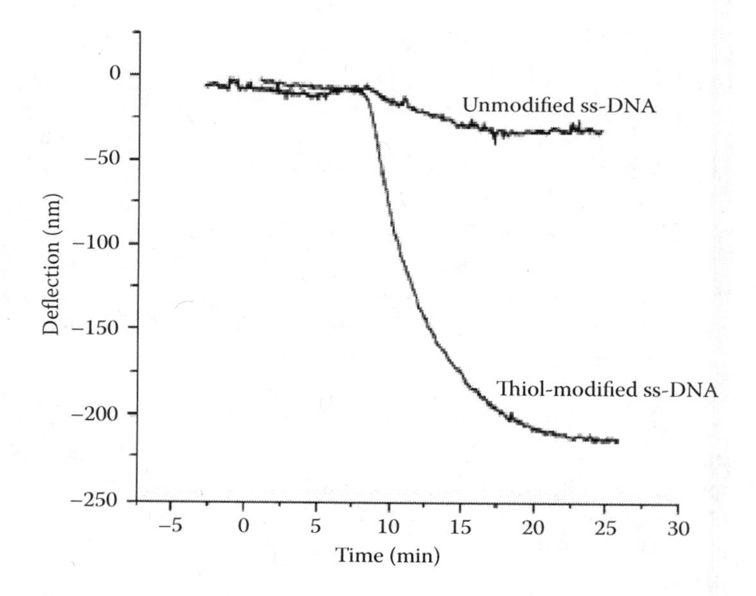

FIGURE 9.87 Comparison of deflection signals of a cantilever between unmodified ss-DNA and thiol-modified ss-DNA. Large deflection of cantilever is observed for thiol-modified ss-DNA, while no deflection observed for unmodified ss-DNA. (Reprinted from Carrascosa et al., *Trends Anal. Chem.*, 25, 196–206, 2006. With permission.)

9.9.6.2.1 *Study of Biomolecular Interaction between TnC and ME[66]*

Troponin C is considered the primary biomarker for myocardial injury, and the study of its interaction with membrane proteins is of considerable importance. In the present work, an optical lever method is employed to study the interaction of rabbit skeletal muscle troponin C (TnC) with a twenty-six-residue membrane protein (ME). The cantilever deflection is measured for different concentrations, as shown in Figures 9.88 and 9.89. These figures present the tip deflection of the cantilever as the interaction progresses, after TnC-ME is allowed to interact on the microcantilever. It is interesting to notice the change in direction for the deflection. The results show the unique pattern of deflection that characterizes the TnC-ME biomechanical interaction. The reversal of deflection also indicates the reversal of stresses produced during biomechanical interactions. One could identify the variation from compressive stress to tensile surface stress during the biointeractions. In addition, the peak deflection could also be used to measure the concentration, as the amplitude of deflection was found to be proportional to the concentration, as shown in Figure 9.90. This deflection pattern becomes the unique signature that is specific for the TnC-ME interaction.

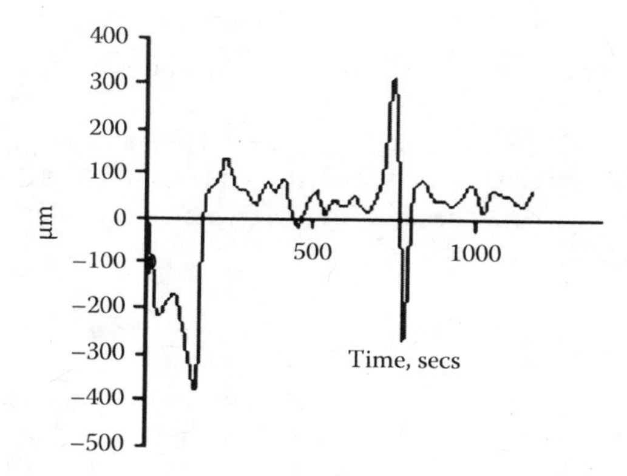

FIGURE 9.88 Cantilever deflection pattern for 20 mg/ml TnC-ME at 0.4 μl. (Reprinted from Amritsar et al., *J. Biomed. Optics*, 11, 021010-1–7, 2006. With permission.)

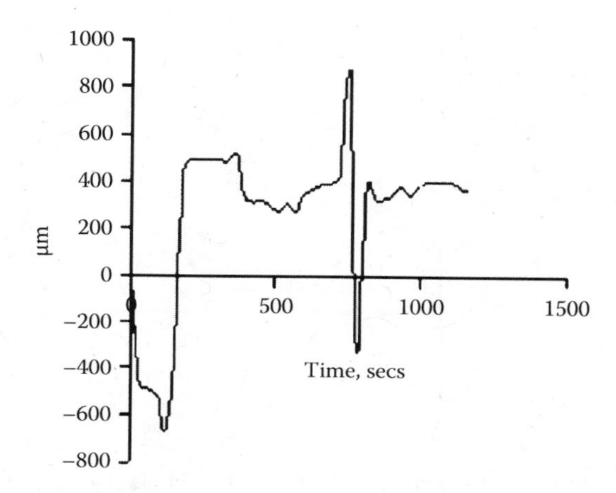

FIGURE 9.89 Cantilever deflection pattern for 40 mg/ml TnC-ME at 0.4 μl. (Reprinted from Amritsar et al., *J. Biomed. Optics*, 11, 021010-1–7, 2006. With permission.)

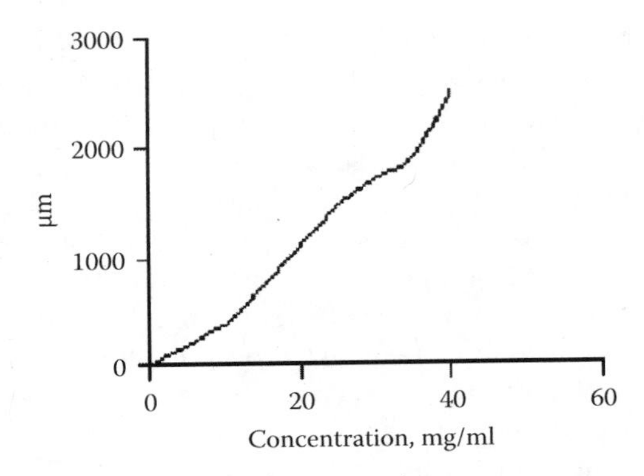

FIGURE 9.90 Cantilever deflection pattern as a function of TnC-ME concentration. (Reprinted from Amritsar et al., *J. Biomed. Optics*, 11, 021010-1–7, 2006. With permission.)

9.9.6.2.2 *Study of Biomolecular Interaction between HRP and* H_2O_2[78]

An enzyme-substrate reaction involves redox cycles that involve electron transfer between an enzyme and its substrate. One of the common enzyme-substrate pairs with high affinity is horseradish peroxidase (HRP) and H_2O_2. The interaction between HRP and H_2O_2 is studied using a microcantilever in the static mode. When the enzyme HRP comes into contact with a selected substrate, it undergoes unique biointeractions. The substrate H_2O_2 is reduced when it interacts with HRP, as given by the following equation:

$$H_2O_2 \rightarrow H_2O + \tfrac{1}{2}\, O_2$$

The recorded deflection pattern of the cantilever during the HRP-H_2O_2 interaction is shown in Figure 9.91. The deflection signature for HRP-H_2O_2 is quite different from the signature of the TnC-ME pair, as the deflection for HRP-H_2O_2 is in only one direction. Hence, one could infer that the microcantilever is able to produce enzyme-specific deflection signatures that can be used for the detection of enzymatic reactions.

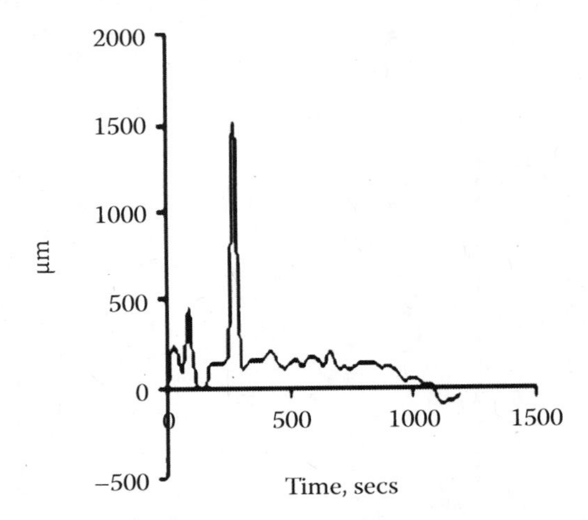

FIGURE 9.91 Cantilever deflection as a function of time for HRP-H_2O_2 interaction. (Adapted from Amritsar et al., "Detection of Acute Myocardial Infarction (AMI) Using BioMEMS," paper presented at the 4th Canadian Workshop on MEMS, Ottawa, Canada, August 19, 2005.)

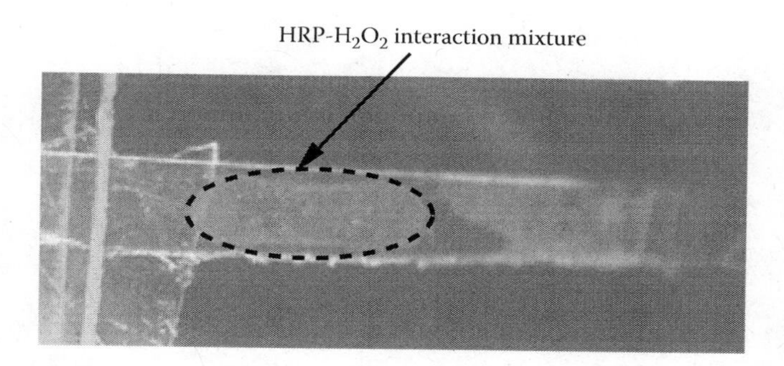

FIGURE 9.92 Mixture of HRP-H_2O_2 on the cantilever beam during interaction. (Adapted from Chandrasekaran and Packirisamy, "Label-Free Detection of Enzymatic Interaction through Dynamic Behavior of Microstructures," paper presented at the International Conference on MEMS IITM, Chennai, India, January 3–5, 2009.)

Similarly, the static mode can also be used to detect cellular behavior and differentiate between live and dead cells. As the cells interact differently on mechanical surfaces, different cells will behave differently and produce unique signatures that can be used for the diagnosis of cell types, cell interactions, number of cells, etc. This technique can also be used to diagnose different cancer cells.[79]

9.9.6.3 Case 3: Detection of Enzymatic Interactions in the Dynamic Mode

In this work, the variation in the natural frequency of the cantilever in response to the interaction between the HRP enzyme and hydrogen peroxide is studied. The natural frequency of the microcantilever during the progress of an enzymatic reaction is measured by laser Doppler velocimetry (LDV).

As this device works on the change in stiffness due to biomechanical stress, it is important to carry out the interaction in the stress-dominant zone of the cantilever, which is close to the root, as shown in Figure 9.92.

The natural frequency variation during the enzymatic reaction is shown in Figure 9.93. This figure shows the three stages of the reaction: (1) the shift in the natural frequency of the microcantilever (around 180 sec) due to the reaction, (2) a variation due to the evaporation of the reagents from the surface (around 700 sec), and (3) the surface stress. This work confirms that by using microcantilevers, important information can be obtained on various types of biointeractions.

Similarly, the dynamic mode technique can also be used to identify the number of cancer cells as frequency changes with the mass of the added cells.[80]

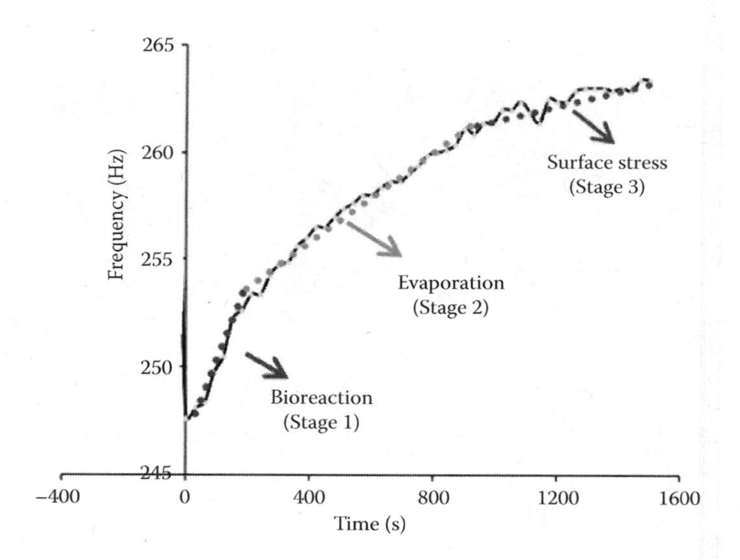

FIGURE 9.93 Variation in natural frequency of the microcantilever due to the biochemical reaction between HRP and H_2O_2. (Adapted from Chandrasekaran and Packirisamy, "Label-Free Detection of Enzymatic Interaction through Dynamic Behavior of Microstructures," paper presented at the International Conference on MEMS IITM, Chennai, India, January 3–5, 2009.)

REFERENCES

1. Büssow, K., Konthur, Z., Lueking, A., Lehrach, H., Walter, G. 2003. Protein array technology. Potential use in medical diagnostics. *Am. J. Pharmacogenomics* 1:1–6.
2. Schena, M., Shalon, D., Davis, R. W., Brown, P. O. 1995. Quantitative monitoring of gene expression patterns with a complementary DNA microarray. *Science* 270(5235):467–70.
3. Lettieri, T. 2006. Recent applications of DNA microarray technology to toxicology and ecotoxicology. *Environ. Health Perspect.* 114:4–9.
4. Templin, M. F., Stoll, D., Schrenk, M., Traub, P. C., Vöringer, C. F., Joos, T. O. 2002. Protein microarray technology. *Trends Biotechnol.* 20:160–66.
5. Gärtner, C., Richard Klemm, R., Becker, H. 2007. Methods and instruments for continuous-flow PCR on a chip. *Proc. SPIE* 6465:646502.
6. Zhang, C., Xing, D. 2007. Miniaturized PCR chips for nucleic acid amplification and analysis: Latest advances and future trends. *Nucleic Acids Res.* 2007:1–15.
7. Wu, J., Cao, W., Wen, W., Choy Chang, D., Sheng, P. 2009. Polydimethylsiloxane microfluidic chip with integrated microheater and thermal sensor. *Biomicrofluidics* 3(1):012005-1–7.
8. Zhang, Y., Ozdemir, P. 2009. Microfludic DNA amplification—A review. *Anal. Chim. Acta* 638:115–25.
9. Kim, H., Dixit, S., Green, C. J. 2009. Nanodroplet real-time PCR system with laser assisted heating. *Opt. Express* 17:218–27.
10. Kiss, M. M., Donnelly, L. O., Reginald Beer, N. R., Warner, J., Bailey, C. G., Colston, B. W., Rothberg, J. M., Link, D. R., Leamon, J. H. 2008. High-throughput quantitative polymerase chain reaction in picoliter droplets. *Anal. Chem.* 80:8975–81.
11. Beer, N. R., Hindson, B. J., Wheeler, E. K., Hall, S. B., Rose, K. A., Kennedy, I. M., Colston, B. W. 2007. On-chip real-time, single-copy polymerase chain reaction in picoliter droplets. *Anal. Chem.* 79:8471–75.
12. Talapatra, A., Rouse, R., Hardiman, G. 2002. Protein microarrays: Challenges and promises. *Pharmacogenomics* 3:1–10.
13. http://wapedia.mobi/en/Antibody_microarray.
14. Hultschig, C., Kreutzberger, J., Seitz, H., Konthur, Z., Buüssow, K., Hans Lehrach, H. 2006. Recent advances of protein microarrays. *Curr. Opin. Chem. Biol.* 10:4–10.
15. Liotta, L. A., Espina, V., Mehta, A. I., Calvert, V., Rosenblatt, K., Geho, D., Munson, P. J., Young, L., Wulfkuhle, J., Petricoin, E. F. 2003. Protein microarrays: Meeting analytical challenges for clinical applications. *Cancer Cell.* 3:317–25.

16. Hall, D. A., Ptacek, J., Snyder, M. 2007. Protein microarray technology. *Mech. Ageing Dev.* 128:161–67.
17. Zhu, H., Snyder, M. 2001. Protein arrays and microarrays. *Curr. Opin. Chem. Biol.* 5:40–45.
18. Spisak, S., Tulassay, Z., Molnar, B., Guttman, A. 2007. Protein microchips in biomedicine and biomarker discovery. *Electrophoresis* 28: 4261-73
19. Hart, T., Alice Zhao, A., Garg, A., Swetha Bolusani1, S., Marcotte, E. M. 2009. Human cell chips: Adapting DNA microarray spotting technology to cell-based imaging assays. *PLoS One* 4(10):e7088.
20. Sauter, G., Simon, R., Hillan, K. 2003. Tissue microarrays in drug discovery. *Nature Rev. Drug Discov.* 2:962–72.
21. Fernandes, T. G., Kwon, S.-J., Lee, M.-Y., Clark, D. S., Joaquim, M. S., Cabral, J. M. S., Dordick, J. S. 2008. On-chip, cell-based microarray immunofluorescence assay for high-throughput analysis of target proteins. *Anal. Chem.* 80:6633–39.
22. Song, H., Tice, J. D., Ismagilov, R. F. 2003. A microfluidic system for controlling reaction networks in time. *Angew. Chem.* 115:792–96.
23. Song, H., Chen, D. L., Ismagilov, R. F. 2006. Reactions in droplets in microfluidic channels. *Angew. Chem. Int. Ed.* 45:7336–56.
24. Kanno, K., Maeda, H., Showta Izumo, S., Ikuno, M., Takeshita, K., Asuka Tashiro, A., Fujiiac, M. 2002. Rapid enzymatic transglycosylation and oligosaccharide synthesis in a microchip reactor. *Lab Chip* 2:15–18.
25. Sotowa, K.-I., Miyoshi, R., Lee, C.-G., Kang, Y., Kusakabe, K. 2005. Mixing and enzyme reactions in a microchannel packed with glass beads. *Kor. J. Chem. Eng.* 22:552–55.
26. Thomsen, M. S., Nidetzky, B. 2008. Microfluidic reactor for continuous flow biotransformations with immobilized enzymes: The example of lactose hydrolysis by a hyperthermophilic *b*-glycoside hydrolase. *Eng. Life Sci.* 8:40–48.
27. Holden, A. M., Jung, S.-Y., Cremer, P. S. 2004. Patterning enzymes inside microfluidic channels via photoattachment chemistry. *Anal. Chem.*76: 1838–43.
28. Hadd, A. G., Raymond, D. E., Halliwell, J. W., Jacobson, S. C., Ramsey, J. M. 1997. Microchip device for performing enzyme assays. *Anal. Chem.* 69:3407–12.
29. Drott, J., Lindström, K., Rosengren, L., Laurell, T. 1997. Porous silicon as the carrier matrix in microstructured enzyme reactors yielding high enzyme activities. *J. Micromech. Microeng.* 7:14–23.
30. Srinivasan, V., Pamula, V. K., Fair, R. B. 2004. Droplet-based microfluidic lab-on-a-chip for glucose detection. *Anal. Chim. Acta* 507:145–50.
31. Song, H., Ismagilov, R. F. 2003. Millisecond kinetics on a microfluidic chip using nanoliters of reagents. *J. Am. Chem. Soc.* 125:14613–19.
32. Hung, L.-H., Lee, A. P. 2007. Microfluidic devices for the synthesis of nanoparticles and biomaterials. *J. Med. Biol. Eng.* 27:1–6.
33. Khan, S. A., Gunther, A., Schmidt, M. A., Jensen, K. F. 2004. Microfluidic synthesis of colloidal silica. *Langmuir* 20:8604–11.
34. Han, A., Han, K.-H., Mohanty, S. K., Frazier, A. B. 2005. Microsystems for whole blood purification and electrophysiological analysis. *J. Semicond. Technol. Sci.* 5:1–9.
35. Tsutsui, H., Ho, C.-M. 2009. Cell separation by non-inertial force fields in microfluidic systems. *Mech. Res. Commun.* 36:92–103.
36. Manz, A., Graber, N., Widmer, H. 1990. Miniaturized total analysis systems: A novel concept for chemical sensors. *Sensors Actuators B* 1:244–48.
37. Krawczyk, S. 2003. Discussion on optical integration in lab-on-a-chip microsystems for medical diagnostics. In *Physica Status Solidi* 0:998–1012.
38. Verpoorte, E., Manz, A., Ludi, H., Widmer, H., van der Schoot, B. H., de Rooij, N. 1991. A novel optical detector cell for use in miniaturized total chemical analysis systems. In *Solid-State Sensors and Actuators, 1991*, Digest of Technical Papers, TRANSDUCERS'91, San Francisco, pp. 796–99.
39. Verpoorte, E., Schoot, B., Jeanneret, S., Manz, A., Widmer, H., Rooij, N. 1994. Three-dimensional micro flow manifolds for miniaturized chemical analysis systems. *J. Micromech. Microeng.* 4:246–56.
40. Jacobson, S. C., Ramsey, J. M. 1996. Integrated microdevice for DNA restriction fragment analysis. *Anal. Chem.* 68:720–23.
41. Guenat, O., Ghiglione, D., Morf, W., de Rooij, N. 2001. Partial electroosmotic pumping in complex capillary systems. Part 2: Fabrication and application of a micro total analysis system (μTAS) suited for continuous volumetric nanotitrations. *Sensors Actuators B Chem.* 72:273–82.
42. Ruano, J. M., Glidle, A., Cleary, A., Walmsley, A., Aitchison, J. S., Cooper, J. M. 2003. Design and fabrication of a silica on silicon integrated optical biochip as a fluorescence microarray platform. *Biosensors Bioelect.* 18:175–84.

43. Ahn, C. H., Choi, J. W., Beaucage, G., Nevin, J., Lee, J. B., Puntambekar, A., Lee, R. 2004. Disposable smart lab on a chip for point-of-care clinical diagnostics. *Proc. IEEE* 92:154–73.
44. Balslev, S., Jørgensen, A. M., Olsen, B. B., Mogensen, K. B., Snakenborg, D., Geschke, O., Kutter, J. P., Kristensen, A. 2006. Lab-on-a-chip with integrated optical transducers. *Lab Chip* 6:213–17.
45. Pal, R., Yang, M., Lin, R., Johnson, B., Srivastava, N., Razzacki, S., Chomistek, K., Heldsinger, D., Haque, R., Ugaz, V. 2005. An integrated microfluidic device for influenza and other genetic analyses. *Lab Chip* 5:1024–32.
46. Minas, G., Wolffenbuttel, R. F., Correia, J. H. 2005. A lab-on-a-chip for spectrophotometric analysis of biological fluids. *Lab Chip* 5:1303–9.
47. Chandrasekaran, A., Packirisamy, M. 2007. Wafer dicing strategic planning technique for clustered BioMEMS devices. *Int. J. Product Dev.* 4:296–309.
48. Chandrasekaran, A., Packirisamy, M. 2008. Enhanced fluorescence-based bio-detection through selective integration of reflectors in microfluidic lab-on-a-chip. *Sens. Rev.* 28:33–38.
49. Ozhikandathil, J., Packirisamy, M. 2010. Silica-on-silicon (SOS)-PDMS platform integrated lab-on-a-chip (LOC) for quantum dot applications. Proceedings of SPIE, Photonics North, Niagara Falls, 2010.
50. Han, J., Lee, S. H., Heo, Y., Hwang, C. J., Chong, H. A. 2005. An on-chip blood serum separator using self-assembled silica microspheres. Paper presented at Filter Transducers: The 13th International Conference on Solid-State Sensors, Korea, June 5–9, 2005.
51. Ray, S., Mehta, G., Srivastava, S. 2010. Label-free detection techniques for protein microarrays: Prospects, merits and challenges. *Proteomics* 10:731–48.
52. Han, M., Gao, X., Su, J. Z., Nie, S. 2001. Quantum-dot-tagged microbeads for multiplexed optical coding of biomolecules. *Nat. Biotechnol.* 19:631–35.
53. Fritz, J., Baller, M. K., Lang, H. P., Rothuizen, H., Vettiger, P., Meyer, E., Güntherodt, H.-J., Gerber, Ch., Gimzewski, J. K. 2000. Translating biomolecular recognition into nanomechanics. *Science* 288:316–18.
54. Raiteri, R., Grattarola, M., Butt, H.-J., Skladal, P. 2001. Micromechanical cantilever-based biosensors. *Sensors Actuators B* 79:115–26.
55. Mullen, R., Mehregany, M., Omar, M., Ko, W. 1991. Theoretical modeling of boundary conditions in microfabricated beams. In *Proceedings of IEEE Micro Electro Mechanical Systems, Conference on Micro Structures, Sensors, Actuators, Machines and Robots*, Nara, Japan, January 30–February 2, 1991, pp. 154–59.
56. Burchman, K. E., Boyd, J. T. 1998. Freestanding, micromachined, multimode silicon optical waveguides at $\lambda = 1.3$ μm for microelectromechanical systems technology. *Appl. Optics* 37:8397–99.
57. Gehring, G. A., Cooke, M. D., Gregory, I. S., Karl, W. J., Watts R. 2000. Cantilever unified theory and optimization for sensors and actuators. *Smart Mater. Struct.* 9:918–31.
58. Kawakatsu, H., Saya, D., Kato, A., Fukushima, K., Toshiyoshi, H., Fujita, H. 2002. Millions of cantilevers for atomic force microscopy. *Rev. Sci. Instrum.* 73:1188–92.
59. Calleja, M., Tamayo, J., Johansson, A., Rasmussen, P., Lechuga, L. M., Boisen, A. 2003. Polymeric cantilever arrays for biosensing applications. *Sensor Lett.* 1:1–5.
60. Amritsar, J., Stiharu, I., Balagopal, G., Li, X., Packirisamy, M. 2004. MOEMS-based cardiac enzymes detector for acute myocardial infarction. *Proc. SPIE* 5578:91–98.
61. Chen, X., Fox, C. H. J., McWilliam, S. 2004. Optimization of a cantilever microswitch with piezoelectric actuation. *J. Intelligent Mater. Syst. Struct.* 15:823–34.
62. Petersen, K. E. 1982. Silicon as a mechanical material. *Proc. IEEE* 70:420–46.
63. Greenwood, J. C. 1988. Silicon in mechanical sensors. *J. Phys. E Sci. Instrum.* 21:1114–28.
64. Pedersen, N. L. 2000. Design of cantilever probes for atomic force microscopy (AFM). *Eng. Optimization* 32:373–92.
65. Rinaldi, G., Packirisamy, M., Stiharu, I. 2005. Geometrical performance conditioning of microstructures. In *Proceedings of ISSS, International Conference on Smart Materials, Structures and Systems*, Bangalore, India, July 28–30, 2005, pp. SD-191–98.
66. Tabard-Cossa, V. 2006. Microcantilever actuation generated by redox-induced surface stress. PhD thesis, McGill University.
67. Amritsar, J., Stiharu, I., Packirisamy, M. 2006. Bioenzymatic detection of troponin C using micro-opto-electro-mechanical systems. *J. Biomed. Optics* 11:021010-1–7.
68. Stoney, G. G. 1909. The tension of metallic films deposited by electrolysis. *Proc. R. Soc. Lond. A Math. Phys. Eng. Sci.* 82:172–75.
69. Goeders, K. M., Colton, J. S., Bottomley, L. A. 2008. Microcantilevers: Sensing chemical interactions via mechanical motion. *Chem. Rev.* 108:522–42.
70. McFarland, A. W., Poggi, M. A., Doyle, M. J., Bottomley, L. A., Colton, J. S. 2005. Influence of surface stress on the resonance behavior of microcantilevers. *Appl. Phys. Lett.* 87:053505.

71. Rinaldi, G., Packirisamy, M., Stiharu, I., Mrad, N. 2009. Simple and versatile micro-cantilever sensors. *Sensors Rev.* 29:44–53.
72. Chandrasekaran, A., Packirisamy, M. 2009. Label-free detection of enzymatic interaction through dynamic behavior of microstructures. Paper presented at the International Conference on MEMS IITM, Chennai, India, January 3–5, 2009.
73. McFarland, A. W. 2005. PhD thesis, Georgia Institute of Technology, Atlanta.
74. Ginsberg, J. 2001. *Mechanical structural vibrations*. New York: John Wiley & Sons.
75. Sader, J. E. 2001. Surface stress induced deflections of cantilever plates with applications to the atomic force microscope: Rectangular plates. *J. Appl. Phys.* 89:2911–21.
76. McFarland, A. W., Colton, J. S. 2005. Role of material microstructure in plate stiffness with relevance to microcantilever sensors. *J. Micromech. Microeng.* 15:1060–67.
77. Rinaldi, G. Packirisamy, M., Stiharu, I. 2007. Dynamic analysis of slotted MEMS cantilevers. *Int. J. COMADEM* 10:13–19.
78. Carrascosa, L. G., Moreno, M., Alvarez, M., Lechuga, L. M. 2006. Nanomechanical biosensors: A new sensing tool. *Trends Anal. Chem.* 25:196–206.
79. Amritsar, J., Stiharu, I., Packirisamy, M. 2005, Detection of acute myocardial infarction (AMI) using BioMEMS. Paper presented at the 4th Canadian Workshop on MEMS, Ottawa, Canada, August 19, 2005.
80. Stiharu, I., Packirisamy, M., Burnier, M., Marshall, J. C. 2008. Method and devices for early detection of cancer cells and types through micromechanical interactions. U.S. Patent application US-2008–0318242-A1.
81. Bhalerao, K. D., Mwenifumbo, S., Soboyejo, A. B. O., Soboyejo, W. O. 2004. Bounds in the sensitivity of BioMEMS devices for cell detection. *Biomed. Microdevices* 6:23–31.

REVIEW QUESTIONS

1. What is a microarray?
2. How can microarrays be classified?
3. What is the core principle behind DNA microarrays? Describe the hybridization process.
4. How is an array experiment usually performed? What are the main steps of a DNA array experiment?
5. What are the main applications of DNA microarrays?
6. What is a polymerase chain reaction (PCR) and what are the main applications of PCR?
7. Describe a continuous flow PCR chip.
8. What is a thermocycler?
9. Describe the schematic structure of a droplet-based PCR chip.
10. What is a protein microarray and what are the applications of protein arrays?
11. How can proteins be immobilized on a substrate?

Index

H

I

K

L

M

N

Q

R

S